Computer-Aided Statics and Strength of Materials

Graham R. Salter

Professor Emeritus
Niagara County Community College
Sanborn, New York

Prentice Hall
Upper Saddle River, New Jersey   Columbus, Ohio

Library of Congress Cataloging-in-Publication Data

Salter, Graham R.
    Computer-aided statics and strength of materials / Graham R.
 Salter.
      p. cm.
    ISBN 0-13-741950-3
    1. Statics—Data processing. 2. Strength of materials—Data
processing. 3. Computer-aided engineering. I. Title.
TA351.S25 2000
620.1'03—dc21

                                      99-17738
                                      CIP

Cover photo: © H. Armstrong Roberts
Editor: Stephen Helba
Production Supervision: Lisa Garboski, bookworks
Design Coordinator: Karrie M. Converse-Jones
Cover Designer: Janoski Advertising Design
Production Manager: Deidra M. Schwartz
Marketing Manager: Chris Bracken

This book was set in Goudy by Maryland Composition Co., Inc. and was printed and bound by R.R. Donnelley & Sons Company. The cover was printed by Phoenix Color Corp.

© 2000 by Prentice-Hall, Inc.
Pearson Education
Upper Saddle River, New Jersey 07458

Printed in the United States of America

10 9 8 7 6 5 4 3 2 1

ISBN: 0-13-741950-3

Prentice-Hall International (UK) Limited, *London*
Prentice-Hall of Australia Pty. Limited, *Sydney*
Prentice-Hall of Canada, Inc., *Toronto*
Prentice-Hall Hispanoamericana, S. A., *Mexico*
Prentice-Hall of India Private Limited, *New Delhi*
Prentice-Hall of Japan, Inc., *Tokyo*
Prentice-Hall (Singapore) Asia Pte. Ltd., *Singapore*
Editora Prentice-Hall do Brasil, Ltda., *Rio de Janeiro*

**This book is dedicated to Maureen
whose encouragement helped far more than she realizes.**

Preface

All phases of engineering, drafting, and design are becoming automated throughout industry. It would be difficult to find any company, no matter how small, that did not depend to some extent on computers. Schools are expanding computer education rapidly, and even primary school children know how to run various programs.

This book is written for students in two-year or four-year engineering technology programs, although technicians in industry may also find the numerous computer programs to be valuable tools. Our aim is to prepare students for industry, and they will not be adequately prepared without a working knowledge of computers. However, running "canned" programs without a thorough understanding of the subject is dangerous and is strongly discouraged.

The objective of this book is to teach the students basic *statics* and *strength of materials*. Worked Example Problems, using a hand-held calculator, cover each topic, and it is not until the end of each chapter that computer programs are introduced. Calculus is not required for a complete understanding of the text, but pages titled "Using Calculus" give more detailed presentations for the students who have an understanding of elementary calculus. A basic understanding of trigonometry is required.

After the students have a thorough understanding of the principles involved and the governing equations, the computer programs not only remove the drudgery of computation, they allow fairly complicated problems to be solved with ease. Students gain a greater understanding of the subject by simply changing input values and seeing what happens. For example, suppose a flange coupling has eight ½-in. bolts. What would happen to the maximum power rating if the design were changed to six $\frac{9}{16}$-in bolts? The answer can be seen immediately.

The 25 original computer programs are written for Microsoft Excel 97, which is currently the most used spreadsheet software in industry. A brief description of each program is given in Appendix B.

The Example Problems following each new concept give the students a clear understanding of the applications and of the principles involved. The numerous problems at the end of each chapter include both calculator and computer computations.

ACKNOWLEDGMENTS

I would like to thank the reviewers of this text: Reuben R. Aronovitz, Delaware County Community College; Edward L. Bernstein, Alabama A&M University; Janak Dave, University of Cincinnati; Ross Lyman, College of Lake County; and Mohammad A. Zahraee, Purdue University-Calumet.

Graham R. Salter

Contents

1   Introduction   1

   1.1   The Nature of Materials   1
   1.2   Units and Constants   2
   1.3   Conversion   7
   1.4   Numerical Computations   8
   1.5   Computer Analyses   10
   1.6   Problems for Chapter 1   11

2   Forces   13

   2.1   Definitions   13
   2.2   Resultants   14
   2.3   Equilibrants   30
   2.4   The Free Body   33
   2.5   Reactions   35
   2.6   Two-Force Members   36
   2.7   Computer Analyses   37
   2.8   Problems for Chapter 2   39

3   Moments   41

   3.1   Moments   41
   3.2   Concentrated Loads   43
   3.3   Uniformly Distributed Loads   49
   3.4   Couples   51
   3.5   Calculating Reactions   53
   3.6   Computer Analyses   72
   3.7   Problems for Chapter 3   74

4   Nonconcurrent Forces, Frames, and Trusses   79

   4.1   Single Members   79
   4.2   Frames   83
   4.3   Space Frames   88
   4.4   Trusses   95

4.5   Computer Analyses   107
4.6   Problems for Chapter 4   116

5   Friction   121

5.1   The Nature of Friction   121
5.2   Coefficient of Friction   122
5.3   Force Required to Move a Body Up a Slope   127
5.4   Wedges   130
5.5   Jackscrews   133
5.6   Belt Friction   136
5.7   Computer Programs for Friction   139
5.8   Problems for Chapter 5   142

6   Direct Stress   145

6.1   Normal, or Axial, Stress   145
6.2   Shear Stress   151
6.3   Bearing Stress   154
6.4   Stress Concentrations   155
6.5   Computer Analyses   159
6.6   Problems for Chapter 6   163

7   Material Properties   167

7.1   The Tension and Compression Test   167
7.2   Strain   169
7.3   Hooke's Law and the Stress-Strain Diagram   170
7.4   Ultimate and Yield Strengths in Shear   174
7.5   Poisson's Ratio   175
7.6   Thermal Deformation   176
7.7   Computer Analyses   177
7.8   Problems for Chapter 7   186

8   Thermal Stress   189

8.1   Single Material   189
8.2   Materials in Parallel   191
8.3   Materials in Series   194
8.4   Computer Analyses   198
8.5   Problems for Chapter 8   201

9 Connections   203

   9.1 Types of Connections   203
   9.2 Welded Connections   203
   9.3 Riveted and Bolted Connections   209
   9.4 Computer Programs for Connections   220
   9.5 Problems for Chapter 9   228

10 Pressure Vessels   231

   10.1 Internal Pressure   231
   10.2 Forces Inside a Cylindrical Pressure Vessel   231
   10.3 Forces Inside a Spherical Pressure Vessel   234
   10.4 Stresses in Pressure Vessels   234
   10.5 Computer Program for Vessels   239
   10.6 Problems for Chapter 10   242

11 Centroids and Moments of Inertia   245

   11.1 Center of Gravity   245
   11.2 Moment of Area   246
   11.3 Centroids of Simple Shapes   249
   11.4 Centroids of Composite Shapes   249
   11.5 Centroids of Any Shape   255
   11.6 Moment of Inertia   257
   11.7 Moment of Inertia of Simple Shapes   258
   11.8 Transfer-of-Axis Theorem   260
   11.9 Moments of Inertia of Composite Shapes   261
   11.10 Moment of Inertia of Any Shape   265
   11.11 Polar Moment of Inertia   268
   11.12 Computer Analyses   270
   11.13 Problems for Chapter 11   282

12 Torsion   285

   12.1 Torque   287
   12.2 Torsional Deformation   287
   12.3 Torsional Shear Stress   289
   12.4 Stress Concentrations   293
   12.5 Power   298
   12.6 Transmission   300
   12.7 Shaft Couplings   302
   12.8 Computer Analyses   309
   12.9 Problems for Chapter 12   316

13   Shear and Bending in Beams   319

13.1   Beam Types   319
13.2   Bending Deformation   320
13.3   The Flexure Formula   325
13.4   Beam Shear Force and Bending Moment Diagrams   327
13.5   Shear Forces   327
13.6   Shear Force Diagrams   327
13.7   Bending Moment Diagrams   349
13.8   Generalized Equations   368
13.9   Moving Loads   381
13.10 Computer Analyses   394
13.11 Computer Program for Moving Loads   409
13.12 Problems for Chapter 13   411

14   Shear and Bending Stresses in Beams   419

14.1   Horizontal Shear Stresses   419
14.2   Bending Stresses   432
14.3   Stress Concentrations   435
14.4   Composite Beams   439
14.5   Problems for Chapter 14   444

15   Deflection of Beams   451

15.1   Beam Deflection Theories   451
15.2   Deflections of Simply Supported Beams   456
15.3   Deflections of Cantilever Beams   464
15.4   Principle of Superposition   471
15.5   Deflection Equations   473
15.6   Computer Analyses   483
15.7   Problems for Chapter 15   489

16   Beam Design   495

16.1   General Considerations   495
16.2   Structural Shapes   498
16.3   Beam Selection   500
16.4   Basic Deflection   504
16.5   Computer Analyses   509
16.6   Problems for Chapter 16   522

17   Statically Indeterminate Beams   527

17.1   The Three-Moment Equation   527
17.2   The Method of Superposition   542
17.3   Fixed Beams   546

17.4   Computer Analyses   550
17.5   Problems for Chapter 17   560

18   Combined Loadings   565

18.1   Pure Tension   565
18.2   Combined Tension-Tension (Stress Fields Normal to Each Other)   571
18.3   Combined Tension and Shear   577
18.4   Principal Planes and Principal Stresses   581
18.5   Combined Torsion and Bending   585
18.6   Combined Bending and Tension or Compression   590
18.7   Eccentric Loads   592
18.8   Computer Analyses   595
18.9   Problems for Chapter 18   603

19   Struts and Columns   605

19.1   Basic Column Theory   605
19.2   Allowable Load   610
19.3   Transition Slenderness Ratio   611
19.4   AISC Column Code   616
19.5   Timber Columns   620
19.6   Eccentric Loads, Initial Curvature, and Stress Reversal   622
19.7   Computer Analyses   625
19.8   Problems for Chapter 19   630

Appendix A   Mathematical Notes   632

A.1   Areas of Figures   632
A.2   Trigonometry   633
A.3   Sigma Notation   637

Appendix B   Computer Program Listings   638

Appendix C   Representative Mechanical Properties of Selected
               Engineering Materials   642

Appendix D   Table of Conversion Factors: U.S. Customary
               Units to SI Units   652

Appendix E   Table of Structural Steel Shapes and Standard
               Timber Sizes   653

References   667

Index   669

1

Introduction

■ *Objectives*

In this chapter you will learn about

1. the structure and nature of engineering materials;
2. the units used in engineering;
3. the meaning of the concept of FORCE;
4. converting from one type of unit to another.

1.1   THE NATURE OF MATERIALS

If you look at a block of solid steel, it might be difficult to imagine that it actually consists of billions of tiny building blocks and contains billions of empty spaces. We call the building blocks *molecules*, and they in turn are made up of atoms, which, themselves, consist of electrons, neutrons, and protons. An atom, which is the smallest piece of an element that possesses the physical and chemical properties of that element, has a radius of about 10^{-8} cm, that is, one hundred-millionth of a centimeter. Stacked edge to edge, it would take 1 million of them to make up the thickness of this page.

Around 1911, Ernest Rutherford (1871–1937) and his colleagues performed experiments whose results indicated that an atom must consist of a tiny, positively charged nucleus containing more than 99.9% of the mass of the atom, surrounded by negatively charged electrons moving in orbits around the nucleus, like planets around the sun (Figure 1.1).

Although we know today that this is a greatly simplified model, it has been—and still is—useful in explaining some of the basic principles of physics. It is generally accepted that the nucleus has a radius of between 10^{-12} and 10^{-13} cm—i.e., somewhere between 1/10,000 and 1/100,000 the size of the atom. Let us try to put this in perspective. If we represent the nucleus with a golf ball (radius 0.84 in.), then the electrons would revolve in a sphere with a radius of 700 ft to 1.3 mi! We are not going to delve into the mysteries of atomic structure, but it is important to understand that even the most solid-looking materials contain lots of space.

We all know that it is easy to stretch a rubber band, and when we let go it will return to its original size. Rubber is an *elastic material*. But so are iron, steel, copper, and zinc—in fact just about all metals and many nonmetals are elastic. They can be stretched and compressed by pulling and squeezing and will return to their original size and shape, provided they are not pulled so hard that they become permanently de-

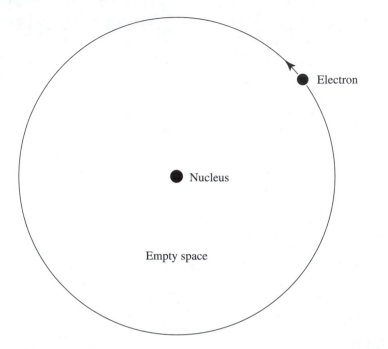

Figure 1.1   Rutherford's model of the atom (1911) (not to scale).

formed. We talk more about this in Chapter 6. Of course, the force needed to stretch a piece of steel is enormous, but steel will stretch. Why will it stretch? As we have seen, it is not just a solid block of metal. It consists of molecules with spaces between them, held together by intermolecular forces. Think of these as being like little strong springs: Pull on them and they will stretch a little but will pull back against you; push and they will compress a little but again will push back against you. Most engineering materials behave like this to some extent, and we will learn the correct words and the underlying equations in the following chapters.

This model also lets us understand thermal expansion and contraction (to be discussed in Chapter 7). When a metal is heated, the molecules move farther apart, and when it is cooled, they move closer together, so the object as a whole expands or contracts.

1.2   UNITS AND CONSTANTS

Units are extremely important for consistency and in defining exactly what we are doing. If we are talking about money and say that we have 100, it makes a big difference whether we mean pennies or dollars. It is good engineering practice to attach the units to *every* number in a calculation. It may seem a bit cumbersome, but it can save a lot of time in checking for mistakes. At the present time in the United States we use two systems of units in engineering. The U.S. Customary System (foot-pound-second system) is gradually being replaced by the meter-kilogram-second system. This is known as SI, or the International System. It is very important that we become familiar with both systems, because it may be a long time before we completely convert to SI.

Apart from being different sets of units, there is a basic and fundamental difference between the two systems. The U.S. customary system is a *gravitational* system, in which *force* or *weight* is the basic unit, whereas in SI, which is an *absolute* system, *mass* is the basic unit. The terms mass and weight are often confused with one another, but it is important to distinguish between them. Mass is a property of the body itself; it is the quantity of matter in the body and is measured by its inertia or resistance to acceleration. The mass of an object remains the same (excluding the effects of relativity) whether the body is on Earth, the Moon, or anywhere else. Weight is a force (a pull) trying to pull the body toward the center of the planet. The "pull" exerted by the Moon is about one-sixth of that exerted by Earth, so an astronaut weighing 180 lb on earth would weigh only 30 lb on the Moon, but his or her mass would not have changed.

Simply, a **force** is a push or a pull. In more scientific terms, it is that which changes, or tends to change, the state of rest of a body or the state of uniform motion of a body in a straight line. Let us read that again. What it says is this: A force

1. can change the state of rest of a body;
2. can *tend* to change the state of rest of a body;
3. can change the state of uniform motion in a straight line;
4. can *tend* to change the state of uniform motion in a straight line.

An example of (1) is a golf ball sitting on a tree. It is at rest. When the golf club hits the ball, there is a force that changes its state of rest.

An example of (2) is a person pushing against the wall of a building. A force is being applied that is tending (trying) to move the wall. Of course it doesn't, but where does that force go? We discussed molecules and the way they can move relative to one another. The person pushing on the wall is pushing the molecules closer together (minutely), and the molecules are pushing back. So the force is actually going into compressing the wall.

In (3) we talk about uniform motion in a straight line. Uniform motion means constant speed. Constant speed in a straight line is called constant *velocity*. In everyday language we may say velocity when we mean speed, but in engineering and science we must be careful to use the correct term. Velocity is a *vector*, which means that both its *magnitude* and *direction* must be defined. We discuss this further in the next section.

A car moving at constant speed along a straight road is a body in a state of uniform motion in a straight line. We can change that state in many ways. For example, we press the gas pedal or the brake pedal to alter the speed. These require the application of a force. We either increase the force on the pistons, which puts more torque on the road wheels, or we create a braking force opposing the wheel rotation. Another way is to turn the steering wheel to change the direction. No matter how we change the state, it requires the application of a force.

In (4), a force acts on the body but does not alter its velocity (neither its speed nor its direction). For example, a gust of wind hits the car, but it has insufficient force to alter its speed or direction.

There are *external* forces and *internal* forces. **External forces** are those applied to a body, such as your weight when you sit on a chair. These are often referred to as *loads*. The chair doesn't collapse because it is pushing back against you with an internal force, which is equal but in the opposite direction to the line of action of your weight. This internal force is created by the intermolecular forces.

Isaac Newton (1642–1727), following the work of Aristotle (384–322 B.C) and Galileo (1564–1642), published his great work on motion, the *Principia*, in 1687. It is Newton's second law that concerns us here. It states that force, *F*, equals mass, *m*, times acceleration, *a*.

$$F = m \cdot a \tag{1.1}$$

Let us consider this. First, what is acceleration? *Acceleration* means that something is going faster and faster as time goes on. A typical example is the motion of a stone dropped from a bridge. Initially it is in your hand and its speed is zero. One second after you let it drop, it is falling at a speed of 32 ft/s. After 2 s its speed is 64 ft/s, and at the 3-s mark it is falling at 96 ft/s. Its speed is increasing by 32 ft/s every second, or, as we say, its *acceleration* is 32 ft/s/s. We usually write this as 32 ft/s². What is making the stone do this? *Gravity*. So we say that the acceleration due to gravity is 32 ft/s².

The interesting fact is that it does not matter whether we drop a stone or a cannonball; the acceleration will be 32 ft/s². This means that if we drop a stone and a cannonball at the same time from the same height, they will hit the ground together (Figure 1.2). Of course, a feather and a stone would drop at different rates because of air resistance, but in a vacuum (for example, on the Moon) they *would* fall together. A more precise value of the gravitational acceleration on Earth is 32.174 ft/s², although it does vary slightly with location on Earth and with altitude. In SI units it is 9.81 m/s². What is the force acting on the stone or the cannonball to make it accelerate? We call this force its *weight*. **Weight** is the force with which Earth (or the Moon or a planet) attracts the body, and it is directed toward the center of the planet. Obviously, the more massive cannonball needs a greater force to accelerate it at 9.81 m/s² than the less massive stone; put another way, the cannonball *weighs more* than the stone. This gravitational acceleration is so important that we give it its own special symbol, *g*.

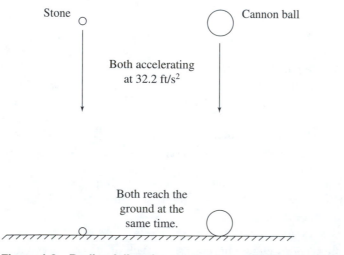

Figure 1.2   **Bodies fall at the same rate regardless of weight.**

$$g = 32.2 \text{ ft/s}^2 \quad \text{or} \quad 9.81 \text{ m/s}^2 \qquad \text{On Earth}$$

On the Moon it is about one-sixth of this value.

We can apply equation (1.1) to our stone-dropping case. The acceleration, a, equals g, and the force, F, equals the weight. So we can write

$$W = m \cdot g \qquad\qquad (1.2)$$

where W represents the weight of the body.

USING CALCULUS

Velocity (v) is the rate of change of displacement (s) and so is given by the derivative of displacement with respect to time (t):

$$v = ds/dt$$

EXAMPLE

Suppose the displacement of a body can be expressed as

$$s = 3t^3 + 4t^2 - 5t + 7 \text{ m}$$

Determine its velocity at time $t = 3$ s.

Solution

$$v = ds/dt = 9t^2 + 8t - 5$$

The velocity at $t = 3$ s is $ds/dt|_{t=3} = 9(3)^2 + 8(3) - 5$.

$$v = 81 \text{ m/s} + 24 \text{ m/s} - 5 \text{ m/s} = 100 \text{ m/s}$$

Answer

The velocity at $t = 3$ s is 100 m/s.

Acceleration (a) is defined as the time rate of change of velocity and is also a vector quantity.

$$a = dv/dt = d^2v/dt^2$$

EXAMPLE

What is the acceleration of the object in the previous example at time $t = 2$ s?

Solution

$$a = dv/dt = d(9t^2 + 8t - 5)/dt = 18t + 8$$

The acceleration at $t = 2$ s is $dv/dt\big|_{t=2} = 36 \text{ m/s}^2 + 8 \text{ m/s}^2 = 44 \text{ m/s}^2$.

Answer

The acceleration is 44 m/s^2.

Now let's consider the basic difference between the two systems of measurement. In the gravitational ft-lb-s system, force or weight is the basic unit expressed in pounds (lb). Acceleration, g, is measured in ft/s^2, so we can determine the units for mass by using equation (1.2). Rearranging that equation gives

$$m = W/g$$

Putting in the units; we have

$$m = \text{lb}/(\text{ft/s}^2)$$

or

$$m = \text{lb s}^2/\text{ft}$$

Pound-seconds squared per foot is too long to say or to write, so we give it the name *slug*.

In the gravitational system, mass is measured in slugs, which are units of lb-s^2/ft.

The basic unit in the absolute system is mass, measured in kilograms (kg). Using equation (1.2),

$$W = m \cdot g \ \text{kg·m/s}^2$$

The unit kg·m/s^2 is called a newton (N).

In the absolute system, weight (or force) is measured in newtons, which are units of kg·m/s^2.

We learn about *stress* in later chapters. **Stress** is defined as force divided by area. Notice that it is *force*/area and not mass/area. For example, a beam can be subjected to a certain amount of stress before it breaks. Suppose a beam can support a 10,000-kg mass on Earth. That is a weight of

$$(10,000 \text{ kg})(9.81 \text{ m/s}^2) = 98,100 \text{ N}$$

Apart from changes in the mechanical properties due to temperature, this beam will be able to support the same weight on the Moon; but on the Moon, the gravitational acceleration is only about 1.7 m/s^2. Therefore,

$$m = \frac{W}{g} = \frac{98,100 \text{ N}}{1.7 \text{ m/s}^2} = 57,700 \text{ kg}$$

A beam that could support a mass of 10,000 kg on Earth would be able to support a mass of 57,700 kg on the Moon.

Quite often, the numbers we use in calculations are very large or very small. For example, a certain type of steel will break at a stress level of 160,000 lb/in.^{2} (psi), and its coefficient of thermal expansion (to be discussed later) is 0.0000065 in./in./°F. In cases like these it is convenient to employ either prefixes or scientific notation. We already know some of them—one centimeter is one-hundredth meter, so the prefix *centi* means $\frac{1}{100}$, or 0.01. In scientific notation $\frac{1}{100} = 10^{-2}$. A millimeter is one-tenth centimeter, or one-thousandth meter, so the prefix *milli* means $\frac{1}{1000}$, or 0.001, or 10^{-3}. Your computer has many megabytes of ram, that is, many *millions* of bytes (10^6), and its hard drive capacity may be measured in gigabytes, which means *billions* (1,000,000,000 = 10^9)) of bytes. The following table lists the prefixes and scientific notation often used in engineering.

Prefix	Symbol	Meaning	Notation
tera	T	1,000,000,000,000	10^{12}
giga	G	1,000,000,000	10^9
mega	M	1,000,000	10^6
kilo	k	1,000	10^3
centi	c	0.01	10^{-2}
milli	m	0.001	10^{-3}
micro	μ	0.000 001	10^{-6}
nano	n	0.000 000 001	10^{-9}

For example, kilometer is written km, and 1 km equals 1000 m, or 10^3 m.

1.3 CONVERSION

Converting from one system of units to the other usually requires conversion factors or sometimes a conversion equation. To convert from feet to meters we simply multiply by 0.3048, but to change from °F to °C we have to use the equation

$$°C = \tfrac{5}{9}(°F - 32) \tag{1.3(a)}$$

To go from °C to °F:

$$°F = 32 + (\tfrac{9}{5})°C \tag{1.3(b)}$$

Remember that the same units in the numerator and denominator of a fraction cancel out, just like numbers. The following very simple example demonstrates this.

EXAMPLE PROBLEM 1.1

Convert 6 yd to inches.

Solution

1 yd = 3 ft, and 1 ft = 12 in.
  So,

$$6 \text{ yd} = (6 \text{ yd}) \times (3 \text{ ft/yd}) \times (12 \text{ in./ft})$$

$$6 \text{ yd} = (6 \times 3 \times 12)(\text{yd})(\text{ft/yd})(\text{in./ft})$$

$$6 \text{ yd} = 216(\text{yd-ft-in.})/(\text{yd-ft})$$

Answer

6 yd = 216 in.

1.4   NUMERICAL COMPUTATIONS

Most of the numbers that we encounter in engineering are *approximate* numbers. For example, a pressure gauge reading is recorded as 3500 psi. Depending on the accuracy of the gauge and the care with which it is read, the true pressure might be anywhere between, say, 3490 and 3510 psi. No measurement can be absolutely precise. Of course there are exact numbers. Most tables have exactly 4 legs, and most cars have exactly 1 steering wheel. We have to be very careful not to arrive at an answer that is more accurate than the given information. The following example should clarify this idea.

You are driving a car that has a digital speedometer, and you maintain a constant speed of 55 mi/h for 1.25 h. How far did you travel?

Most of us would be tempted to do the following calculation:

$$\text{distance} = \text{speed} \times \text{time} = 55 \text{ mi/h} \times 1.25 \text{ h} = 68.75 \text{ mi}$$

Answer
Distance = 68.75 mi.

Let us think about this. The speedometer would show 55 mi/h at a speed of anywhere between 54.5 and 55.4 mi/h. It would round up to 55 from 54.5 and down to 55 at 55.4 mi/h. Also, how accurate is your watch as you measure 1.25 h? But suppose the time reading is exact for this demonstration.

The following table shows the range of possible distances:

Actual Speed (mi/h)	Distance (mi)
54.5	68.125
54.6	68.25
54.7	68.375
54.8	68.5
54.9	68.625
55.0	68.75
55.1	68.875
55.2	69.0
55.3	69.125
55.4	69.25

The first three answers would round to 68 mi, and the last seven would round to 69 mi. What is the answer? Probably 69 miles, but we will discuss the rules for numerical computations a little later.

1.4.1   Significant Digits

Significant means meaningful. Clearly the digits following the decimal points in the previous table of distances are not meaningful.

Significant Digits

All the digits in an approximate number are considered significant digits except zeros used as positioning zeros (placeholders).

Expressing a pressure as 3500 psi means that the pressure is three thousand five hundred pounds per square inch, not three hundred fifty or thirty-five. The zeros simply place the 3 and the 5 in the correct columns; they are placeholders. The number 3500 has two significant digits. Similarly, 0.00065 means 65 hundred-thousandths. It also has two significant digits. However, 3505 psi indicates that we know the pressure to four significant digits, the zero being one of them. Again, 0.10065 indicates that we know the two zeros are not merely placeholders, so this number has five significant digits.

1.4.2 Rule for Multiplying and Dividing

When approximate numbers are multiplied or divided, the result should be expressed with the *same number of significant digits as the given one with the least significant digits*. Put another way, the result should be expressed with the *same accuracy as the least accurate given number*.

Returning to the problem of the distance traveled, the car speed was given to two significant digits and the time to three significant digits. The least accurate is the speed, and so the answer should be expressed with two significant digits.

$$\text{distance} = 68.75 \qquad \text{rounded up to 69 miles}$$

Answer
Distance = 69 mi.

1.4.3 Precision

The **precision** of an approximate number refers to the position of the last significant digit. For example, 3.141 is precise to the nearest one-thousandth, and 3500 is precise to the nearest one hundred.

1.4.4 Rule for Adding and Subtracting

When approximate numbers are added or subtracted, the result should be expressed with the *same precision as the least precise given number*. As an example, the sum of 3.1415926 + 1.32453 + 3.27 is 7.7361226, but it should be expressed as 7.74.

EXAMPLE PROBLEM 1.2
Evaluate the following expression:

$$x = \frac{(2.345)(13.64)}{6.735 - 3.63}$$

Solution
Both numbers in the numerator have four significant digits, so their product should be expressed with four significant digits:

$$(2.345)(13.64) = 31.9858 \qquad \text{rounded to 31.99}$$

The least precise number in the denominator is 3.63, so

$$(6.735 - 3.63) = 3.105 \qquad \text{rounded to } 3.11$$

In the division 31.99/3.11 the least accurate number (3.11) has three significant digits. (Notice, however, that the numbers happen to have the same precision.) Using a calculator:

$$31.99/3.11 = 10.28617363$$

which must be expressed to three significant digits as 10.3.

Answer
$x = 10.3$

1.5   COMPUTER ANALYSES

The programs in this book are written for *Microsoft Excel 97* and are included in the software package.

1.5.1   Program 'Convert'

This program allows you to convert from customary units to SI or from SI to customary units quickly and easily.

EXAMPLE PROBLEM 1.3

Convert 1000 ft^3 to m^3.

Solution
1. Open the Excel program 'Convert'.
2. Enter US in cell D5.
3. Enter 7 in cell I5.
4. Enter 1000 in cell I22.
5. The answer, 28.3168, appears in cell I24.

Answer
1000 ft^3 = 30 m^3.

EXAMPLE PROBLEM 1.4

Convert 15°C to °F.

1. Enter SI in cell D5.
2. Enter 9 in cell I5.
3. Enter 15 in cell I22.
4. The answer, 59, appears in cell I24.

Answer
15°C = 59°F

1.6   PROBLEMS FOR CHAPTER 1

Hand Calculations

1. A body weighs 100 lb on Earth.
   a. What is its mass, in slugs?
   b. What would it weigh on a planet where the acceleration due to gravity is 64.4 ft/s^2?
2. What would a mass of 100 kg weigh on a planet where the acceleration due to gravity is 5 m/s^2? Give your answer in newtons.
3. An object weighs 500 N on a planet where the acceleration due to gravity is 25 ft/s^2. What will it weigh, in lb, on Earth?
4. Convert 10 slugs to kg, using the information that 1 lb = 4.448 N.
5. Given the fact that 1 ft = 0.3048 m, convert 1000 ft/min to m/s.
6. 6.4 G bytes = 6.4 × 10^n k bytes. What is the value of n?
7. 1000 bytes = 1 × 10^n M bytes. What is the value of n?
8. Calculate the force required to accelerate a 44-lb shopping cart on a level frictionless surface from rest to 1.7 ft/s in 2.0 s.
9. A 1.0 × 10^{-4}-kg spider is descending on a strand with an acceleration of 4.21 m/s^2. What is the magnitude of the supporting force given by the strand?
10. Calculate the acceleration of a 160-lb skydiver if air resistance exerts a force of 60 lb.

11. A 0.322-lb baseball strikes the catcher's mitt at 78.3 mi/h. The mitt recoils backward, bringing the ball to rest in 6.20 ms. What was the average force exerted by the ball on the glove?
12. If a body moves at a constant velocity, what is the highest power of t (time) in the displacement equation?
13. If a body moves with constant acceleration, what is the highest power of t (time) in the displacement equation?
14. The distance s feet that a body moved in time t seconds was determined experimentally to be given by the equation

$$s = 16.1t^2 + 12.6t - 8$$

   a. How far had it traveled in the first 3 s?
   b. What was its speed at time $t = 2$ s?
   c. What was its acceleration?
15. Evaluate the expression $A = \pi(6.23 \text{ mm})^2/4$.

Note: The 4 in this expression is exact.
16. Evaluate the expression

$$y = \frac{(31.2345)(0.0070)}{3.005 - 2.8}$$

Computer Calculations

17. Convert
   a. 35 ft to m
   b. 1500 in.^{2} to m^2
   c. 45° to rad
   d. 60 km/h to mi/h
18. Convert:
   a. 250 N to lb
   b. 2500 lb/in.^{2} to MPa (megapascals)
   c. 55 hp to W (watts)
   d. 50 kg to slugs
   e. 75 lb-ft to N·m
   f. −40°C to °F
19. Write a computer program to calculate the mass of a body and its weight on a planet,

given its weight on Earth and the gravitational acceleration on the planet. Use either U.S. customary or SI units.
20. Write a computer program to convert from degrees Celsius (°C) to degrees Fahrenheit (°F), degrees Rankine (°R), and kelvins (K) using the following equations:

$$T(°C) = \tfrac{5}{9}(T(°F) - 32)$$
$$T(K) = T(°C) + 273.15$$
$$T(°R) = 1.8(T(K))$$

2

Forces

■ **Objectives**

In this chapter you will learn about *resultants* and *equilibrants* of multiple forces acting on a body.

2.1   DEFINITIONS

We discussed the concept of *force* in Chapter 1, but we review it here before considering what happens when more than one force act on a body.

We gave the following definition: A force is that which changes, or tends to change, the state of rest of a body or the state of uniform motion of a body in a straight line.

That is, a force

1. can change the state of rest of a body;
2. can *tend* to change the state of rest of a body;
3. can change the state of uniform motion in a straight line;
4. can *tend* to change the state of uniform motion in a straight line.

When multiple forces act on a body, they can be

1. collinear or noncollinear;
2. coplanar or noncoplanar;
3. concurrent or nonconcurrent.

We explain these terms below.

2.1.1   Collinear Forces

Collinear means *in a straight line*. For example, in a tug of war, the two teams are exerting collinear forces in opposite directions.

2.1.2   Coplanar Forces

Coplanar means *in the same plane*. Coplanar forces are illustrated by a bow string when an archer shoots an arrow (see Figure 2.1). The two halves of the string and the arrow are all in one plane (usually a vertical plane). The forces are the tensions in the string and the opposing force created by the archer's pull.

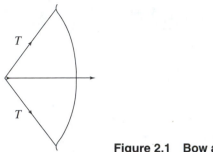

Figure 2.1   Bow and arrow.

2.1.3   Concurrent Forces

Concurrent means *acting through a common point*. In the bow-and-arrow example, the forces created by the tensions on each half of the string and the pull exerted by the archer all pass through the common point where the notch in the arrow fits onto the string.

2.2   RESULTANTS

A **resultant** is a *single force* that has the same effect on a body as any number of *concurrent* forces have on that body. For example, when the archer lets go, the two equal forces, T, in Figure 2.1 pull on the arrow at equal angles and thrust the arrow straight ahead. It is as though there were a single horizontal force, the *resultant*, pushing on the arrow.

2.2.1   Resultant of Collinear Forces

The resultant of collinear forces is calculated by simple addition or subtraction. We show a block in Figure 2.2(a) being pushed by a 75-N force and pulled by a 100-N force. The effect is the same whether we *push* it with a 175-N force or *pull* it with a 175-N force.

This brings us to the **transmissibility** of a force: We can consider a force to be acting at *any point on its line of action*, provided, of course, that we don't change its magnitude or direction.

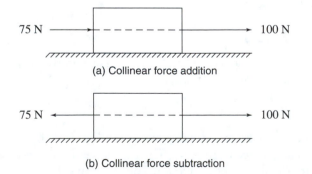

Figure 2.2

We have to be careful here. The principle of transmissibility of a force can be applied when we are dealing with *overall effects* on a body, such as its acceleration. When dealing with local effects, however, we must realize that a push will tend to compress the body, whereas a pull will tend to stretch it.

In Figure 2.2(b) both forces are pulling on the body but in opposite directions, resulting in a combined force (resultant) of 25 N to the right.

2.2.2   Resultant of Coplanar Concurrent Forces

We obtain the resultant of two concurrent forces by *completing the parallelogram*, as in Figure 2.3(a). We draw both forces to scale and at the correct angles. For example, if F_1 is a horizontal force of 50 lb and $F_2 = 30$ lb at 60°, we could draw a horizontal line 5 in. long and a 3-in. line at an angle of 60°. Then we complete the parallelogram *ABCD* and draw the diagonal *AC*. *AC* is the resultant, **R.**

Notice that we obtain exactly the same result if we draw a 5-in. horizontal line to represent F_1 and then a 3-in. line at an angle of 60° but *shifted so that its tail starts at the head of* F_1. The line from the *tail* of F_1 to the *head* of F_2 is **R,** the resultant. We show this in Figure 2.3(b). Notice that the given forces flow from tail to head, tail to head, and the resultant goes from the *tail of the first* to the *head of the second force*. Think of it this way: You drive east 50 mi from A to B; then you turn 60° left and drive another 30 mi to arrive at C. You could have arrived at the same destination by driving from A to C.

The forces F_1 and F_2 both have magnitude and direction; they are **vectors,** and finding their resultant is known as *vector addition*. If the angle between the two forces is 90°, we can use the Pythagorean theorem to find the magnitude of the resultant and then trigonometry to determine its angle. If the angle between them is not 90°, we can use the law of sines and/or the law of cosines to calculate the resultant. These laws are discussed in Appendix 1.

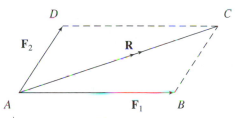

(a) Resultant determined by completing the parallelogram

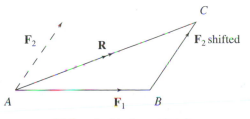

(b) Resultant by force triangle

Figure 2.3

EXAMPLE PROBLEM 2.1

A body is acted upon by a horizontal force of 40 N to the right and a vertical upward force of 70 N. What is the resultant force on the body?

Solution

Make a simple sketch. See Figure 2.4(a). Now shift the vertical force so that its tail touches the head of the horizontal force (Figure 2.4(b)). It makes no difference which force we move. We could move the tail of the 40-N force to the head of the 70-N one. The result would be the same.

By the Pythagorean theorem;

$$a^2 = b^2 + c^2$$

where a = the resultant, $\mathbf{R}$
      b = the 40-N force
      c = the 70-N force

$$R^2 = (40\ \text{N})^2 + (70\ \text{N})^2 = 1600\ \text{N}^2 + 4900\ \text{N}^2 = 6500\ \text{N}^2$$

$$R = 81\ \text{N}$$

Let the angle between $\mathbf{R}$ and the horizontal be α.

$$\tan \alpha = \frac{\text{opposite side}}{\text{adjacent side}}$$

$$= \frac{70\ \text{N}}{40\ \text{N}} = 1.75$$

so α = arctan (1.75) = 60.3°.

Answer

The resultant has a magnitude of 81 N and a direction of 60.3° to the horizontal.

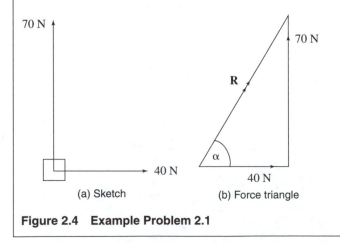

(a) Sketch   (b) Force triangle

Figure 2.4   Example Problem 2.1

EXAMPLE PROBLEM 2.2

Figure 2.5(a) shows a body being acted upon by a horizontal push of 200 lb and a 500-lb pull at 45° from the horizontal. Determine the resultant.

Solution

In this case, the force triangle is already drawn. We do not have to shift one of the forces because they are tail to head in the first sketch, but it is better to draw a separate force triangle, as in Figure 2.5(b). The internal angle at B is $\alpha = (180° - 45°) = 135°$. So we have a triangle with one side of 200 lb and another side of 500 lb; the angle between them is 135°. Applying the law of cosines (see Figure 2.5(b)):

$$R^2 = (AB)^2 + (BC)^2 - 2(AB)(BC)\cos \alpha$$

$$AB = 200 \text{ lb}, \qquad BC = 500 \text{ lb}, \quad \alpha = 135°$$

So,
$$R^2 = (200 \text{ lb})^2 + (500 \text{ lb})^2 - 2(200 \text{ lb})(500 \text{ lb})\cos 135°$$

$$= 4 \times 10^4 \text{ lb}^2 + 25 \times 10^4 \text{ lb}^2 - (20 \times 10^4 \text{ lb}^2)(-0.7071)$$

$$= 431{,}420 \text{ lb}^2$$

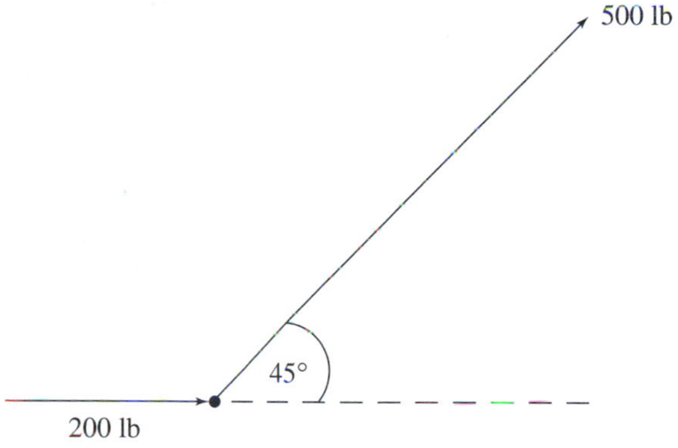

(a) Sketch

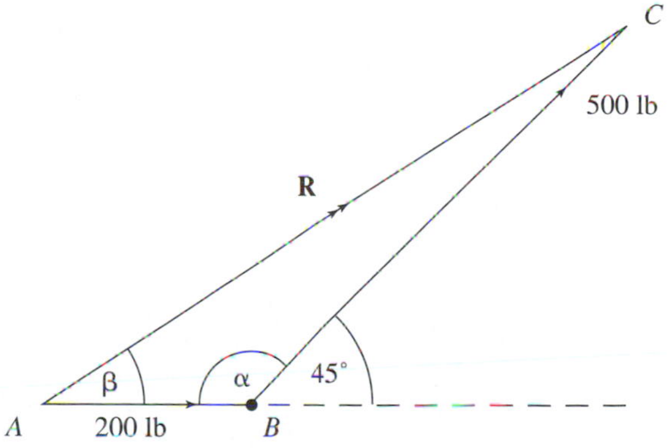

(b) Force triangle

Figure 2.5 Example Problem 2.2

Thus, $R = 656.8$ lb.

Applying the law of sines (see Figure 2.5(b)):

$$\frac{R}{\sin \alpha} = \frac{BC}{\sin \beta}$$

$$= \frac{500 \text{ lb}}{\sin \beta}$$

So

$$\sin \beta = \frac{(500 \text{ lb})(\sin \alpha)}{R \text{ lb}}$$

$$= \frac{(500 \text{ lb})(\sin 135°)}{656.8 \text{ lb}}$$

$$= \frac{(500 \text{ lb})(0.7071)}{656.8 \text{ lb}} = 0.5383$$

So $\beta = \arcsin (0.5383) = 32.6°$

Answer

The resultant has a magnitude of 700 lb and a direction of 32.6° to the horizontal.

For more than two coplanar concurrent forces, we can find the resultant by *completing the polygon*. This is an extension of the force triangle. The three forces, F_1, F_2, and F_3, in Figure 2.6(a) are in the same plane (coplanar) and pass through a common point (concurrent). We draw F_1 to scale and at the correct angle (see Figure 2.6(b)); we then draw F_2, *putting its tail at the head of* F_1. Then we draw F_3 with its tail at the head of F_2. The line, R, from the tail of F_1 to the head of F_3 is the resultant.

We can find the resultant this way for any number of coplanar concurrent forces. Unfortunately, this method does not lend itself readily to mathematical analysis. A more useful method is to *resolve* the forces into horizontal and vertical components.

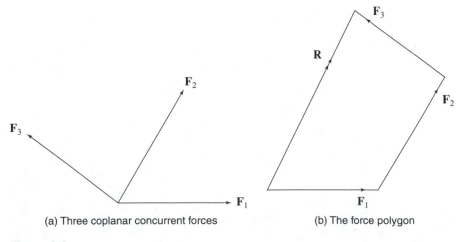

(a) Three coplanar concurrent forces   (b) The force polygon

Figure 2.6

2.2.3   Resolution of Forces

Resolution of a force means to *resolve* it (divide it) into two components, usually one horizontal and one vertical. This is the reverse of finding the resultant. Just as the resultant **R** in Figure 2.4(b) can replace the 40-N and the 70-N forces in Figure 2.4(a), so could the 40-N and the 70-N forces replace the single force (resultant) **R**. We say that we have *resolved* **R** into its horizontal and vertical components. The following example shows how to find the resultant of three coplanar concurrent forces by resolution. The method can be applied to any number of coplanar concurrent forces.

EXAMPLE PROBLEM 2.3

Determine the resultant of the three coplanar concurrent forces shown in Figure 2.7(a)

Solution

Figure 2.7(b), (c), and (d) shows the forces resolved into their horizontal and vertical components. Because F_1 (Figure 2.7(b)), is a horizontal force, it needs no resolution. F_2, the 40-N force (Figure 2.7(c)), is resolved into a horizontal force component, F_{2h}, and a vertical force component, F_{2v}. F_2 is actually the resultant of F_{2h} and F_{2v}, so the forces should run tail to head, tail to head, and the resultant should run from the tail of one to the head of the other, which they do. We now have a right-angle triangle with a 30° angle, an opposite side of F_{2v}, an adjacent side of F_{2h}, and a hypotenuse of $F_2 = 40$ N.

Using trigonometry:

$$F_{2h} = F_2\cos 30° = (40 \text{ N})(0.866) = 34.6 \text{ N}$$

$$F_{2v} = F_2\sin 30° = (40 \text{ N})(0.5) = 20 \text{ N}$$

We show the 50-N force in Figure 2.7(d).

Note: In order to keep the positive and negative signs correct, always measure the angles *counterclockwise, with east as zero*. The angle between the horizontal and the 50-N force is $(180° - 60°) = 120°$.

The horizontal component is

$$F_{3h} = F_3\cos 120° = (50 \text{ N})(-0.5) = -25 \text{ N}$$

The minus sign indicates that the force direction is to the left.

$$F_{3v} = F_3\sin 120° = (50 \text{ N})(0.866) = 43.3 \text{ N}$$

We now have three horizontal components, 30 N, 34.6 N, and -25 N, and two vertical components, 20 N, and 43.3 N, as shown in Figure 2.7(e).

The horizontal components are collinear, so we can add them algebraically (algebraically means taking account of the plus signs and the minus signs):

$$F_h = (30 \text{ N}) + (34.6 \text{ N}) + (-25 \text{ N}) = 39.6 \text{ N}$$

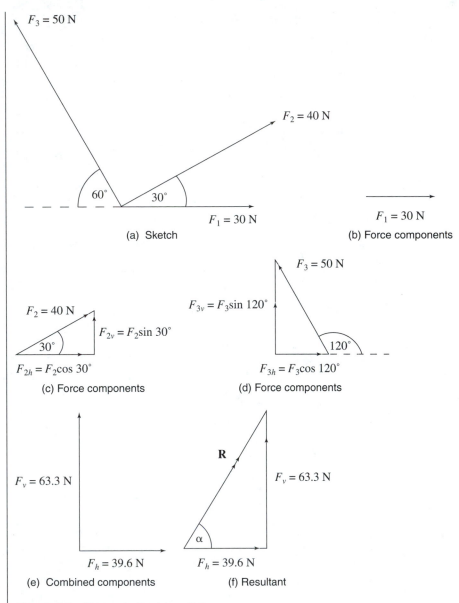

Figure 2.7   Example Problem 2.3

Similarly, for the vertical force components,

$$F_v = (20 \text{ N}) + (43.3 \text{ N}) = 63.3 \text{ N}$$

The resultant of these is found by drawing the force triangle shown in Figure 2.7(f).

$$R^2 = F_v^2 + F_h^2$$

$$F_v = 63.3 \text{ N}, \qquad F_h = 39.6 \text{ N}$$

Therefore,

$$R^2 = (63.3\ \text{N})^2 + (39.6\ \text{N})^2 = 5775\ \text{N}^2$$

So, $\qquad\qquad R = 74.7\ \text{N}$

$$\tan \alpha = \frac{63.3\ \text{N}}{39.6\ \text{N}} = 1.598$$

The original forces have only one significant digit, so the answer must be given to one significant digit also.

$$R = 70\ \text{N}$$

Thus, $\alpha = \arctan(1.598) = 58.0°$.

Answer

The resultant has a magnitude of 70 N and is at 58.0° to the horizontal.

2.2.4 Resultant of Coplanar Concurrent Forces: General Equations

We will now put these ideas together in a general form in preparation for writing a computer program. If we have N coplanar concurrent forces, $\mathbf{F}_1, \mathbf{F}_2, \mathbf{F}_3, \ldots, \mathbf{F}_N$, and they are at angles $\alpha_1, \alpha_2, \alpha_3, \ldots, \alpha_N$ to the horizontal (measured counterclockwise with east as zero), then the horizontal components are

$$F_{1h} = F_1 \cos \alpha_1, \qquad F_{2h} = F_2 \cos \alpha_2, \qquad F_{3h} = F_3 \cos \alpha_3, \ldots, F_{Nh} = F_N \cos \alpha_N$$

and the vertical components are

$$F_{1v} = F_1 \sin \alpha_1, \qquad F_{2v} = F_2 \sin \alpha_2, \qquad F_{3v} = F_3 \sin \alpha_3, \ldots, F_{Nv} = F_N \sin \alpha_N$$

The horizontal components are collinear, so we can add them algebraically:

$$F_h = \sum_{n=1}^{N} F_n \cos \alpha_n \qquad (2.1(a))$$

Similarly, for the vertical components,

$$F_v = \sum_{n=1}^{N} F_n \sin \alpha_n \qquad (2.1(b))$$

Refer to Appendix 1 if these symbols are unfamiliar to you. F_h and F_v are the horizontal and vertical legs of a right-angle triangle, as in Figure 2.7(f), so we can determine the resultant by the Pythagorean theorem:

$$R = \sqrt{F_h^2 + F_v^2} \qquad (2.2(a))$$

and the angle, α, that the resultant makes with the horizontal is

$$\alpha = \arctan(F_v/F_h) \qquad (2.2(b))$$

EXAMPLE PROBLEM 2.4

Determine the resultant of the following five coplanar concurrent forces. All angles are measured counterclockwise with east as zero.

$$F_1 = 15 \text{ lb} \quad \text{at} \quad \alpha_1 = 20°, \qquad F_2 = 30 \text{ lb} \quad \text{at} \quad \alpha_2 = 65°$$
$$F_3 = 40 \text{ lb} \quad \text{at} \quad \alpha_3 = 135°, \qquad F_4 = 15 \text{ lb} \quad \text{at} \quad \alpha_4 = 225°$$
$$F_5 = 55 \text{ lb} \quad \text{at} \quad \alpha_5 = 320°.$$

Solution

From equation (2.1 (a)), $F_h = \sum_{n=1}^{N} F_n \cos \alpha_n$

Here $N = 5$, and equation (2.1 (a)) becomes

$$F_h = F_1 \cos \alpha_1 + F_2 \cos \alpha_2 + F_3 \cos \alpha_3 + F_4 \cos \alpha_4 + F_5 \cos \alpha_5$$

So $\quad F_h = (15 \text{ lb}) \cos 20° + (30 \text{ lb}) \cos 65° + (40 \text{ lb}) \cos 135°$
$$+ (15 \text{ lb}) \cos 225° + (55 \text{ lb}) \cos 320°$$
$$F_h = (15 \text{ lb})(0.9397) + (30 \text{ lb})(0.4226) + (40 \text{ lb})(-0.7071)$$
$$+ (15 \text{ lb})(-0.7071) + (55 \text{ lb})(0.7660).$$
$$F_h = 30.0 \text{ lb}$$

Using equation (2.1 (b)),

$$F_v = \sum_{n=1}^{N} F_n \sin \alpha_n$$

$$F_v = F_1 \sin \alpha_1 + F_2 \sin \alpha_2 + F_3 \sin \alpha_3 + F_4 \sin \alpha_4 + F_5 \sin \alpha_5$$

So $\quad F_v = (15 \text{ lb}) \sin 20° + (30 \text{ lb}) \sin 65° + (40 \text{ lb}) \sin 135°$
$$+ (15 \text{ lb}) \sin 225° + (55 \text{ lb}) \sin 320°$$
$$= (15 \text{ lb})(0.3420) + (30 \text{ lb})(0.9063) + (40 \text{ lb})(0.7071)$$
$$+ (15 \text{ lb})(-0.7071) + (55 \text{ lb})(-0.6428)$$
$$= 14.6 \text{ lb}$$
$$R = \sqrt{F_h^2 + F_v^2} \qquad \text{from equation (2.2(a))}$$
$$= \sqrt{(30.0 \text{ lb})^2 + (14.6 \text{ lb})^2}$$
$$= \sqrt{1113.2 \text{ lb}^2}$$

So, $\quad R = 33.4 \text{ lb}$
$$\alpha = \arctan (F_v/F_h) \qquad \text{from equation (2.2(b))}$$
$$= \arctan (14.6 \text{ lb}/30.0 \text{ lb})$$
$$= \arctan (0.4867)$$

So $\alpha = 26.0°$.

Answer

The resultant has a magnitude of 30 lb and is at 26.0° to the horizontal.

2.2.5   Resultant of Noncoplanar Concurrent Forces

An example of noncoplanar (not all in the same plane) concurrent (all acting through a common point) forces is a tripod. See Figure 2.8. Finding the resultant is the three-dimensional equivalent of the coplanar case. Each force is resolved into not two components, but three, along *mutually perpendicular axes*. Mutually perpendicular axes are axes that are all at 90° to each other, like the three edges of a shoe box at one corner. Before getting to the details of resolving the forces into their three components, let us first consider the problem of determining the resultant of three mutually perpendicular forces.

Resultant of Mutually Perpendicular Concurrent Forces

Figure 2.9 shows three mutually perpendicular forces, $\mathbf{F}_x$, $\mathbf{F}_y$, and $\mathbf{F}_z$. In the same way as we complete the parallelogram for two forces, we complete the *parallelepiped* (the shoe box) for three forces. The forces $\mathbf{F}_x$, $\mathbf{F}_y$, and $\mathbf{F}_z$ are directed along three mutually perpendicular axes, AB, AC, AD. If we consider just the two forces $\mathbf{F}_x$ and $\mathbf{F}_z$, their resultant is the diagonal AH, and by the Pythagorean theorem,

$$(AH)^2 = F_x^2 + F_z^2$$

Now we find the resultant of AH and F_y by drawing the diagonal AG. Again, by the Pythagorean theorem,

$$R^2 = (AG)^2 = (AH)^2 + F_y^2$$

Substituting for $(AH)^2$ gives

$$R^2 = F_x^2 + F_y^2 + F_z^2 \qquad (2.3)$$

Let the angles between the forces and the resultant (see Figure 2.10) be α_x, α_y, and α_z. You can see from the diagrams in Figure 2.10 that

$$\cos \alpha_x = \frac{F_x}{R} \qquad \cos \alpha_y = \frac{F_y}{R} \qquad \cos \alpha_z = \frac{F_z}{R} \qquad (2.4)$$

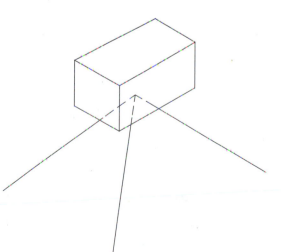

Figure 2.8 Tripod: Example of noncoplanar concurrent forces.

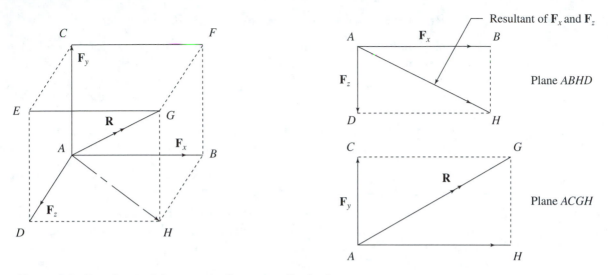

Figure 2.9   Resultant of three mutually perpendicular forces.

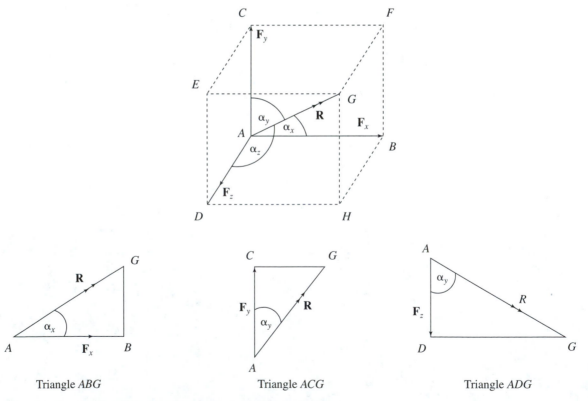

Figure 2.10   Angles between forces and resultant.

These are referred to as the **direction cosines.** We square each one,

$$\cos^2\alpha_x = \frac{F_x^2}{R^2} \qquad \cos^2\alpha_y = \frac{F_y^2}{R^2} \qquad \cos^2\alpha_z = \frac{F_z^2}{R^2}$$

and add them together:

$$\cos^2\alpha_x + \cos^2\alpha_y + \cos^2\alpha_z = \frac{F_x^2}{R^2} + \frac{F_y^2}{R^2} + \frac{F_z^2}{R^2}$$

$$= \frac{F_x^2 + F_y^2 + F_z^2}{R^2}$$

But $F_x^2 + F_y^2 + F_z^2 = R^2$, so

$$\cos^2\alpha_x + \cos^2\alpha_y + \cos^2\alpha_z = R^2/R^2$$

or $\qquad\qquad\qquad \cos^2\alpha_x + \cos^2\alpha_y + \cos^2\alpha_z = 1 \qquad\qquad\qquad\qquad (2.5)$

The sum of the squares of the direction cosines is 1.0.

EXAMPLE PROBLEM 2.5

Determine the resultant of the following three mutually perpendicular forces.

$$F_x = 50 \text{ N} \qquad F_y = 30 \text{ N} \qquad F_z = 60 \text{ N}$$

Solution

$$R^2 = F_x^2 + F_y^2 + F_z^2 \qquad \text{from equation (2.3)}$$

$$F_x = 50 \text{ N} \qquad F_y = 30 \text{ N} \qquad F_z = 60 \text{ N}$$

So, $\qquad\qquad R^2 = (50 \text{ N})^2 + (30 \text{ N})^2 + (60 \text{ N})^2 = 7000 \text{ N}^2$

Therefore, $R = 83.666$ N. Using equation (2.4):

$$\cos\alpha_x = \frac{F_x}{R} \qquad \cos\alpha_y = \frac{F_y}{R} \qquad \cos\alpha_z = \frac{F_z}{R}$$

$$\cos\alpha_x = \frac{50 \text{ N}}{83.666 \text{ N}} = 0.5976$$

$$\alpha_x = 53.3°$$

$$\cos\alpha_y = \frac{30 \text{ N}}{83.666 \text{ N}} = 0.3586$$

$$\alpha_y = 69.0°$$

$$\cos\alpha_z = \frac{60 \text{ N}}{83.666 \text{ N}} = 0.7171$$

$$\alpha_z = 44.2°$$

Check

Is the sum of the squares of the direction cosines equal to 1.0?

$$\cos \alpha_x = 0.5976 \qquad \cos \alpha_y = 0.3586 \qquad \cos \alpha_z = 0.7171$$

$$\cos^2\alpha_x + \cos^2\alpha_y + \cos^2\alpha_z = (0.5976)^2 + (0.3586)^2 + (0.7171)^2$$

$$= 0.3571 + 0.1286 + 0.5142$$

$$= 0.9999$$

The slight discrepancy is due to round-off error.

Answer

$$R = 80 \text{ N}, \qquad \alpha_x = 53.3°, \qquad \alpha_y = 69.0°, \qquad \alpha_z = 44.2°.$$

Now we can approach the problem the other way, that is, resolve a given force into three mutually perpendicular forces.

EXAMPLE PROBLEM 2.6

A force of 100 lb makes the following angles with two of the three axes:

$$\alpha_x = 60° \qquad \alpha_y = 45°$$

What are its three components?

Solution

The given force, call it **F,** is like the resultant, **R,** of the three forces we want to calculate, so using equation (2.4):

$$\cos \alpha_x = F_x/R \qquad \cos \alpha_y = F_y/R \qquad \cos \alpha_z = F_z/R$$

or $\qquad\qquad \cos \alpha_x = F_x/F \qquad \cos \alpha_y = F_y/F \qquad \cos \alpha_z = F_z/F$

So, we can write:

$$F_x = F \cos \alpha_x \qquad F_y = F \cos \alpha_y \qquad F_z = F \cos \alpha_z$$

But, we don't know α_z. We do know, however, that $\cos^2\alpha_x + \cos^2\alpha_y + \cos^2\alpha_z = 1.0$.

$$\alpha_x = 60° \qquad \alpha_y = 45°$$

So, $\qquad\qquad \cos^2 60° + \cos^2 45° + \cos^2\alpha_z = 1.0$

$$(0.5)^2 + (0.7071)^2 + \cos^2\alpha_z = 1.0$$

$$0.25 + 0.50 + \cos^2\alpha_z = 1.0$$

Thus

$$\cos^2 \alpha_z = 1.0 - 0.25 - 0.50 = 0.25$$

$$\cos \alpha_z = \sqrt{0.25} = 0.50$$

Therefore, $\alpha_z = \arccos (0.50) = 60°$.

Now $F_x = F \cos \alpha_x$, where $F = 100$ lb and $\alpha_x = 60°$.

So
$$F_x = (100 \text{ lb})(\cos 60°)$$
$$= (100 \text{ lb})(0.50)$$
$$= 50.0 \text{ lb}$$
$$F_y = F \cos \alpha_y$$
$$F = 100 \text{ lb} \quad \text{and} \quad \alpha_y = 45°$$
$$F_y = (100 \text{ lb})(\cos 45°)$$
$$= (100 \text{ lb})(0.7071)$$
$$= 70.7 \text{ lb}$$
$$F_z = F \cos \alpha_z$$
$$F = 100 \text{ lb} \quad \text{and} \quad \alpha_z = 60°$$
$$F_z = (100 \text{ lb})(\cos 60°)$$
$$= (100 \text{ lb})(0.50)$$
$$= 50.0 \text{ lb}$$

Answer

The three components of the given force are

$$F_x = 50 \text{ lb} \qquad F_y = 71 \text{ lb} \qquad F_z = 50 \text{ lb}$$

Resultant of Nonmutually Perpendicular Concurrent Forces

The problem of determining the resultant of noncoplanar concurrent forces that are *not* mutually perpendicular consists of first resolving *each* of the given forces into three components that are mutually perpendicular, algebraically adding the forces in the *x*-direction, the *y*-direction, and the *z*-direction, then finding the resultant of those three force components.

EXAMPLE PROBLEM 2.7

The magnitudes of the forces shown in Figure 2.11(a) are

$$F_1 = 100 \text{ N} \qquad F_2 = 300 \text{ N} \qquad F_3 = 200 \text{ N}$$

Three mutually perpendicular axes, *x*, *y*, and *z*, are drawn through their common point, *A*.

F_1 (Figure 2.11(b)) makes angles

$$\alpha_{1x} = 70° \qquad \alpha_{1y} = 45° \qquad \alpha_{1z} = 51.77°$$

with the three axes.

F_2 (Figure 2.11(c)) makes angles

$$\alpha_{2x} = 135° \qquad \alpha_{2y} = 60° \qquad \alpha_{2z} = 60°$$

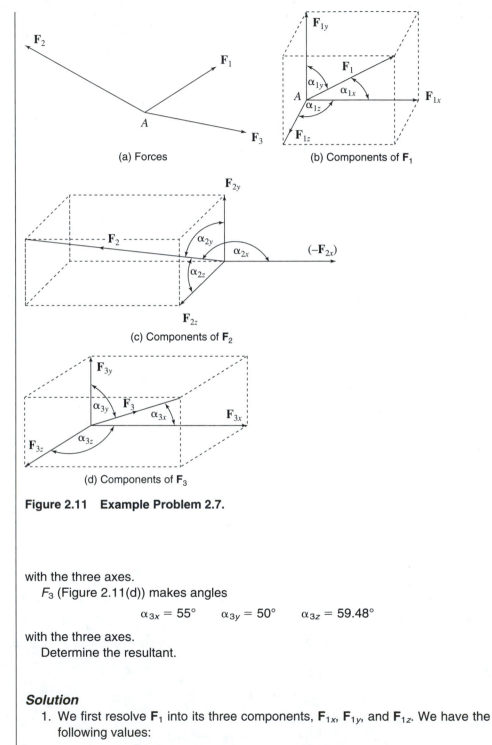

(a) Forces

(b) Components of $\mathbf{F}_1$

(c) Components of $\mathbf{F}_2$

(d) Components of $\mathbf{F}_3$

Figure 2.11   Example Problem 2.7.

with the three axes.

F_3 (Figure 2.11(d)) makes angles

$$\alpha_{3x} = 55° \qquad \alpha_{3y} = 50° \qquad \alpha_{3z} = 59.48°$$

with the three axes.

Determine the resultant.

Solution

1. We first resolve $\mathbf{F}_1$ into its three components, $\mathbf{F}_{1x}$, $\mathbf{F}_{1y}$, and $\mathbf{F}_{1z}$. We have the following values:

$$F_1 = 100 \text{ N} \qquad \alpha_{1x} = 70° \qquad \alpha_{1y} = 45° \qquad \alpha_{1z} = 51.77°$$

Therefore,

$$F_{1x} = F_1\cos\alpha_{1x} = (100\ N)(\cos 70°) = 34.2\ N$$
$$F_{1y} = F_1\cos\alpha_{1y} = (100\ N)(\cos 45°) = 70.7\ N$$
$$F_{1z} = F_1\cos\alpha_{1z} = (100\ N)(\cos 51.77°) = 61.9\ N$$

2. Similarly, for $\mathbf{F_2}$,

$$F_2 = 300\ N \qquad \alpha_{2x} = 135° \qquad \alpha_{2y} = 60° \qquad \alpha_{2z} = 60°$$

Therefore,

$$F_{2x} = F_2\cos\alpha_{2x} = (300\ N)(\cos 135°) = -212.1\ N$$
$$F_{2y} = F_2\cos\alpha_{2y} = (300\ N)(\cos 60°) = 150.0\ N$$
$$F_{2z} = F_2\cos\alpha_{2z} = (300\ N)(\cos 60°) = 150.0\ N$$

3. For F_3,

$$F_3 = 200\ N \qquad \alpha_{3x} = 55° \qquad \alpha_{3y} = 50° \qquad \alpha_{3z} = 59.48°$$

Therefore,

$$F_{3x} = F_3\cos\alpha_{3x} = (200\ N)(\cos 55°) = 114.7\ N$$
$$F_{3y} = F_3\cos\alpha_{3y} = (200\ N)(\cos 50°) = 128.6\ N$$
$$F_{3z} = F_3\cos\alpha_{3z} = (200\ N)(\cos 59.48°) = 101.6\ N$$

4. Add the components algebraically:

$$F_x = F_{1x} + F_{2x} + F_{3x} = (34.2\ N) + (-212.1\ N) + (114.7\ N)$$
$$= -63.2\ N$$
$$F_y = F_{1y} + F_{2y} + F_{3y} = (70.7\ N) + (150.0\ N) + (128.6\ N)$$
$$= 349.3\ N$$

$$F_z = F_{1z} + F_{2z} + F_{3z} = (61.9\ N) + (150.0\ N) + (101.6\ N)$$
$$= 313.5\ N$$

5. Now find the resultant of the components F_x, F_y, F_z.

$$R^2 = F_x^2 + F_y^2 + F_z^2 \qquad \text{from equation (2.3)}$$
$$= (-63.2\ N)^2 + (349.3\ N)^2 + (313.5\ N)^2$$
$$= 22.43 \times 10^4\ N^2$$

So, $R = 473.6\ N$.

Using equation (2.4) to determine the angles that the resultant makes with the three axes gives

$$\cos\alpha_x = F_x/R \qquad \cos\alpha_y = F_y/R \qquad \cos\alpha_z = F_z/R$$

where $F_x = -63.2\ N$, $F_y = 349.3\ N$, $F_z = 313.5\ N$, and $R = 473.6\ N$. So

$$\cos\alpha_x = \frac{-63.2\ N}{473.6\ N} = -0.1334$$
$$\alpha_x = \arccos(-0.1334) = 97.7°$$

$$\cos \alpha_y = \frac{349.3 \text{ N}}{473.6 \text{ N}} = 0.7375$$

$$\alpha_y = \text{arccoss} \ (0.7375) = 42.5°$$

$$\cos \alpha_z = \frac{313.5 \text{ N}}{473.6 \text{ N}} = 0.6620$$

$$\alpha_z = \text{arccos} \ (0.6620) = 48.6°$$

6. As a check, $\cos^2\alpha_x + \cos^2\alpha_y + \cos^2\alpha_z$ should equal 1.0.

$$(-0.1334)^2 + (0.7375)^2 + (0.6620)^2 = 1 \qquad \text{Check}$$

Answer

The resultant has a magnitude of 500 N and is at an angle of 97.7° to the *x*-axis, 42.5° to the *y*-axis, and 48.6° to the *z*-axis.

2.3 EQUILIBRANTS

An **equilibrant** is a single force that opposes a force or a group of forces, such that the whole system—equilibrant plus forces—is in equilibrium. Simply, the equilibrant is the *balancing* force. For example, in the case of an archer pulling on the bow string (see Figure 2.12), the two equal tensions, T, have a resultant, R, that will propel the arrow straight forward. Before letting go, the archer is pulling back with a force E, keeping the system in balance. E is the equilibrant. We see, then, that the equilibrant of a system of forces is equal in magnitude to the resultant and in exactly the opposite direction.

2.3.1 Equilibrant of Collinear Forces

The two collinear (in a straight line) forces acting on the box in Figure 2.13(a) have a resultant of 100 N acting to the right (Figure 2.13(b)). To keep the box from moving we need a force, the equilibrant, of 100 N acting to the left, as shown in Figure 2.13(c).

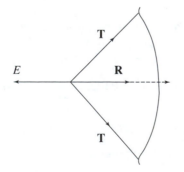

Figure 2.12 Equilibrant due to archer's pull.

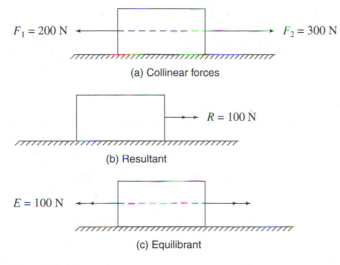

(a) Collinear forces

(b) Resultant

(c) Equilibrant

Figure 2.13

2.3.2 Equilibrant of Coplanar Concurrent Forces

To determine the *resultant*, R, of two coplanar concurrent forces, F_1 and F_2, we made a force triangle by shifting the tail of F_2 to the head of F_1; then we drew the resultant, $\overrightarrow{AC}$ from the tail of F_1 to the head of F_2 (see Figure 2.3(b)).

The equilibrant is *equal in magnitude* but *opposite in direction to the resultant*, so the equilibrant must be the vector $\overrightarrow{CA}$. Figure 2.14(a) shows two coplanar concurrent forces, F_1 and F_2; their equilibrant, **E,** is shown in Figure 2.14(b). Notice that the directions of the forces and the equilibrant *follow each other around the triangle*.

For more than two coplanar concurrent forces we can find the equilibrant by completing the polygon but drawing the "resultant" in the opposite direction. It is more convenient, however, first to determine the resultant by dividing each force into its rectangular components and then equating the equilibrant to the negative of the resultant.

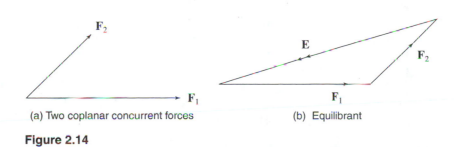

(a) Two coplanar concurrent forces

(b) Equilibrant

Figure 2.14

EXAMPLE PROBLEM 2.8

The two strings in Figure 2.15(a) make angles of 45° and 60° with the vertical. They support a mass of 75 kg. What is the tension in each string?

Solution

First resolve tension T_1 into its horizontal and vertical components. See Figure 2.15(b).

$$T_{1h} = T_1\sin 45° = T_1(0.7071)$$
$$T_{1v} = T_1\cos 45° = T_1(0.7071)$$

Resolve T_2 into its horizontal and vertical components.

$$T_{2h} = T_2\sin 60° = T_2(0.8660)$$
$$T_{2v} = T_2\cos 60° = T_2(0.50)$$

The mass of 75 kg is hanging straight down. *Be careful:* We have to covert this into a *force* (its weight). The weight, *W,* is

$$W = (75 \text{ kg})(9.81 \text{ m/s}^2) = 735.5 \text{ N}$$

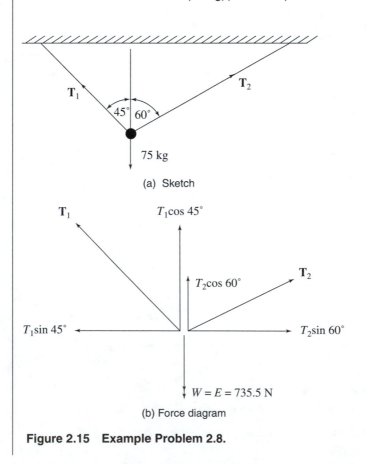

(a) Sketch

(b) Force diagram

Figure 2.15   Example Problem 2.8.

Because the system is in equilibrium (the object isn't moving up or down), the upward pull of the resultant must be equal in magnitude to the weight. That is,

$$T_{1v} + T_{2v} = 735.5 \text{ N}$$

So
$$0.7071 T_1 + 0.50 T_2 = 735.5 \text{ N} \tag{1}$$

But since the object is in equilibrium, it is also not moving sideways, so the two horizontal components must be equal:

$$T_{1h} = T_{2h}$$
$$0.7071 T_1 = 0.8660 T_2 \tag{2}$$

Substituting $(0.8660 T_2)$ for $(0.7071 T_1)$ in equation (1) gives

$$0.8660 T_2 + 0.50 T_2 = 735.5 \text{ N}$$
$$1.3660 T_2 = 735.5 \text{ N}$$
$$T_2 = 538.4 \text{ N}$$

Substituting for T_2 in equation (2):

$$0.7071 T_1 = (0.8660)(538.4 \text{ N}) = 466.3 \text{ N}$$
$$T_1 = \frac{466.3 \text{ N}}{0.7071} = 659.4 \text{ N}$$

Answer
The tensions in the strings are 660 N and 540 N.

Does it make sense that the steeper string should carry more of the weight than the more horizontal string? Yes. Take them to the limit. Imagine that string 1 is vertical and string 2 is horizontal. Then string 1 would take *all* the weight, and string 2 would take none.

2.4   THE FREE BODY

The only external forces acting on the 75-kg mass shown in Figure 2.15(a) are its weight, **W**, and the two tensions, **T**₁ and **T**₂. For calculation purposes we can ignore the ceiling to which the strings are attached and draw just the forces acting on the body. The result is Figure 2.16, the **free-body diagram.**

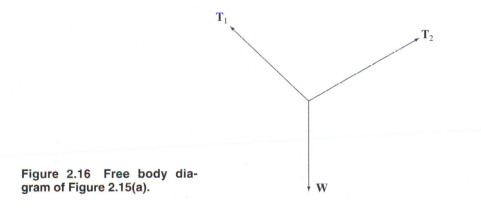

Figure 2.16 Free body diagram of Figure 2.15(a).

In Figure 2.17(a) we show a ladder against a vertical wall. The external forces acting on the ladder are its weight, **W,** a ground force, $\mathbf{F}_g$ (called a reaction), which is preventing it from sinking and from sliding, and a wall force (another reaction), $\mathbf{F}_w$, that is holding it up and preventing it from sliding down the wall. We set the ladder out as a free body, as in Figure 2.17(b), where all the external forces acting on the ladder are

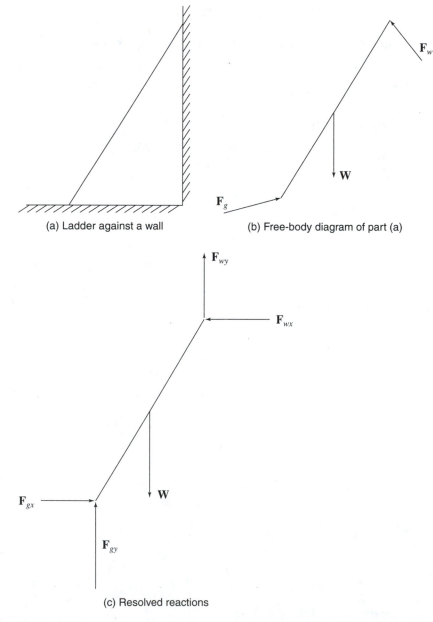

(a) Ladder against a wall   (b) Free-body diagram of part (a)

(c) Resolved reactions

Figure 2.17

included. We then resolve the ground reaction, $\mathbf{F}_g$, and the wall reaction, $\mathbf{F}_w$, into their horizontal and vertical components, as in Figure 2.17(c).

The sum of the forces acting downward must be equal in magnitude to the sum of the forces acting upward. If they were not equal, the ladder would sink or rise. Because downward-acting forces are negative and upward acting forces are positive, their algebraic sum must be zero. These are all in the y-direction, so in compact form we can write

$$\Sigma F_y = 0$$

Also, for equilibrium the sum of the x-components would have to be zero. If they were not, the ladder would move left or right. So,

$$\Sigma F_x = 0$$

Similarly, for three-dimensional cases, the sum of the z-components must also be zero for equilibrium.

$$\Sigma F_z = 0$$

In general, then, for a body to be in equilibrium, the following conditions must be satisfied:

$$\Sigma F_x = 0 \qquad\qquad (2.6(\text{a}))$$
$$\Sigma F_y = 0 \qquad\qquad (2.6(\text{b}))$$
$$\Sigma F_z = 0 \qquad\qquad (2.6(\text{c}))$$

These are *necessary*, but *not sufficient*, conditions for a body to be in equilibrium. We will see in the next chapter that the sum of the *moments* must also be zero.

2.5   REACTIONS

The ladder in Figure 2.17(a) is not sinking into the ground, pushing the wall over, or sliding. (We hope!) The forces acting at the ends of the ladder, $\mathbf{F}_g$ and $\mathbf{F}_w$ (Figure 2.17(b)), are called the *reactions*. Since $\mathbf{W}$, $\mathbf{F}_g$, and $\mathbf{F}_w$ are the only external forces acting on the ladder and since the ladder is in equilibrium, then *the force triangle must close*. This is because the resultant of $\mathbf{F}_g$ and $\mathbf{F}_w$ must balance the weight, or $\mathbf{W}$ is the equilibrant of $\mathbf{F}_g$ and $\mathbf{F}_w$. We show this force triangle in Figure 2.18.

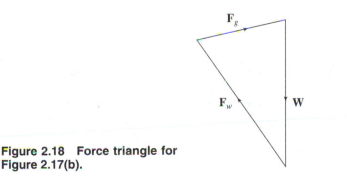

Figure 2.18   Force triangle for Figure 2.17(b).

2.6   TWO-FORCE MEMBERS

A **two-force member** is a member that is subjected solely to axial forces or to forces whose resultants are purely axial. Two-force members are found in many bridges, roof trusses, and other structures. For example, let us think about the simple truss shown in Figure 2.19(a). It consists of two members, AB and BC, pin-jointed to each other and to the wall. The weight, W, is trying to compress BC and stretch AB.

Imagine that we remove member AB. Member BC would swing and hang straight down from point C. Similarly, if we were to remove BC, AB would swing and hang straight down from point A. There are no forces trying to bend either AB or BC. The forces acting on them are purely axial. AB and BC are two-force members. If, for example, we firmly fix member AB to the wall by concreting it in, then if we remove BC, weight W would try to bend member AB. In that case AB is *not* a two-force member.

Figure 2.19(b) shows the free-body diagram for the truss, and since the forces are in equilibrium, the force triangle must close, as in Figure 2.19(c).

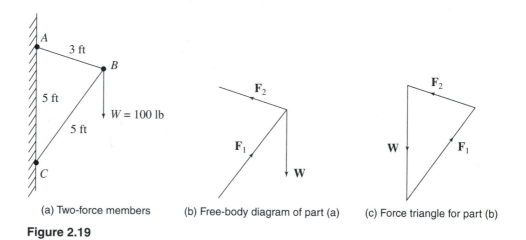

(a) Two-force members   (b) Free-body diagram of part (a)   (c) Force triangle for part (b)

Figure 2.19

EXAMPLE PROBLEM 2.9

Determine the forces in *AB* and *BC* in Figure 2.19(a).

Solution

1. Draw the force triangle, as in Figure 2.19(c). We know that the triangle will close because the system is in equilibrium. First, draw a vertical line, say, 10 cm long, to represent the 100-lb weight. Because *BC* is a two-force member, the force acting on it, the compression, is pushing in the direction *C → B*, opposing the external force that is compressing it. So we can draw a line from the head of *W* upward at the same angle that *CB* is to the vertical. We don't know how long to make it, but we do know that the tension force in *AB* is pulling up in the direction *B → A* and that the force triangle must close. All we have to do,

then, to complete the force triangle is to draw F_2 backward from the tail of W to meet F_1.

2. Calculate the forces. The force triangle, Figure 2.19(c), is identical in shape to the physical triangle formed by the members. *This is so only because we are dealing with two-force members.* We say the triangles are *similar triangles.* In geometry, *similar* means having the same shape but not the same size.

   The lengths of the sides of the force triangle (Figure 2.19(c)) represent the magnitudes of the forces.

   We have represented the 100-lb weight by a vertical line 10 cm long, so 1 cm represents a force of (100 lb/10 cm) × 1 cm = 10 lb. In the similar triangle (the physical triangle) in Figure 2.19(a), the vertical line *AC* is like the line **W** in the force triangle and is 5 ft long. The member *BC* is also 5 ft long, so the line representing force **F₁** in Figure 2.19(c) must also be 10 cm long, so F_1 must be 100 lb. Since *AB* is 3 ft long and 5 ft is represented by 10 cm, the line representing force **F₂** must be (10 cm/5 ft) (3 ft) = 6 cm long. Because 1 cm represents 10 lb, 6 cm represents 60 lb.

Answer
The forces are, therefore,

$$F_1 = 100 \text{ lb} \quad \text{(compression)} \qquad F_2 = 60 \text{ lb} \quad \text{(tension)}$$

2.7 COMPUTER ANALYSES

2.7.1 Program 'Resultant'

The program 'Resultant' calculates the resultant of up to 10 concurrent forces. The forces may be *coplanar* or *noncoplanar*. For the coplanar case, you enter the magnitudes of the forces and their angles in degrees measured in a counterclockwise direction, with east being zero. For the noncoplanar case, you enter the magnitudes of the forces and their angles from the x- and y-axes. The angles from the z-axis are automatically determined.

EXAMPLE PROBLEM 2.10

Determine the resultant of the following three coplanar concurrent forces.

$$F_1 = 300 \text{ N at } 30°, \qquad F_2 = 500 \text{ N at } 75°, \qquad F_3 = 850 \text{ N at } 125°$$

Solution
1. Open program 'Resultant.'
2. Enter SI in cell C4.
3. Enter the forces 300, 500, 850 in the Force column under *Coplanar.*
4. Enter the corresponding angles, 30, 75, 125.

The answer is shown as

Resultant = 1332.9 N

At an angle of 94.2 degrees

EXAMPLE PROBLEM 2.11

Determine the resultant of the following six coplanar concurrent forces.

$F_1 = 50$ lb at $25°$, $F_2 = 75$ lb at $57°$, $F_3 = 123$ lb at $17°$

$F_4 = 146$ lb at $45°$, $F_5 = 185$ lb at $320°$, $F_6 = 212$ lb at $197°$

Solution
1. Enter US in cell C4.
2. Enter 50, 75, 123, 146, 185, 212 in the Force column
3. Enter 25, 57, 17, 45, 320, 197 in the Angle column.

The answer is given as

Resultant $= 249.6$ lb

At an angle of 9.76 degrees

EXAMPLE PROBLEM 2.12

This example problem repeats Example Problem 2.7 to demonstrate its solution using program 'Resultant.'

The magnitudes of the forces are

$F_1 = 100$ N, $F_2 = 300$ N, $F_3 = 200$ N

F_1 makes angles

$\alpha_{1x} = 70°$, $\alpha_{1y} = 45°$, $\alpha_{1z} = 51.77°$

with the three axes. F_2 makes angles

$\alpha_{2x} = 135°$, $\alpha_{2y} = 60°$, $\alpha_{2z} = 60°$

with the three axes, and F_3 makes angles

$\alpha_{3x} = 55°$, $\alpha_{3y} = 50°$, $\alpha_{3z} = 59.48°$

with the three axes. Determine the resultant.

Solution
1. Enter SI in cell C4.
2. Go to the *Noncoplanar* section of the screen.
3. Enter 100, 300, 200 in the Force column.
4. Enter 70, 135, 55 in the Angle *x* column.
5. Enter 45, 60, 50 in the Angle *y* column.

The answer is given as

Resultant $= 473.5$ N

At angles of 97.7 degrees to the *x*-axis

42.5 degrees to the *y*-axis

48.6 degrees to the z axis

These results agree with the hand calculation in Example Problem 2.7.

2.8   PROBLEMS FOR CHAPTER 2

Hand Calculations

1. A body is acted upon by a horizontal force of 120 N to the right and a vertical downward force of 100 N. What is the resultant force on the body?

2. Determine the resultant of a horizontal force of 300 lb to the right and a force of 500 lb at an angle of 135° to the horizontal, taking east as zero.

3. Two forces that are equal in magnitude act on a body. One is pointing in a northwesterly direction, and the resultant of the two forces points west.

   a. In which direction is the second force acting?
   b. Show that if the magnitude of each force is F pounds, then the magnitude of the resultant is $\sqrt{2} \times F$ pounds.

4. A vertical force of 200 N is resolved into two components, one of 500 N and the other of 400 N. Calculate the angle each makes with the horizontal.

5. Determine the resultant of the three coplanar concurrent forces shown in Figure 2.20.

6. Find the equilibrant for the force system shown in Figure 2.20.

7. Determine the resultant of the following five coplanar concurrent forces. All angles are measured counterclockwise with east as zero.

$$F_1 = 25 \text{ lb at } \alpha_1 = 25°,$$
$$F_2 = 30 \text{ lb at } \alpha_2 = 30°,$$
$$F_3 = 40 \text{ lb at } \alpha_3 = 95°,$$
$$F_4 = 75 \text{ lb at } \alpha_4 = 180°,$$
$$F_5 = 55 \text{ lb at } \alpha_5 = 220°$$

8. Determine the resultant of the following three mutually perpendicular forces.

$$F_x = 45 \text{ kN}, \qquad F_y = 70 \text{ kN}, \qquad F_z = 25 \text{ kN}$$

9. A force of 250 lb makes the following angles with two of the three axes:

$$\alpha_x = 55°, \qquad \alpha_y = 35°$$

What are its three components?

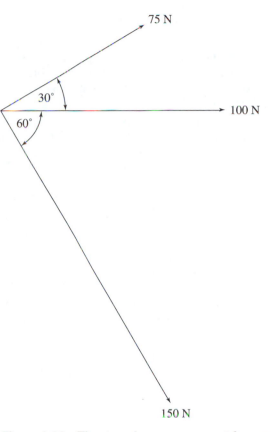

Figure 2.20   Three coplanar concurrent forces.

Hint: There is a quick way to do this. Notice that the sum of the two given angles is 90°; sin $\alpha = \cos(90° - \alpha)$; and $\sin^2\alpha + \cos^2\alpha = 1$.

10. Resolve a force of 60 N into two components, one of which is 15 N and makes an angle of 75° with the other force.

Hint: Draw the force diagram, use the cosine rule, and then use the sine rule.

11. An engine block is suspended by two ropes attached to a ceiling beam. Both ropes make angles of 50° with the vertical, and the tension in each rope is 865 N. What is the *mass* of the engine block?

12. A concrete block weighing 700 N is suspended by a 3.5-m-long rope. It is to be loaded on to a truck by pushing it horizontally a distance of 1.0 m. How much force is required, and what will be the tension in the rope?

13. If the person pushing the block in the previous question can exert a maximum thrust of 650 N, how far can the block be pushed horizontally?

14. The ends of a wire are to be attached to two hooks 2 ft apart on the same level, and a weight of 300 lb is to be suspended from the middle of the wire. If the wire will break with

a direct pull of 200 lb, what minimum length of wire should be used?

15. In Problem 14, if the distance between the hooks is d feet, the suspended weight is W pounds, and the breaking tension is T pounds, show that the minimum length of wire is given by the equation

$$L = \frac{2Td}{\sqrt{4T^2 - W^2}}$$

Hint: Let the angle between the wire and the vertical be α. Express $\cos\alpha$ in terms of W and T. Express $\sin\alpha$ in terms of d and L. Use $\cos^2\alpha + \sin^2\alpha = 1$, and simplify.

Computer Calculations

16. Repeat Problem 5.

17. Repeat Problem 7.

18. Repeat Problem 8.

Note: all the input angles are zero or 90°.

19. Determine the resultant of the following nine coplanar concurrent forces.

$F_1 = 55$ N at $\alpha_1 = 15°$,
$F_2 = 30$ N at $\alpha_2 = 230°$

$F_3 = 47$ N at $\alpha_3 = 75°$,
$F_4 = 75$ N at $\alpha_4 = 173°$

$F_5 = 152$ N at $\alpha_5 = 102°$,
$F_6 = 68$ N at $\alpha_6 = 180°$

$F_7 = 85$ N at $\alpha_7 = 225°$,
$F_8 = 175$ N at $\alpha_8 = 12°$

$F_9 = 105$ N at $\alpha_9 = 340°$

20. The magnitudes of three noncoplanar concurrent forces are

$F_1 = 550$ lb, $F_2 = 300$ lb, $F_3 = 465$ lb

F_1 makes angles $\alpha_{1x} = 85°$, $\alpha_{1y} = 35°$, $\alpha_{1z} = 55.47°$ with the three axes. F_2 makes angles $\alpha_{2x} = 45°$, $\alpha_{2y} = 60°$, $\alpha_{2z} = 60°$ with the three axes. F_3 makes angles $\alpha_{3x} = 53°$, $\alpha_{3y} = 90°$, $\alpha_{3z} = 37°$ with the three axes. Determine the resultant.

21. Write a program to determine the angle between a force and the z-axis given the angles between the force and the x-axis and the force and the y-axis. Include a test to check that the two given angles are possible. If they are not, display a warning that there is an input error.

3

Moments

■ Objectives

In this chapter you will learn about moments and couples and how to determine reactions for beams carrying concentrated and distributed loads.

3.1 MOMENTS

It is common experience that it is easier to loosen a nut by pushing near the end of a wrench (point 1 in Figure 3.1(a)) than by pushing close to the nut (point 2). In fact, it is much easier if you slide a long tube over the wrench and push at the end of the tube (see Figure 3.1(b)). We say that this gives us more *leverage*. In engineering terms, we are talking about *moments*.

The moment of a force about an axis is equal to the magnitude of the force times the perpendicular distance from the line of action of the force to the axis.

In Figure 3.2(a) we show a 10-lb force pushing on a 9-in.-long wrench at 90°. The perpendicular distance, l, from the line of action of the force to the axis P, the center of the nut, is 9 in. So the moment about P is

$$M_P = F \times l = (10 \text{ lb})(9 \text{ in.}) = 90 \text{ lb-in.}$$

The same 10-lb force is acting on the same 9-in. wrench in Figure 3.2(b), but this time at 30° to the wrench handle. The distance l is now $(9 \text{ in.})(\sin 30°) = (9 \text{ in.})(0.5) = 4.5$ in., so the moment about P is

$$M_P = (10 \text{ lb})(4.5 \text{ in.}) = 45 \text{ lb-in.}$$

The "twisting" on the nut is only half of what it was when the force was exerted at a right angle to the wrench handle.

Applying the force right along the axis of the handle will push only on the nut (Figure 3.2(c)). The distance l will be zero, and there will be no moment about P.

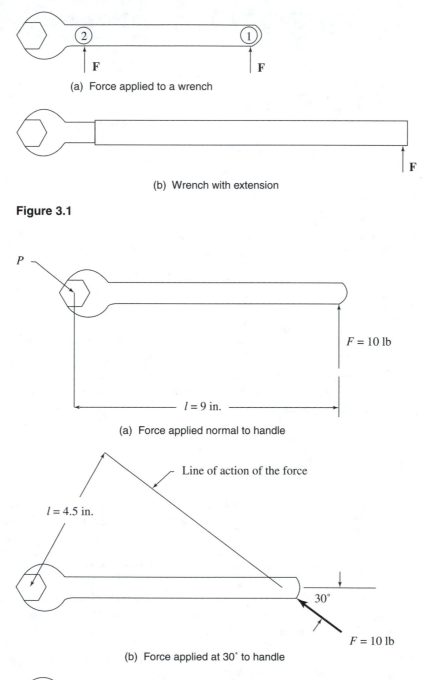

(a) Force applied to a wrench

(b) Wrench with extension

Figure 3.1

P

$F = 10$ lb

$l = 9$ in.

(a) Force applied normal to handle

Line of action of the force

$l = 4.5$ in.

$30°$

$F = 10$ lb

(b) Force applied at 30° to handle

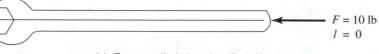

$F = 10$ lb
$l = 0$

(c) Force applied along handle axis

Figure 3.2

EXAMPLE PROBLEM 3.1

What is the moment about P in Figure 3.2(b) if the force is 100 N, the wrench handle is 23 cm, and the angle is 45°?

Solution

The perpendicular distance, l, from the line of action of the force to the axis, P, is $l = (230 \text{ mm})(\sin 45°) = 162.6 \text{ mm}$.

$$M_P = F \times l$$
$$= (100 \text{ N})(162.6 \text{ mm}) = 16{,}260 \text{ N·mm} = 16.3 \text{ N·m}$$

Answer

The moment about P is 16 N·m.

We are going to use moments and forces to determine the *reactions* acting on bodies. Reactions are forces that arise to balance external forces. For example, a ladder leaning against a wall would push the wall over if the wall were not pushing back with an equal force—the wall reaction. Also, it would sink into the ground if there were not a ground reaction opposing the weight. A plank lying across two sawhorses is being held up by the reactions arising in the supports.

3.2   CONCENTRATED LOADS

A **concentrated load** is an external force that acts essentially at a point. The two children at each end of a teeter-totter (seesaw), as in Figure 3.3(a), are concentrated loads. The loads are their weights, W_1 and W_2, and they act through their centers of gravity (Figure 3.3(b)).

The children and the plank are being supported by the reaction, R. If the lengths from the support to the children, l_1 and l_2, are equal, then we know by experience that W_1 must equal W_2 for the plank to balance. The moment of W_1 about the support is

$$M_1 = W_1 \cdot l_1$$

This is a counterclockwise moment, trying to turn the plank in the opposite direction to the hands of a clock.

The moment due to W_2 is $M_2 = W_2 \cdot l_2$, which is a clockwise moment.

We need a sign convention. In normal Cartesian (x, y) coordinates, going vertically upward is positive and going vertically downward is negative. The weights, W_1 and W_2, are acting vertically downward (negative). Also, moving to the right is positive, and moving to the left is negative. For weight W_1, the distance *from the support to the weight* is in the negative direction, so the moment, M_1, is the product of two negative values and is, therefore, positive. This means, then, that *counterclockwise mo-*

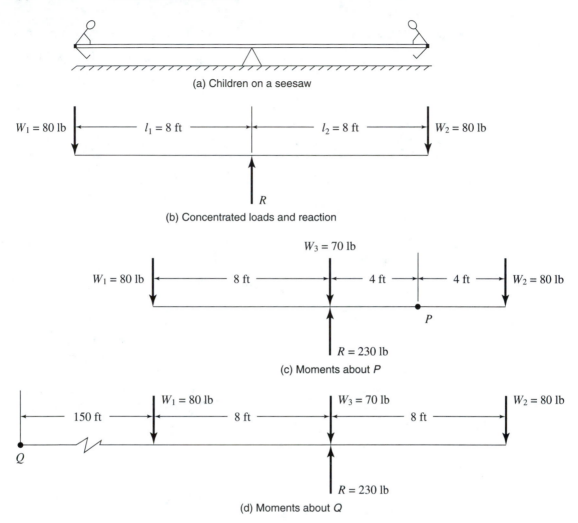

(a) Children on a seesaw

(b) Concentrated loads and reaction

(c) Moments about *P*

(d) Moments about *Q*

Figure 3.3

ments *are positive*. Similarly, W_2 is negative and l_2 is positive. Their product is negative, showing that *clockwise moments are negative*.

Counterclockwise moments are positive.

Clockwise moments are negative.

Suppose that, in Figure 3.3(b), $l_1 = l_2 = 8$ ft and that both children weigh 80 lb. Then the counterclockwise moment, M_1, is

$$M_1 = (80 \text{ lb})(8 \text{ ft}) = 640 \text{ lb-ft}$$

and the clockwise moment, M_2, is

$$M_2 = -(80 \text{ lb})(8 \text{ ft}) = -640 \text{ lb-ft}$$

Notice that the reaction R, which is a force, has no moment about the support because the distance from R to R is zero.

For the seesaw to balance, the algebraic sum of the moments must be zero:

$$M_1 + M_2 = 0$$
$$640 \text{ lb-ft} + (-640 \text{ lb-ft}) = 0.$$

Let us now sum up the moments about any other station, say P in Figure 3.3(c). The reaction R *will* have a moment about point P, so we will have to determine the magnitude of R before we can calculate its moment. We will call the weight of the plank W_3, and this will act at its center of gravity, which is where the support is. If the magnitude of R were less than $(W_1 + W_2 + W_3)$, the plank would sink to the ground. If R were greater than $(W_1 + W_2 + W_3)$, the plank would rise off the ground. We must conclude, therefore, that $R = (W_1 + W_2 + W_3)$

Suppose the plank weighs 70 lb; then $R = (80 \text{ lb} + 80 \text{ lb} + 70 \text{ lb}) = 230 \text{ lb}$. This is positive because R is acting in an upward direction.

Next, we take moments about P (Figure 3.3(c)). The moment due to W_1 is a counterclockwise moment:

$$M_1 = (80 \text{ lb})(12 \text{ ft}) = 960 \text{ lb-ft}$$

The moment due to W_2 is a clockwise moment:

$$M_2 = -(80 \text{ lb})(4 \text{ ft}) = -320 \text{ lb-ft}$$

The moment due to W_3 is a counterclockwise moment:

$$M_3 = (70 \text{ lb})(4 \text{ ft}) = 280 \text{ lb-ft}$$

The moment due to R is a clockwise moment:

$$M_R = -(230 \text{ lb})(4 \text{ ft}) = -920 \text{ lb-ft}$$

The algebraic sum of the moments about P is

$$M_1 + M_2 + M_3 + M_R = (960 \text{ lb-ft}) + (-320 \text{ lb-ft}) + (280 \text{ lb-ft})$$
$$+ (-920 \text{ lb ft})$$
$$= 0$$

This is a general rule that can be applied to any number of moments.

For a body to be in equilibrium, the algebraic sum of the moments about any point acting on the body must be zero.

$$\Sigma M)_P = 0 \tag{3.1}$$

To emphasize that we do mean *any* point, let us take moments about point Q (Figure 3.3(d)), which is 150 ft to the left of the left end of the plank.

$$M_1 = -(80 \text{ lb})(150 \text{ ft}) = -12{,}000 \text{ lb-ft}$$
$$M_2 = -(80 \text{ lb})(150 \text{ ft} + 8 \text{ ft} + 8 \text{ ft}) = -(80 \text{ lb})(166 \text{ ft})$$
$$= -13{,}280 \text{ lb-ft}$$
$$M_3 = -(70 \text{ lb})(150 \text{ ft} + 8 \text{ ft}) = -(70 \text{ lb})(158 \text{ ft})$$
$$= -11{,}060 \text{ lb-ft}$$
$$M_R = (230 \text{ lb})(158 \text{ ft}) = 36{,}340 \text{ lb-ft}$$

The sum of the moments is

$$M_1 + M_2 + M_3 + M_R = -12,000 \text{ lb-ft} + (-13,280 \text{ lb-ft})$$
$$+ (-11,060 \text{ lb-ft}) + 36,340 \text{ lb-ft}$$
$$= 0$$

Equation (3.1), together with equations (2.6(a), (b), and (c), are the four *conditions to be satisfied for a body to be in equilibrium*.

$$\Sigma M)_P = 0 \qquad \text{where } P \text{ is any point}$$
$$\Sigma F_x = 0$$
$$\Sigma F_y = 0$$
$$\Sigma F_z = 0$$

EXAMPLE PROBLEM 3.2

A 7-m beam (Figure 3.4) is supported at a point 2 m from its left end. There are external loads of 500 N at the left end and 100 N that is 2 m from the right end. What *mass, W,* at the right end of the beam is needed to balance the beam?

Solution

We could solve this by taking moments about any point, but the easy way is to take moments about the support; then we don't have to know the reaction, *R.* For equilibrium,

$$\Sigma M)_R = 0$$

where
$$\Sigma M)_R = M_1 + M_2 + M_3.$$
$$M_1 \text{ (ccw)} = +(500 \text{ N})(2 \text{ m}) = 1000 \text{ N·m}$$
$$M_2 \text{ (cw)} = -(100 \text{ N})(3 \text{ m}) = -300 \text{ N·m}$$
$$M_3 = -(W \text{ N})(5 \text{ m}) = -5W \text{ N·m}$$

So

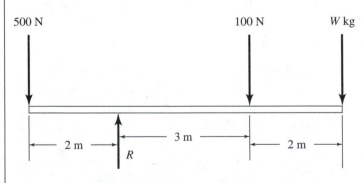

Figure 3.4   Example Problem 3.2.

$$\Sigma M)_R = 1000 \text{ N·m} + (-300 \text{ N·m}) + (-5W \text{ N·m}) = 0$$

Therefore,

$$5W = 1000 \text{ N·m} - 300 \text{ N·m} = 700 \text{ N·m}$$

$$W = \frac{700 \text{ N·m}}{5 \text{ m}} = 140 \text{ N}$$

We now have to convert this to kg:

$$W \text{ (kg)} = \frac{W \text{ (N)}}{9.81 \text{ m/s}^2}$$

So $W = (140 \text{ N})/(9.81 \text{ m/s}^2) = 14.3$ kg.

Answer
The required mass is 14 kg.

EXAMPLE PROBLEM 3.3

A 15-ft beam is supported on a frictionless roller that is positioned at 6 ft from its left end (see Figure 3.5(a)). A 50-lb force is applied to the left end of the beam at an angle of 45° to the horizontal. What force, at what angle, applied to the right end of the beam is necessary to keep the beam in equilibrium?

Solution
There are three conditions to be satisfied to keep the beam in equilibrium:

1. The algebraic sum of the vertical force components must equal zero.
2. The algebraic sum of the horizontal force components must equal zero.
3. The algebraic sum of the moments about any point must be zero.

We will resolve the two forces into their horizontal and vertical components (Figure 3.5(b)) and then apply the preceding conditions.

The horizontal component of the 50-lb force is a force pointing to the right:

$$F_{1h} = (50 \text{ lb})(\cos 45°) = (50 \text{ lb})(0.7071)$$

$$= 35.36 \text{ lb}$$

Because the roller is frictionless, it cannot prevent the beam from sliding, so the only horizontal forces acting on the beam are the two horizontal-end force components. The right-end horizontal component of force, which is $F_2 \cos \alpha$, must therefore be equal to 35.36 lb:

$$F_2 \cos \alpha = 35.36 \text{ lb}$$

The vertical component of the 50-lb force is

$$F_{1v} = (50 \text{ lb})(\sin 45°) = (50 \text{ lb})(0.7071) = 35.36 \text{ lb}$$

If we take moments about the roller, the reaction, **R**, is not required because it has no moment contribution, so we do not have to include the calculation of balancing vertical forces.

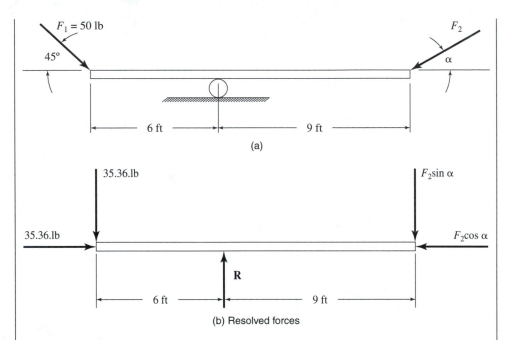

(a)

(b) Resolved forces

Figure 3.5   Example Problem 3.3

We next take moments about the roller. The vertical component of force at the left end (35.36 lb) creates a counterclockwise moment about the roller, and the vertical component of force at the right end ($F_2\sin\alpha$) creates a clockwise moment about the roller. For the beam to balance (to be in equilibrium), the algebraic sum of the two moments created by these force components must be zero.

$$-(F_2\sin\alpha)(9\text{ ft}) + (35.36\text{ lb})(6\text{ ft}) = 0$$

$$-(F_2\sin\alpha)(9\text{ ft}) + 212.16\text{ lb-ft} = 0$$

So
$$F_2\sin\alpha = \frac{212.16\text{ lb-ft}}{9\text{ ft}}$$

$$F_2\sin\alpha = 23.57\text{ lb}$$

We now have the following values for $F_2\cos\alpha$ and $F_2\sin\alpha$:

$$F_2\cos\alpha = 35.36\text{ lb} \qquad F_2\sin\alpha = 23.57\text{ lb}$$

Remembering that $\sin\alpha/\cos\alpha = \tan\alpha$,

$$\frac{F_2\sin\alpha}{F_2\cos\alpha} = \tan\alpha \qquad \text{(because the } F_2\text{'s cancel)}$$

So
$$\tan\alpha = 23.57/35.36 = 0.6667$$

Therefore, $\alpha = \arctan(0.6667) = 33.69°$ and $F_2\cos\alpha = 35.36$ lb, so $F_2 = (35.36$ lb$)/(\cos 33.69°) = 42.5$ lb.

Answer
The required force has a magnitude of 43 lb, and it is pointing down at 33.7°.

3.3 UNIFORMLY DISTRIBUTED LOADS

An example of a *uniformly distributed load* is a layer of snow on a roof or a layer of concrete (the roadway) on a bridge. In diagrams, we show a uniformly distributed load as
a rectangle, as in Figure 3.6(a), with its weight per foot or weight per meter inside the
rectangle.

The distributed load in Figure 3.6(a) is 100 N/m and extends a distance of 5 m. The
total load is (100 N/m)(5 m) = 500 N. Imagine that we split this load into five 100-
N blocks and treat each block as a concentrated load, as shown in Figure 3.6(b). Label them 1 through 5 from left to right. Now take moments about A.

For block 1 l_1 = 10.5 m, because its weight acts at its center, which is 10 m +
0.5 m from the left end of the beam. Its weight is 100 N, and it is creating a clockwise
(negative) moment about A:

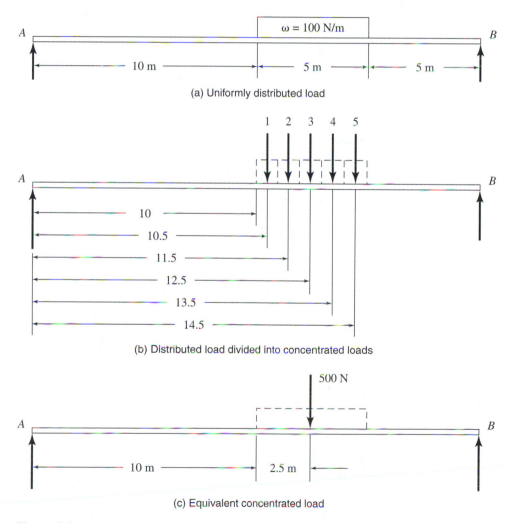

(a) Uniformly distributed load

(b) Distributed load divided into concentrated loads

(c) Equivalent concentrated load

Figure 3.6

$$M_{A1} = -(100 \text{ N})(10.5 \text{ m}) = -1050 \text{ N·m}$$

Let us tabulate the calculation:

Block No.	Dist. From A (m)	Load (N)	Moment About A (N·m)
1	10.5	100	−1050
2	11.5	100	−1150
3	12.5	100	−1250
4	13.5	100	−1350
5	14.5	100	−1450

The total moment about A is the sum of the last column, which is −6250 N·m.

In Figure 3.6(c) we have replaced the distributed load with a single concentrated load located at the center of gravity of the distributed load. Its weight is 500 N, and it is 12.5 m from A. Its moment about A is

$$M_A = -(500 \text{ N})(12.5 \text{ m}) = -6250 \text{ N·m}, \quad \text{as before}$$

We see, therefore, that to calculate moments due to distributed loads, we simply replace each load by its equivalent concentrated load and take moments as before.

EXAMPLE PROBLEM 3.4

Calculate the moment about A due to the external loads shown in Figure 3.7(a).

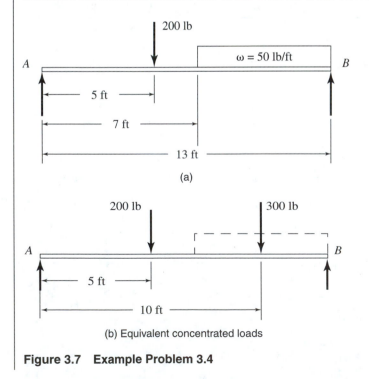

(a)

(b) Equivalent concentrated loads

Figure 3.7   Example Problem 3.4

Solution

The distributed load extends from 7 ft to 13 ft, so it is 6 ft long. Because it weighs 50 lb/ft, its total weight is 300 lb. This will be acting at its center of gravity, which is 7 ft + 3 ft, i.e., 10 ft from A. The equivalent concentrated loading diagram is shown in Figure 3.7(b).

Taking moments about *A*:

$$M_A = -(200 \text{ lb})(5 \text{ ft}) - (300 \text{ lb})(10 \text{ ft})$$
$$= -(1000 \text{ lb-ft}) - (3000 \text{ lb-ft})$$

So, $M_A = -4000$ lb-ft.

Answer

The moment about *A* is 4000 lb-ft (clockwise).

3.4 COUPLES

A couple is a special type of moment created by two parallel forces that are equal in magnitude but opposite in direction.

A couple is being applied to the steering wheel in Figure 3.8. If the wheel radius is 0.8 ft, then the upward-directed 25-lb force on the right creates a moment about the center of

$$M_C = (25 \text{ lb})(0.8 \text{ ft}) = 20 \text{ lb-ft} \qquad \text{(counterclockwise)}$$

The downward-directed 25-lb force on the left creates a moment about the center of

$$M_C = (20 \text{ lb})(0.8 \text{ ft}) = 20 \text{ lb-ft} \qquad \text{(counterclockwise)}$$

So the total moment about C is 40 lb-ft. This is the same as taking one of the forces (25 lb) and multiplying it by the perpendicular distance between the forces (1.6 ft): 25 lb × 1.6 ft = 40 lb-ft.

The perpendicular distance between the two forces of a couple is called the **arm.**

The moment of a couple, or the torque, is equal to one of the forces times the arm.

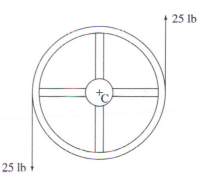

Figure 3.8   Couple applied to a steering wheel.

25 lb

25 lb

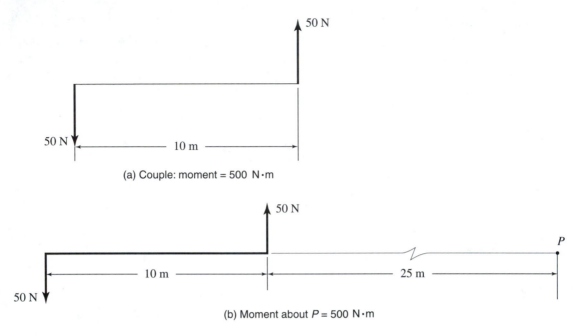

(a) Couple: moment = 500 N·m

(b) Moment about P = 500 N·m

Figure 3.9

We show a couple in Figure 3.9(a) whose forces are 50 N and whose arm is 10 m. Its moment, or torque, is $(50\,\text{N})(10\,\text{m}) = 500\,\text{N·m}$. If we calculate its moment about, say, point P in Figure 3.9(b):

$$M_P = -(50\,\text{N})(25\,\text{m}) + (50\,\text{N})(35\,\text{m}) = -1250\,\text{N·m} + 1750\,\text{N·m}$$

or $M_p = 500\,\text{N·m}$, as before.

The moment of a couple is a constant.

EXAMPLE PROBLEM 3.5

Do the forces in Figure 3.10(a) constitute a couple? If so, what is its moment?

Solution
The forces at each end of the arm are collinear, so they can be added algebraically, resulting in Figure 3.10(b).

This *is* a couple because it comprises two parallel forces that are equal in magnitude and opposite in direction.

The moment is $M = (30\,\text{lb})(12\,\text{ft}) = 360\,\text{lb-ft}$.

Answer
Yes, they do form a couple. Its moment is 360 lb-ft (ccw).

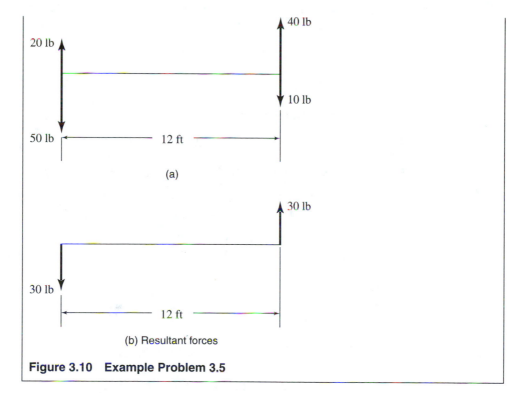

Figure 3.10   Example Problem 3.5

3.5   CALCULATING REACTIONS

Reactions are the forces that oppose external loads.

3.5.1   Calculating Reactions for One Concentrated Load

Figure 3.11 shows a single concentrated load of 100 lb at 4 ft from the left end of a 10-ft beam. The beam is resting on two supports, one at each end. The question is, What are the reactions R_1 and R_2? Before answering that question, let us ask the reasonable question, Why do we care? The reason is that once we know the reactions, which are the forces in the supports, we can determine the stresses and so design them so that they will not fail.

If we ignore the weight of the beam, which is often a good approximation because the beam weight is usually less than 10% of the total load, and then take moments about A:

$$\Sigma M)_A = -(100 \text{ lb})(4 \text{ ft}) + (R_2)(10 \text{ ft}) = 0$$

So
$$R_2 (10 \text{ ft}) = 400 \text{ lb-ft}$$
$$R_2 = 40 \text{ lb}$$

Now taking moments about B gives

$$\Sigma M)_B = (100 \text{ lb})(10 \text{ ft} - 4 \text{ ft}) - (R_1)(10 \text{ ft}) = 0$$

So,
$$R_1 (10 \text{ ft}) = 600 \text{ lb-ft}$$
$$R_1 = 60 \text{ lb}$$

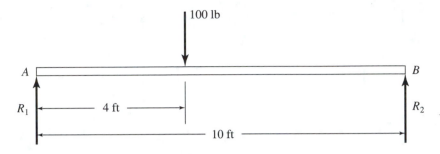

Figure 3.11   Single concentrated load.

As a check, does the sum of the vertical forces equal zero?
$$\Sigma F_y = R_1 + (-100 \text{ lb}) + R_2$$
$$= (60 \text{ lb}) + (-100 \text{ lb}) + (40 \text{ lb}) = 0 \qquad \text{Check}$$

Answer
The reactions are $R_1 = 60$ lb and $R_2 = 40$ lb.

EXAMPLE PROBLEM 3.6

Calculate the reactions for the loading shown in Figure 3.12.

Solution
In this case the beam overhangs beyond the supports, but the calculation procedure is exactly the same.
    Taking moments about C (the *left support,* not the end of the beam):
$$\Sigma M)_C = -(100 \text{ kN})(6 \text{ m} - 3 \text{ m}) + (R_2)(10 \text{ m} - 3 \text{ m}) = 0$$
$$-300 \text{ kN·m} + (R_2)(7 \text{ m}) = 0$$
$$R_2 = 42.9 \text{ kN}$$

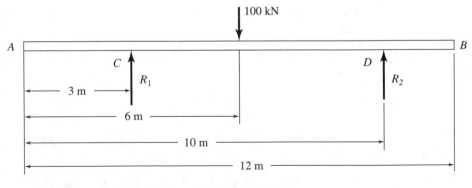

Figure 3.12   Loading for Example Problem 3.6.

Now take moments about D (the right support):

$$\Sigma M)_D = (100 \text{ kN})(10 \text{ m} - 6 \text{ m}) - (R_1)(10 \text{ m} - 3 \text{ m}) = 0$$

$$400 \text{ kN·m} - (R_1)(7 \text{ m}) = 0$$

$$R_1 = 57.1 \text{ kN}$$

Check

$$\Sigma F_y = R_1 + (-100 \text{ kN}) + R_2 = 0$$

$$= 57.1 \text{ kN} + (-100 \text{ kN}) + 42.9 \text{ kN} = 0 \qquad \text{Check}$$

Answer

The reactions are $R_1 = 57$ kN and $R_2 = 43$ kN.

3.5.2 Calculating Reactions for Two Concentrated Loads

Figure 3.13 shows a beam supporting two concentrated loads. Again, we want to calculate the reactions R_1 and R_2. Taking moments about A,

$$\Sigma M)_A = -(100 \text{ lb})(3 \text{ ft}) - (200 \text{ lb})(7 \text{ ft}) + (R_2)(10 \text{ ft}) = 0$$

$$(R_2)(10 \text{ ft}) = 1700 \text{ lb-ft}$$

$$R_2 = 170 \text{ lb}$$

Taking moments about B,

$$\Sigma M)_B = (200 \text{ lb})(10 \text{ ft} - 7 \text{ ft}) + (100 \text{ lb})(10 \text{ ft} - 3 \text{ ft}) - (R_1)(10 \text{ ft}) = 0$$

$$600 \text{ lb ft} + 700 \text{ lb-ft} - (R_1)(10 \text{ ft}) = 0$$

$$(R_1)(10 \text{ ft}) = 1300 \text{ lb ft}$$

$$R_1 = 130 \text{ lb}$$

Check

$$\Sigma F_y = R_1 + (-100 \text{ lb}) + (-200 \text{ lb}) + R_2 = 0$$

$$130 \text{ lb} - 100 \text{ lb} - 200 \text{ lb} + 170 \text{ lb} = 0 \qquad \text{Check}$$

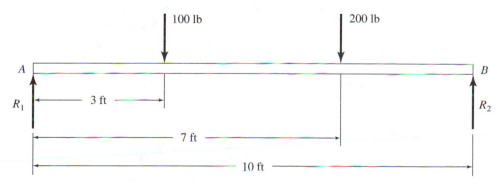

Figure 3.13   Two concentrated loads.

Answer
The reactions are $R_1 = 130$ lb and $R_2 = 170$ lb.

3.5.3   Calculating Reactions for Any Number of Concentrated Loads

Example Problem 3.7 extends the procedure just outlined.

EXAMPLE PROBLEM 3.7

Calculate the reactions for the loading shown in Figure 3.14.

Solution
We take moments about reaction R_1:

$$\Sigma M)_{R_1} = (100 \text{ N})(4 \text{ m} - 2 \text{ m}) - (500 \text{ N})(5 \text{ m} - 4 \text{ m}) + (300 \text{ N})(7 \text{ m} - 4 \text{ m})$$
$$- (400 \text{ N})(9 \text{ m} - 4 \text{ m}) + (R_2)(10 \text{ m} - 4 \text{ m}) - (200 \text{ N})(12 \text{ m} - 4 \text{ m})$$
$$= 200 \text{ N·m} - 500 \text{ N·m} + 900 \text{ N·m} - 2000 \text{ N·m} + (R_2)(6 \text{ m})$$
$$- 1600 \text{ N·m}$$

So $\Sigma M)_{R_1} = -3000 \text{ N·m} + (R_2)(6 \text{ m})$

For equilibrium, this must be zero; therefore,

$$(R_2)(6 \text{ m}) = 3000 \text{ N·m}$$
$$R_2 = \frac{3000 \text{ N·m}}{6 \text{ m}} = 500 \text{N}$$

Now taking moments about reaction R_2:

$$\Sigma M)_{R_2} = (100 \text{ N})(10 \text{ m} - 2 \text{ m}) - (R_1)(10 \text{ m} - 4 \text{ m}) + (500 \text{ N})(10 \text{ m} - 5 \text{ m})$$
$$- (300 \text{ N})(10 \text{ m} - 7 \text{ m}) + (400 \text{ N})(10 \text{ m} - 9 \text{ m}) - (200 \text{ N})$$
$$\times (12 \text{ m} - 10 \text{ m})$$
$$= 800 \text{ N·m} - (R_1)(6 \text{ m}) + 2500 \text{ N·m} - 900 \text{ N·m} + 400 \text{ N·m}$$
$$- 400 \text{ N·m}$$

So $\Sigma M)_{r_2} = 2400 \text{ N·m} - (R_1)(6 \text{ m})$

For equilibrium, this must be zero; therefore,

$$(R_1)(6 \text{ m}) = 2400 \text{ N·m}$$
$$R_1 = \frac{2400 \text{ N·m}}{6 \text{ m}} = 400 \text{ N}$$

Check
Do the vertical forces add up to zero, i.e, is $L_1 + L_2 + L_3 + L_4 + L_5 + R_1 + R_2 = 0$?

$$(-100 \text{ N}) + (-500 \text{ N}) + (300 \text{ N}) + (-400 \text{ N}) + (-200 \text{ N}) + (400 \text{ N}) + (500 \text{ N})$$
$$= 0 \quad \text{Check}$$

Answer
The reactions are $R_1 = 400$ N and $R_2 = 500$ N.

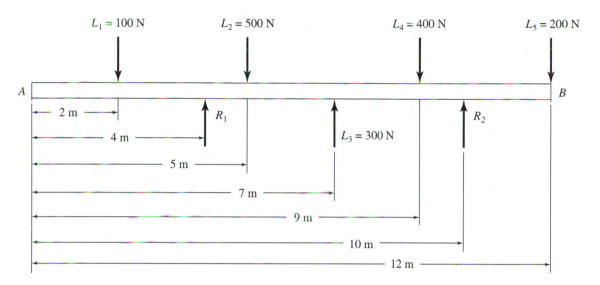

Figure 3.14   Loading for Example Problem 3.7.

3.5.4   Calculating Reactions for One Distributed Load

Figure 3.15(a) shows a 10-ft beam carrying a distributed load of 50 lb/ft, starting at 2 ft from A, the left end of the beam, and ending at 6 ft from A.

First draw the equivalent concentrated loading diagram, as in Figure 3.15(b). To do this, replace the distributed load of 50 lb/ft by a (50-lb/ft × 4-ft = 200-lb) load at 4 ft from A, which is the location of the center of gravity of the distributed load. Then, taking moments about A,

$$\Sigma M)_A = -(200 \text{ lb})(4 \text{ ft}) + (R_2)(10 \text{ ft}) = 0$$
$$-(800 \text{ lb-ft}) + (R_2)(10 \text{ ft}) = 0.$$
$$(R_2)(10 \text{ ft}) = 800 \text{ lb-ft}$$
$$R_2 = \frac{800 \text{ lb-ft}}{10 \text{ ft}} = 80 \text{ lb}$$

Taking moments about B:

$$\Sigma M)_B = (200 \text{ lb})(10 \text{ ft} - 4 \text{ ft}) - (R_1)(10 \text{ ft}) = 0$$
$$(1200 \text{ lb-ft}) - (R_1)(10 \text{ ft}) = 0$$
$$(R_1)(10 \text{ ft}) = 1200 \text{ lb-ft}$$
$$R_1 = \frac{1200 \text{ lb-ft}}{10 \text{ ft}} = 120 \text{ lb}$$

Check

$$\Sigma F_y = R_1 + (-200 \text{ lb}) + R_2$$
$$= 120 \text{ lb} + (-200 \text{ lb}) + 80 \text{ lb} = 0 \qquad \text{Check}$$

Answer

The reactions are $R_1 = 120$ lb and $R_2 = 80$ lb.

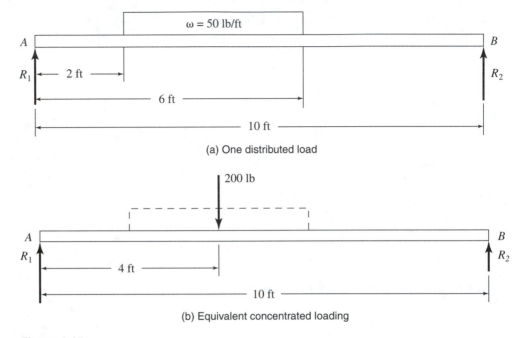

(a) One distributed load

(b) Equivalent concentrated loading

Figure 3.15

EXAMPLE PROBLEM 3.8

The beam in Figure 3.16(a) overhangs its supports and carries a distributed load of 100 N/m starting at 1 m from A and extending to 7 m. The two reactions are at 3 m and 9 m from A. Determine the reactions.

Solution
Draw the equivalent concentrated loading diagram (Figure 3.16(b)). Taking moments about C, which is at the left support, yields

$$\Sigma M)_C = -(600 \text{ N})(4 \text{ m} - 3 \text{ m}) + (R_2)(9 \text{ m} - 3 \text{ m})$$
$$= -600 \text{ N·m} + (R_2)(6 \text{ m}) = 0$$
$$(R_2)(6 \text{ m}) = 600 \text{ N·m}$$

Therefore, $R_2 = (600 \text{ N·m})/(6 \text{ m}) = 100 \text{ N}$.
Taking moments about D,

$$\Sigma M)_D = (600 \text{ N})(9 \text{ m} - 4 \text{ m}) - (R_1)(9 \text{ m} - 3 \text{ m})$$
$$= (3000 \text{ N·m}) - (R_1)(6 \text{ m}) = 0$$
$$(R_1)(6 \text{ m}) = 3000 \text{ N·m}$$
$$R_1 = \frac{3000 \text{ N·m}}{6 \text{ m}} = 500 \text{ N}$$

Check

$$\Sigma F_y = R_1 + (-600 \text{ N}) + R_2$$
$$= 500 \text{ N} + (-600 \text{ N}) + 100 \text{ N} = 0 \quad \text{Check}$$

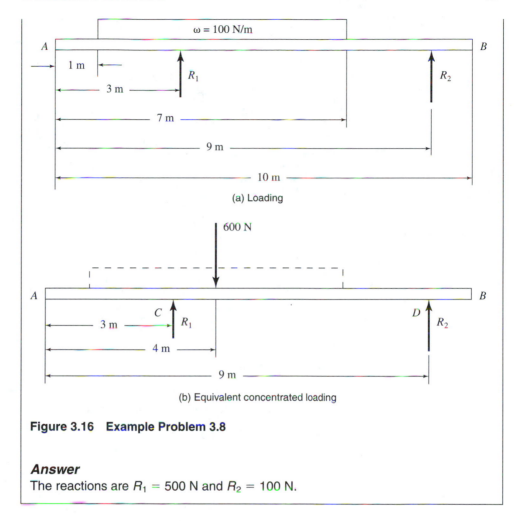

(a) Loading

(b) Equivalent concentrated loading

Figure 3.16   Example Problem 3.8

Answer

The reactions are $R_1 = 500$ N and $R_2 = 100$ N.

3.5.5   Calculating Reactions for Two Distributed Loads

EXAMPLE PROBLEM 3.9

We show two distributed loads in Figure 3.17(a). The first distributed load of 100 lb/ft extends from 1 ft to 3 ft. The second, 50 lb/ft, extends from 5 ft to 8 ft. The supports are at 2 ft and 10 ft. Determine the reactions.

Solution

Draw the equivalent concentrated loading diagram (Figure 3.17(b)). The first distributed load is equivalent to a concentrated load of (100 lb/ft) (3 ft − 1 ft) = 200 lb. The second is equivalent to (50 lb/ft) (8 ft − 5 ft) = 150 lb. We take moments about *C*, the location of the left support. The 200-lb equivalent load is right at the support and so has no moment about it.

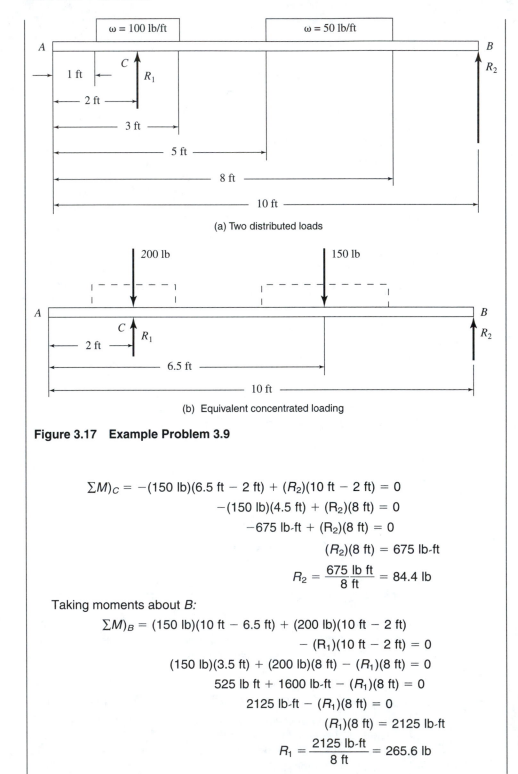

(a) Two distributed loads

(b) Equivalent concentrated loading

Figure 3.17   Example Problem 3.9

$$\Sigma M)_C = -(150 \text{ lb})(6.5 \text{ ft} - 2 \text{ ft}) + (R_2)(10 \text{ ft} - 2 \text{ ft}) = 0$$
$$-(150 \text{ lb})(4.5 \text{ ft}) + (R_2)(8 \text{ ft}) = 0$$
$$-675 \text{ lb-ft} + (R_2)(8 \text{ ft}) = 0$$
$$(R_2)(8 \text{ ft}) = 675 \text{ lb-ft}$$
$$R_2 = \frac{675 \text{ lb ft}}{8 \text{ ft}} = 84.4 \text{ lb}$$

Taking moments about *B:*
$$\Sigma M)_B = (150 \text{ lb})(10 \text{ ft} - 6.5 \text{ ft}) + (200 \text{ lb})(10 \text{ ft} - 2 \text{ ft})$$
$$- (R_1)(10 \text{ ft} - 2 \text{ ft}) = 0$$
$$(150 \text{ lb})(3.5 \text{ ft}) + (200 \text{ lb})(8 \text{ ft}) - (R_1)(8 \text{ ft}) = 0$$
$$525 \text{ lb ft} + 1600 \text{ lb-ft} - (R_1)(8 \text{ ft}) = 0$$
$$2125 \text{ lb-ft} - (R_1)(8 \text{ ft}) = 0$$
$$(R_1)(8 \text{ ft}) = 2125 \text{ lb-ft}$$
$$R_1 = \frac{2125 \text{ lb-ft}}{8 \text{ ft}} = 265.6 \text{ lb}$$

Check

$$\Sigma F_y = R_1 + (-200 \text{ lb}) + (-50 \text{ lb}) + R_2$$
$$= 265.6 \text{ lb} - 200 \text{ lb} - 150 \text{ lb} + 84.4 \text{ lb} = 0 \quad \text{Check}$$

Answer

The reactions are $R_1 = 270$ lb and $R_2 = 80$ lb.

3.5.6 Calculating Reactions for Any Number of Distributed Loads

Example Problem 3.10 extends the procedure outlined previously.

EXAMPLE PROBLEM 3.10

Calculate the reactions for the loading shown in Figure 3.18(a).

Solution

The equivalent concentrated loads and their distances from the left end of the beam are (see Figure 3.18(b)):

$$L_1 = (200 \text{ N/m})(7 \text{ m} - 2 \text{ m}) = 1000 \text{ N} \quad \text{at } 4.5 \text{ m}$$
$$L_2 = (100 \text{ N/m})(14 \text{ m} - 11 \text{ m}) = 300 \text{ N} \quad \text{at } 12.5 \text{ m}$$
$$L_3 = (350 \text{ N/m})(23 \text{ m} - 18 \text{ m}) = 1750 \text{ N} \quad \text{at } 20.5 \text{ m}$$

Taking moments about C, which is the location of the left support:

$$\Sigma M)_C = -(1000 \text{ N})(4.5 \text{ m} - 4 \text{ m}) - (300 \text{ N})(12.5 \text{ m} - 4 \text{ m})$$
$$- (1750 \text{ N})(20.5 \text{ m} - 4 \text{ m}) + (R_2)(22 \text{ m} - 4 \text{ m})$$
$$= -500 \text{ N·m} - 2550 \text{ N·m} - 28,875 \text{ N·m} + (R_2)(18 \text{ m}).$$
$$= -31,925 \text{ N·m} + (R_2)(18 \text{ m})$$

For equilibrium this must be zero:

$$-31,925 \text{ N·m} + (R_2)(18 \text{ m}) = 0$$
$$(R_2)(18 \text{ m}) = 31,925 \text{ N·m}$$
$$R_2 = \frac{31,925 \text{ N·m}}{18 \text{ m}} = 1773 \text{ N}$$

Taking moments about D, which is the location of the right support:

$$\Sigma M)_D = -(R_1)(22 \text{ m} - 4 \text{ m}) + (1000 \text{ N})(22 \text{ m} - 4.5 \text{ m})$$
$$+ (300 \text{ N})(22 \text{ m} - 12.5 \text{ m}) + (1750 \text{ N})(22 \text{ m} - 20.5 \text{ m})$$
$$= -(R_1)(18 \text{ m}) + 17,500 \text{ N·m} + 2850 \text{ N·m} + 2625 \text{ N·m}$$
$$= -(R_1)(18 \text{ m}) + 22,975 \text{ N·m}$$

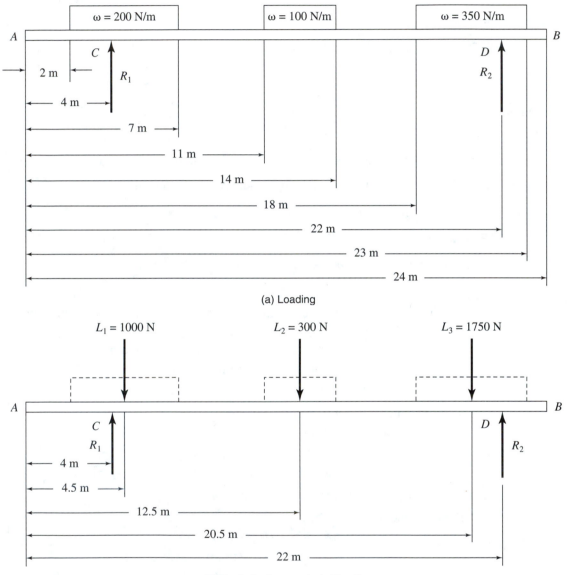

(a) Loading

(b) Equivalent concentrated loading

Figure 3.18   Example Problem 3.10

For equilibrium this must be zero:

$$-(R_1)(18 \text{ m}) + 22{,}975 \text{ N·m} = 0$$

$$(R_1)(18 \text{ m}) = 22{,}975 \text{ N·m}$$

$$R_1 = \frac{22{,}975 \text{ N·m}}{18 \text{ m}} = 1276 \text{ N}$$

Check

$$\Sigma F_y = R_1 + (-1000 \text{ N}) + (-300 \text{ N}) + (-17{,}550 \text{ N}) + R_2$$

$$= 1276 \text{ N} + (-1000 \text{ N}) + (-300 \text{ N}) + (-1750 \text{ N}) + 1774 \text{ N} = 0 \qquad \text{Check}$$

Answer

The reactions are $R_1 = 1300$ N and $R_2 = 1800$ N.

3.5.7 Calculation of Reactions for Any Number of Concentrated and Distributed Loads

The method of calculation is simply an extension of the preceding examples. First we replace all the distributed loads by their concentrated load equivalents; then we include the given concentrated loads and solve as before.

EXAMPLE PROBLEM 3.11

Calculate the reactions for the loading shown in Figure 3.19(a).

Solution

The given concentrated loads are

$$L_1 = 100 \text{ lb}, \qquad L_2 = 300 \text{ lb}, \qquad L_3 = 50 \text{ lb}$$

The given distributed loads are

$$\omega_1 = 10 \text{ lb/ft}, \qquad \omega_2 = 25 \text{ lb/ft}$$

which have equivalent concentrated loads of (see Figure 3.19(b)):

$$(10 \text{ lb/ft})(6 \text{ ft}) = 60 \text{ lb} \qquad \text{at 3 ft, and}$$

$$(25 \text{ lb/ft})(21 \text{ ft} - 13 \text{ ft}) = 200 \text{ lb} \qquad \text{at 17 ft}$$

Taking moments about C, which is the location of the left support:

$$\Sigma M)_C = (60 \text{ lb})(4 \text{ ft} - 3 \text{ ft}) - (100 \text{ lb})(8 \text{ ft} - 4 \text{ ft})$$

$$- (300 \text{ lb})(15 \text{ ft} - 4 \text{ ft}) - (200 \text{ lb})(17 \text{ ft} - 4 \text{ ft})$$

$$- (50 \text{ lb})(24 \text{ ft} - 4 \text{ ft}) + (R_2)(24 \text{ ft} - 4 \text{ ft})$$

$$= 60 \text{ lb-ft} - 400 \text{ lb-ft} - 3300 \text{ lb-ft} - 2600 \text{ lb-ft}$$

$$- 1000 \text{ lb-ft} + (R_2)(20 \text{ ft})$$

$$= -7240 \text{ lb-ft} + (R_2)(20 \text{ ft})$$

This must be equal to zero for equilibrium,

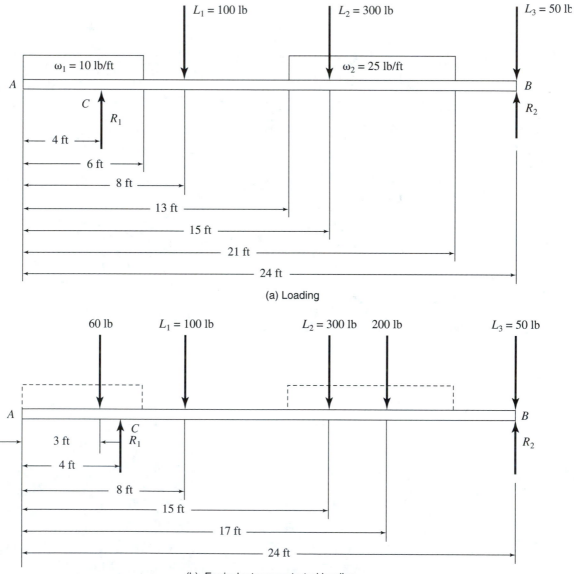

(a) Loading

(b) Equivalent concentrated loading

Figure 3.19   Example Problem 3.11

$$(R_2)(20 \text{ ft}) - 7240 \text{ lb-ft} = 0$$

$$(R_2)(20 \text{ ft}) = 7240 \text{ lb-ft}$$

$$R_2 = \frac{7240 \text{ lb-ft}}{20 \text{ ft}} = 362 \text{ lb}$$

Taking moments about B:

$$\Sigma M)_B = (60 \text{ lb})(24 \text{ ft} - 3 \text{ ft}) - (R_1)(24 \text{ ft} - 4 \text{ ft}) + (100 \text{ lb})(24 \text{ ft} - 8 \text{ ft})$$

$$+ (300 \text{ lb})(24 \text{ ft} - 15 \text{ ft}) + (200 \text{ lb})(24 \text{ ft} - 17 \text{ ft})$$

$$= 1260 \text{ lb-ft} - (R_1)(20 \text{ ft}) + 1600 \text{ lb-ft} + 2700 \text{ lb-ft} + 1400 \text{ lb-ft}$$

$$= -(R_1)(20 \text{ ft}) + 6960 \text{ lb-ft}$$

This must be equal to zero for equilibrium.

$$-(R_1)(20 \text{ ft}) + 6960 \text{ lb-ft} = 0$$

$$(R_1)(20 \text{ ft}) = 6960 \text{ lb-ft}$$

$$R_1 = \frac{6960 \text{ lb-ft}}{20 \text{ ft}} = 348 \text{ lb}$$

Check

$$\Sigma F_y = (-60 \text{ lb}) + R_1 + (-100 \text{ lb}) + (-300 \text{ lb}) + (-200 \text{ lb}) + (-50 \text{ lb}) + R_2$$

$$= (-60 \text{ lb}) + 348 \text{ lb-ft} + (-100 \text{ lb}) + (-300 \text{ lb})$$

$$+ (-200 \text{ lb}) + (-50 \text{ lb}) + 362 \text{ lb}$$

$$= 0 \qquad \text{Check}$$

Answer

$R_1 = 350 \text{ lb}$ and $R_2 = 360 \text{ lb}$.

3.5.8   General Case for One Concentrated Load

We can express the foregoing calculations in the form of general equations in preparation for writing computer programs. In Figure 3.20 we show a concentrated load, L_1, at a distance x_{L1} from A, the left end of the beam. The reactions R_1 and R_2 are at x_{R1} and x_{R2} from A, respectively. Taking moments about C, as we did in Example Problem 3.5,

$$\Sigma M)_C = (L_1)(x_{L1} - x_{R1}) + (R_2)(x_{R2} - x_{R1}) = 0$$

Note: Notice that we are entering the load as L_1, *not* $-L_1$. In some cases, where there are multiple loads (to be considered later), some of the loads may be acting in an upward direction and others in a downward direction. When we enter *numbers* into the equations, downloads will be negative and uploads will be positive.

So,

$$R_2 = -\frac{(L_1)(x_{L1} - x_{R1})}{x_{R2} - x_{R1}}$$

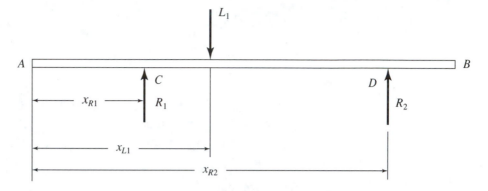

Figure 3.20   General loading for one concentrated load.

Taking moments about D:

$$\Sigma M)_D = (L_1)(x_{L1} - x_{R2}) + (R_1)(x_{R1} - x_{R2}) = 0$$

$$(R_1)(x_{R1} - x_{R2}) = -(L_1)(x_{L1} - x_{R2})$$

$$R_1 = \frac{L_1\,(x_{L1} - x_{R2})}{x_{R2} - x_{R1}}$$

So the two equations for the reactions are

$$R_1 = \frac{L_1\,(x_{L1} - x_{R2})}{x_{R2} - x_{R1}} \qquad (3.2(a))$$

$$R_2 = \frac{-L_1\,(x_{L1} - x_{R1})}{x_{R2} - x_{R1}} \qquad (3.2(b))$$

3.5.9   General Case for N Concentrated Loads

We now consider the general case of N concentrated loads. See Figure 3.21. The reaction R_1 is at a distance x_{R1} from A, the left end of the beam, and load L_1 is at a distance x_{L1} from A. Similarly, L_2 is at a distance x_{L2} from A; in general, L_n is x_{Ln} from A. The last load, L_N, is x_{LN} from A, and the reaction R_2 is at a distance x_{R2}.

Taking moments about C:

$$\Sigma M)_C = (L_1)(x_{L1} - x_{R1}) + (L_2)(x_{L2} - x_{R1}) + \cdots + (L_n)(x_{Ln} - x_{R1})$$
$$+ \cdots + (L_N)(x_{LN} - x_{R1}) + (R_2)(x_{R2} - x_{R1})$$
$$= 0$$

We can write this in compact notation (see Appendix 1) as

$$\Sigma M)_C = \sum_{n=1}^{N} L_n\,(x_{Ln} - x_{R1}) + (R_2)(x_{R2} - x_{R1}) = 0$$

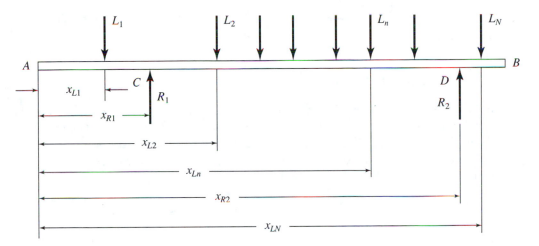

Figure 3.21 *N concentrated loads.*

Therefore,
$$R_2 = -\sum_{n=1}^{N} L_n \frac{(x_{Ln} - x_{R1})}{x_{R2} - x_{R1}}$$

Taking moments about D (the location of R_2 in Figure 3.21):
$$\Sigma M)_D = (L_1)(x_{L1} - x_{R2}) + (L_2)(x_{L2} - x_{R2}) + \cdots + (L_n)(x_{Ln} - x_{R2})$$
$$+ \cdots + (L_N)(x_{LN} - x_{R2}) + (R_1)(x_{R1} - x_{R2})$$
$$= 0$$

In compact notation,
$$\Sigma M)_D = \sum_{n=1}^{N} L_n (x_{Ln} - x_{R2}) + (R_1)(x_{R1} - x_{R2}) = 0$$

Therefore,
$$R_1 = -\sum_{n=1}^{N} L_n \frac{(x_{Ln} - x_{R2})}{x_{R1} - x_{R2}}$$

or
$$R_1 = \sum_{n=1}^{N} L_n \frac{(x_{Ln} - x_{R2})}{x_{R2} - x_{R1}}$$

So the equations for the reactions are
$$R_1 = \sum_{n=1}^{N} L_n \frac{(x_{Ln} - x_{R2})}{x_{R2} - x_{R1}} \tag{3.2(c)}$$

$$R_2 = -\sum_{n=1}^{N} L_n \frac{(x_{Ln} - x_{R1})}{x_{R2} - x_{R1}} \tag{3.2(d)}$$

Notice the similarity between equations (3.2(a)) and (3.2(c)), and between (3.2(b)) and (3.2(d)).

3.5.10   General Case for One Distributed Load

In Figure 3.22(a) we show a distributed load, ω (this is load per unit length), starting at a distance x_{Ds} from A, the left end of the beam, and extending to x_{De}. The reactions R_1 and R_2 are at x_{R1} and x_{R2}, respectively. Figure 3.22(b) shows the equivalent concentrated loading. The equivalent concentrated load is equal to the load per unit length, which is ω, multiplied by the load length, which is $(x_{De} - x_{Ds})$. So

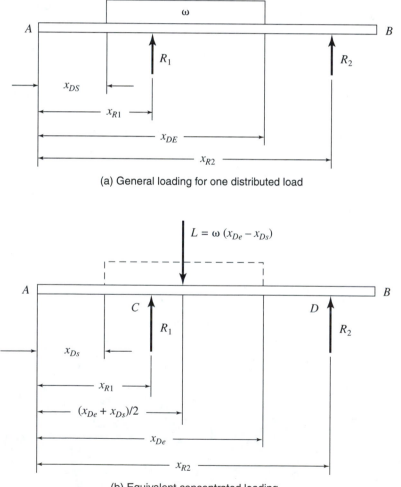

(a) General loading for one distributed load

(b) Equivalent concentrated loading

Figure 3.22

$$L = \omega(x_{De} - x_{Ds})$$

The distance from A to L is $(x_{De} + x_{Ds})/2$, i.e., to the center of the distribution.

We take moments about C (Figure 3.22(b)). The distance from C to the equivalent load is $[(x_{De} + x_{Ds})/2 - x_{R1}]$, and the distance from C to R_2 is $(x_{R2} - x_{R1})$. Therefore,

$$\Sigma M)_C = \omega(x_{De} - x_{Ds}) [(x_{De} + x_{Ds})/2 - x_{R1}] + R_2(x_{R2} - x_{R1}) = 0$$

So

$$R_2(x_{R2} - x_{R1}) = -\omega(x_{De} - x_{Ds}) [(x_{De} + x_{Ds})/2 - x_{R1}]$$

or

$$R_2 = -\omega(x_{De} - x_{Ds}) [(x_{De} + x_{Ds})/2 - x_{R1}]/(x_{R2} - x_{R1})$$

Similarly, we take moments about D. The distance from D to the equivalent load is $[(x_{De} + x_{Ds})/2 - x_{R1}]$, and the distance from D to R_1 is $(x_{R1} - x_{R2})$. Therefore,

$$\Sigma M)_C = \omega(x_{De} - x_{Ds}) [(x_{De} + x_{Ds})/2 - x_{R2}] + R_1(x_{R1} - x_{R2}) = 0$$

So

$$R_1(x_{R1} - x_{R2}) = -\omega(x_{De} - x_{Ds}) [(x_{De} + x_{Ds})/2 - x_{R2}]$$

or

$$R_1 = \omega(x_{De} - x_{Ds}) [(x_{De} + x_{Ds})/2 - x_{R2}]/(x_{R2} - x_{R1})$$

Thus, the reactions for the case of one distributed load are

$$R_1 = \frac{\omega(x_{De} - x_{Ds}) [(x_{De} + x_{Ds})/2 - x_{R2}]}{x_{R2} - x_{R1}} \qquad (3.3(a))$$

$$R_2 = \frac{-\omega(x_{De} - x_{Ds}) [(x_{De} + x_{Ds})/2 - x_{R1}]}{x_{R2} - x_{R1}} \qquad (3.3(b))$$

3.5.11   General Case for M Distributed Loads

Notice that we had a general case for N concentrated loads (Section 3.5.9); now we develop a general case for M distributed loads. The reason for this is that in a case where there are both concentrated and distributed loads, there may not be the same number of each. Figure 3.23(a) shows M distributed loads, $\omega_1, \omega_2, \ldots, \omega_n, \ldots, \omega_M$. The first starts at x_{D1s} and extends to x_{D1e}. The nth starts at x_{Dns} and extends to x_{Dne}. All distances are measured from A, the left end of the beam. The supports are at x_{R1} and x_{R2}.

We show the equivalent concentrated loading in Figure 3.23(b). The load L_1 is equal to the distributed load ω_1 times its length, $(x_{D1e} - x_{D1s})$. Similar statements are true for all the loads. So, in general we have

$$L_n = \omega_n(x_{Dne} - x_{Dns})$$

The first equivalent concentrated load is at a distance $(x_{D1e} + x_{D1s})/2$ from A. Similar statements are true for all the loads. So, in general, L_n is at station $(x_{Dne} + x_{Dns})/2$.

Taking moments about C in Figure 3.23(b),

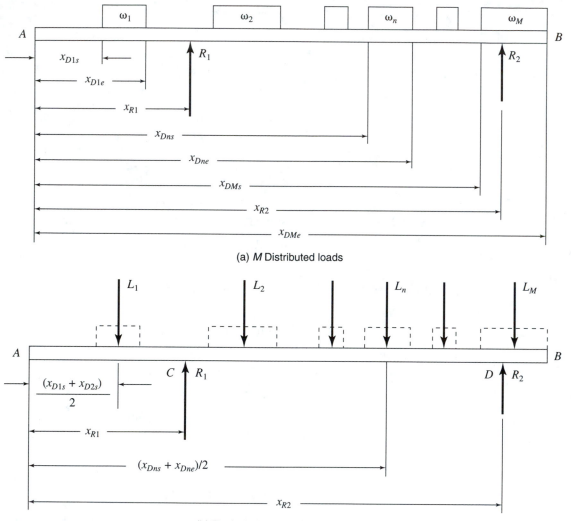

(a) *M* Distributed loads

(b) Equivalent concentrated loading

Figure 3.23

$$
\begin{aligned}
\sum M)_C = 0 = {} & (L_1)\left[(x_{D1e} + x_{D1s})/2 - x_{R1}\right] \\
& + (L_2)\left[(x_{D2e} + x_{D2s})/2 - x_{R1}\right] \\
& + \cdots \\
& + (L_n)\left[(x_{Dne} + x_{Dns})/2 - x_{R1}\right] \\
& + \cdots \\
& + (L_M)\left[(x_{DMe} + x_{DMs})/2 - x_{R1}\right] \\
& + (R_2)(x_{R2} - x_{R1})
\end{aligned}
$$

Writing this in compact form (see Appendix 1) gives

$$\Sigma M)_C = 0 = \sum_{n=1}^{M} L_n \left[(x_{Dne} + x_{Dns})/2 - x_{R1}\right] + (R_2)(x_{R_2} - x_{R_1})$$

or

$$R_2 = -\sum_{n=1}^{M} L_n[(x_{Dne} + x_{Dns})/2 - x_{R1}]/(x_{R_2} - x_{R_1})$$

But,

$$L_n = \omega_n(x_{Dne} - x_{Dns})$$

So

$$R_2 = -\sum_{n=1}^{M} \frac{\omega_n(x_{Dne} - x_{Dns}) \left[(x_{Dne} + x_{Dns})/2 - x_{R1}\right]}{x_{R2} - x_{R1}}$$

Taking moments about D gives

$$\Sigma M)_D = 0 = (L_1) \left[(x_{D1e} + x_{D1s})/2 - x_{R2}\right]$$
$$+ (L_2) \left[(x_{D2e} + x_{D2s})/2 - x_{R2}\right]$$
$$+ \cdots$$
$$+ (L_n) \left[(x_{Dne} + x_{Dns})/2 - x_{R2}\right]$$
$$+ \cdots$$
$$+ (L_M) \left[(x_{DMe} + x_{DMs})/2 - x_{R2}\right]$$
$$+ (R_1)(x_{R1} - x_{R2})$$

In compact form,

$$\Sigma M)_D = 0 = \sum_{n=1}^{M} L_n \left[(x_{Dne} + x_{Dns})/2 - x_{R2}\right] + (R_1)(x_{R1} - x_{R2})$$

or

$$R_1 = -\sum_{n=1}^{M} L_n \left[(x_{Dne} + x_{Dns})/2 - x_{R2}\right]/(x_{R1} - x_{R2})$$

$$= \sum_{n=1}^{M} L_n \left[(x_{Dne} + x_{Dns})/2 - x_{R2}\right]/(x_{R2} - x_{R1})$$

Substituting for L_n,

$$R_1 = \sum_{n=1}^{M} \frac{\omega_n(x_{Dne} - x_{Dns}) \left[(x_{Dne} + x_{Dns})/2 - x_{R2}\right]}{x_{R2} - x_{R1}}$$

So the equations for the reactions are

$$R_1 = \sum_{n=1}^{M} \frac{\omega_n(x_{Dne} - x_{Dns}) \left[(x_{Dne} + x_{Dns})/2 - x_{R2}\right]}{x_{R2} - x_{R1}} \qquad (3.3(c))$$

$$R_2 = -\sum_{n=1}^{M} \frac{\omega_n(x_{Dne} - x_{Dns}) \left[(x_{Dne} + x_{Dns})/2 - x_{R1}\right]}{x_{R2} - x_{R1}} \qquad (3.3(d))$$

3.5.12   General Case for N Concentrated and M Distributed Loads

Deriving the reaction equations for the case of concentrated *and* distributed loads is simply a matter of adding equations (3.2(c)) and (3.3(c)) for R_1 and (3.2(d)) and (3.3(d)) for R_2.

Repeating these equations,

$$R_1 = \sum_{n=1}^{N} \frac{L_n(x_{Ln} - x_{R2})}{x_{R2} - x_{R1}} \tag{3.2(c)}$$

$$R_2 = -\sum_{n=1}^{N} \frac{L_n(x_{Ln} - x_{R1})}{x_{R2} - x_{R1}} \tag{3.2(d)}$$

$$R_1 = \sum_{n=1}^{M} \frac{\omega_n(x_{Dne} - x_{Dns})\left[(x_{Dne} + x_{Dns})/2 - x_{R2}\right]}{x_{R2} - x_{R1}} \tag{3.3(c)}$$

$$R_2 = -\sum_{n=1}^{M} \frac{\omega_n(x_{Dne} - x_{Dns})\left[(x_{Dne} + x_{Dns})/2 - x_{R1}\right]}{x_{R2} - x_{R1}} \tag{3.3(d)}$$

So for R_1,

$$R_1 = \sum_{n=1}^{N} \frac{L_n(x_{Ln} - x_{R2})}{x_{R2} - x_{R1}} + \sum_{n=1}^{M} \frac{\omega_n(x_{Dne} - x_{Dns})\left[(x_{Dne} + x_{Dns})/2 - x_{R2}\right]}{x_{R2} - x_{R1}} \tag{3.4(a)}$$

and for R_2,

$$R_2 = -\sum_{n=1}^{N} \frac{L_n(x_{Ln} - x_{R1})}{x_{R2} - x_{R1}} - \sum_{n=1}^{M} \frac{\omega_n(x_{Dne} - x_{Dns})\left[(x_{Dne} + x_{Dns})/2 - x_{R1}\right]}{x_{R2} - x_{R1}} \tag{3.4(b)}$$

3.6   COMPUTER ANALYSES

3.6.1   Program 'Shr&mmt'

'Shr&mmt' stands for shear and moment. We use this program later to determine the shear force and bending moment diagrams for a loaded beam. Before we can determine shear forces and bending moments, we have to know the reactions, so we can use this program to calculate the reactions.

Input Data

1. Enter the positions x_1 and x_2 of the supports as distances from the left end of the beam.
2. Enter the beam length.
3. Enter the concentrated loads and their distances from the left end of the beam.

Note: *Downward-acting loads are entered as negative values, and upward-acting loads are positive.*

4. Enter the distributed loads (downward-acting negative, upward-acting positive) and the positions of the starts and the ends of each.
5. Enter the units for concentrated loads, distributed loads, and distances in cells L3, L4, and L5, respectively.

Output Data

The support reactions, R_1 and R_2, are displayed. R_1 and R_2 are the left and right reactions, respectively.

EXAMPLE PROBLEM 3.12

An 18-ft-long simply supported beam carries a load of 75,000 lb at station 11 ft. Determine the reactions.

Solution

1. Open program 'Shr&mmt'.
2. Because this is not an overhanging beam, which is a beam with loads to the left of the left support or to the right of the right support, $x_1 = 0$ and $x_2 = 18$.
3. Enter the beam length of 18.
4. There is only one concentrated load, and there are no distributed loads, so enter $-75,000$ for the concentrated load and 11 for x.
5. Enter the units for the concentrated load and the distances, lb and ft.

Answer

The results are given as follows:

The left reaction is $R_1 = 29.2$ lb.
The right reaction is $R_2 = 45.8$ lb.

EXAMPLE PROBLEM 3.13

Calculate the reactions for the loading shown in Figure 3.24.

Solution

1. Enter the support locations, 7 and 22.
2. Enter the beam length, 24.
3. Enter the concentrated loads, $-50, -200, -150, -100$.
4. Enter their locations, 0, 8, 15, 19.

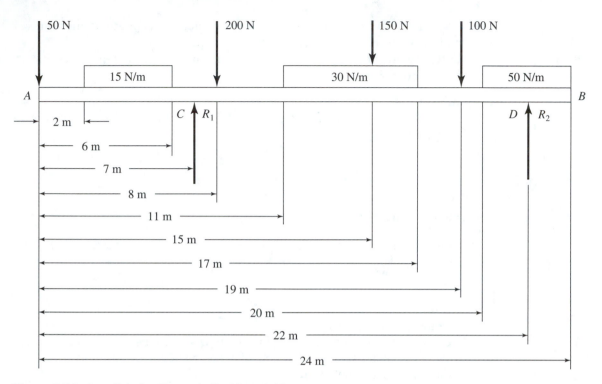

Figure 3.24   Loading for Example Problem 3.13.

5. Enter the distributed loads, -15, -30, -50.
6. Enter their start locations, 2, 11, 20.
7. Enter their end locations, 6, 17, 24.
8. Enter the units, N, N/m, m.

Answer

The reactions are given as follows.

The left reaction is $R_1 = 518$ N.
The right reaction is $R_2 = 422$ N.

3.7   PROBLEMS FOR CHAPTER 3

Hand Calculations

1. A 12-ft beam (Figure 3.25) is supported at a point 5 ft from its left end. There is an external load of 350 lb at the left end and a mass of 10 slugs at the right end. What load, L (lb), at a point 3 ft from the left end of the beam is needed to balance the beam? What is the reaction?

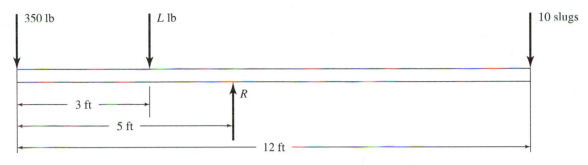

Figure 3.25

2. A 7-m beam is supported on a frictionless roller that is positioned at 3 m from its left end. (See Figure 3.26). A 35-kN force is applied at a point 1 m from the right end of the beam at an angle of 60° to the horizontal. What force, at what angle, applied to the left end of the beam is necessary to keep it in equilibrium? *Include the weight of the beam, which is 10 kN.*

3. Show that the forces in Figure 3.27 constitute a couple. What is its torque?

4. Calculate the reactions for the loading shown in Figure 3.28.

5. Calculate the reactions for the loading shown in Figure 3.29.

6. Calculate the reactions for the loading shown in Figure 3.30.

7. Calculate the reactions for the loading shown in Figure 3.31.

8. A floor is constructed of 16-ft joists simply supported at their ends and covered with plywood sheets. The joists are spaced on 4-ft centers, and the total load, including the weight of the floor, is 1 lb/in². Calculate the end reactions for the joists.

9. The wheel base of a truck is 3.5 m, and its rear axle has a load limit of 30 kN. How far ahead of the rear axle must a 5500-kg mass be placed to avoid damaging the rear axle?

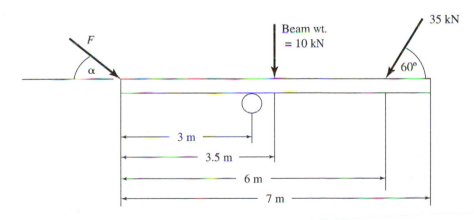

Figure 3.26

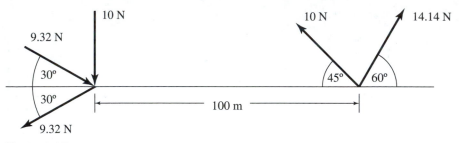

Figure 3.27

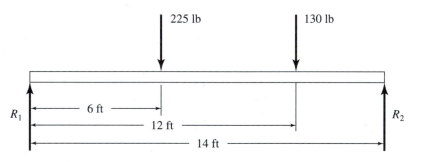

Figure 3.28

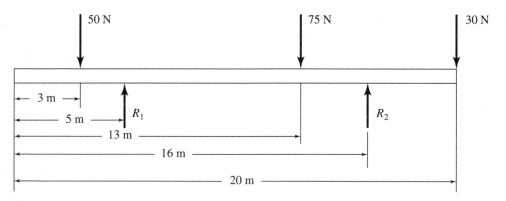

Figure 3.29

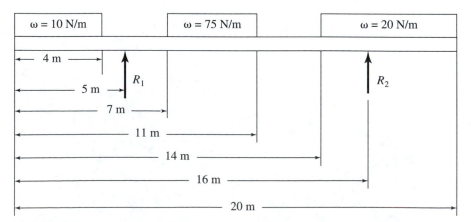

Figure 3.30

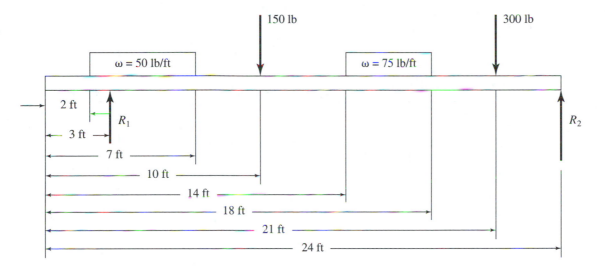

Figure 3.31

Computer Calculations

10. Repeat Problem 7.
11. Calculate the reactions for the loading shown in Figure 3.32.
12. Calculate the reactions for the loading shown in Figure 3.33.
13. Return to Example Problem 3.13 and Figure

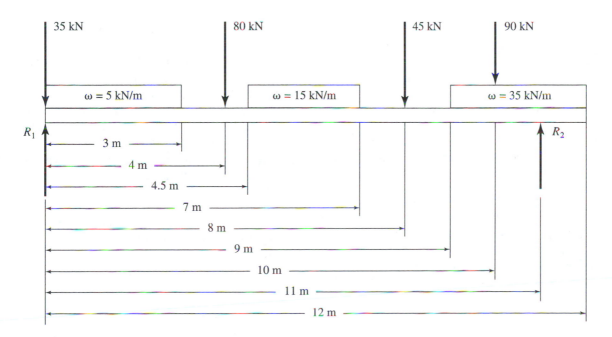

Figure 3.32

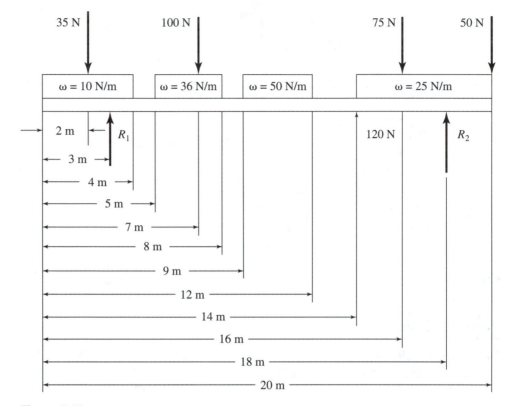

Figure 3.33

3.24 and move the supports to the ends of the beam. What are the new reactions?

14. Keeping the left support at $x_1 = 0$ in Problem 13, move the right support inboard until the beam is on the point of tipping. What value of x_2, to the nearest 0.1 m, will just prevent tipping?

15. Returning to Example Problem 3.13 and Figure 3.24, move the supports inboard, keeping them both the same distance from each end, until the beam is on the point of tipping. What are the values of x_1 and x_2, to the nearest 0.1 m, that will just prevent tipping?

16. Returning to Example Problem 3.13 and Figure 3.24, by how much would the 50-N load on the left end of the beam have to be in-

creased to just tip the beam? Give your answer to the nearest 1 N.

17. Write a computer program to determine the reactions for a simply supported beam carrying up to three concentrated loads.

18. Write a computer program to determine the reactions for an overhanging beam carrying up to three concentrated loads.

19. Write a computer program to determine the reactions for a simply supported beam carrying up to three distributed loads.

20. Write a computer program to determine the reactions for an overhanging beam carrying up to two concentrated loads and two distributed loads.

4

Nonconcurrent Forces, Frames, and Trusses

■ Objectives

In this chapter you will learn about *frames*, *space frames*, and *trusses*.

Nonconcurrent forces are forces that do not act through a common point. We considered some cases in Chapter 3. In the present chapter we expand the study to include frames, space frames, and trusses.

To solve these types of problems we generally have to use some or all of the conditions for equilibrium laid out in the previous chapter:

1. $\Sigma M)_p = 0$ where P is any point
2. $\Sigma F_x = 0$
3. $\Sigma F_y = 0$
4. $\Sigma F_z = 0$

In words, the following hold for equilibrium:

1. The algebraic sum of the moments about any point of all the external forces acting on a body must be zero.
2. The algebraic sum of the *x*-components of all the external forces acting on a body must be zero.
3. The algebraic sum of the *y*-components of all the external forces acting on a body must be zero.
4. The algebraic sum of the *z*-components of all the external forces acting on a body must be zero.

4.1 SINGLE MEMBERS

Let us consider the case of a ladder against a smooth vertical wall (Figure 4.1(a)). Choosing a smooth wall simplifies the problem. If it were rough, we would have to know something about the friction between the ladder and the wall. A smooth wall can only push against the ladder. It cannot prevent it from slipping, so the wall reaction, $\mathbf{F}_{wx}$, must be a *horizontal force*. Figure 4.1(b) shows the ladder set out as a free body. The ladder weighs 30 lb, and this acts vertically down at the center of the ladder. The person standing at the 8-ft height weighs 150 lb. These two weights are be-

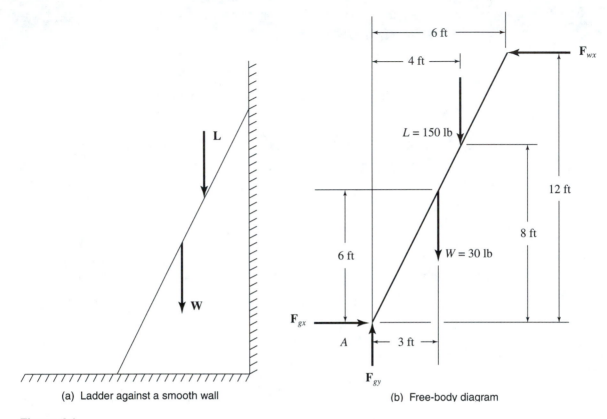

(a) Ladder against a smooth wall

(b) Free-body diagram

Figure 4.1

ing supported by the vertical component, $\mathbf{F}_{gy}$, of the ground reaction, which—of course—must be acting in an upward direction. The ladder is being prevented from slipping by the horizontal component, $\mathbf{F}_{gx}$, of the ground reaction, acting to the right.

We can readily find $\mathbf{F}_{gy}$ by using the equilibrium condition that the sum of the vertical forces must be zero:

$$\Sigma F_y = 0$$
$$(-30 \text{ lb}) + (-150 \text{ lb}) + F_{gy} = 0$$
$$-180 \text{ lb} + F_{gy} = 0$$

So

$$F_{gy} = 180 \text{ lb}$$

We also know that the sum of the horizontal forces must be zero:

$$\Sigma F_x = 0$$
$$F_{gx} + F_{wx} = 0$$

So

$$F_{gx} = -F_{wx}$$

Thus the magnitude of $\mathbf{F}_{gx}$ equals the magnitude of $\mathbf{F}_{wx}$ (and the negative sign indicates that they act in opposite directions), but we do not know this magnitude yet. We do know, however, that the sum of the moments about any point must be zero for equilibrium.

Taking moments about A, recall that clockwise moments are negative and counterclockwise moments are positive. (Remember to use the force times the distance that is perpendicular to that force.)

$$\Sigma M)_A = -(30 \text{ lb})(3 \text{ ft}) - (150 \text{ lb})(4 \text{ ft}) + (F_{wx})(12 \text{ ft}) = 0$$

So,

$$-90 \text{ lb-ft} - 600 \text{ lb-ft} + (F_{wx})(12 \text{ ft}) = 0$$
$$-690 \text{ lb-ft} + (F_{wx})(12 \text{ ft}) = 0$$
$$F_{wx} = (690 \text{ lb-ft})/(12 \text{ ft})$$
$$F_{wx} = 57.5 \text{ lb}$$

We now have the complete solution:

$$F_{gx} = 57.5 \text{ lb}, \qquad F_{gy} = 180 \text{ lb}, \qquad F_{wx} = 57.5 \text{ lb}$$

Note: Keeping to our sign convention, F_{wx} should be negative because it is acting from right to left. However, simple logic told us its direction, and we drew it that way. In more complicated cases it is not always intuitively obvious which way a certain force acts. In these cases draw the force in the positive direction. The final solution will indicate whether it is negative or positive. This is very important when writing computer programs.

As an example, let us repeat the preceding problem, but this time we write all the *unknown* forces as positive forces. F_{gx} and F_{gy} are already in the positive direction, so all we have to do is reverse the direction of F_{wx}.

Taking moments about A, all the moments will be clockwise and, therefore, negative:

$$\Sigma M)_A = -(30 \text{ lb})(3 \text{ ft}) - (150 \text{ lb})(4 \text{ ft}) - (F_{wx})(12 \text{ ft}) = 0$$
$$-90 \text{ lb-ft} - 600 \text{ lb-ft} - (F_{wx})(12 \text{ ft}) = 0$$

So

$$F_{wx} = (-690 \text{ lb-ft})/12 \text{ ft}$$
$$= -57.5 \text{ lb}$$

indicating that F_{wx} acts from right to left.

EXAMPLE PROBLEM 4.1

The 10-m-long bar shown in Figure 4.2(a) weighs 50 kN. It is hinged to the wall at A and held in position by a wire attached to its upper end, C, and to the wall at B. The bar makes an angle of 30° to the wall, and the wire makes an angle of 60° to the wall. Determine the tension in the wire and the components of the reaction at A.

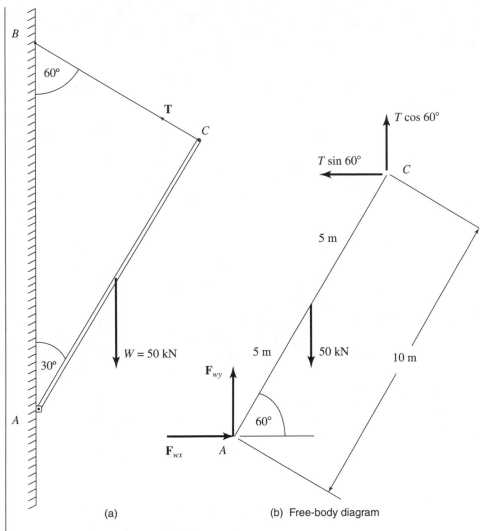

(a)   (b) Free-body diagram

Figure 4.2   Example Problem 4.1

Solution

Draw the free-body diagram (Figure 4.2(b)). First, at *A*, there will be an upward component, $\mathbf{F}_{wy}$, of the reaction to prevent the bar from sliding down the wall. There will also be a horizontal component, $\mathbf{F}_{wx}$, in the positive direction, pushing against the bar. At *C*, we can resolve the tension into a horizontal force, $T \sin 60°$, and a vertical force, $T \cos 60°$.

Taking moments about *A*,

$$\Sigma M)_A = 0$$

The moment arm for the bar weight of 50 kN (see Figure 4.2(b)) is (5 m) (cos 60°), so the moment of this weight about *A* is

$$-(50 \text{ kN})(5 \text{ m})(\cos 60°) = -(50 \text{ kN})(2.5 \text{ m}) = -125 \text{ kN·m}$$

The moment arm for the vertical component at C ($T \cos 60$), about A is (10 m) (cos 60°), so its moment is

$$(T \cos 60°)[(10 \text{ m})(\cos 60°)] = (0.5T)(5 \text{ m}) = 2.5T \text{ m}$$

The moment arm for the horizontal component at C ($T \sin 60°$), about A is (10 m) (sin 60°), so its moment is

$$(T \sin 60°)[(10 \text{ m})(\sin 60°)] = (0.8660 T)(8.66 \text{ m}) = 7.50 T \text{ m}$$

For the bar to be in equilibrium, the algebraic sum of these moments must be zero:

$$-125 \text{ kN·m} + 2.5T \text{ m} + 7.50T \text{ m} = 0$$
$$-125 \text{ kN·m} + 10T \text{ m} = 0$$
$$T = 12.5 \text{ kN}$$

Summing the horizontal forces,

$$F_{wx} - T \sin 60° = 0$$

therefore,

$$F_{wx} = T \sin 60° = (12.5 \text{ kN})(0.866) = 10.83 \text{ kN}$$

Summing the vertical forces,

$$F_{wy} + (-50 \text{ kN}) + T \cos 60° = 0$$

Therefore,

$$F_{wy} = 50 \text{ kN} - (12.5 \text{ kN})(0.5)$$
$$F_{wy} = 50 \text{ kN} - 6.25 \text{ kN}$$
$$F_{wy} = 43.75 \text{ kN}$$

Answer

The tension in the wire is 13 kN.
The horizontal force component at A is 11 kN.
The vertical force component at A is 44 kN.

4.2   FRAMES

We generally refer to two-dimensional frames—that is, frames with all the members and forces in one plane—simply as **frames.** Three-dimensional frames, which we discuss in the next section, are called **space frames.** The A-frame shown in Figure 4.3(a) is a typical frame. It is pin-jointed at B, C, and D and carries a load on the cross bar. To design a frame like this, we would need to determine the pin reactions at B, C, and D in order to make them strong enough.

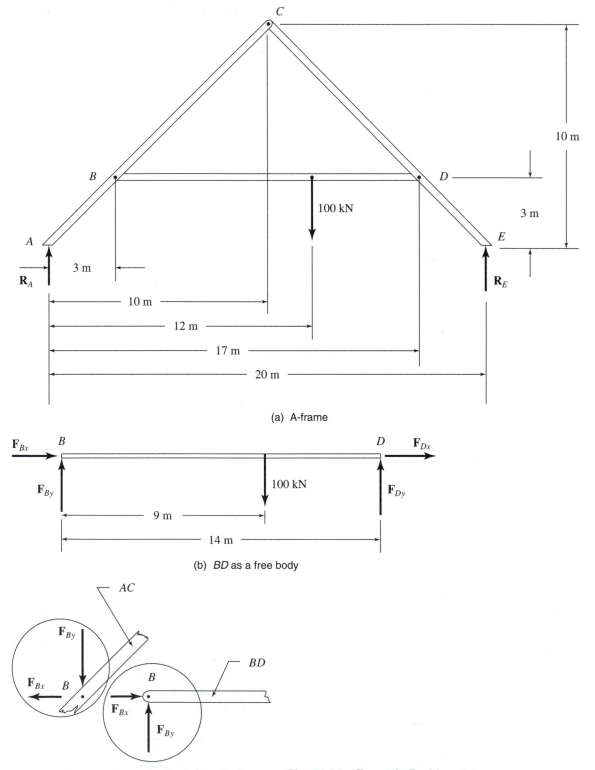

(a) A-frame

(b) *BD* as a free body

(c) Equal and opposite forces on pin *B*   **Figure 4.3   Example Problem 4.2**

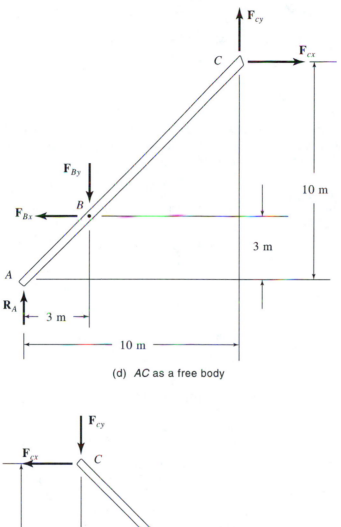

(d) *AC* as a free body

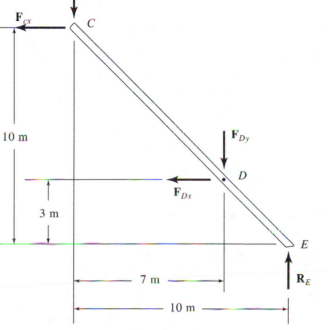

(e) *CE* as a free body

EXAMPLE PROBLEM 4.2

Calculate the support reactions at *A* and *E* and the pin reactions at *B, C,* and *D* for the frame shown in Figure 4.3(a).

Solution
Because the only load acts vertically downward, the reactions at *A* and *E* must be vertically upward, there being no tendency for sideways movement.
Taking moments about *A* (Figure 4.3(a)):

$$\Sigma M)_A = 0$$

The only two external forces creating moments about *A* are the load and the reaction at *E*, R_E.

$$-(100 \text{ kN})(12 \text{ m}) + (R_E)(20 \text{ m}) = 0$$

$$-1200 \text{ kN·m} + (R_E)(20 \text{ m}) = 0$$

$$R_E = 1200 \text{ kN·m}/20 \text{ m}$$

$$= 60 \text{ kN}$$

Summing the vertical forces gives

$$\Sigma F_y = 0$$

$$R_A + (-100 \text{ kN}) + R_E = 0$$

$$R_A - 100 \text{ kN} + 60 \text{ kN} = 0$$

$$R_A - 40 \text{ kN} = 0$$

$$= 40 \text{ kN}$$

Now set out the crossbar, *BD,* as a free body (Figure 4.3(b)). The horizontal and vertical components of the force at *B* are F_{Bx} and F_{By}, respectively. Similarly, at *D* they are F_{Dx} and F_{Dy}. Rather than making assumptions about the directions of these unknown forces, we make them all positive. If the solution shows a force to be negative, then we know it is in the opposite direction. Taking moments about *B* (Figure 4.3(b)) gives

$$\Sigma M)_B = 0$$

$$-(100 \text{ kN})(9 \text{ m}) + (F_{Dy})(14 \text{ m}) = 0$$

$$(F_{Dy})(14 \text{ m}) = 900 \text{ kN·m}$$

$$F_{Dy} = 900 \text{ kN·m}/14 \text{ m} = 64.3 \text{ kN}$$

Summing the vertical forces gives

$$\Sigma F_y = 0$$

$$F_{By} + (-100 \text{ kN}) + F_{Dy} = 0$$

$$F_{By} + (-100 \text{ kN}) + 64.3 \text{ kN} = 0$$

$$F_{By} - 35.7 \text{ kN} = 0$$

$$= 35.7 \text{ kN}$$

Summing the horizontal forces yields

$$\Sigma F_x = 0$$

$$F_{Bx} + F_{Dx} = 0$$

So $F_{Bx} = -F_{Dx}$, showing that their magnitudes are equal but in opposite directions.

We are now going to set out AC as a free body (Figure 4.3(d)), but before doing so let us carefully consider the force directions. Isolating the pin B as a free body (see Figure 4.3(c)), the horizontal forces acting on it must be equal and opposite. If not, it would be moving sideways. Similarly, the vertical forces must balance. Therefore, when we draw AC as a free body, we must reverse the directions of F_{Bx} and F_{By}. Returning now to Figure 4.3(d) and taking moments about C gives

$$\Sigma M)c = 0$$

$$-(R_A)(10 \text{ m}) + (F_{By})(10 \text{ m} - 3 \text{ m}) - (F_{Bx})(10 \text{ m} - 3 \text{ m}) = 0$$

But $R_A = 40$ kN and $F_{By} = 35.7$ kN, so

$$-(40 \text{ kN})(10 \text{ m}) + (35.7 \text{ kN})(7 \text{ m}) - (F_{Bx})(7 \text{ m}) = 0$$

$$-400 \text{ kN·m} + 249.9 \text{ kN·m} - (F_{Bx})(7 \text{ m}) = 0$$

$$-150.1 \text{ kN·m} - (F_{Bx})(7 \text{ m}) = 0$$

$$F_{Bx} = -150.1 \text{ kN·m}/7 \text{ m}$$

$$= -21.4 \text{ kN}$$

Since $F_{Bx} = -F_{Dx}$,

$$F_{Dx} = 21.4 \text{ kN}$$

Summing the horizontal forces in Figure 4.3(d):

$$\Sigma F_x = 0$$

$$F_{Bx} + F_{Cx} = 0$$

So

$$F_{Cx} = -F_{Bx} = 21.4 \text{ kN}$$

Summing the vertical forces gives

$$\Sigma F_y = 0$$

$$R_A + F_{By} + F_{Cy} = 0$$

$$40 \text{ kN} - 35.7 \text{ kN} + F_{Cy} = 0$$

$$4.3 \text{ kN} + F_{Cy} = 0$$

$$F_{Cy} = -4.3 \text{ kN}$$

We can summarize the *magnitudes* of the forces, but their directions depend on which member we are considering:

$R_A = 40$ kN	$R_E = 60$ kN
$F_{Bx} = 92.8$ kN	$F_{By} = 35.7$ kN
$F_{Cx} = 21.4$ kN	$F_{Cy} = 4.3$ kN
$F_{Dx} = 21.4$ kN	$F_{Dy} = 64.3$ kN

Check: Set out CE as a free body (Figure 4.3(e)), making sure that the force directions at C and D are opposite to those at C for AC and D for BD.

Taking moments about *C* yields

$$\Sigma M)_C = 0$$

$$(R_E)(10 \text{ m}) - (F_{Dy})(7 \text{ m}) - (F_{Dx})(10 \text{ m} - 3 \text{ m}) = 0$$

$$(60 \text{ kN})(10 \text{ m}) - (64.3 \text{ kN})(7 \text{ m}) - (21.4 \text{ kN})(7 \text{ m}) = 0$$

$$600 \text{ kN·m} - 450.1 \text{ kN·m} - 149.8 \text{ kN·m} = 0$$

(It is actually 0.1 due to round-off error.)

Answer

$R_A = 40$ kN	$R_E = 60$ kN
$F_{Bx} = 93$ kN	$F_{By} = 36$ kN
$F_{Cx} = 21$ kN	$F_{Cy} = 4$ kN
$F_{Dx} = 21$ kN	$F_{Dy} = 64$ kN

4.3 SPACE FRAMES

Space frames are three-dimensional frames. Probably the most convenient method of determining the loads in the members of a space frame is that known as the *method of tension coefficients*. The tension coefficient of a member AB of a frame is defined as the *tension per unit length* and is denoted by t_{AB}. The member AB in Figure 4.4 is of length *L* and has a tensile load *T*.

$$t_{AB} = T/L \tag{4.1}$$

If the force in a member is compressive, the tension coefficient for that member is negative.

First we draw a set of mutually perpendicular axes, *x*, *y*, and *z*. The projected lengths of *L* on the *x*-, *y*-, and *z*-axes are L_x, L_y, and L_z, respectively. Instead of drawing L_y and L_z on the *y*- and *z*-axes, we have transferred those lengths to the end of the member. Now we have two right-angled triangles, ABC and ACD. Let angle BAC be θ and angle CAD be ϕ.

A force *T* along AB can be resolved into *T* sin θ along A*y* and *T* cos θ along AC. The force *T* cos θ along AC can be resolved further into

$$T \cos \theta \cdot \cos \phi \quad \text{along A}x, \quad \text{and}$$

$$T \cos \theta \cdot \sin \phi \quad \text{along A}z$$

What we have so far is this:

The component of **T** along the *x*-axis is *T* cos θ·cos ϕ.
The component of **T** along the *y*-axis is *T* sin θ.
The component of **T** along the *z*-axis = *T* cos θ·sin ϕ

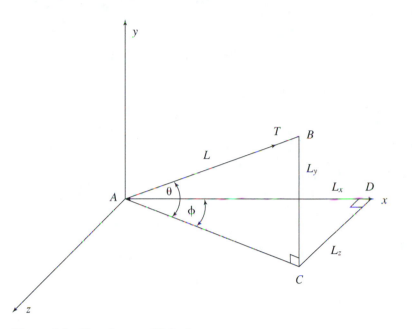

Figure 4.4   Tension coefficients.

Now let us look at the lengths:

$$AC = L \cos \theta,$$

so
$$L_x = AC \cos \phi = L \cos \theta \cdot \cos \phi$$
$$L_y = BC = L \sin \theta$$
$$L_z = CD = AC \sin \phi = L \cos \theta \cdot \sin \phi$$

We can write these as

$$\cos \theta \cos \phi = L_x/L$$
$$\sin \theta = L_y/L$$
$$\cos \theta \sin \phi = L_z/L$$

Going back to the force components, we see that

The component of $\mathbf{T}$ along the x-axis $= TL_x/L = (T/L)L_x = t_{AB}L_x$
The component of $\mathbf{T}$ along the y-axis $= TL_y/L = (T/Ly)L_y = t_{AB}L_y$
The component of $\mathbf{T}$ along the z-axis $= TL_z/L = (T/L)L_z = t_{AB}L_z$

What we see from this is that the component of the force in a member in any direction is the tension coefficient times the projected length of the member in that direction. This is particularly useful because the projected lengths are given in the three views of an engineering drawing.

Suppose now that we have a space frame made of three members, AB, AC, and AD, as in Figure 4.5. The lengths of the members are L_{AB}, L_{AC}, and L_{AD}. We draw a set of mutually perpendicular axes through point A and denote the projections of AB on to

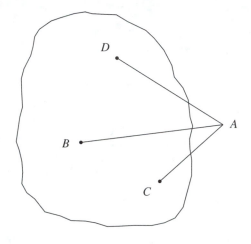

Figure 4.5   Space frame.

these axes as L_{ABx}, L_{ABy}, and L_{ABz}. Similarly, the projections of AC on the axes are L_{ACx}, L_{ACy}, and L_{ACz}, and the projections of AD are L_{ADx}, L_{ADy}, and L_{ADz}.

If there is a tensile force of T_{AB}, its components are

> In the x-direction, $t_{AB} \cdot L_{ABx}$,
> In the y-direction, $t_{AB} \cdot L_{ABy}$,
> In the z-direction, $t_{AB} \cdot L_{ABz}$,

where $t_{AB} = T_{AB}/L_{AB}$, the tension coefficient for member AB.

Similarly, if there is a tensile force of T_{AC} in AC, its components are

> In the x-direction, $t_{AC} \cdot L_{ACx}$,
> In the y-direction, $t_{AC} \cdot L_{ACy}$,
> In the z-direction, $t_{AC} \cdot L_{ACz}$.

And if there is a tensile force of T_{AD} in AD, its components are

> In the x-direction, $t_{AD} \cdot L_{ADx}$,
> In the y-direction, $t_{AD} \cdot L_{ADy}$,
> In the z-direction, $t_{AD} \cdot L_{ADz}$.

So the total component of the forces in the x-direction is

$$t_{AB} \cdot L_{ABx} + t_{AC} \cdot L_{ACx} + t_{AD} \cdot L_{ADx}$$

If there are external forces whose total component is F_x in the x-direction, then for equilibrium, the sum of the forces in the x direction must be zero, or

$$t_{AB} \cdot L_{ABx} + t_{AC} \cdot L_{ACx} + t_{AD}.L_{ADx} + F_x = 0 \qquad (4.2(a))$$

Similarly, if the total y-component of the external forces is F_y,

$$t_{AB} \cdot L_{ABy} + t_{AC} \cdot L_{ACy} + t_{AD} \cdot L_{ADy} + F_y = 0 \qquad (4.2(b))$$

and for the z-component,

$$t_{AB} \cdot L_{ABz} + t_{AC} \cdot L_{ACz} + t_{AD} \cdot L_{ADz} + F_z = 0 \qquad (4.2(c))$$

In an actual problem, we would know the projected lengths of the members from the three-view drawing. We would also know the external forces, so there would be three unknown values, t_{AB}, t_{AC}, and t_{AD}, and three equations. Having solved for the tension coefficients, we would then determine the tensile forces in the members by multiplying the tension coefficients by the member lengths. The length of a member is given by

$$L = \sqrt{L_x^2 + L_y^2 + L_z^2} \tag{4.3}$$

EXAMPLE PROBLEM 4.3

A three-view of a space frame is shown in Figure 4.6. The dimensions are in millimeters. Determine the forces in the members AB, AC, and AD.

Solution

The elevation shows the xy-plane, the plan shows the xz-plane, and the right-side view shows the yz-plane. AE in the top view is the vertical plane xy. AE in the front view is the horizontal plane xz. Because the point A is the axes' origin, all the x-values will be negative. The z-component of AD will be negative, as will be the y-components of AB and AC. The remainder will be positive.

Collecting these together, we have

$$L_{ABx} = L_{ACx} = L_{ADx} = -350 \text{ mm}$$

$$L_{ABy} = -100 \text{ mm}, \qquad L_{ACy} = -150 \text{ mm}, \qquad L_{ADy} = 100 \text{ mm}$$

$$L_{ABz} = 150 \text{ mm}, \qquad L_{ACz} = 50 \text{ mm}, \qquad L_{ADz} = -50 \text{ mm}$$

$$F_x = 2.8 \text{ kN}, \qquad F_y = -1.0 \text{ kN}, \qquad F_z = 1.6 \text{ kN}$$

Inserting these values into equations (4.2 (a)), (b), and (c) gives

$$t_{AB}(-350 \text{ mm}) + t_{AC}(-350 \text{ mm}) + t_{AD}(-350 \text{ mm}) + 2.8 \text{ kN} = 0 \tag{1}$$

$$t_{AB}(-100 \text{ mm}) + t_{AC}(-150 \text{ mm}) + t_{AD}(100 \text{ mm}) + (-1.0 \text{ kN}) = 0 \tag{2}$$

$$t_{AB}(150 \text{ mm}) + t_{AC}(50 \text{ mm}) + t_{AD}(-50 \text{ mm}) + 1.6 \text{ kN} = 0 \tag{3}$$

Multiply (2) by 3.5:

$$t_{AB}(-350 \text{ mm}) + t_{AC}(-525 \text{ mm}) + t_{AD}(350 \text{ mm}) + (-3.5 \text{ kN}) = 0$$

Add this to (1):

$$t_{AB}(-700 \text{ mm}) + t_{AC}(-875 \text{ mm}) + (-0.7 \text{ kN}) = 0 \tag{4}$$

Now multiply (3) by 7:

$$t_{AB}(1050 \text{ mm}) + t_{AC}(350 \text{ mm}) + t_{AD}(-350 \text{ mm}) + 11.2 \text{ kN} = 0$$

Subtract this from (1):

$$t_{AB}(-1400 \text{ mm}) + t_{AC}(-700 \text{ mm}) + (-8.4 \text{ kN}) = 0 \tag{5}$$

Multiply (4) by 2:

$$t_{AB}(-1400 \text{ mm}) + t_{AC}(-1750 \text{ mm}) + (-1.4 \text{ kN}) = 0$$

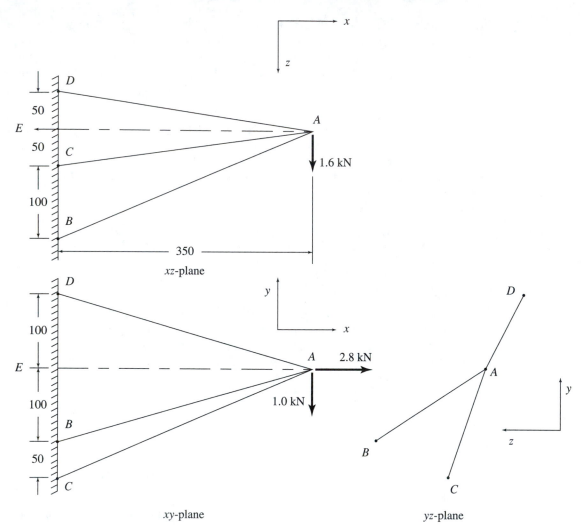

Figure 4.6   Space frame for Example Problem 4.3.

Subtract this from (5):

$$t_{AC}(1050 \text{ mm}) + (-7.0 \text{ kN}) = 0$$

So $t_{AC} = 7.0 \text{ kN}/1050 \text{ mm} = 6.67 \times 10^{-3} \text{ kN/mm}$

Substitute this value into (5):

$$t_{AB}(-1400 \text{ mm}) + (6.67 \times 10^{-3} \text{ kN/mm})(-700 \text{ mm}) + (-8.4 \text{ kN}) = 0$$

$$t_{AB}(-1400 \text{ mm}) + (-4.67 \text{ kN}) + (-8.4 \text{ kN}) = 0$$

$$t_{AB}(-1400 \text{ mm}) = 13.07 \text{ kN}$$

So $t_{AB} = 13.07 \text{ kN}/(-1400 \text{ mm}) = -9.34 \times 10^{-3} \text{ kN/mm}$

Substitute the t_{AB} and t_{AC} values into any of equations (1), (2), or (3). For example, substituting into (3) gives

$$(-9.34 \times 10^{-3} \text{ kN/mm})(150 \text{ mm}) + (6.67 \times 10^{-3} \text{ kN/mm})(50 \text{ mm}) +$$
$$t_{AD}(-50 \text{ mm}) + 1.6 \text{ kN} = 0$$
$$(-1.40 \text{ kN}) + (0.33 \text{ kN}) + t_{AD}(-50 \text{ mm}) + 1.6 \text{ kN} = 0$$
$$t_{AD}(-50 \text{ mm}) + 0.53 \text{ kN} = 0$$

So $\qquad t_{AD} = (-0.53 \text{ kN})/(-50 \text{ mm}) = 10.6 \times 10^{-3} \text{ kN/mm}$

Collecting these results gives

$$t_{AB} = -9.34 \times 10^{-3} \text{ kN/mm}$$
$$t_{AC} = 6.67 \times 10^{-3} \text{ kN/mm}$$
$$t_{AD} = 10.6 \times 10^{-3} \text{ kN/mm}$$

We now have to multiply these tension coefficients by the lengths of the members. From equation (4.3),

$$L = \sqrt{L_x^2 + L_y^2 + L_z^2}$$

So
$$L_{AB} = \sqrt{L_{ABx}^2 + L_{ABy}^2 + L_{ABz}^2}$$
$$= \sqrt{(-350 \text{ mm})^2 + (-100 \text{ mm})^2 + (150 \text{ mm})^2)}$$
$$= 393.7 \text{ mm}$$
$$L_{AC} = \sqrt{L_{ACx}^2 + L_{ACy}^2 + L_{ACz}^2}$$
$$= \sqrt{(-350 \text{ mm})^2 + (-150 \text{ mm})^2 + (50 \text{ mm})^2)}$$
$$= 384.1 \text{ mm}$$
$$L_{AD} = \sqrt{L_{ADx}^2 + L_{ADy}^2 + L_{ADz}^2}$$
$$= \sqrt{(-350 \text{ mm})^2 + (100 \text{ mm})^2 + (-50 \text{ mm})^2)}$$
$$= 367.4 \text{ mm}$$

We can now determine the forces in each member by using $t = T/L$, from which

$$T = tL$$

For *AB*, $\qquad t_{AB} = -9.34 \times 10^{-3} \text{ kN/mm}$ and $L_{AB} = 393.7 \text{ mm}$

Therefore, $\quad T_{AB} = (-9.34 \times 10^{-3} \text{ kN/mm})(393.7 \text{ mm}) = -3.68 \text{ kN}$

For *AC*, $\qquad t_{AC} = 6.67 \times 10^{-3} \text{ kN/mm}$ and $L_{AC} = 384.1 \text{ mm}$

Therefore, $\quad T_{AC} = (6.67 \times 10^{-3} \text{ kN/mm})(384.1 \text{ mm}) = 2.56 \text{ kN}$

For *AD*, $\qquad t_{AD} = 10.6 \times 10^{-3} \text{ kN/mm}$ and $L_{AD} = 367.4 \text{ mm}$

Therefore, $\quad T_{AD} = (10.6 \times 10^{-3} \text{ kN/mm})(367.4 \text{ mm}) = 3.89 \text{ kN}$

Answer

$$T_{AB} = 3.7 \text{ kN} \quad \text{(compression)} \qquad \text{because it was negative}$$
$$T_{AC} = 2.6 \text{ kN} \quad \text{(tension)}$$
$$T_{AD} = 3.9 \text{ kN} \quad \text{(tension)}$$

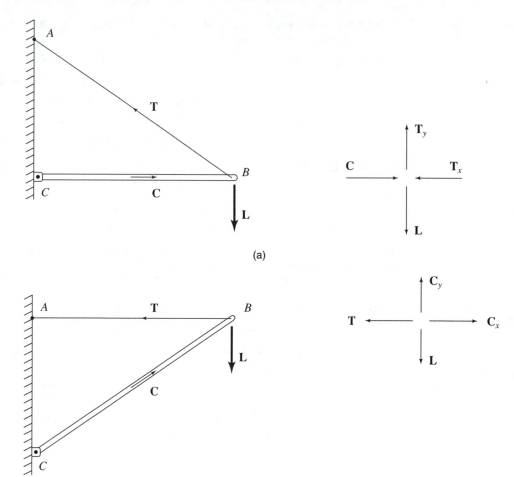

(a)

(b)

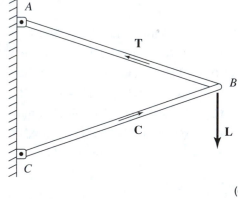

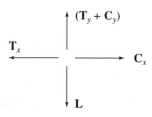

(c)

Figure 4.7   Trusses: Single triangles.

4.4   TRUSSES

Trusses are structures that consist of triangles. They are found in roof trusses, bridge structures, airplane wings, cranes, and many more structures.

4.4.1   The Single Triangle

Figure 4.7(a), (b), and (c) shows three simple trusses, each being used to support a load **L**.

In (a), the load is transmitted to the wall by the tension member *AB*. The compression member *BC* balances out the horizontal component of the tension.

In (b), the load is transmitted to the wall by the compression member *BC*. The tension member *AB* balances out the horizontal component of the compression.

In (c), the load is supported by the sum of the vertical components of the tensile and the compressive forces, and the horizontal components of these forces balance out.

4.4.2   Analysis of a Single-Triangle Truss

We calculated the forces in the members of a single-triangle truss in Example Problem 2.9. To do that, we drew the force triangle, but an alternative approach would be to balance the *x*-components and the *y*-components. Because this is the method of analysis used for more complex trusses and also in computer analyses of trusses, we will consider a simple example.

EXAMPLE PROBLEM 4.4

Find the forces in the truss members shown in Figure 4.8(a).

Solution

Set out point Q as a free body (Figure 4.8(b)), showing the tensile force, **T**, and the compressive force, **C**. Then resolve the compressive force C into its x- and y-components. The x-component will be ($C \cos 50°$), and the y-component will be ($C \sin 50°$). The horizontal forces must balance, so

$$T = C \cos 50° = 0.643C$$

Also, the vertical forces must balance, so

$$C \sin 50° = 200 \text{ N}$$
$$0.766C = 200 \text{ N}$$
$$C = 200 \text{ N}/0.766 = 261 \text{ N}$$

Since $T = 0.643C$,

$$T = (0.643)(261 \text{ N}) = 168 \text{ N}$$

Answer

The tensile force is 170 N and the compressive force is 260 N.

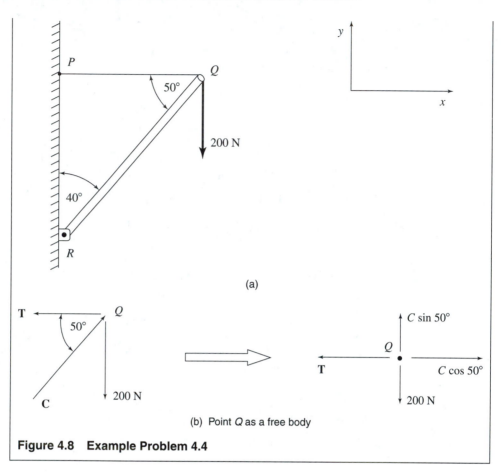

(a)

(b) Point Q as a free body

Figure 4.8   Example Problem 4.4

EXAMPLE PROBLEM 4.5

Repeat Example Problem 4.4 using tension coefficients.

Solution

Instead of *AB, AC,* and *AD,* as in Figure 4.6, the members are *QP, QR,* and 0 (it is a two-dimensional problem). Let the length of *QR* (Figure 4.8(a)) be L_{QR}. Then its projected length on the *y*-axis (the vertical wall) is (be careful of the signs)

$$L_{QRy} = -L_{QR}\sin 50° = -0.7660L_{QR}$$

and its projected length on the *x*-axis is

$$L_{QRx} = -L_{QR}\cos 50° = -0.6428L_{QR}$$

Let the length of *QP* be L_{QP}. Its projected length on the *y*-axis is $L_{QPy} = 0$, and its projected length on the *x*-axis is simply

$$L_{QPx} = -L_{QP}$$

The external force components are $F_x = 0$ and $F_y = -200$ N. This is a two-dimensional problem, so all z-components are zero. Applying equations (4.2(a)) and (4.2(b)):

$$t_{QR}L_{QRx} + t_{QP}L_{QPx} + F_x = 0$$
$$t_{QR}L_{QRy} + t_{QP}L_{QPy} + F_y = 0$$

But,
$$L_{QRx} = -0.6428L_{QR} \qquad L_{QPx} = -L_{QP}$$
$$L_{QRy} = -0.7660L_{QR} \qquad L_{QPy} = 0$$
$$F_x = 0 \qquad\qquad F_y = -200 \text{ N}$$

Substituting these into the equations,

$$t_{QR}(-0.6428L_{QR}) + t_{QP}(-L_{QP}) = 0$$
$$t_{QR}(-0.7660L_{QR}) + (-200 \text{ N}) = 0$$

The second of these equations yields

$$t_{QR}L_{QR} = (200 \text{ N})/(-0.7660) = -261.1 \text{ N}$$

But $t_{QR}L_{QR} = T_{QR}$, the tension in member QR. Because this is a negative value, it is a compressive force. Substituting this value into the first of the preceding equations gives

$$(-261.1 \text{ N})(-0.6428) + t_{QP}(-L_{QP}) = 0$$

So $t_{QP}L_{QP} = (261.1 \text{ N})(0.6428) = 167.8$ N. Also, $t_{QP}L_{QP} = T_{QP}$, the tension in member QP.

Answer
As before, the solution is

$$T_{QR} = 260 \text{ N (compression)}$$
$$T_{QP} = 170 \text{ N (tension)}$$

4.4.3   Development of the Truss

Imagine that we are designing an airplane wing spar. It has to be long and slender and must fit into the space shown in Figure 4.9(a). We could design it as a single triangle, as in Figure 4.9(b), but the long, thin compression member would probably buckle. What we have to do is to transmit the lift load back to the fuselage through a series of small triangles, as in Figure 4.9(c). This structure is known as the *Pratt truss*. Another type, shown in Figure 4.9(d), is the *Warren truss*.

Other examples of trusses are the *Fink* and *compound Fink*, used as roof trusses (Figure 4.10(a) and (b)), and the crane truss, which is basically a Warren truss, in Figure 4.10(c).

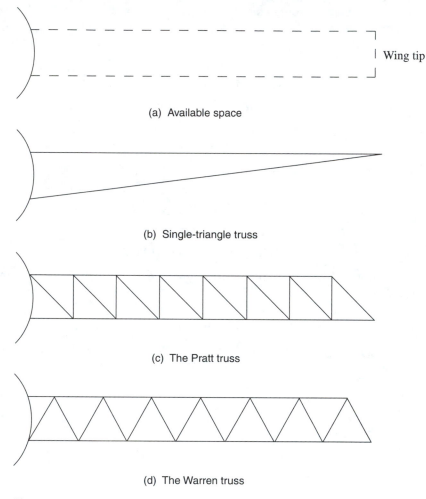

(a) Available space

(b) Single-triangle truss

(c) The Pratt truss

(d) The Warren truss

Figure 4.9

4.4.4 Method of Joints

The members of a truss are assumed to be two-force members (only axial forces act on each strut). In the *method of joints*, each joint is considered separately as a free body, and the forces acting on the joint are balanced in the x-direction and in the y-direction. As an example, suppose AB and AC are members of a truss joined at joint A (Figure 4.11). The angle between the members is 30°, and there is a vertical reaction of 100 N at A.

First consider the force directions. The member AB must be pushing down to balance the vertical reaction, but because it is at an angle to the horizontal it must also be pushing to the left. The only force to balance this component is the force in AC. Therefore $\mathbf{F}_{AC}$ must be acting to the right.

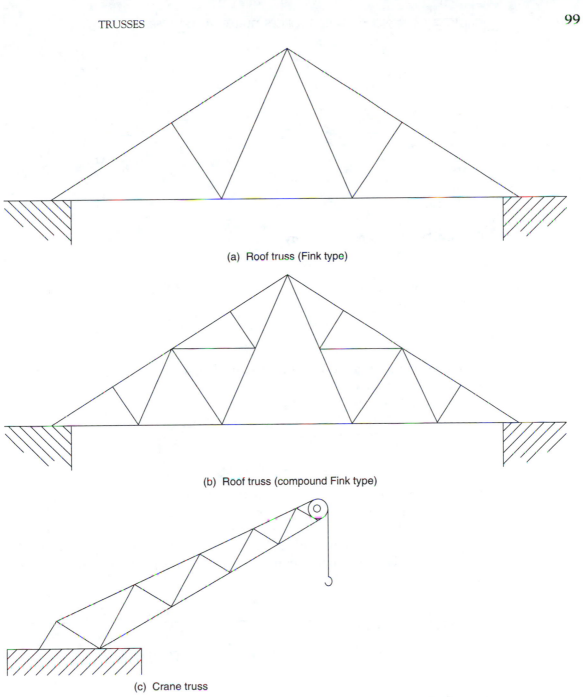

(a) Roof truss (Fink type)

(b) Roof truss (compound Fink type)

(c) Crane truss

Figure 4.10

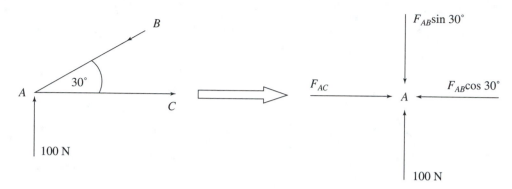

Figure 4.11   Joint *A* as a free body.

The components of $\mathbf{F}_{AB}$ are $F_{AB}\sin 30°$ and $F_{AB}\cos 30°$. Balancing the forces in the y-direction gives

$$100 \text{ N} = F_{AB}\sin 30° = 0.5F_{AB}$$

So $$F_{AB} = 200 \text{ N}$$

This is *pushing* against joint *A*. A *strut* could do that, whereas a *string* could not. So $\mathbf{F}_{AB}$ is a *compressive* force.

Balancing the forces in the x-direction gives

$$F_{AC} = F_{AB}\cos 30° = 0.8660F_{AB}$$

So

$$F_{AC} = (0.8860)(200 \text{ N}) = 177.2 \text{ N}$$

This is *pulling* against joint *A*. A *string* could do that, so $\mathbf{F}_{AC}$ is a *tensile* force.

In an actual problem, you would then move to the next joint, and so on, until all the forces had been determined. Notice, however, that we have only two equations, so we can analyze only joints with two unknown forces or one unknown force. This does not mean that we cannot determine *all* the forces; we just have to do them in the correct order.

EXAMPLE PROBLEM 4.6

Determine the forces in all the members in the truss shown in Figure 4.12(a).

Solution

1. Calculate the reactions. Taking moments about *A* gives

$$\Sigma M)_A = 0$$

$$(20 \text{ m})(R_E) - (15 \text{ m})(5000 \text{ N}) = 0$$

$$R_E = \frac{75{,}000 \text{ N·m}}{20 \text{ m}} = 3750 \text{ N}$$

Taking moments about E yields

$$\sum M)_E = 0$$

$$(5\text{ m})(5000\text{ N}) - (20\text{ m})(R_A) = 0$$

$$R_A = \frac{25,000\text{ N·m}}{20\text{ m}} = 1250\text{ N}$$

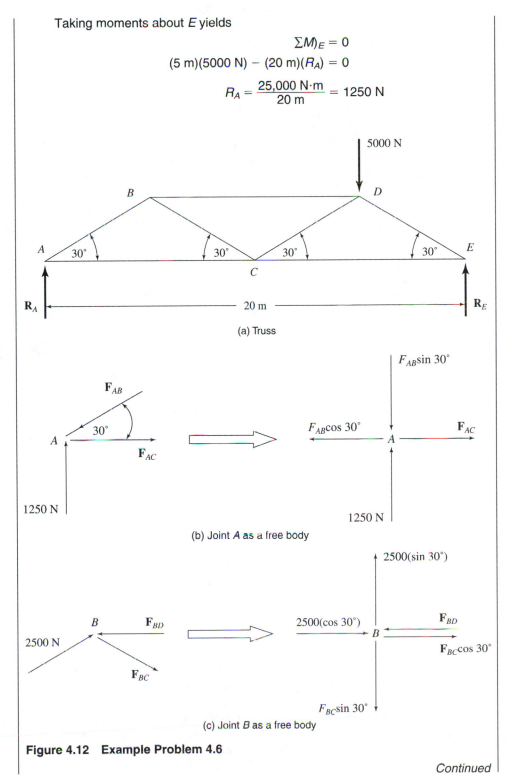

(a) Truss

(b) Joint *A* as a free body

(c) Joint *B* as a free body

Figure 4.12   Example Problem 4.6

Continued

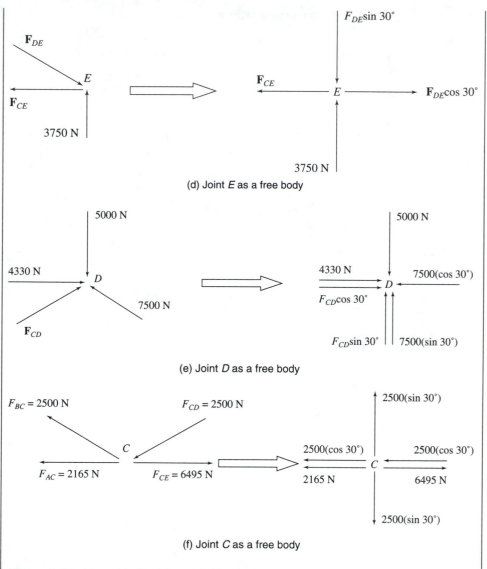

(d) Joint *E* as a free body

(e) Joint *D* as a free body

(f) Joint *C* as a free body

Figure 4.12   Example Problem 4.6 (*Continued*)

As a check, the sum of the external vertical forces should equal zero, which it is:

$$3750 \text{ N} + 1250 \text{ N} - 5000 \text{ N} = 0$$

2. Set out joint *A* as a free body (Figure 4.12(b)).

There is an upward vertical reaction of 1250 N, which must be balanced by the vertical component of F_{AB}. So F_{AB} must act downward and, consequently, to the left, making F_{AC} act to the right.

Balancing the vertical forces at *A*,

$$1250 \text{ N} - F_{AB}\sin 30° = 0$$

So

$$F_{AB}(0.50) = 1250 \text{ N}$$
$$= 2500 \text{ N}$$

Balancing the horizontal forces at *A*,

$$F_{AC} - F_{AB}\cos 30° = 0$$
$$F_{AC} = (2500 \text{ N})(0.8660) = 2165 \text{ N}$$

3. Set out joint *B* as a free body (Figure 4.12(c)). First consider the member *AB*. We know that the force $\mathbf{F}_{AB}$ is pushing down against joint *A*. With regard to joint *B*, there must be an equal and opposite force to stabilize the member *AB*. So there is an *upward* force $\mathbf{F}_{AB}$ acting at joint *B*. Now consider member *BC*. Its vertical force component must be acting down to balance the upward component of $\mathbf{F}_{AB}$. Because the horizontal components of both $\mathbf{F}_{AB}$ and $\mathbf{F}_{BC}$ are acting to the right, $\mathbf{F}_{BD}$ must act to the *left*.

Balancing the vertical forces at B gives

$$(2500 \text{ N})(\sin 30°) - F_{BC}\sin 30° = 0$$

So

$$F_{BC} = 2500 \text{ N}$$

Balancing the horizontal forces yields

$$(2500 \text{ N})(\cos 30°) + F_{BC}\cos 30° - F_{BD} = 0$$
$$(2500 \text{ N})(0.8660) + (2500 \text{ N})(0.8660) - F_{BD} = 0$$

So

$$F_{BD} = 4330 \text{ N}$$

4. Instead of going to joint *C* next, we leave it as the final check, and go to joint *E*. (We *could* use joint *C* next. It is just a matter of choice.) Set out joint *E* as a free body (Figure 4.12(d)). Using the same arguments as before, $\mathbf{F}_{DE}$ acts down and $\mathbf{F}_{CE}$ acts to the left.

Balancing the vertical forces at E,

$$3750 \text{ N} - F_{DE}\sin 30° = 0$$

So

$$F_{DE} = \frac{3750 \text{ N}}{0.50} = 7500 \text{ N}$$

Balancing the horizontal forces,

$$F_{DE}\cos 30° - F_{CE} = 0$$
$$F_{CE} = (7500 \text{ N})(0.8660) = 6495 \text{ N}$$

5. Set out joint D as a free body (Figure 4.12(e)). Balancing the vertical forces at D,

$$F_{CD}\sin 30° + (7500 \text{ N})(\sin 30°) - 5000 \text{ N} = 0$$

$$0.50F_{CD} + 3750 \text{ N} - 5000 \text{ N} = 0$$

$$F_{CD} = \frac{1250 \text{ N}}{0.50} = 2500 \text{ N}$$

6. Summary of the forces:

F_{AB} = 2500 N	(compressive)	F_{BC} = 2500 N	(tensile)
F_{CD} = 2500 N	(compressive)	F_{DE} = 7500 N	(compressive)
F_{AC} = 2165 N	(tensile)	F_{BD} = 4330 N	(compressive)
F_{CE} = 6495 N	(tensile)		

7. As a check, set out joint C as a free body (Figure 4.12(f)).

Balancing the vertical forces at C,

$$(2500 \text{ N})(\sin 30°) - (2500 \text{ N})(\sin 30°) \stackrel{?}{=} 0 \qquad \text{Yes}$$

Balancing the horizontal forces at C,

$$6495 \text{ N} - (2500 \text{ N})(\cos 30°) - (2500 \text{ N})(\cos 30°) - 2165 \text{ N} \stackrel{?}{=} 0$$

$$6495 \text{ N} - (2500 \text{ N})(0.8660) - (2500 \text{ N})(0.8660) - 2165 \text{ N} \stackrel{?}{=} 0$$

$$6495 \text{ N} - 2165 \text{ N} - 2165 \text{ N} - 2165 \text{ N} \stackrel{?}{=} 0? \qquad \text{Yes}$$

Answer

F_{AB} = 2500 N	(compressive)	F_{BC} = 2500 N	(tensile)
F_{CD} = 2500 N	(compressive)	F_{DE} = 7500 N	(compressive)
F_{AC} = 2200 N	(tensile)	F_{BD} = 4300 N	(compressive)
F_{CE} = 6500 N	(tensile)		

4.4.5 Method of Sections

The *method of sections* allows us to determine the force in a member without having to calculate joint by joint. It is very useful if we need to know the forces in only a few members of a multimembered truss. As an example, we will determine the forces in BD and CE of the truss in Figure 4.12(a). We have already determined these to be F_{BD} = 4330 N and F_{CE} = 6495 N by the method of joints in Example Problem 4.6

To determine the forces, cut the truss as shown in Figure 4.13(a).

To find $\mathbf{F}_{BD}$, the force in member BD, consider the section to the left of the cut (Figure 4.13(b)) as a free body, and take moments about joint C. We have already determined that the reaction at A is 1250 N from Example Problem 4.6. Also, the truss height is $(5 \text{ m}) (\tan 30°) = (5 \text{ m}) (0.5774) = 2.887$ m. For equilibrium,

$$\Sigma M)_C = 0$$

$$(2.887 \text{ m})(F_{BD}) - (10 \text{ m})(1250 \text{ N}) = 0$$

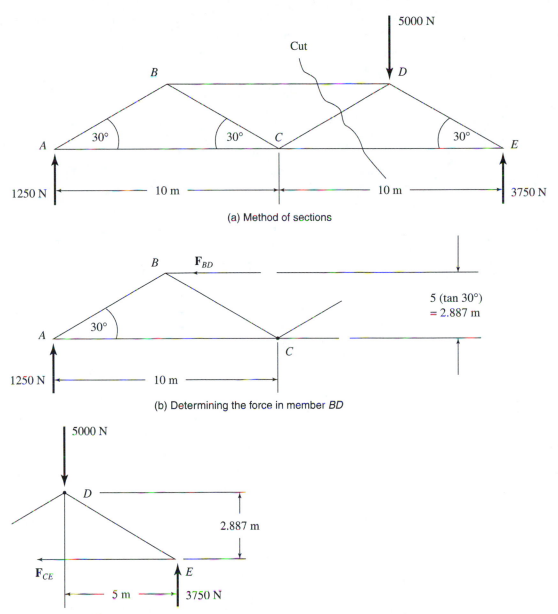

(a) Method of sections

(b) Determining the force in member *BD*

(c) Determining the force in member *CE*

Figure 4.13

Note: We see immediately that $\mathbf{F}_{BD}$ must be acting to the left to balance the moment about C created by the reaction at A.

$$(2.887 \text{ m})(F_{BD}) - 12500 \text{ N·m} = 0$$

$$F_{BD} = \frac{12{,}500 \text{ N}}{2.887 \text{ m}} = 4330 \text{ N}$$

To find $\mathbf{F}_{CE}$, the force in member CE, consider the section to the right of the cut (Figure 4.13(c)), and take moments about D.
For equilibrium,

$$\Sigma M)_D = 0$$
$$(5 \text{ m})(3750 \text{ N}) - (2.887 \text{ m})(F_{CE}) = 0$$
$$18{,}750 \text{ N·m} - (2.887 \text{ m})(F_{CE}) = 0$$
$$F_{CE} = \frac{18{,}750 \text{ N·m}}{2.887 \text{ m}} = 6495 \text{ N}$$

Notice that this method alone could not be used to determine the forces in the slant members.

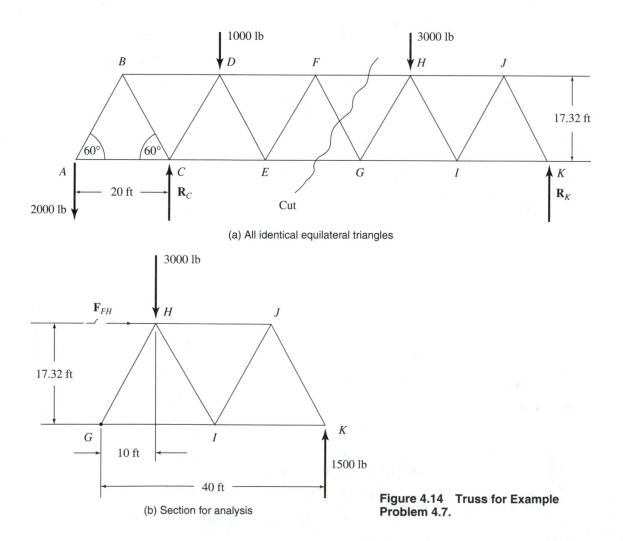

(a) All identical equilateral triangles

(b) Section for analysis

Figure 4.14   Truss for Example Problem 4.7.

EXAMPLE PROBLEM 4.7

Determine the force in member *FH* in Figure 4.14(a).

Solution

1. Calculate the reaction, R_K, at joint *K* by taking moments about *C* and equating their algebraic sum to zero:

$$\Sigma M)_C = 0$$

$$(2000 \text{ lb})(20 \text{ ft}) + (R_K)(80 \text{ ft}) - (1000 \text{ lb})(10 \text{ ft}) - (3000 \text{ lb})(50 \text{ ft}) = 0$$

$$4 \times 10^4 \text{ lb ft} + (R_K)(80 \text{ ft}) - 1 \times 10^4 \text{ lb-ft} - 15 \times 10^4 \text{ lb-ft} = 0$$

$$(R_K)(80 \text{ ft}) - 12 \times 10^4 \text{ lb-ft} = 0$$

$$R_K = \frac{12 \times 10^4 \text{ lb-ft}}{80 \text{ ft}}$$

$$= 1500 \text{ lb}$$

2. Cut the truss through members *FH, FG,* and *EG,* and consider the section on the right (Figure 4.14(b)). We do not know the direction of the force in *FH,* so assume it is to the right. If it acts in the opposite direction, the final result will be a negative value.

   Taking moments about joint *G* (to exclude the forces in *FG* and in *EG* from the moment balance) and equating their algebraic sum to zero, we have

$$\Sigma M)_G = 0$$

$$(1500 \text{ lb})(40 \text{ ft}) - (3000 \text{ lb})(10 \text{ ft}) - (F_{FH})(17.32 \text{ ft}) = 0$$

$$6 \times 10^4 \text{ lb-ft} - 3 \times 10^4 \text{ lb-ft} - (F_{FH})(17.32 \text{ ft}) = 0$$

$$3 \times 10^4 \text{ lb-ft} - (F_{FH})(17.32 \text{ ft}) = 0$$

$$F_{FH} = \frac{3 \times 10^4 \text{ lb-ft}}{17.32 \text{ ft}}$$

$$F_{FH} = 1732 \text{ lb}$$

The result is a positive value, so our assumption concerning its direction was correct. *FH* is pushing against joint *H,* and so it must be a compressive force.

Answer

The force in member FH is 1700 lb (compressive).

4.5 COMPUTER ANALYSES

We present two programs in this section. The first, program 'Truss', is for solving single-triangle truss problems. The second, program 'Spcframe', is for solving space frame problems.

4.5.1   Program 'Truss'

This program calculates the forces in the members of single-triangle trusses. The input data consists of the angles of the members, the load or external force, and its angle. The following examples demonstrate how to use the program.

Caution: When writing programs containing angles, note that most programs use angles in *radians*, not degrees. Suppose your input contains an angle ALPHA degrees and you need to find its cosine. In Excel, =cos (ALPHA) will give an incorrect value. You have to write =cos (ALPHA*PI()/180), or =cos (RADIANS (ALPHA)). This will convert ALPHA to radians; then you can find the cosine.

EXAMPLE PROBLEM 4.8

Determine the forces in members *AB* and *BC* in Figure 4.15.

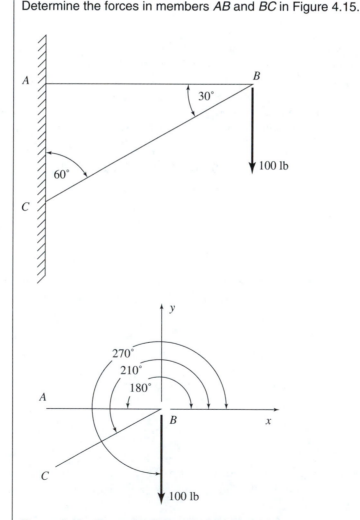

Figure 4.15   Truss for Example Problem 4.8.

'Truss' for single-triangle calculations

Member AB

Angle	0

Member BC

Angle	0

Load

Force	0
Angle	0

Load Units

lb, klb, N, kN | lb |

Results

Member AB	#DIV/0! lb	#DIV/0!
Member BC	#DIV/0! lb	#DIV/0!

Figure 4.16   Input/output sheet for program 'Truss'.

Solution

1. Determine the angles. All angles are measured *counterclockwise, with east as zero and the intersection point of the three forces as the origin.*

   Member *AB* is at angle 180°.
   Member *BC* is at angle (180 + 30)° = 210°.
   The load of 100 lb is at angle 270°.

2. Open program 'Truss'. Your screen should look like Figure 4.16.

   Enter 180 in cell B4   (Member *AB*)

   Enter 210 in cell E4   (Member *BC*)

   Enter 100 in cell B7   (Load, Force)

   [Note that this is entered as a *positive* value, although it is acting in a downward direction because the angle of 270° directs it downward. We could enter an angle of 90° and a load of −100 lb.]

   Enter 270 in cell B8   (Load, Angle)

   Enter lb in cell E7   (Load Units)

'Truss' for single-triangle calculations

Member AB

Angle	180

Member BC

Angle	210

Load

Force	100
Angle	270

Load Units

lb, klb, N, kN | lb |

Results

Member AB	−173.205 lb	Tension
Member BC	200.00 lb	Compression

Figure 4.17   Input/output sheet for Example Problem 4.8.

Answer

The results are (Figure 4.17):

$$F_{AB} = 173\text{-lb tension}$$
$$F_{BC} = 200\text{-lb compression}$$

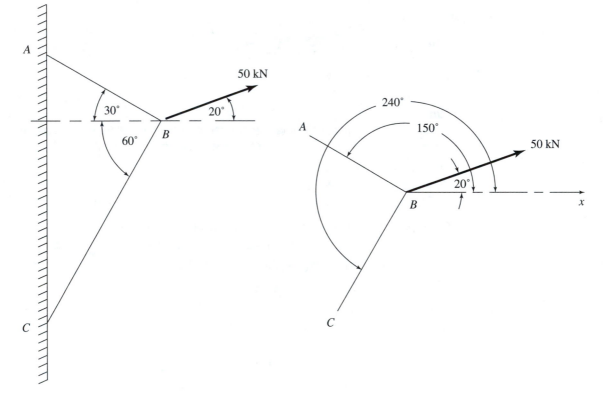

Figure 4.18   Truss for Example Problem 4.9.

EXAMPLE PROBLEM 4.9

Determine the forces in members *AB* and *BC* in Figure 4.18.

Solution

1. Determine the angles.

   AB is at $(180 - 30)° = 150°$.
   BC is at $(180 + 60)° = 240°$.

   The external force is at 20°.

'Truss' for single-triangle calculations

Member AB

Angle	150

Member BC

Angle	240

Load

Force	50
Angle	20

Load Units

lb, klb, N, kN	kN

Results

Member AB	−32.1394 kN	Tension
Member BC	−38.30 kN	Tension

Figure 4.19   Input/output sheet for Example Problem 4.9.

2. Input the following values.

   Enter 150 in cell B4 (see Figure 4.19).
   Enter 240 in cell E4.
   Enter 50 in cell B7.
   Enter 20 in cell B8.
   Enter kN in cell E7.

Answer

The results are shown in Figure 4.19.

$$F_{AB} = 32\text{-kN tension}$$

$$F_{BC} = 38\text{-kN tension}$$

The components of the 50-kN load are

$$F_x = 47 \text{ kN}$$

$$F_y = 17 \text{ kN}$$

4.5.1   Program 'Spcframe'

Program 'Spcframe' calculates the forces in the members of a space frame. The input data consist of the projected lengths of the members on each axis and the x-, y-, and z-components of the external force. The following examples demonstrate how to use the program.

EXAMPLE PROBLEM 4.10

Calculate the *stresses* in the members of the space frame shown in Figure 4.20. Each member is 30 mm in diameter.

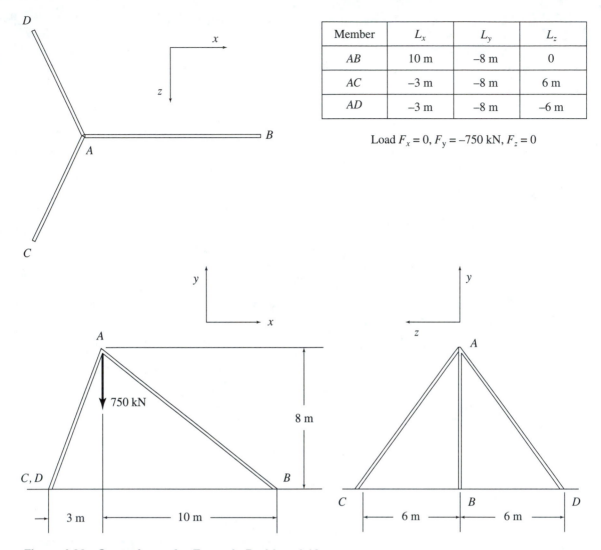

Member	L_x	L_y	L_z
AB	10 m	–8 m	0
AC	–3 m	–8 m	6 m
AD	–3 m	–8 m	–6 m

Load $F_x = 0$, $F_y = -750$ kN, $F_z = 0$

Figure 4.20   Space frame for Example Problem 4.10.

Note: The *stress* in a member is the force divided by its cross-sectional area. We discuss this in detail in Chapter 6.

Solution

1. Set up the axes. Remember that the origin is the point where the members meet. In this case, it is point *A*. Draw the *x*- and *y*-directions for the front view (see Figure 4.20). The z-direction is determined by the *right-hand screw rule*. Imagine that you are holding a screwdriver in your right hand. You are going to turn a screw so that it goes *into* a piece of wood, and the direction of rota-

tion is from *x* to *y*. The screwdriver is pointing in the *z*-direction. Draw the *x*- and *z*-directions on the top view and the *y*- and *z*-directions on the end view (Figure 4.20).

We show the projected lengths in the table of Figure 4.20. Make sure you understand each one. For example, the projected length L_y for member *AB* is negative because it is measured from the origin, *A,* in a downward (negative *y*) direction.

2. Open program 'Spcframe'. Your screen should look like Figure 4.21.

Enter the following data (see Figure 4.22)

10 in cell B4	−3 in cell E4	−3 in cell H4
−8 in cell B5	−8 in cell E5	−8 in cell H5
0 in cell B6	6 in cell E6	−6 in cell H6
0 in cell B9	−750 in cell B10	0 in cell B11
kN in cell B13		

3. The results are (see Figure 4.22)

$$F_{AB} = 277.1 \text{ kN} \quad \text{compression}$$
$$F_{AC} = 376.5 \text{ kN} \quad \text{compression}$$
$$F_{AD} = 376.5 \text{ kN} \quad \text{compression}$$

4. Convert these values to stresses. Each member has a cross-sectional area of

$$A = \pi d^2/4 = \pi (30 \text{ mm})^2/4 = 706.9 \text{ mm}^2$$

The stress is $s = F/A$, so

$$S_{AB} = \frac{277.1 \text{ kN}}{706.9 \text{ mm}^2} = \frac{277,100 \text{ N}}{706.9 \text{ mm}^2}$$

Therefore, $\quad S_{AB} = 392 \text{ MPa}$

Space frame 'Spcframe'

Member AB

Lx	0
Ly	0
Lz	0

Member AC

Lx	0
Ly	0
Lz	0

Member AD

Lx	0
Ly	0
Lz	0

Load

Fx	0
Fy	0
Fz	0

Tension Coefficients

Member AB	#DIV/0!
Member AC	#DIV/0!
Member AD	#DIV/0!

Member Lengths

Member AB	0
Member AC	0
Member AD	0

Enter load units
lb, klb, N, kN   lb

Member Forces

Member AB	#DIV/0!	lb	#DIV/0!
Member AC	#DIV/0!	lb	#DIV/0!
Member AD	#DIV/0!	lb	#DIV/0!

Figure 4.21   Input/output sheet for program 'Spcframe'.

Space frame 'Spcframe'

Member AB	
Lx	10
Ly	−8
Lz	0

Member AC	
Lx	−3
Ly	−8
Lz	6

Member AD	
Lx	−3
Ly	−8
Lz	−6

Load	
Fx	0
Fy	−750
Fz	0

Tension Coefficients	
Member AB	−21.6346
Member AC	−36.0577
Member AD	−36.0577

Member Lengths	
Member AB	12.806
Member AC	10.440
Member AD	10.440

Enter load units
lb, klb, N, kN   kN

Member Forces		
Member AB	277.06 kN	COMPRESSION
Member AC	376.45 kN	COMPRESSION
Member AD	376.45 kN	COMPRESSION

Figure 4.22   Input/output sheet for Example Problem 4.10.

$$S_{AC} = S_{AD} = \frac{376.5 \text{ kN}}{706.9 \text{ mm}^2} = \frac{376{,}500 \text{ N}}{706.9 \text{ mm}^2}$$

Therefore, $S_{AC} = S_{AD} = 533$ MPa

Answer
The stresses are

$$S_{AB} = 390 \text{ MPa}, \qquad S_{AC} = S_{AD} = 530 \text{ MPa}$$

EXAMPLE PROBLEM 4.11

Calculate the forces in the members of the space frame shown in Figure 4.23.

Solution
1. Set up the axes, and tabulate the projected lengths, as shown in Figure 4.23.
2. Enter the data, as in Figure 4.24.

Answer

$$F_{AB} = 3700 \text{ lb} \qquad \text{compression}$$
$$F_{AC} = 2600 \text{ lb} \qquad \text{tension}$$
$$F_{AD} = 3900 \text{ lb} \qquad \text{tension}$$

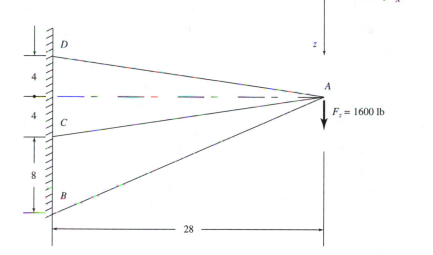

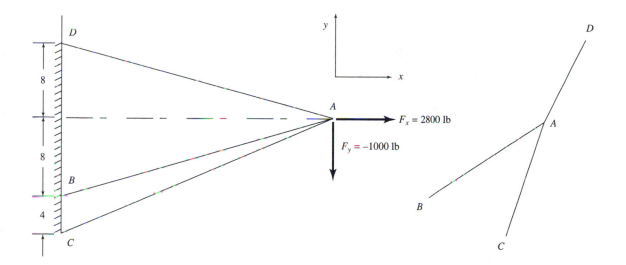

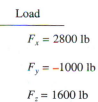

Member	L_x	L_y	L_z
AB	−28	−8	12
AC	−28	−12	4
AD	−28	8	−4

Load
$F_x = 2800$ lb
$F_y = -1000$ lb
$F_z = 1600$ lb

Figure 4.23   Space frame for Example Problem 4.11.

Space frame 'Spcframe'

Member AB

Lx	−28
Ly	−8
Lz	12

Member AC

Lx	−28
Ly	−12
Lz	4

Member AD

Lx	−28
Ly	8
Lz	−4

Load

Fx	2800
Fy	−1000
Fz	1600

Tension Coefficients

Member AB	−116.667
Member AC	83.3333
Member AD	133.333

Member Lengths

Member AB	31.496
Member AC	30.725
Member AD	29.394

Enter load units

lb, klb, N, kN | lb |

Member Forces

Member AB	3674.54	lb	COMPRESSION
Member AC	2560.38	lb	TENSION
Member AD	3919.18	lb	TENSION

Figure 4.24   Input/output sheet for Example Problem 4.11.

4.6   PROBLEMS FOR CHAPTER 4

Hand Calculations

1. The 15-ft-long bar shown in Figure 4.25 weighs 100 lb. It is hinged to the wall at *A* and held in position by a wire attached to its upper end, *C*, and to the wall at *B*. The bar makes an angle of 45° to the wall, and the wire makes an angle of 30° to the wall. Determine the tension in the wire and the components of the reaction at *A*.

2. Calculate the support reactions at *A* and *E* and the pin reactions at *B*, *C*, and *D* for the frame shown in Figure 4.26.

3. Determine the forces in members *AB*, *AC*, and *AD* shown in Figure 4.27.

4. Calculate the forces in the strut *AB* and the wire *BC*, shown in Figure 4.28.

5. Calculate the forces in the strut *AB* and the wire *BC*, shown in Figure 4.29.

6. Determine the forces in all the members in the truss shown in Figure 4.30, and indicate for each member whether it is in tension or compression.

7. Calculate the force in member *CE* shown in Figure 4.31. Is it a tensile or a compressive force?

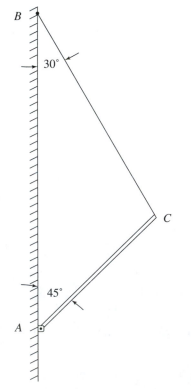

Figure 4.25

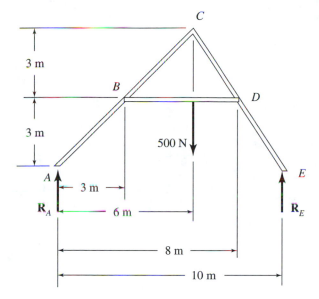

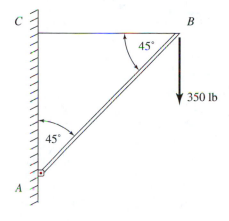

Figure 4.28

Figure 4.26

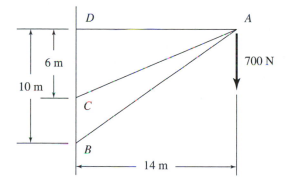

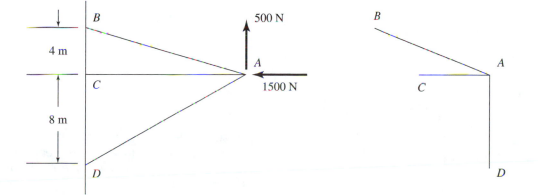

Figure 4.27

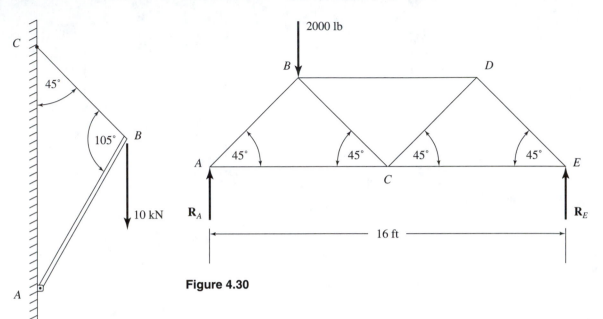

Figure 4.30

Figure 4.29

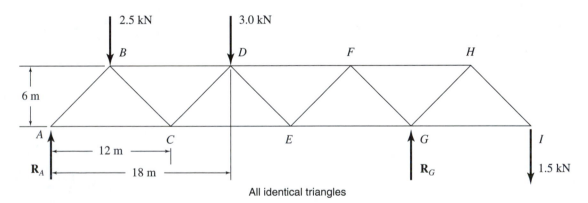

All identical triangles

Figure 4.31

Computer Calculations

8. Repeat Problem 3.

9. Repeat Problem 4.

10. Repeat Problem 5.

11. Determine the forces in members *AB* and *BC* shown in Figure 4.32.

12. At what angle should the 75-N force in Figure 4.32 act in order to create an equal *tension* in both members? What is the magnitude of this tension?

13. Determine the forces in members *AB* and *BC* shown in Figure 4.33.

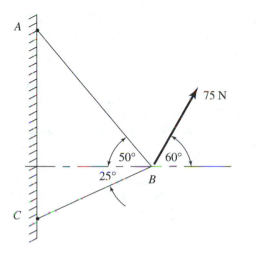

Figure 4.32

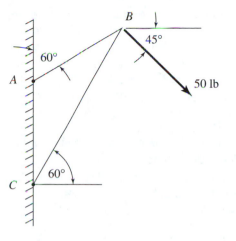

Figure 4.33

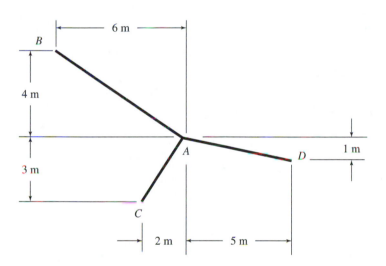

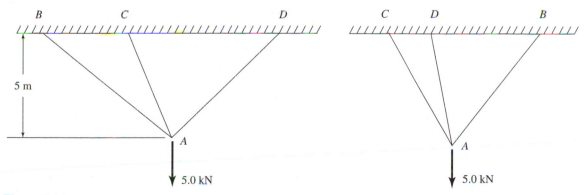

Figure 4.34

14. If the 50-lb load in Figure 4.33 were to act vertically downward, what would the forces be in the members?

15. Calculate the stresses in members *AB, AC,* and *AD* shown in Figure 4.34. The diameter of each member is 10 mm.

16. Calculate the forces in members *AB, AC,* and *AD* shown in Figure 4.34 if the 5.0-kN force acts horizontally to the right instead of vertically downward.

17. In Figure 4.34, the plane *BCD* is rotated 90° so that point *C* is the highest point. The 5.0-kN load still acts vertically downward. Determine the forces in the members. (*Hint:* Just change the applied force components.)

18. In Figure 4.34, the plane *BCD* is rotated 90° so that point *D* is the highest point. The 5.0-kN load still acts vertically downward. Determine the forces in the members.

19. In Figure 4.20, Example Problem 4.10, plane *BCD* is rotated 90° so that *B* is the highest point. The 750-kN load still acts vertically downward. Determine the forces in the members.

20. In Figure 4.20, Example Problem 4.10, plane *BCD* is rotated 90° so that *D* is the highest point. The 750-kN load still acts vertically downward. Determine the forces in the members.

5

Friction

■ Objectives

In this chapter you will learn about static friction and kinetic friction and the application of friction theory to wedges, jackscrews, and belt friction.

5.1 THE NATURE OF FRICTION

Imagine a brick lying on a flat horizontal table. Push it gently with your finger parallel to the table top. What happens? Apparently nothing. It is not until you push harder that the brick begins to slide. We say *apparently* nothing happens prior to the brick moving, but, of course, an equal and opposite force must be developed to balance the force created by your finger. This force, acting between the brick and the table top, is the *static friction* force. It is called static because the brick is at rest.

Before we begin to push the brick, there is no friction force. If we push with a force of $\frac{1}{2}$ lb and the brick does not slide, then a friction force of $\frac{1}{2}$ lb is being developed. If a 1-lb force is not sufficient to move the brick, then a 1-lb friction force is being developed, and so on, until the brick commences to slide. If, instead of pushing horizontally, we push at an angle of, say, 30° to the horizontal, then the frictional force resisting motion is equal and opposite to the **horizontal component** of the applied force. See Figure 5.1.

Once the brick begins to slide, the friction force reduces a little and then remains essentially constant, independent of the applied force. (This is not quite true. High speeds create frictional heating, which affects the frictional force.) The friction developed during motion is called *kinetic friction*. Figure 5.2 shows a graph of friction force plotted against the component of applied force parallel to the interface between the two surfaces.

It should be understood that the type of friction we are discussing is called *dry* friction, or Coulomb friction. The French physicist Charles Coulomb (1736–1806) conducted friction research and developed the empirical equations. If there is a lubricant between the surfaces, we have a condition known as *viscous* friction, which is analyzed in a different manner. Viscous friction involves molecular attraction and momentum transfer and is beyond the scope of this text.

Although dry friction is to be avoided in most engineering designs and operations, it is sometimes required, as in the case of car brakes, for example. Also, one or two turns of rope around a capstan can hold a boat to the dock.

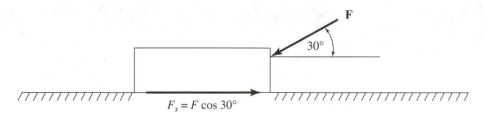

Figure 5.1   Static friction force.

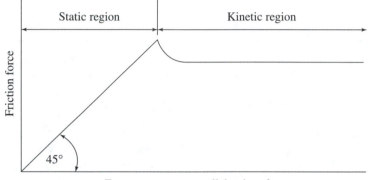

Figure 5.2   Static and kinetic friction forces.

One of the perhaps surprising results of Coulomb's experiments is that the maximum friction force that can be developed is *independent of the contact area*. You can show this by placing the brick on its edge and observing that the same force is required to move it. He found that the friction force is *proportional to the normal force* between the surfaces. In our case of the brick on the table, the normal force is simply the weight of the brick.

5.2 COEFFICIENT OF FRICTION

We saw in the previous section that for any two surfaces, say steel on steel or rope on wood, the maximum frictional force that can be developed is dependent only on the normal force and is proportional to it, so we can write

$$F_s = \mu_s N \qquad (5.1)$$

where

F_s = maximum frictional resistance

N = normal force between the surfaces

μ_s is the proportionality *constant*, called the *coefficient of static friction*

We write *constant* to emphasize that μ_s is dependent on the two materials and on their finishes. Approximate values of μ_s, taken from various sources, are tabulated in the following table.

Surfaces	Coefficient of Static Friction, μ_s
Hard steel on hard steel	0.42
Mild steel on mild steel	0.65
Brass on steel	0.51
Copper on steel	0.53
Aluminum on steel	0.50
Wood on steel	0.40
Wood on wood	0.45
Rope on wood	0.70
Rubber on solids	1 to 4
Teflon on steel	0.04
Teflon on Teflon	0.04
Nylon 6/6 on nylon 6/6	0.09

EXAMPLE PROBLEM 5.1

What is the magnitude of the horizontal force required to start a 10-kg block of aluminum to slide on a horizontal steel tabletop?

Solution

The required force is equal to the maximum static frictional resistance.

$$F = F_s = \mu_s N$$

N is the normal force between the surfaces, which is the weight (*not* the mass) of the block.

$$N = (10 \text{ kg})(9.81 \text{ m/s}^2) = 98.1 \text{ N}$$

The coefficient of static friction for aluminum on steel is $\mu_s = 0.50$. Therefore,

$$F = (98.1 \text{ N})(0.50) = 49 \text{ N}.$$

Answer

The required force is 49 N.

EXAMPLE PROBLEM 5.2

The table in the previous problem is tilted until the block just begins to slide. What is the angle?

Solution

Referring to Figure 5.3, the component of the weight acting normal to the surface is

$$N = W \cos \alpha$$

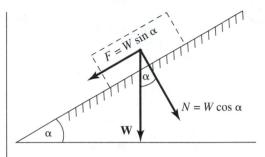

$$F = W \sin \alpha$$

$$N = W \cos \alpha$$

Figure 5.3   Block on inclined plane.

where $W = 98.1$ N and α is the tilt angle; the component of the weight acting along the surface is

$$F = W \sin \alpha$$

Because F is the force required to just start the block moving, it is equal to the maximum static frictional force, $F_{\mathbf{s}}$, which is equal to $\mu_s N$

So $F = W \sin \alpha = \mu_s N$

But $N = W \cos \alpha$

Therefore, $W \sin \alpha = \mu_s (W \cos \alpha)$

$$\frac{\sin \alpha}{\cos \alpha} = \mu_s$$

But $(\sin \alpha)/(\cos \alpha) = \tan \alpha$, so

$$\tan \alpha = \mu_s$$

In this example $\mu_s = 0.50$; therefore, $\alpha = 26.6°$.

Answer
The required tilt angle is 26.6°.

Example Problem 5.2 indicates a simple method of determining the coefficient of static friction between two surfaces. Simply tilt the surface until the object just begins to slide. Measure the tilt angle, and take the tangent of it. This angle is known as the

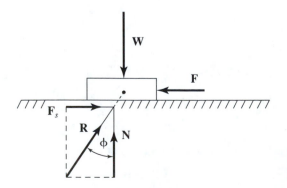

Figure 5.4   Angle of static friction.

angle of static friction and is usually represented by the Greek letter φ (phi) with subscript s. Thus, an alternative form of equation (5.1) is

$$\tan \phi_s = F_s/N = \mu_s \qquad (5.2)$$

Returning now to the case of a block on a horizontal plane (Figure 5.4), we can combine the friction force F_s and the normal force N into one resultant force R, which acts at angle ϕ_s from the normal force.

EXAMPLE PROBLEM 5.3

The member shown in Figure 5.5 is free to slide on the vertical 80-mm-diameter shaft. If the weight is too far from the center line of the shaft, the member will bind on the shaft and will not slide. What is the maximum distance that the weight can be from the shaft centerline and still permit the member to slide? Both the shaft and the member are made of hard steel.

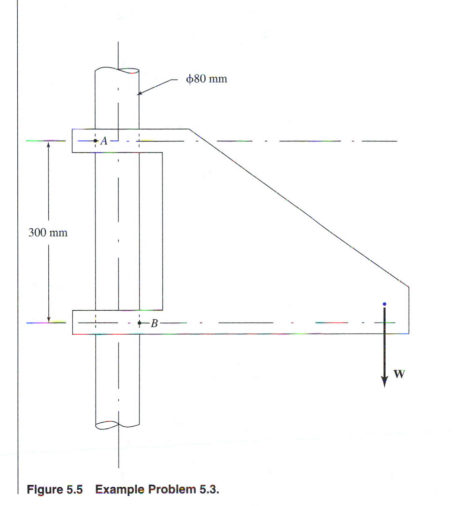

Figure 5.5   Example Problem 5.3.

Solution

The weight **W** will create normal forces $\mathbf{N}_A$ and $\mathbf{N}_B$ at points A and B, respectively. (See Figure 5.6.) There will also be friction forces $\mathbf{F}_{sA}$ and $\mathbf{F}_{sB}$ at points A and B. The resultants of these forces are $\mathbf{R}_A$ at A and $\mathbf{R}_B$ at B. Because the member is on the point of slipping (impending motion), the angle between the resultants and the normal forces is ϕ_s, the friction angle. *For the three forces $\mathbf{R}_A$, $\mathbf{R}_B$, and **W** to be in equilibrium, their lines of action must cross at a common point.* (The forces must be concurrent; see Chapter 2.)

Let the horizontal distance from the shaft centerline to the line of action of the weight be x millimeters, and the vertical distance from point B to the point of intersection of the three forces be y millimeters. Then, referring to Figure 5.6,

$$y/(x - 40 \text{ mm}) = \tan \phi_s$$

But $\tan \phi_s = \mu_s$, the coefficient of static friction, which for hard steel on hard steel is 0.42.

Therefore,
$$y/(x - 40 \text{ mm}) = 0.42$$

$$y = 0.42x - 16.8 \text{ mm} \tag{1}$$

Carrying out the same calculation for point A,
$$(300 \text{ mm} - y)/(x + 40 \text{ mm}) = 0.42$$
$$300 \text{ mm} - y = 0.42x + 16.8 \text{ mm} \tag{2}$$

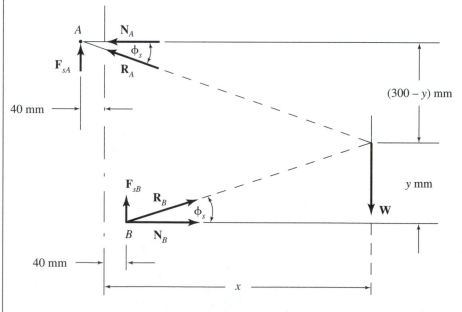

Figure 5.6   Force diagram for Example Problem 5.3.

Substitute for y from equation (1):
$$300 \text{ mm} - (0.42x - 16.8 \text{ mm}) = 0.42x + 16.8 \text{ mm}$$
$$300 \text{ mm} - 0.42x + 16.8 \text{ mm} = 0.42x + 16.8 \text{ mm}$$
$$0.84x = 300 \text{ mm}$$
$$x = 357 \text{ mm}$$

Answer
The distance of the weight from the shaft centerline must be less than 357 mm for the member to slide.

Notice that the solution is independent of the weight.

5.3   FORCE REQUIRED TO MOVE A BODY UP A SLOPE

Figure 5.7 shows a block of weight W, length L, and height H on a plane inclined at angle α to the horizontal. What force $\mathbf{F}$, at angle θ to the plane, is necessary to start the block moving up the plane? As the magnitude of $\mathbf{F}$ is increased from zero, will the block lift or slip first?

Resolving the forces into components along and normal to the plane (see Figure 5.8), we have a net component pulling up, parallel to the plane of $(F \cos \theta - W \sin \alpha)$, and a net force normal to the plane of $\mathbf{N} = (W \cos \alpha - F \sin \theta)$.

For sliding, the net force parallel to the plane is $\mathbf{F}_s$, the static friction force. But $\mathbf{F}_s = \mu_s N = \mu_s(W \cos \alpha - F \sin \theta)$. Therefore,
$$(F \cos \theta - W \sin \alpha) = \mu_s(W \cos \alpha - F \sin \theta)$$
Rearranging gives $\quad F(\cos \theta + \mu_s \sin \theta) = W(\mu_s \cos \alpha + \sin \alpha)$
$$F = \frac{W(\mu_s \cos \alpha + \sin \alpha)}{\cos \theta + \mu_s \sin \theta} \tag{5.3}$$

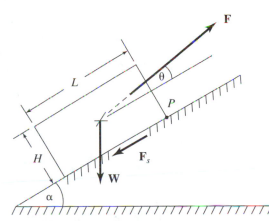

Figure 5.7   Block being pulled up an inclined plane.

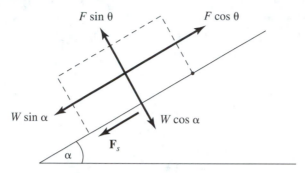

Figure 5.8   Free-body diagram of Figure 5.7.

For tipping, the clockwise moment about the point P must be greater than the counterclockwise moment about that point. See Figure 5.7.

$$\frac{(F \cos \theta - W \sin \alpha)H}{2} > \frac{(W \cos \alpha - F \sin \theta)L}{2}$$

$$F(L \sin \theta + H \cos \theta) > W(L \cos \alpha + H \sin \alpha)$$

$$F > \frac{W(L \cos \alpha + H \sin \alpha)}{L \sin \theta + H \cos \theta} \qquad (5.4)$$

EXAMPLE PROBLEM 5.4

What force F is required (see Figure 5.7) to just move the block if the block weighs 100 lb, $\alpha = 20°$, $\theta = 45°$, $L = 10$ in., H = 5 in., and $\mu_s = 0.30$? Will it tilt or slide first?

Solution

For sliding:
$$F = \frac{W(\mu_s \cos \alpha + \sin \alpha)}{\cos \theta + \mu_s \sin \theta}$$

$$= \frac{(100 \text{ lb})(0.30 \cos 20° + \sin 20°)}{\cos 45° + 0.30 \sin 45°}$$

$$= \frac{(100 \text{ lb})(0.282 + 0.342)}{0.707 + 0.212}$$

$$= 67.9 \text{ lb}$$

For tipping:
$$F > \frac{W(L \cos \alpha + H \sin \alpha)}{L \sin \theta + H \cos \theta}$$

$$> \frac{(100 \text{ lb})[(10 \text{ in.}) \cos 20° + (5 \text{ in.})\sin 20°]}{(5 \text{ in.})\sin 45° + (10 \text{ in.})\cos 45°}$$

$$> \frac{(100 \text{ lb})(9.40 \text{ in.} + 1.71 \text{ in.})}{3.54 \text{ in.} + 7.07 \text{ in}}$$

$$> 104.7 \text{ lb}$$

Answer
The block will *slide* when the force reaches 67.9 lb.

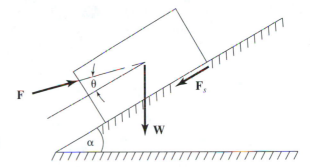

Figure 5.9 Block being pushed up an inclined plane.

In Figure 5.9 the applied force pushes the block at angle θ to the plane. This increases the normal force, which now becomes

$$N = \mu_s(W \cos \alpha + F \sin \theta) \qquad \text{See Figure 5.10.}$$

Following the same analysis as for the case of pulling the block, the force required to start the block sliding up the plane is

$$F = \frac{W(\mu_s \cos \alpha + \sin \alpha)}{\cos \theta - \mu_s \sin \theta} \qquad (5.5)$$

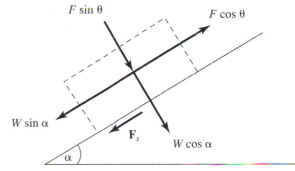

Figure 5.10 Free-body diagram of Figure 5.9.

EXAMPLE PROBLEM 5.5

Compare the forces required to start a block sliding up an inclined plane by pulling or pushing at angle θ to the plane. What is the ratio of pulling force to pushing force for $\theta = 45°$ and $\mu_s = 0.25$?

Solution
For pulling,

$$F_{\text{pull}} = W(\mu_s \cos \alpha + \sin \alpha)/(\cos \theta + \mu_s \sin \theta) \qquad (5.3)$$

For pushing:

$$F_{\text{push}} = W(\mu_s \cos \alpha + \sin \alpha)/(\cos \theta - \mu_s \sin \theta) \qquad (5.5)$$

So the ratio F_{pull}/F_{push} is $(\cos\theta - \mu_s\sin\theta)/(\cos\theta + \mu_s\sin\theta)$. Dividing top and bottom by $\cos\theta$,

$$F_{pull}/F_{push} = (1 - \mu_s\tan\theta)/(1 + \mu_s\tan\theta) \qquad (5.6)$$

which is independent of both the weight and the slope of the plane. For $\theta = 45°$ and $\mu_s = 0.25$,

$$F_{pull}/F_{push} = (1 - 0.25\tan 45°)/(1 + 0.25\tan 45°)$$
$$\tan 45° = 1$$
$$F_{pull}/F_{push} = (1 - 0.25)/(1 + 0.25)$$
$$= 0.60$$

Answer
The pulling force is equal to 60% of the pushing force.

5.4   WEDGES

A wedge is a device that converts a force in one direction into a larger force at right angles to that direction. Figure 5.11 shows a force **F** acting on a wedge to lift a block of weight W_b. The wedge angle is $\alpha°$, and the weight of the wedge is W_w. The angles of static friction and the resultant forces between the surfaces are

ϕ_1 and R_1 between the block and the wedge
ϕ_2 and R_2 between the block and the wall
ϕ_3 and R_3 between the wedge and the ground.

The free-body diagram for the block is shown in Figure 5.12. Summing the forces in the x-direction:

$$\Sigma F_x = R_1\sin(\phi_1 + \alpha) - R_2\cos\phi_2 = 0$$

Therefore,
$$R_2 = R_1\sin(\phi_1 + \alpha)/\cos\phi_2 \qquad (5.7)$$

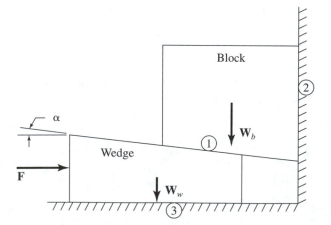

Figure 5.11   Wedge lifting a block.

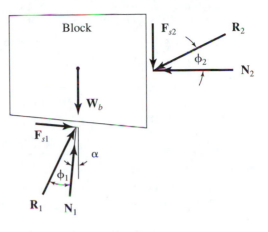

Figure 5.12 Free-body diagram for the block.

Summing the forces in the y-direction:

$$\Sigma F_y = R_1 \cos(\phi_1 + \alpha) - W_b - R_2 \sin \phi_2 = 0 \qquad (5.8)$$

Substituting for R_2 from equation (5.7) gives

$$R_1 \cos(\phi_1 + \alpha) - W_b - R_1 \sin(\phi_1 + \alpha) \sin \phi_2 / \cos \phi_2 = 0$$

or

$$R_1[\cos(\phi_1 + \alpha) - \sin(\phi_1 + \alpha)\tan \phi_2] = W_b$$

$$R_1 = \frac{W_b}{\cos(\phi_1 + \alpha) - \sin(\phi_1 + \alpha)\tan \phi_2} \qquad (5.9)$$

Referring now to the free-body diagram for the wedge (Figure 5.13):

$$\Sigma F_x = F - R_1 \sin(\phi_1 + \alpha) - R_3 \sin \phi_3 = 0 \qquad (5.10)$$

and

$$\Sigma F_y = R_3 \cos \phi_3 - R_1 \cos(\phi_1 + \alpha) - W_w = 0 \qquad (5.11)$$

From equation (5.11)

$$R_3 = \frac{R_1 \cos(\phi_1 + \alpha) + W_w}{\cos \phi_3}$$

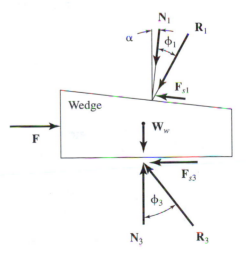

Figure 5.13 Free-body diagram for the wedge.

Substituting into equation (5.10),

$$F - R_1\sin(\phi_1 + \alpha) - [R_1\cos(\phi_1 + \alpha) + W_w]\tan \phi_3 = 0$$
$$F - R_1[\sin(\phi_1 + \alpha) + \cos(\phi_1 + \alpha)\tan \phi_3] - W_w\tan \phi_3 = 0$$

So

$$F = R_1[\sin(\phi_1 + \alpha) + \cos(\phi_1 + \alpha)\tan \phi_3] + W_w\tan \phi_3$$

Substituting for R_1 from equation (5.9),

$$F = \frac{W_b[\sin(\phi_1 + \alpha) + \cos(\phi_1 + \alpha)\tan \phi_3]}{\cos(\phi_1 + \alpha) - \sin(\phi_1 + \alpha)\tan \phi_2} + W_w\tan \phi_3 \qquad (5.12)$$

Equation (5.12) is the general form of the wedge equation.

EXAMPLE PROBLEM 5.6

The block shown in Figure 5.11 weighs 3 kN and the wedge weighs 200 N. The wedge angle is 8° and the following coefficients of friction apply:

$$\mu_1 = 0.20 \text{ between the block and the wedge}$$
$$\mu_2 = 0.30 \text{ between the block and the wall}$$
$$\mu_3 = 0.15 \text{ between the wedge and the ground}$$

What force **F** is required to start lifting the block?

Solution

$$\phi_1 = \arctan \mu_1 = \arctan 0.20 = 11.31°$$
$$\phi_2 = \arctan \mu_2 = \arctan 0.30 = 16.70°$$
$$\phi_3 = \arctan \mu_3 = \arctan 0.15 = 24.23°$$
$$(\phi_1 + \alpha) = 11.31° + 8° = 19.31°$$
$$\sin(\phi_1 + \alpha) = 0.331, \qquad \cos(\phi_1 + \alpha) = 0.944,$$
$$\tan \phi_2 = 0.30, \qquad\qquad \tan \phi_3 = 0.15$$

From equation (5.12):

$$F = \frac{W_b[\sin(\phi_1 + \alpha) + \cos(\phi_1 + \alpha)\tan \phi_3]}{\cos(\phi_1 + \alpha) - \sin(\phi_1 + \alpha)\tan \phi_2} + W_w\tan \phi_3$$

$$F = \frac{(3000 \text{ N})(0.331 + 0.944 \times 0.15)}{0.944 - 0.331 \times 0.30} + (200 \text{ N})(0.15)$$

$$F = \frac{(3000 \text{ N})(0.473)}{0.845} + 30 \text{ N}$$

Therefore, $F = 1700$ N.

Answer
The required force is 1.7 kN.

EXAMPLE PROBLEM 5.7

What force would be required if the wedge angle were changed to 6° in the Example 5.6?

Solution

Because $\alpha = 6°$, $(\phi_1 + \alpha) = 11.31° + 6° = 17.31°$.

$$\sin(\phi_1 + \alpha) = 0.298, \qquad \cos(\phi_1 + \alpha) = 0.955$$

$$F = \frac{W_b[\sin(\phi_1 + \alpha) + \cos(\phi_1 + \alpha)\tan\phi_3]}{\cos(\phi_1 + \alpha) - \sin(\phi_1 + \alpha)\tan\phi_2} + W_w\tan\phi_3$$

$$F = \frac{(3000\ \text{N})(0.298 + 0.955 \times 0.15)}{0.955 - 0.298 \times 0.30} + (200\ \text{N})(0.15)$$

$$F = \frac{(3000\ \text{N})(0.441)}{0.866} + 30\ \text{N}$$

Therefore, $F = 1558$ N.

Answer

The required force is 1.6 kN.

5.5 JACKSCREWS

A jackscrew, used in vises, car jacks, etc., usually has a square, threaded screw that turns in a fixed base. (See Figure 5.14.) The rotary motion applied to the handle is converted to an axial motion, and because of the wedge action and the moment arm of the handle, the output force is far greater than the applied force.

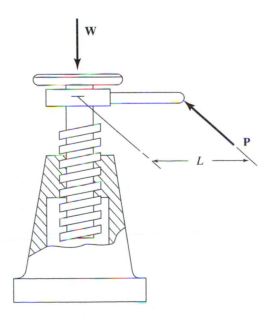

Figure 5.14 Jackscrew.

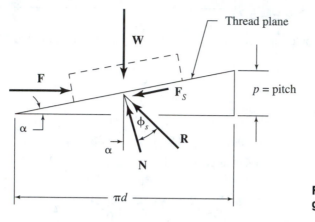

Figure 5.15 Free-body diagram for jackscrew.

The circumference of the thread is πd, where d is the mean thread diameter, and the axial distance moved in one revolution is p, the thread pitch. Figure 5.15 shows one revolution of the thread "unwound" and the forces acting on the mating thread of the base.

The lead angle is $\alpha = \arctan(p/\pi d)$. Denoting the angle of static friction as ϕ_s and the resultant as **R**, we have the following. Summing the forces in the x-direction,

$$\Sigma F_x = F - R\sin(\phi_s + \alpha) = 0 \tag{5.13}$$

where **F** is the force on the thread in the x-direction. Summing the forces in the y-direction,

$$\Sigma F_y = R\cos(\phi_s + \alpha) - W = 0$$

(where W is the weight being lifted), from which

$$R = \frac{W}{\cos(\phi_s + \alpha)}$$

Substituting for R in equation (5.13):

$$F = \frac{W\sin(\phi_s + \alpha)}{\cos(\phi_s + \alpha)}$$

So,

$$F = W\tan(\phi_s + \alpha) \tag{5.14}$$

To find the force, **P**, required on the handle, whose length is L, we take moments about the thread centerline. Noting that the force **F** acts at a distance $d/2$ from the centerline,

$$PL = Fd/2$$
$$P = Fd/2L$$

Substituting for F from equation (5.14),

$$P = Wd\tan(\phi_s + \alpha)/2L \qquad \text{lifting} \tag{5.15}$$

When the weight is being lowered, the friction force acts in the opposite direction (see Figure 5.16), and the resultant shifts to the left of the normal force, **N**. The angle between **R** and the vertical then becomes $(\phi_s - \alpha)$, and equation (5.15) changes to

$$P = Wd\tan(\phi_s - \alpha)/2L \qquad \text{lowering} \tag{5.16}$$

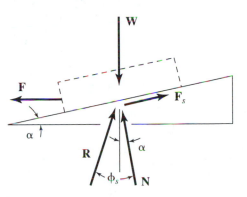

Figure 5.16 Free body diagram for jackscrew lowering.

Notice that if the angle of static friction, ϕ_s, is less than the lead angle α, then P will be negative. This means that the weight will lower itself without any applied force; that is, the jackscrew will not be *self-locking*.

EXAMPLE PROBLEM 5.8

The mean diameter of a jackscrew is 90 mm, and the thread pitch is 15 mm. If the coefficient of static friction is 0.13, what force applied to the end of a handle 600 mm long would raise a mass of 2000 kg? What force would be required to lower the mass?

Solution

The lead angle is $\alpha = \arctan(p/\pi d) = \arctan(15 \text{ mm}/90\pi \text{ mm})$:

$$\alpha = \arctan(0.0531) = 3.04°$$

The angle of static friction is $\phi_s = \arctan(0.13) = 7.41°$. So,

$$\tan(\phi_s + \alpha) = \tan(10.45°) = 0.184$$

and

$$\tan(\phi_s - \alpha) = \tan(4.37°) = 0.076$$

The weight is $W = (2000 \text{ kg})(9.81 \text{ m/s}^2) = 19.6$ kN.
For lifting,

$$P = Wd\tan(\phi_s + \alpha)/2L$$

where $W = 19.62$ kN, $d = 90$ mm, $\tan(\phi_s + \alpha) = 0.184$, and $L = 600$ mm.

$$P = (19.62 \text{ kN})(90 \text{ mm})(0.184)/1200 \text{ mm}$$
$$= 271 \text{ N}$$

For lowering,

$$P = Wd\tan(\phi_s - \alpha)/2L$$

where $\tan(\phi_s - \alpha) = 0.076$.

$$P = (19.62 \text{ kN})(90 \text{ mm})(0.076)/1200 \text{ mm}$$
$$= 112 \text{ N}$$

Answers
The force to lift the load is 270 N, and the force to lower the load is 110 N.

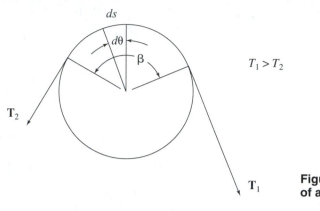

Figure 5.17   Belt around portion of a drum.

5.6   BELT FRICTION

Figure 5.17 shows a flexible belt wrapped around part of a fixed drum, with the angle between the belt and the drum being β radians. Taking the impending motion of the belt to be clockwise relative to the drum, it can be shown that (see Using Calculus at the end of this section)

$$T_1/T_2 = e^{\mu\beta} \tag{5.17}$$

where

$$T_1 = \text{the larger tension}$$
$$T_2 = \text{the smaller tension}$$
$$e = \text{the base of the natural logarithm } (e = 2.7183)$$
$$\mu = \text{the coefficient of static friction}$$

EXAMPLE PROBLEM 5.9

An experiment was conducted (see Figure 5.18) to determine the coefficient of static friction between the belt and the drum. It was observed that the belt just began to slip when weight W_1 was twice weight W_2. Calculate the coefficient of static friction.

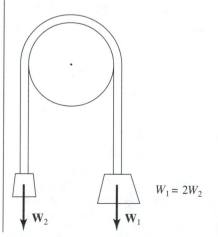

$W_1 = 2W_2$

Figure 5.18   Example Problem 5.16.

Solution

$$T_1/T_2 = e^{\mu\beta}$$

Taking the natural logarithm of both sides,

$$\ln(T_1/T_2) = \mu\beta$$

Therefore,

$$\mu = \ln(T_1/T_2)/\beta$$

$$T_1/T_2 = 2 \qquad \beta = 90° = \pi \text{ rad}$$

So,

$$\mu = \ln(2)/\pi = 0.22$$

Answer

The coefficient of static friction between the belt and the drum is 0.22.

EXAMPLE PROBLEM 5.10

A line is wrapped around a capstan on the dock to secure a ship. It makes $3\frac{1}{2}$ turns. The coefficient of static friction between the line and the capstan is 0.30, and the pull exerted by the ship is 22,000 lb. How much force must a sailor apply to the free end of the line to hold the ship?

Solution

$$T_1/T_2 = e^{\mu\beta}$$

$$T_1 = 22,000 \text{ lb}, \qquad \mu = 0.30, \qquad \beta = (3.5)(2\pi) = 7\pi \text{ rad}$$

Therefore,

$$\mu\beta = (0.30)(7\pi) = 6.597$$

So,

$$T_1/T_2 = e^{6.597} = 733$$

$$T_2 = T_1/733 = (22,000 \text{ lb})/733 = 30 \text{ lb}$$

Answer

A 30-lb force is required to hold the ship!

EXAMPLE PROBLEM 5.11

In Example Problem 5.10, the sailor can pull on the line with a maximum sustained force of 80 lb, and the coefficient of static friction between the capstan and the line is 0.20. How many turns are necessary for the sailor to hold the ship?

Solution

$$T_1/T_2 = e^{\mu\beta}$$

Taking the natural logarithm of both sides,

$$\ln(T_1/T_2) = \mu\beta$$

Therefore,

$$\beta = \ln(T_1/T_2)/\mu$$
$$T_1 = 22{,}000 \text{ lb}, \qquad T_2 = 80 \text{ lb}, \qquad \mu = 0.20$$
$$\beta = \ln(22{,}000 \text{ lb}/80 \text{ lb})/0.20$$
$$= \ln(275)/0.20$$
$$= 5.62/0.20 = 28.1 \text{ rad}$$

There are 2π rad in one turn, so the number of turns required is $28.1/2\pi = 4.47$ turns.

Answer

$4\frac{1}{2}$ turns are required.

USING CALCULUS

Figure 5.19 shows an expanded view of the element ds in Figure 5.17. The belt tensions on each side of the element are T and $(T + dT)$, acting at angles $d\theta/2$, as shown. Summing forces in the x-direction gives

$$\Sigma F_x = (T + dT)\cos d\theta/2 - T\cos d\theta/2 - R\sin \phi_s = 0$$

Therefore, $\qquad dT\cos d\theta/2 = R\sin \phi_s$ $\qquad\qquad$ (1)

The cosine of very small angles is approximately 1, so equation (1) reduces to

$$dT = R\sin \phi_s \qquad\qquad (1a)$$

Summing forces in the y-direction yields

$$\Sigma F_y = R\cos \phi_s - T\sin d\theta/2 - (T + dT)\sin d\theta/2 = 0$$

Therefore, $\qquad R\cos \phi_s - 2T\sin d\theta/2 - dT\sin d\theta/2 = 0$

The sine of very small angles is approximately equal to the angle in radians, so

$$R\cos \phi_s - 2T\, d\theta/2 - dT\cdot d\theta/2 = 0$$

But the term involving the product of two infinitesimals ($dT\cdot d\theta/2$) is negligible compared to the other terms in the equation. Therefore,

$$R\cos \phi_s - \frac{2T\, d\theta}{2} = 0$$

or $\qquad\qquad\qquad\qquad\qquad T\, d\theta = R\cos \phi_s \qquad\qquad (2)$

Dividing equation (1a) by equation (2) gives

$$dT/T = \tan \phi_s\, d\theta$$

But $\qquad\qquad\qquad\qquad \tan \phi_s = \mu_s, \text{ so}$

$$dT/T = \mu_s\, d\theta$$

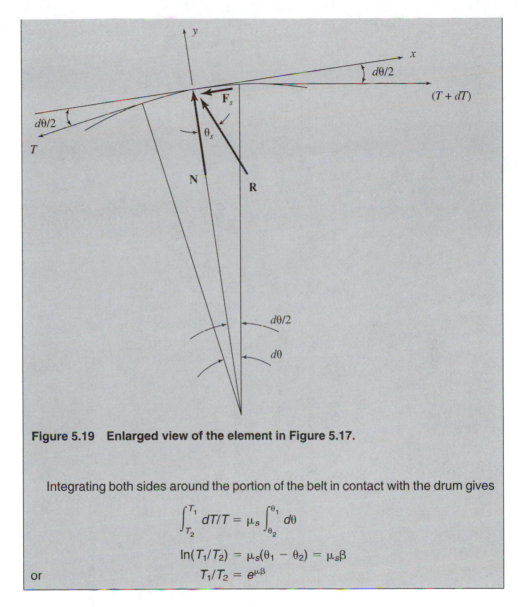

Figure 5.19   Enlarged view of the element in Figure 5.17.

Integrating both sides around the portion of the belt in contact with the drum gives

$$\int_{T_2}^{T_1} dT/T = \mu_s \int_{\theta_2}^{\theta_1} d\theta$$

$$\ln(T_1/T_2) = \mu_s(\theta_1 - \theta_2) = \mu_s\beta$$

or
$$T_1/T_2 = e^{\mu\beta}$$

5.7   COMPUTER PROGRAMS FOR FRICTION

Program 'Friction' is actually three programs on one sheet. These programs are for calculations concerning wedges, jackscrews, and belts. If you call up program 'Friction' you will see the headings WEDGES, JACKSCREWS, and BELTS and the required input variables for each. Each program needs the input US or SI (uppercase) to define the units being used. The following examples explain how to use these programs.

Program 'Friction'

WEDGES

Enter US or SI [US]

Input

Weight to be lifted, Wb	2000	lb
Weight of wedge, Ww	50	lb
Wedge angle, alpha	6	degrees

Coeffs of static friction:

between block and wedge	0.15
between block and wall	0.2
between wedge and ground	0.18

Output

| Required Force | 935 | lb |

JACKSCREWS

Enter US or SI [US]

Input

Weight to be lifted, Wb	1000	lb
Mean thread diameter	3.5	in.
Thread pitch	0.5	in.
Coeff of static friction	0.2
Length of handle	24	in.

Output

| Required Lifting Force | 18 | lb |
| Required Lowering Force | 11 | lb |

BELTS

Enter US or SI [SI]

Input

Larger tension, T1	8000	N
Coeff of static friction	0.3
Wrap angle, beta	720	degrees

Output

| Smaller tension, T2 | 184 | N |

Figure 5.20   Input/output for Example Problems 5.12, 5.13, and 5.14

EXAMPLE PROBLEM 5.12

A 50-lb wedge with a wedge angle of 6° is being used to lift a 2000-lb weight. The following coefficients of static friction apply:

Between the block and the wedge: 0.15
Between the block and the wall: 0.20
Between the wedge and the ground: 0.18

What is the magnitude of the required horizontal force?

Solution
Call up program 'Friction' and enter the values under WEDGES shown on the left of Figure 5.20. Do not forget to enter US for the units. The required force is shown to be 935 lb.

Answer
The required force is 935 lb.

EXAMPLE PROBLEM 5.13

A jackscrew with a mean thread diameter of 3.5 in. and having a 2-ft-long handle is being used to lift a 1000-lb weight. The thread pitch is 0.5 in. and the coefficient of static friction is 0.20. What force is required to lift the load? What force is required to lower the load?

Solution
Go to JACKSCREWS and enter the variables, as shown in Figure 5.20.
    The output shows that the required lifting force is 18 lb and the lowering force is 11 lb.

Answer
The required lifting force is 18 lb, and the required lowering force is 11 lb.

EXAMPLE PROBLEM 5.14

A mass of 815.5 kg is hanging from a rope that is wound around a horizontal round shaft. If the rope makes two complete turns around the shaft and the coefficient of static friction is 0.3, what force is required on the free end of the rope to hold the weight?

Solution

First convert the mass to weight, which will be the tension T_1 in the rope:

$$T_1 = (815.5 \text{ kg})(9.81 \text{ m/s}^2) = 8000 \text{ N}$$

The wrap angle is $2 \times 360° = 720°$. The inputs are shown in Figure 5.20 under BELTS, and the output is seen to be 184 N.

Answer

The required force is 184 N.

Note: Any of these programs (as with all spreadsheet programs) may be used to determine one of the input variables given the output. For example, if the maximum available force were 150 N in the previous example, how many turns of rope would be required? *Do not enter 150 in the output cell!* Change the wrap angle until the output equals 150 N. We leave this as an exercise for the student.

5.8   PROBLEMS FOR CHAPTER 5

Hand Calculations

1. A block of Teflon lies on a steel tabletop. The table is tilted until the block begins to slide. What is the tilt angle?

2. A block of mild steel lies on a mild steel table top. The table is tilted until the block begins to slide. What is the tilt angle?

3. What horizontal force is required to start a 100-lb block of wood sliding on a horizontal wooden floor?

4. What force would be required in Problem 3 if the force were pushing down at an angle of 20° to the horizontal?

5. What force would be required (assuming the block does not tilt) in Problem 3 if the force were pulling up at an angle of 20° to the horizontal?

6. The block in Problem 5 is 2 ft long (*L*) and $4\frac{1}{2}$ ft high (*H*). Will it slide or tilt?

7. One end of a 12-ft-long plank of wood rests on the tailgate of a truck, 3 ft high. The other end rests on the ground. It is being used as an inclined plane to load a 120-lb block of steel on to the truck. What force does the person loading the truck have to exert parallel to the plank?

8. Show that, in the absence of any friction forces, the horizontal force required to raise a load of weight *W* with a wedge of angle α is $F = W \tan \alpha$.

9. A block of copper is being lifted with a wedge of steel. The vertical wall is steel and the floor is wood. The copper block has a mass of 250 kg and the wedge has a mass of 25 kg. If the

wedge angle is 5°, what horizontal force is required? Is this an efficient system?

10. If you think that the wedge system in Problem 9 is not efficient, is there a wedge angle that would make it efficient?

11. Show that if the ratio of the horizontal force on a wedge to the block weight is some value f, then ignoring the weight of the wedge:

$$\tan(\phi_1 + \alpha) = (f - \mu_3)/(1 + f\mu_2)$$

where α is the wedge angle, ϕ_1 is the angle of static friction between the block and the wedge, μ_2 is the coefficient of static friction between the block and the wall, and μ_3 is the coefficient of static friction between the wedge and the wall.

12. A block is to lifted by a wedge by the application of a horizontal force equal to one-half the weight of the block. What wedge angle is required if the coefficients of static friction are as follows?

0.15 between the block and the wedge

0.20 between the block and the wall

0.25 between the wedge and the ground

Neglect the weight of the wedge.

13. The mean diameter of a jackscrew is 3.5 in., and the thread pitch is 0.65 in. If the coefficient of static friction is 0.12, what force applied to the end of a handle 18 in. long would raise a weight of 4000 lb? What force would be required to lower the load?

14. A steel block weighing 1500 lb is to held in a vise that has a mean thread diameter of 2.0 in., a thread pitch of 0.30 in., a handle length of 8 in., and a coefficient of static friction of 0.15. The coefficient of static friction between the vise jaws and the block is 0.60. What is the minimum force required on the handle?

15. A farmer is leading a steer on a rope when suddenly the steer pulls on the rope with a force of 700 lb. The farmer manages to wrap the rope one-half turn around a tree. Is the farmer able to hold the steer if he/she can exert a sustained pull of 90 lb?

16. A belt is wrapped one-half turn over a drum, and it is noted that the belt just begins to slip when the ratio of the tensions is 1 to 3. With the same belt on the same drum, what is the ratio of the tensions if the belt wraps one-fourth turn over the drum?

Computer Calculations

17. Repeat Problem 9.

18. Repeat Problem 12. Enter any weight for the block, and enter the coefficients of static friction. Try values of wedge angle until the required force is equal to half the weight of the block.

19. Repeat Problem 18 and find the wedge angle required for the force to be three-fourths of the load.

20. Repeat Problem 13.

21. Repeat Problem 14.

22. Repeat Problem 15.

23. Write a computer program to calculate the forces required to push and to pull a block of weight W up an inclined plane of angle α. The block length is L and its height is H. The angle of the force to the plane is θ, and the coefficient of static friction between the block and the plane is μ_s. Also determine the forces required for tipping. Write the program such that either U.S. customary units or SI units may be used. See equations (5.3), (5.4), and (5.5).

24. Use the program developed in Problem 23 to redo Problems 3, 4, 5, 6, and 7.

6

Direct Stress

■ Objectives

In this chapter you will learn about *simple*, or *direct*, stress, which is categorized as *normal stress* (tensile or compressive), *shear stress*, or *bearing stress*.

First, what do we mean by **stress**? Stress is a measure of the *intensity of a force*—the amount of force that acts on a unit of area. Simply, it is force divided by the area over which it is acting. If you know what *pressure* is, then you will see that it appears to be the same thing. The difference between pressure and stress is that pressure refers to an *external* force, or a load, acting on an area, whereas stress refers to an *internal* force acting on an area.

We saw in Chapter 1 that when an external force acts on a body, the molecules that make up the body resist the force, trying to prevent deformation. It is this internal resistive force divided by the area over which it acts that we refer to as stress.

6.1 NORMAL, OR AXIAL, STRESS

Normal, or **axial**, **stresses** arise when two equal and opposite forces, on the same straight line of action, pull (tension) or push (compression) on a body. See Figure 6.1(a) and (b). These stresses are called *tensile stresses* and *compressive stresses*, respectively. Notice that the stress is different at different locations if the body is not of the same cross-sectional area throughout, being higher in the regions of smaller area. In the case of normal stresses, the stress in any plane normal to the line of action of the force is (very nearly) the same at all points. That is, the stress distribution is flat. We will see later that this is not so when the body contains notches, holes, grooves, etc.

6.1.1 Tensile Stress

Tensile stress is one of the stresses categorized as *simple*, or *direct*, stress. Other stresses in this category are compressive, shear, and bearing stresses. In later chapters we will learn about *indirect* stresses, which are bending and torsional, and *combined* stresses, which are combinations of direct and indirect stresses.

We pointed out in Chapter 1 that solids, although they look solid, actually contain billions of spaces, and because of that they can be deformed. Admittedly, it takes thousands of pounds of force to stretch a steel bar just a little, but it *will* stretch. The word *tensile* means stretch, as opposed to *compressive*, which means squeeze.

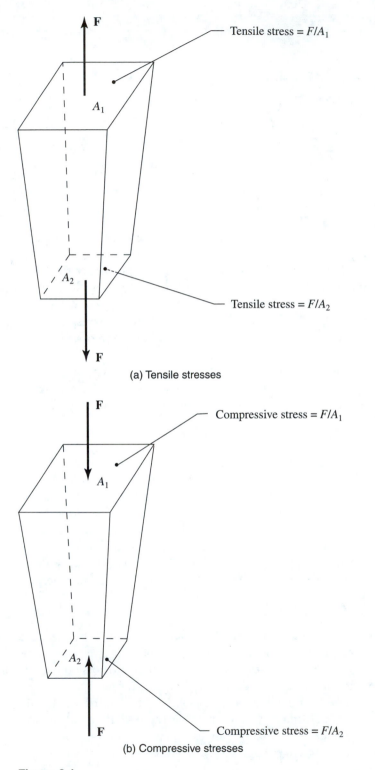

(a) Tensile stresses

(b) Compressive stresses

Figure 6.1

Imagine that we have a 10-in.-long steel bar. It is square in cross section, measuring 1 in. by 1 in. We grip each end of the bar in very strong jaws and apply a force (using a special machine) to stretch the bar. The machine is called a *tensile testing* machine; we learn more about it in Chapter 7. Suppose it takes a force of 30,000 lb to stretch the bar 0.01 in. (see Figure 6.2(a)).

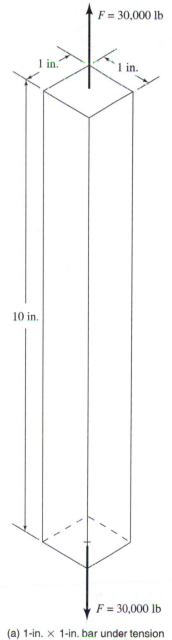

$F = 30,000$ lb

1 in.   1 in.

10 in.

$F = 30,000$ lb

Figure 6.2   (a) 1-in. × 1-in. bar under tension

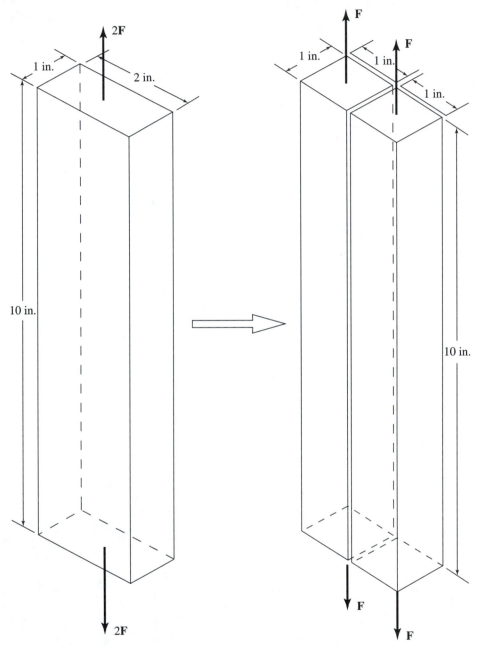

(b) 1-in. × 2-in. bar under tension

Figure 6.2   (*Continued*)

Note: Equal and opposite forces of 30,000 lb each on an object, as in Figure 6.2(a), create a force of 30,000 lb on the object—*not* 60,000 lb. If we applied a force to one end only, the body would move and experience no internal force. Whether we pull on both ends with equal forces of 30,000 lb or anchor one end and pull on the other with 30,000 lb, the bar resists with an internal fore of 30,000 lb.

Now take a bar of the same material, also 10 in. long, but with a 2-in. by 1-in. cross section (Figure 6.2(b)). Can we figure out what force would be required to stretch *this* bar 0.01 in.? Remember in Chapter 1 we said that it is as if the molecules were held together with strong, tiny springs. There are two times as many in the second bar as in the first, so it would be reasonable to assume that two times the force would be necessary—that is, 60,000 lb. It is similar to having two 1-in. square bars side by side (Figure 6.2(b)). Doing the same experiment again with a 10-in. bar of the same material, but whose cross section measures 2 in. by 2 in., what force would we expect to have to apply to stretch it, again, 0.01 in.? The cross-sectional area of this bar is 2 in. $\times$ 2 in. = 4 in.^{2}, which is four times the area of the first bar, so it should require 4 $\times$ 30,000 lb = 120,000 lb.

What we are seeing here is that, in the first case, the required force divided by the cross-sectional area of the bar equals 30,000 lb/1 in.^{2}, which we write as 30,000 psi. In the second case, force/area = 60,000 lb/2 in.^{2} = 30,000 psi. In the third case the force/area = 120,000 lb/4 in.^{2} = 30,000 psi.

We conclude from this that bars of the same material and the same length will be stretched the same amount if the ratio of the applied force to the cross-sectional area remains the same. Obviously, then, this is a very important ratio. We call it *tensile stress*.

Tensile stress = applied force/cross-sectional area

$$s = F/A \qquad\qquad (6.1)$$

Looking at this another way, take a rubber band and stretch it 1 in. It will require a certain pull (a force), say, 0.1 lb. Now put 10 identical bands together and stretch them 1 in. Each will need 0.1 lb, for a total force of 10 $\times$ 0.1 lb = 1 lb. The total cross-sectional area of the 10 bands is 10 times that of one band, so we see in both cases that the force divided by the area, i.e., the tensile stress, is the same.

EXAMPLE PROBLEM 6.1

A 0.5-in.-diameter bar has a tensile stress of 55,000 psi. What tensile force is being applied?

Solution
First, we need the cross-sectional area.

$$A = \pi(\text{diameter})^2/4$$
$$= \pi(0.5 \text{ in.})^2/4 = 0.1963 \text{ in.}^2$$

Now rearrange equation (6.1) to read

$$F = sA$$

Inserting the numbers $s = 55{,}000$ lb/in.^{2} and $A = 0.1963$ in.^{2} gives

$$F = (55{,}000 \text{ lb/in.}^2)(0.1963 \text{ in.}^2) = 10{,}800 \text{ lb}$$

Answer
The applied force is 10,800 lb.

EXAMPLE PROBLEM 6.2

A 35-kg mass is suspended from a 1-mm-diameter wire. What is the tensile stress in the wire?

Solution
Be careful! Stress is defined as *force/area*, not *mass/area*. First we have to convert 35 kg to newtons.

$$F = 35 \text{ kg} \times 9.81 \text{ m/s}^2 = 343 \text{ N}$$

The cross-sectional area of the wire is

$$A = \pi(1 \text{ mm})^2/4 = 0.785 \text{ mm}^2$$

Therefore, the tensile stress is

$$s = F/A = 343 \text{ N}/0.785 \text{ mm}^2 = 437 \text{ N/mm}^2$$

$$\mathbf{1 \text{ N/mm}^2 = 1 \text{ megapascal}} \quad \textbf{(MPa)}$$

$$s = 437 \text{ MPa}$$

Answer
The stress in the wire is 440 MPa.

6.1.2   Compressive Stress

For the same reasons that materials can be stretched, they can be compressed, and again, the definition of compressive stress is applied force divided by cross-sectional area; of course, the applied force is one that tends to compress the material. One major difference arises between tensile and compressive stresses. If the piece being compressed is long and slender (a column) it may buckle rather than just compress. This condition is handled in an entirely different manner, as we will see in Chapter 19.

EXAMPLE PROBLEM 6.3

A structure having a mass of 250 metric tons is supported equally by 25 short posts, 200 mm in diameter. What is the average compressive stress in each post?

Solution

$$1 \text{ metric ton} = 1000 \text{ kg} = 10^3 \text{ kg}$$
$$250 \text{ metric tons} = 250 \times 10^3 \text{ kg} = 2.5 \times 10^5 \text{ kg}$$

First, convert the mass to a force:

$$F = 2.5 \times 10^5 \text{ kg} \times 9.81 \text{ m/s}^2 = 2.45 \times 10^6 \text{ N}$$

This force is shared equally by 25 posts, so the force on each post is

$$F = 2.45 \times 10^6 \text{ N}/25 = 98,100 \text{ N}$$

The cross-sectional area of one post is

$$A = \pi(200 \text{ mm})^2/4 = 31,420 \text{ mm}^2$$

So the compressive stress is $s = F/A = 98,100 \text{ N}/31,420 \text{ mm}^2 = 3.12 \text{ MPa}$.

Answer

The compressive stress in each post is 3.12 MPa.

6.2 SHEAR STRESS

In the cases of direct tension and direct compression, the areas resisting the forces are *perpendicular* to the direction of the forces. When a member is subjected to direct shear the resisting area is *parallel* to the force. Figure 6.3(a), (b), and (c) shows three examples of shear.

We show a rivet being sheared in Figure 6.3(a), a plate being sheared in Figure 6.3(b), and a hole being punched in Figure 6.3(c). In the first case, the shear area is the cross-sectional area of the rivet. This area is parallel to the applied force. It is the total of all the molecules that are being torn away from, and sliding over, their neighboring molecules. In Figure 6.3(b) the shear area is the cross-sectional area of the plate. Again, it is the area that is sliding over its adjacent area. In the case of the hole being punched, Figure 6.3(c), the area that is being torn away and sliding over its adjacent area is a cylindrical surface whose diameter is the punch diameter and whose depth is the depth of the plate.

Shear stress is defined as the applied force divided by the shear area.

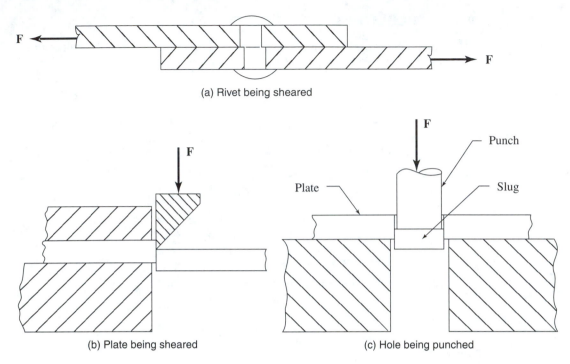

(a) Rivet being sheared

(b) Plate being sheared

(c) Hole being punched

Figure 6.3

EXAMPLE PROBLEM 6.4

The aluminum alloy plate shown in Figure 6.3(b) is 100 cm long and 3 mm wide. The shearing force is 620 kN. What is the shear stress?

Solution
The shear area is A = (100 cm $\times$ 10 mm/cm) (3 mm) = 3000 mm^2 The force is F = 620 kN = 620 $\times$ 10^3 N. The shear stress is

$$s = F/A = (620 \times 10^3 \text{ N})/3000 \text{ mm} = 207 \text{ MPa}$$

Answer
The shear stress is 207 MPa.

EXAMPLE PROBLEM 6.5

A two-inch-diameter disk is to be punched out of a 1/4-in. copper alloy plate. See Figure 6.3(c). The value of the stress at which the copper alloy will shear (known as the *ultimate* shear strength) is 25 ksi.

  a. What force is required?
  b. What is the compressive stress in the punch?

Solution

  a. The shear area, A, is the area of a cylindrical surface with a 2-in. diameter and a height of 1/4 in.

$$A = \pi \times \text{diameter} \times \text{height}$$
$$= \pi(2 \text{ in.})(0.25 \text{ in.}) = 1.57 \text{ in.}^2$$

The shear stress is $s = 25$ ksi, which means 25,000 lb/in². So the required force is $F = sA = (25,000 \text{ lb/in.}^2)(1.57 \text{ in.}^2) = 39,250$ lb.

  b. The compressive stress in the punch is based on the cross-sectional area of the punch, which is $A = \pi \times (\text{diameter})^2/4$:

$$A = \pi(2 \text{ in.})^2/4 = 3.14 \text{ in.}^2$$

The compressive stress is $s = F/A = 39,250 \text{ lb}/3.14 \text{ in.}^2 = 12,500$ psi.

Answer

The required force is 39,300 lb. The compressive stress in the punch is 12,500 psi.

We show an example of what is called *double shear* in Figure 6.4. The rivet is being sheared in two places; i.e., there are two areas of the rivet resisting the shear force.

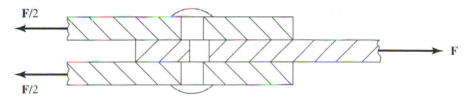

Figure 6.4   Rivet in double shear.

EXAMPLE PROBLEM 6.6

The rivet in Figure 6.4 has a diameter of 3/8 in., and the ultimate shear strength of the rivet material is 17,500 psi. At what force will the rivet fail?

Solution

The cross-sectional area of the rivet is π (0.375 in.)^2/4 = 0.110 in.^{2}, but this is only one-half of the resisting area because the rivet is in double shear. The shear area is, therefore, A = 0.220 in.^{2} The force required to shear the rivet is

$$F = sA = (17,500 \text{ lb/in.}^2)(0.220 \text{ in.}^2) = 3850 \text{ lb}$$

Answer

The required shearing force is 3850 lb.

6.3   BEARING STRESS

Bearing stress is the term given to the stress developed when one solid body presses on another. An example of this is shown in Figure 6.5, where the rivets are *bearing* against the plates, and the plates are bearing against the rivets. Whether the rivets will fail in shear or in bearing or whether the plates will fail in bearing (or possibly in tension) are questions that we address in Chapter 9. Thinking about the bearing of the rivets on the plates, we see that maximum stress must be developed at points labeled (a) shown in Figure 6.5, falling off to zero at points labeled (b). As a simplification of the actual stress distribution, we take the bearing area as the projected area, which is the rivet diameter times the plate thickness, and the force as the total force. The following example should clarify this.

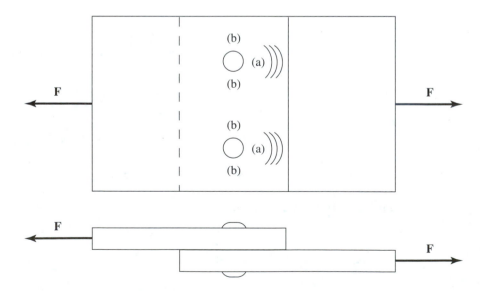

Figure 6.5   Bearing stress.

EXAMPLE PROBLEM 6.7

The upper plate in Figure 6.5 is 8 mm thick, and the lower plate is 12 mm thick. The rivet diameter is 6 mm, and the applied force is 4 kN. What are the bearing stresses on the upper and lower plates?

Solution

The two rivets share the total load equally, so the force on one rivet is $F = 4$ kN/2 $= 2000$ N.

Upper plate: The bearing area of one rivet against the plate is

$$A = \text{(rivet diameter)} \times \text{(plate thickness)}$$
$$= (6 \text{ mm})(8 \text{ mm}) = 48 \text{ mm}^2$$

The bearing stress is

$$s = F/A = 2000 \text{ N}/48 \text{ mm}^2 = 41.7 \text{ MPa}$$

Lower plate:

$$A = (6 \text{ mm})(12 \text{ mm}) = 72 \text{ mm}^2$$
$$s = 2000 \text{ N}/72 \text{ mm}^2 = 27.8 \text{ MPa}$$

Answer

The upper plate bearing stress is 42 MPa, and the lower plate bearing stress is 28 MPa.

6.4   STRESS CONCENTRATIONS

Equation (6.1), $s = F/A$, is valid only for members having uniform cross sections. Suppose we are applying a tensile load of 10,000 lb to the bar shown in Figure 6.6(a). This bar has an abrupt change in diameter from 0.5 in. to 1 in. The tensile stress in the smaller section is $s = F/A = 10{,}000 \text{ lb}/[\pi(0.5 \text{ in.})^2/4)] = 51{,}000$ psi. This stress will remain constant from the end of the bar up to about one diameter (0.5 in.) from the shoulder. The stress will then rise sharply, reaching a peak at the shoulder, and then fall to the stress level in the larger diameter ($s = 10{,}000 \text{ lb}/0.785 \text{ in.}^2 = 12{,}740$ psi) at about 1 in. (one diameter) from the shoulder. It will remain at this level to the end of the bar. This phenomenon is known as *Saint-Venant's principle*.

The maximum stress reached depends on the ratio of the diameters, D/d, and also on the fillet-to-diameter ratio, r/d. In this example, if the fillet radius is 0.05 in., the maximum stress would be two times that in the smaller section; i.e., it would be 102,000 psi.

We have to be very aware of this in any design work.

6.4.1   Stress-Concentration Factors

A convenient way to account for stress concentrations is to use experimentally determined *stress-concentration factors*. We first calculate the higher stress, which is the stress based on the minimum cross-sectional area of the member, and then apply the stress concentration factor, k, for that case.

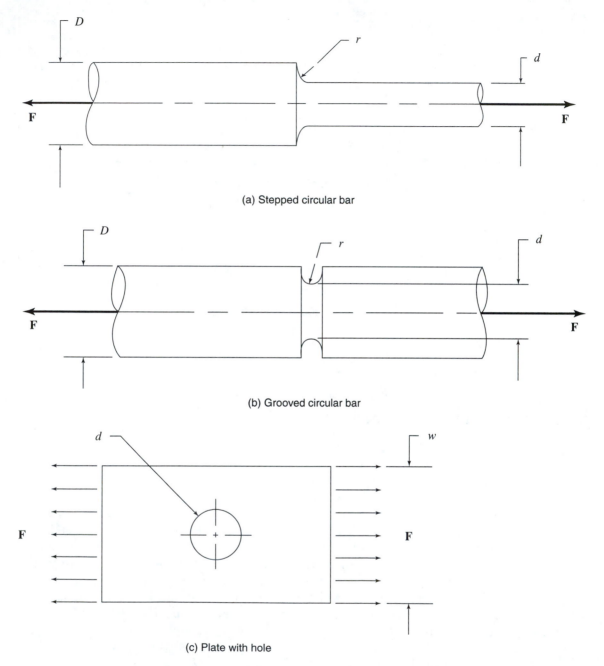

(a) Stepped circular bar

(b) Grooved circular bar

(c) Plate with hole

Figure 6.6

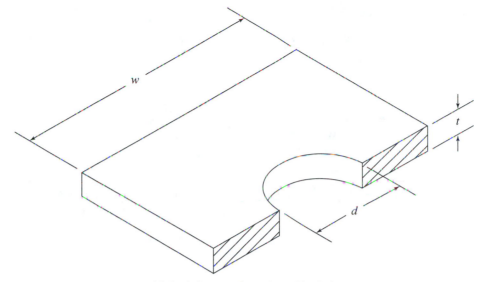

(d) Resisting area for a plate with a hole

Figure 6.6   (*Continued*)

For a stepped circular bar, like the one shown in Figure 6.6(a), it has been determined that the stress concentration factor, k, depends on the ratio D/d and on the ratio r/d. The experimental data for this case can be expressed by the equation

$$k = 1.052/(r/d)^n \tag{6.2(a)}$$

where

$$n = -0.0265(D/d)^2 + 0.1551(D/d) + 0.0752 \tag{6.2(b)}$$

for $3.0 \geq (D/d) \geq 1.2$, which means for the ratio D/d equal to or less than 3.0 and equal to or greater than 1.2, i.e., lying between 1.2 and 3.0.

Similarly, for the grooved circular bar shown in Figure 6.6(b), the experimental data can be correlated as

$$k = 1.041/(r/d)^n \tag{6.3(a)}$$

where

$$n = 2.2704(D/d)^3 - 10.3515(D/d)^2 + 15.4(D/d) - 7.1479 \tag{6.3(b)}$$

for $2.0 \geq (D/d) \geq 1.05$.

Another situation that often arises is that shown in Figure 6.6(c), which is a plate with a circular hole under tension. The minimum cross-sectional area resisting the force—and, therefore, the area of maximum stress—is the shaded area shown in Figure 6.6(d), which is $A = (w - d)t$. This is the area we would use to calculate the maximum stress in the plate, *but* stress concentrations arise due to the discontinuity (the hole).

The equation for this case is

$$k = 2.11(w/d - 1)^{0.0982} \tag{6.4}$$

EXAMPLE PROBLEM 6.8

A stepped circular bar (Figure 6.2(a)) has a large diameter of 70 mm, a small diameter of 40 mm, and a fillet radius of 6 mm. A tensile load of 180 kN is applied. What is the maximum stress in the bar?

Solution

1. First calculate the maximum stress, ignoring stress concentrations. The minimum cross-sectional area is

$$A = \pi(40 \text{ mm})^2/4 = 1257 \text{ mm}^2$$

So the maximum stress (ignoring stress concentrations) is

$$s = 180{,}000 \text{ N}/1257 \text{ mm}^2 = 143.2 \text{ MPa}$$

2. Calculate the stress-concentration factor. For a stepped circular bar we have the equation

$$k = 1.052/(r/d)^n \qquad\qquad (6.2(a))$$

where

$$n = -0.0265(D/d)^2 + 0.1551(D/d) + 0.0752 \qquad\qquad (6.2(b))$$

for $3.0 \geq (D/d) \geq 1.2$. In the present case, $D/d = 70 \text{ mm}/40 \text{ mm} = 1.75$, which is within the range.

   Calculate n:

$$n = -0.0265(1.75)^2 + 0.1551(1.75) + 0.0752$$
$$= -0.0812 + 0.2714 + 0.0752$$
$$= 0.2654$$

   Calculate k:

$$k = 1.052/(6 \text{ mm}/40 \text{ mm})^{0.2654}$$
$$= 1.052/(0.15)^{0.2654}$$
$$= 1.052/0.6044$$
$$= 1.74$$

3. Calculate the maximum stress:

$$s = 143.2 \text{ MPa} \times 1.74 = 249 \text{ MPa}$$

Answer

The maximum stress is 250 MPa.

EXAMPLE PROBLEM 6.9

A 2.5-in.-wide, 0.5-in.-thick plate has a 0.75-in.-diameter hole drilled through it (Figure 6.6(c)). The plate material will break if the tensile stress exceeds 95 ksi. It is to be used in an application where a tensile force of 45,000 lb will be applied. Will it fail?

Solution

1. First, calculate the stress, ignoring the stress-concentration factor. The resisting area (see Figure 6.6(d)) is

$$A = (w - d)t = (2.5 \text{ in.} - 0.75 \text{ in.}) \times 0.5 \text{ in.} = 0.875 \text{ in.}^2$$
$$s = F/A = 45,000 \text{ lb}/0.875 \text{ in.}^2 = 51,430 \text{ psi} = 51.4 \text{ ksi}$$

At this point in the calculation we may think that it is a safe design, *but* we must find the stress-concentration factor.

2. Calculate the stress-concentration factor. For this case we have the equation:

$$k = 2.11(w/d - 1)^{0.0982} \tag{6.4}$$

Therefore,
$$k = 2.11(2.5 \text{ in.}/0.75 \text{ in.} - 1)^{0.0982}$$
$$= 2.11(3.33 - 1)^{0.0982}$$
$$= 2.11(2.33)^{0.0982}$$
$$= 2.11 \times 1.09$$
$$= 2.3$$

3. Calculate the *actual* maximum stress:

$$s_{max} = 2.3 \times 51.4 \text{ ksi} = 118.2 \text{ ksi}$$

Answer

The plate *will* fail.

6.5   COMPUTER ANALYSES

6.5.1   Program 'Dstress'

Program 'Dstress' is actually two programs. One is for the calculation of stress or force for tensile or compressive loads, and the other is for the calculation of stress or force for shear loads.

General Input Data

1. Enter US or SI for the type of units.
2. Enter S if you want to calculate stress or F if you want to calculate force.

Tensile or Compressive Stress Calculations

The members from which you may choose are a

1. Round continuous bar,
2. Round, stepped bar,
3. Round, grooved bar,
4. Rectangular bar,
5. Plate,
6. Plate containing a circular hole.

To select one of these, highlight its number and, holding down the Ctrl key, press *t*. Depending on your selection, you will be asked to enter certain data. The stress or force will be determined.

Shear Stress Calculations

First enter S for single shear or D for double shear. The members from which you may choose are a

1. Round bar, rivet, bolt, or similar member,
2. Tube or pipe,
3. Rectangular bar or plate.

Or, you may want to determine the force or stress involved with punching a hole in a plate. In that case, select item 4. To select one of these items highlight its number and, holding down the Ctrl key, press s. Again, depending on your selection, you will be asked to enter certain data. The stress or force will be determined.

EXAMPLE PROBLEM 6.10

A 50-mm-diameter bar has a 40-mm-diameter groove. Both fillet radii are 6 mm. What maximum stress will be developed under a tensile load of 200 kN?

Solution
1. Open program 'Dstress'.
2. This is a tensile stress calculation, so we will use the left side of the screen. Enter SI for the units and S for a stress calculation.
3. Highlight 3 in the selection column; holding down the Ctrl key, press *t*.
4. The active cell will now be the data entry cell for the bar diameter. Enter 50.
5. Press the Enter key and type 40.
6. Press the Enter key and type 6.
7. Press the Enter key and type 200000.

Answer
The stress is 329 MPa.

EXAMPLE PROBLEM 6.11

A 3-in.-wide, 0.25-in.-thick steel plate contains a 1-in.-diameter hole. What is the maximum stress in the plate under a tensile load of 6000 lb?

Solution

1. Again, this is a tensile stress calculation, so we will use the left side of the screen. Enter US for the units and S for a stress calculation.
2. Highlight 6 in the selection column and, holding down the Ctrl key, press t.
3. The active cell will now be the data entry cell for the plate width. Enter 3.
4. Press the Enter key and type 0.25.
5. Press the Enter key and type 1.
6. Press the Enter key and type 6000.

Answer

The stress is 27,100 psi.

EXAMPLE PROBLEM 6.12

A 45-kg mass is suspended from a yellow brass, stepped round bar (see Figure 6.7). The bar has diameters 10 cm and 5 cm, and the ultimate tensile strength of yellow brass is 420 MPa.

a. Calculate the smallest fillet radius that could be used without the bar breaking.
b. If the fillet radius calculated in (a) is doubled, how much mass could the bar hold before breaking?

Solution

a. We will determine the fillet radius by entering different values until the stress is 420 MPa.
1. Enter SI for the units and S for a stress calculation.
2. Highlight number 2 in the selection column and press Ctrl-t (hold down the Ctrl key and press t).
3. Enter 100 for the bar's larger diameter (*Note:* It must be in mm.)
4. Enter 50 for the bar's smaller diameter.
5. Enter 10 for the fillet radius *as an initial guess.*
6. Note that we are given a *mass* of 45 kg. Changing this to a force gives

$$F = 45 \text{ kg} \times 9.81 \text{ m/s}^2 = 441{,}300 \text{ N}$$

Type 441300 and press Enter.

The stress is given as 371 MPa.

The ultimate tensile strength of the material is 420 MPa, so the fillet radius can be reduced. Reducing the fillet radius will increase the stress-concentration factor, thereby increasing the stress for the same load.

Change the fillet radius to 5 (5 mm). Now the stress is 450 MPa, which is too high.

Try a fillet radius of 7 mm. The answer, 410 MPa, is too low.

Try 6 mm. This is close; stress = 428 MPa.

Try 6.5 mm. The answer this time is stress = 418 MPa, which is close enough.

b. Enter the filet radius: 2 × 6.5 mm = 13 mm.

1. Change to a force calculation by entering F in the appropriate cell.
2. Enter the stress value of 420.

The answer is $F = 538{,}000$ N. Converting this to mass,

$$m = \frac{538{,}000 \text{ N}}{9.81 \text{ m/s}^2} = 54.8 \text{ kg}$$

So the bar can support 21% more mass if the fillet radius is increased from 6.5 mm to 13 mm.

Answer

The required minimum fillet radius is 6.5 mm. Increasing the fillet radius to 13 mm allows the bar to support 21% more mass.

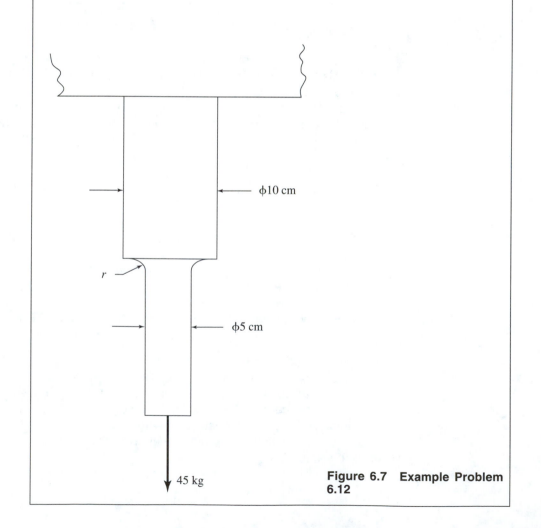

φ10 cm

r

φ5 cm

45 kg

Figure 6.7   Example Problem 6.12

EXAMPLE PROBLEM 6.13

This problem is a repeat of Example Problem 6.5. A two-inch-diameter disk is to be punched out of a 1/4-in. copper alloy plate. See Figure 6.3(c). The value of the stress at which the copper alloy will shear (known as the *ultimate shear strength*) is 25 ksi.
    a. What force is required?
    b. What is the compressive stress in the punch?

Solution

Part (a) is a shear *force* calculation, and part (b) is a compressive *stress* calculation.

    a. For shear force:
      1. Enter US for the units and F for force.
      2. Move to the right side of the screen for a shear calculation, and enter S for single shear.
      3. Highlight 4 in the selection column for a punched hole, and press Ctrl-s.
      4. Enter the plate thickness: 0.25.
      5. Enter the hole diameter: 2.
      6. Enter the stress: 25000.

Answer

The required force is 39,270 lb.

    b. For stress in the punch:
      1. Move to the left side of the screen, and change to a stress calculation by entering S.
      2. Highlight 1 in the selection column and press Ctrl-t.
      3. Enter the punch diameter: 2.
      4. Enter the force: 39270.

Answer

The stress in the punch is 12,500 psi.

6.6   PROBLEMS FOR CHAPTER 6

Hand Calculations

1. A short, round bar is being compressed by a 200-kN force. The diameter of the bar is 50 mm. Calculate the stress.

2. Six 0.06-in.-diameter wires hang vertically from the ceiling and support a steel plate weighing 600 lb. What is the stress in each wire?

3. Three short white pine posts are to support a mass of 1000 slugs. The allowable compressive stress for the wood is 725 psi. What diameter posts are required?

4. Calculate the shear stress on the pin shown in Figure 6.8. The pin diameter is 0.375 in. and the tensile force is 8000 lb.

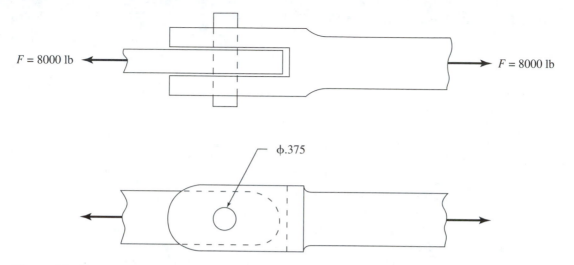

$F = 8000$ lb

$F = 8000$ lb

$\phi.375$

Figure 6.8   Problems 4 and 5

5. The pin in Figure 6.8 is replaced by a tube having an outside diameter of 0.375 in. and an inside diameter of 0.188 in. Will the 8000-lb force cause failure if the tube material will shear at a shear stress of 45 ksi?

6. What force is required to punch a hole 2 in. in diameter through a 0.75-in.-thick plate if a stress of 40 ksi will cause the plate material to shear?

7. A steel plate 8 cm wide and 5 mm thick contains a 2.5-cm hole. What maximum tensile force can be applied if the maximum stress in the plate is not to exceed 300 MPa?

Computer Calculations:

8. Repeat Problem 6.

9. Repeat Problem 7.

10. A steel plate 8 cm wide and 5 mm thick contains a 2.5-cm hole. A tensile force of 30 kN is applied to the ends of the plate. Determine the stress in the plate.

11. If the maximum allowable stress in Problem 10 is 156 MPa, how thick should the plate be if the width and the hole diameter remain the same?

12. Calculate the maximum stress in a stepped round bar whose diameters are 2.25 in. and 1.25 in. if the fillet radius is 0.125 in. The tensile load is 15,000 lb.

13. The same maximum stress calculated in Problem 12 is required, but the tensile load has been increased to 20,000 lb. What fillet radii is necessary?

14. A tensile force of 150 kN is applied to a 60-mm-diameter bar that has a 50-mm-diameter groove with fillet radii of 6 mm. What maximum stress will be developed?

15. If the bar in Problem 14 can withstand two times the calculated stress, what minimum fillet radius could be used?

16. An aluminum alloy tube has an outside diameter of 4.0 cm and is in double shear with an applied force of 200 kN. The allowable shear stress is 93 MPa. What is the maximum safe inside diameter?

17. What is the solid rod diameter of the same material that could be used to replace the tube in Problem 16?

18. The ultimate shear strength of aluminum alloy 2014-T6 is 42 ksi. What force is required

to punch a 1.0-in.-diameter hole in a 1.0-in.-thick plate of 2014-T6?

19. The hole diameter in Problem 18 is increased to 2.0 in. What force is now required? Explain why this answer is exactly two times that of the previous problem.

20. Write a computer program to calculate the maximum stress in a stepped round bar, given the diameters, fillet radius, and the applied load.

21. Write a computer program to calculate the shear stress in a round bar, a rectangular bar, or a tube, given the dimensions and the shear force. *Include a macro for selection of the member.*

7

Material Properties

■ *Objectives*

In this chapter you will learn about the following:

1. How materials are tested to determine their ultimate and yield strengths
2. The concept of strain
3. Hooke's law and the stress-strain diagram
4. Allowable stress
5. Poisson's ratio
6. Thermal expansion

7.1   THE TENSION AND COMPRESSION TEST

We know from experience or intuition that if we apply tensile forces (pull) to two bars of identical geometry, where one is made of steel and the other copper, the copper one will break at a lower force than the steel. Steel and copper have different *properties:* in this case, different *ultimate tensile strengths*.

It is extremely important to know the properties of all materials we deal with, and the only reliable way to determine them is by conducting tests. For this reason, the American Society for Testing and Materials (ASTM) and other groups have defined standard tests that may be performed to determine, for example, ultimate tensile strength and a property known as *yield strength*. Let us imagine that we are conducting a test to determine these two properties for 6061-T6 aluminum alloy, which is a metal used quite frequently in aircraft construction.

First, we need a specimen to test. The standard specimen is either a round or rectangular bar. We show a typical round-bar specimen in Figure 7.1. The parallel part in the middle is the test specimen; the ends are for gripping in a testing machine. We attach an *extensometer* (an instrument for measuring extremely small movements) between two points on the specimen. These points are accurately located, usually 2.000 in. or 50.0 mm apart.

The tensile test machine, also known as a universal testing machine (see Figure 7.2), has a stationary jaw for holding one end of the specimen and a movable one for the other end. The tensile test machine is designed to produce tremendous forces and to record the force continually during a test. As the bar is gradually stretched, the force being applied, F, and the extension, δ, caused by the force, are recorded. Having carefully measured the diameter, d, of the specimen with a micrometer before the test, we

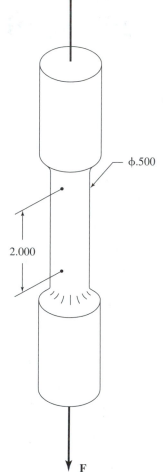

φ.500

2.000

F

F

Figure 7.1   Typical tensile test specimen.

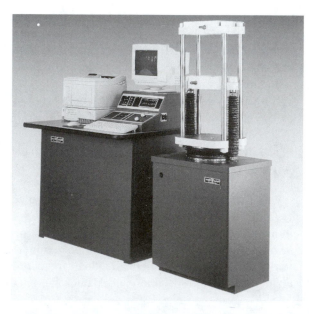

Figure 7.2   Tinius Olsen Super "L" 60,000-lb-capacity Universal Testing Machine Co., Inc., Willow Grove, Penn.

calculate its cross-sectional area, $A = \pi d^2/4$, and can, therefore, determine the tensile stress at any time during the test by $s = F/A$. We have also measured the extension (how much the bar has stretched) throughout the test. This brings us to the concept of *strain*.

7.2 STRAIN

Thinking back to Chapter 1, where we said that it is as if the molecules in a material are held together by tiny springs, we see that as a bar is stretched, each tiny spring is stretched. (There are no actual springs, of course, but the intermolecular forces *act* like springs). If there are a billion of these springs stacked end to end in a 10-cm-long bar, then there are 2 billion in a 20-cm bar. Logically, then, a 20-cm bar would stretch two times the distance that a 10-cm one would for the same applied force.

Suppose we apply a load of 6000 lb to a 0.500-in.-diameter, 2.000-in.-long specimen, and it stretches to 2.0052 in. We calculate the cross-sectional area: $A = \pi(0.500$ in.$)^2/4 = 0.196$ in^2. So, the tensile stress is $s = F/A = (6000$ lb$)/(0.196$ in.$^2) = 30{,}610$ psi. The extension is $\delta = 2.0052$ in. $- 2.0000$ in. $= 0.0052$ in.

Now we repeat the test with a 4.000-in.-long specimen of the same material and the same diameter (0.500 in.). Under the same loading of 6000 lb, the bar will stretch to 4.0104 in. Its extension, δ, will be 0.0104 in., two times that of the 2.000-in. bar.

If we repeat the test with a 6.000-in. specimen of the same material and diameter, δ would be three times that of the 2.000-in. bar, i.e., 0.0156 in.

The constant parameter in all this is the ratio of the extension to the original length. For the 2-in. bar, that ratio is 0.0052 in./2.000 in. = 0.0026 in/in. For the 4-in. bar it is 0.0104 in./4.000 in. = 0.0026 in./in., and for the 6-in. bar it is 0.0156 in./6.000 in = 0.0026 in./in.

This ratio is called the *strain*.

$$\text{strain, } \varepsilon = \delta/l \tag{7.1}$$

where δ is the extension (final length $-$ original length) and l is the original length, both, of course, in the same units.

EXAMPLE PROBLEM 7.1

A 10-cm-long rectangular steel bar, 20 mm $\times$ 30 mm, is being compressed end to end by a force of 200 kN. It compresses 0.16 mm. Calculate the compressive stress and the strain.

Solution

The cross-sectional area is $A = 20$ mm $\times$ 30 mm $= 600$ mm^2.

$$\text{stress} = F/A = 200{,}000 \text{ N}/600 \text{ mm}^2 = 333 \text{ MPa}$$

$$\text{strain} = \delta/l = 0.16 \text{ mm}/100 \text{ mm} = 0.0016 \text{ mm/mm}$$

Answer

Stress = 333 MPa, strain = 0.0016 mm/mm.

7.3   HOOKE'S LAW AND THE STRESS-STRAIN DIAGRAM

Robert Hooke (1635–1703), laid the foundations of stress analysis with his famous law that "most strong materials show a constant ratio between load and deflection."

$$F/\delta = \text{a constant}$$

the value of the constant depending on the dimensions and the type of material. We know, however, that if we replace the force with stress and the extension with strain, then the law can be made independent of dimensions. In this more general form

$$\text{stress/strain} = \text{constant}$$

$$s/\varepsilon = \text{constant}$$

Now the value of the constant depends only on the type of material (except at extreme temperatures). It is usually given the symbol E and is known as the **modulus of elasticity**. Some people call it the *elastic modulus,* and some call it *Young's modulus:*

$$s/\varepsilon = E \qquad\qquad (7.2)$$

What this tells us is that if we double the stress, we double the strain; if we triple the stress, we triple the strain, and so on. So, if we plot a graph of stress versus strain, we should see a straight line. This does actually happen, but only to a certain point. This linear relationship holds only in what is called the *elastic range.* Part of the stress-strain curve for 6061-T6 aluminum alloy is shown in Figure 7.3. We see that for this material Hooke's law holds up to a strain of 0.00135 in./in. That is, the stress is directly proportional to the strain from $\varepsilon = 0$ to 0.00135 in./in.

This point is called the *proportional limit,* or the elastic limit. There is a small difference between the proportional and the elastic limits, but it is insignificant for most engineering work.

While conducting our test, if we stopped the machine at, say, the fifth data point on Figure 7.3 (around a stress of 12,500 psi) and then decreased the load back to zero, the bar would contract back to its original length. This is because we did not exceed the elastic range. Beyond the elastic range, however, the material gets a *permanent* set, which means that even when all the load is removed, it will be slightly longer than its original length. This, of course, is undesirable because it weakens the part, and its permanent deformation may result in interference with adjacent parts. Many materials show a kink in the stress-strain curve just beyond the elastic limit. This is the *yield* point. Beyond yield, the material stretches more easily.

7.3.1   Ultimate and Yield Strengths in Tension and Compression

For design purposes the yield stress (value of the stress at yield) is more important than ultimate stress (the breaking stress), because you should never design a reusable part to work outside of the elastic range.

We defined the elastic modulus as $E = s/\varepsilon$. We must now modify this to read $E = s/\varepsilon$ *in the elastic range.* Returning to Figure 7.3, we can calculate E for 6061-T6 aluminum alloy. In the elastic range, the strain goes from 0 to 0.00135 in./in. for a stress increase of (17,600 psi − 2500 psi) = 15,100 psi. So $E = s/\varepsilon = (15,100 \text{ psi})/(0.00135 \text{ in./in.}) = 11.2 \times 10^6$ psi.

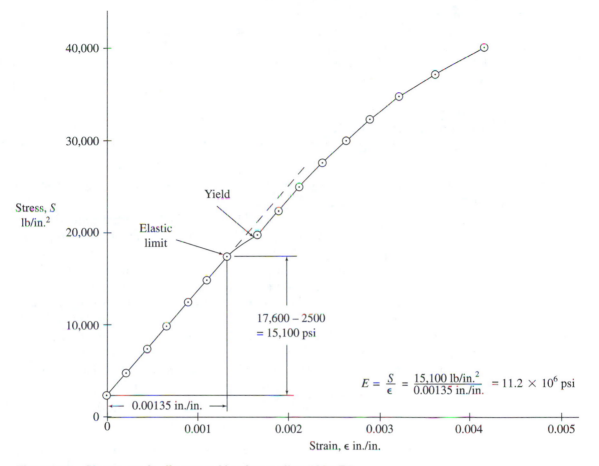

Figure 7.3   Stress-strain diagram: Aluminum alloy 6061-T6

The average ultimate (breaking) strength and yield strength in tension and in compression have been determined and tabulated for most engineering materials. Program 'Proptab', to be introduced at the end of this chapter, allows you to quickly find these values for more than 100 materials.

EXAMPLE PROBLEM 7.2

A 1.6-m-long steel piano wire has a 2-mm diameter.

a. What is the tensile stress in the wire if it stretches 3 mm when tightened?
b. What force is being applied to the wire? Young's modulus for steel is 200×10^3 MPa.

> **Solution**
> The wire diameter is $d = \pi(2 \text{ mm})^2/4 = 3.14 \text{ mm}^2$. The strain in the wire is $\varepsilon = \delta/1$
> $= 3 \text{ mm}/1600 \text{ mm} = 1.88 \times 10^{-3}$ mm/mm. Since $E = s/\varepsilon$, $s = E \cdot \varepsilon = (200 \times 10^3$
> MPa) $(1.88 \times 10^{-3} \text{ mm})$. So, $s = 376$ MPa. The applied force is $F = s \cdot A = (375$
> N/mm)(3.14 mm) = 1178 N.
>
> **Answer**
> The applied force is 1180 N.

Some metals stretch and neck down (see Figure 7.4) before they break. These are referred to as *ductile* materials.

Others, known as *brittle* materials, fail suddenly with little plastic deformation.

Ductile and brittle materials show quite different types of stress-strain curves. We compare the two types in Figures 7.5(a) and 7.5(b). The ductile-material stress-strain curve shows a definite yield point, whereas the brittle-material curve does not. Even so, it is desirable to define a yield point for brittle materials as a stress limit not to exceed. A popular way to do this is the *point two percent offset method*. Referring to Figure 7.5(b), we draw a line parallel to the straight portion of the curve and offset 0.0020 in./in. or 0.0020 mm/mm; where it intersects the stress-strain curve is the accepted *yield* point for that material.

We can take the two parts of a failed tensile test specimen, fit them back together, and measure the final length. The **percent elongation** is defined as:

$$\% \text{ elongation} = \frac{(\text{final length} - \text{original length}) \times 100}{\text{original length}}$$

Obviously the percent elongation will be greater for ductile materials than for brittle ones. In fact, a rough rule of thumb is that ductile materials have a percent elongation greater than 5%, and brittle materials have a percent elongation less than 5%.

7.3.2   Allowable Stress for Tension and Compression

Obviously we don't want to design parts using ultimate stress, because that is the breaking stress. We don't even want to design to yield stress, so how do we choose what stress level to design to? What *allowable* stress should we use?

Many engineering companies have their own ways of determining allowable stress based on their experience and test results. One reliable method is to use **safety factors,** which are numbers by which the ultimate or yield stresses are divided. These factors

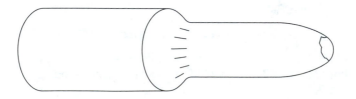

Figure 7.4   Failed ductile material specimen.

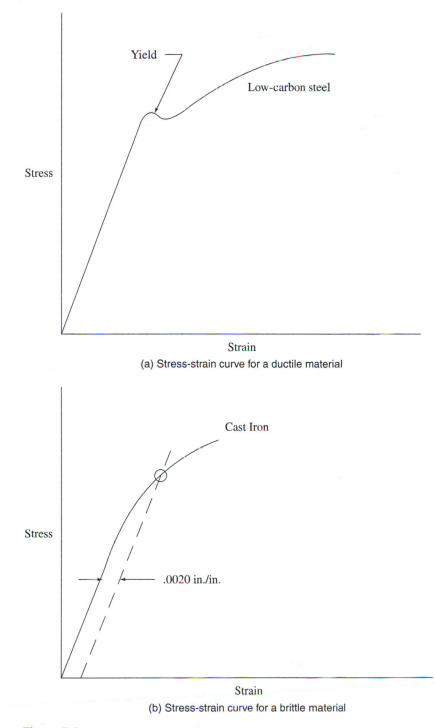

(a) Stress-strain curve for a ductile material

(b) Stress-strain curve for a brittle material

Figure 7.5

depend on whether the material is ductile or brittle and on the type of loading being applied. If the load is to be applied fairly gradually and the material is ductile, then one-half of the *yield* stress may be used safely. In this case the safety factor equals 2. On the other hand, if the loading is to be shock loading and the material is brittle, then the recommended allowable stress is one-twentieth of the *ultimate* stress, i.e., a safety factor of 20. Recommended values are as follows.

Loading Method	Ductile Material	Brittle Material
Steady load	$s_{yield}/2$	$s_{ultimate}/7$
Repeated load	$s_{ultimate}/8$	$s_{ultimate}/10$
Shock load	$s_{ultimate}/12$	$s_{ultimate}/20$

EXAMPLE PROBLEM 7.3

A short, square-cross-section Grade 20 gray iron support carries a constant compressive load of 50,000 lb. The ultimate compressive strength of Grade 20 gray iron is 20 ksi, and its percent elongation is less than 1%. What would you recommend its cross-section dimensions should be?

Solution
Since its percent elongation is less than 1%, it is a brittle material. The loading is steady, so from the safety factor chart:

$$\text{allowable stress} = s_{ultimate}/7$$

$$= 20{,}000 \text{ psi}/7 = 2860 \text{ psi}$$

The cross-sectional area, *A,* is *F/s:*

$$A = 50{,}000 \text{ lb}/2860 \text{ psi} = 17.5 \text{ in.}^2$$

So the length of one side is $l = \sqrt{17.5 \text{ in.}^2} = 4.2$ in.

Answer
The support should be 4.2 in. $\times$ 4.2 in.

7.4   ULTIMATE AND YIELD STRENGTHS IN SHEAR

If experimentally determined values are available, they should be used, but it is often difficult to find these values for shear. The following guidelines give good, conservative, estimates.

The following symbols are used:

$$s_u = \text{ultimate strength in tension}$$
$$s_{us} = \text{ultimate strength in shear}$$
$$s_y = \text{yield strength in tension}$$
$$s_{ys} = \text{yield strength in shear}$$

Material	Ultimate shear strength (s_{us})
Steel	$0.82s_u$
Gray cast iron	$1.30s_u$
Malleable iron	$0.90s_u$
Aluminum alloys	$0.65s_u$
Copper alloys	$0.90s_u$

For yield strength in shear:

$$s_{ys} = s_y/2$$

7.4.1 Allowable Stress for Shear

There are very few published data for brittle materials. Tests should be conducted whenever possible.

Loading Method	Ductile Material
Steady load	$s_{ys}/2$
Repeated load	$s_{ys}/4$
Shock load	$s_{ys}/6$

7.5 POISSON'S RATIO

When a material is stretched end to end, its cross-sectional dimensions decrease slightly, and when compressed, they increase. There is, therefore, a *transverse* strain in addition to the longitudinal strain.

Simeon Poisson (1781–1840) found that the ratio of transverse strain to longitudinal strain was a constant value for a given material; i.e., it is a property of the material. It is called Poisson's ratio and is denoted by μ.

$$\mu = \frac{\text{transverse strain}}{\text{longitudinal strain}}$$

When a body is being deformed by tensile or compressive forces, the forces create stresses and strains in the material. We have defined the ratio of stress to strain (in the elastic range) as the elastic modulus, E. In a similar manner, when shear forces act on a body, shear stresses and shear strains are created, so we can define an *elastic modulus in shear*. It is usually called the **modulus of rigidity** and is denoted by G.

$$G = \frac{\text{shear stress}}{\text{shear strain}}$$

We use G in Chapter 12, which deals with torsional shear stresses.

The modulus of rigidity, G, can be determined from the elastic modulus, E, and Poisson's ratio through the equation

$$G = \frac{E}{2(1 + \mu)} \tag{7.3}$$

EXAMPLE PROBLEM 7.4

302 stainless steel has an elastic modulus of 28×10^6 psi and a Poisson's ratio of 0.305. Calculate its modulus of rigidity.

Solution
Using equation (7.3):

$$G = E/[2(1 + \mu)]$$
$$= 28 \times 10^6 \text{ psi}/[2(1 + 0.305)]$$
$$= 28 \times 10^6 \text{ psi}/2.61 = 10.7 \times 10^6 \text{ psi}$$

Answer
The modulus of rigidity for 302 stainless steel is 10.7×10^6 psi.

7.6   THERMAL DEFORMATION

Most materials expand when heated and contract when cooled, but the amount of expansion or contraction depends on the material, its original length, and the amount of the temperature change.

If we denote the change in length by δ, the temperature change by ΔT, and the original length by l, then

$$\delta = \alpha \cdot l \cdot \Delta T \tag{7.4}$$

where α is a constant for a given material and is called the **coefficient of linear expansion**.

The values of α for some common materials are given next.

	Coefficient of Linear Expansion	
Material	per °C	per °F
Aluminum alloy	23.0×10^{-6}	12.8×10^{-6}
Brass	18.7×10^{-6}	10.4×10^{-6}
Bronze	18.0×10^{-6}	10.0×10^{-6}
Cast iron	11.3×10^{-6}	6.3×10^{-6}
Concrete	9.9×10^{-6}	5.5×10^{-6}
Copper	16.6×10^{-6}	9.2×10^{-6}
Glass (ordinary)	9.0×10^{-6}	5.0×10^{-6}
Glass (Pyrex)	3.0×10^{-6}	1.7×10^{-6}
Lead	29.0×10^{-6}	16.1×10^{-6}
Monel	14.0×10^{-6}	7.8×10^{-6}
Nylon 6/6	72.0×10^{-6}	40.0×10^{-6}
Phenolic	72.0×10^{-6}	40.0×10^{-6}
Polyester	90.0×10^{-6}	50.0×10^{-6}
Polyethylene	216×10^{-6}	120×10^{-6}
Quartz	0.4×10^{-6}	0.2×10^{-6}
SAN	65.0×10^{-6}	36.0×10^{-6}
Steel	11.7×10^{-6}	6.5×10^{-6}
Wrought iron	11.5×10^{-6}	6.4×10^{-6}

EXAMPLE PROBLEM 7.5

A 2-m brass rod is heated from 15°C to 150°C. What is its final length?

Solution
The original length is $l = 2$ m $= 2 \times 10^3$ mm. The temperature rise is $\Delta T = (150°C - 15°C) = 135°C$. The coefficient of linear expansion for brass is $\alpha = 18.7 \times 10^{-6}$ per °C.
 Using equation (7.4),

$$\delta = \alpha \cdot l \cdot \Delta T$$
$$= (18.7 \times 10^{-6}\ C^{-1})(2 \times 10^3\ mm)(135°C)$$
$$= 5.0\ mm$$

The final length of the bar is

$$(1 + \delta) = (2000\ mm + 5.0\ mm) = 2005\ mm$$

Answer
The final length is 2005 mm.

EXAMPLE PROBLEM 7.6

A steel rod, measuring 3 ft at 50°F, is placed in a bath of liquid nitrogen at −250°F. How much will it contract?

Solution
The temperature *drop* is (50°F − (−250°F)) = 300°F.
 The coefficient of linear contraction is the same as the coefficient of linear expansion, which for steel is

$$\alpha = 6.5 \times 10^{-6}\ F^{-1}$$

The original length of the rod is $l = 3$ ft $= 36$ in. Applying equation (7.4):

$$\delta = \alpha \cdot l \cdot \Delta T$$
$$= (6.5 \times 10^{-6}\ F^{-1})(36\ in.)(300°F) = 0.070\ in.$$

Answer
The steel rod contracts 0.070 in.

7.7 COMPUTER ANALYSES

7.7.1 Spreadsheet Programs Quattro Pro and Excel

Two widely used, excellent spreadsheet programs are Quattro Pro and Excel. We use Quattro Pro in Example Problem 7.7 to calculate and plot a stress-strain curve. Then we use Excel in Example Problem 7.8 to calculate and plot the same curve. In this way you will see the similarities and the differences between the two programs.

Note: If you are familiar with writing computer programs, you may want to skip to Section 7.7.2.

EXAMPLE PROBLEM 7.7

Write a program to calculate and plot a stress-strain curve using Quattro Pro, and apply it to the following data from a tension test of yellow brass.

original specimen gage length = 2.000 in.

original specimen diameter = 0.5048 in.

final specimen gage length = 2.438 in.

Load (kips) (*Note:* 1 kip = 10^3 lb)	Load (lb)	Extension (mil) (*Note:* 1 mil = 0.001 in.)
0.5	500	0
1.0	1000	0.4
1.5	1500	0.7
2.0	2000	1.1
2.5	2500	1.5
3.0	3000	1.8
3.5	3500	2.2
4.0	4000	2.6
4.5	4500	3.0
5.0	5000	3.4
5.5	5500	3.9
6.0	6000	4.4
6.5	6500	4.9
7.0	7000	5.5
7.5	7500	6.3
8.0	8000	7.2
8.5	8500	8.1
9.0	9000	9.2
9.5	9500	14.0

Solution

1. Click on the Quattro Pro icon in the Windows environment. The spreadsheet will appear on your screen with A, B, C, D, . . . across the top and 1, 2, 3, 4, . . . down the left side. In this way each cell is identified by a letter and a number. For example, the cell that is 3 across and 5 down is cell C5.

2. Highlight cell A1 by moving your mouse to the top left corner cell, and type STRESS-STRAIN DIAGRAM FOR; then press Enter.

3. Highlight cell D1, type YELLOW BRASS, and press Enter.

4. The cells and their entries are shown in Figure 7.6(a).

Note: A:A1 means cell A1 of page A.

You do not have to type the single quote (`) in front of the text; Quattro Pro will do that for you if you are entering text. Numbers do not have the leading quote mark. Referring to Figure 7.6(a) and 7.6(b), enter the text and the numbers into the cells down to and including F7. Two cells that may need some explanation are H3 and H4.

H3 is 0.7854*C5^2, which calculates the cross-sectional area of the specimen whose diameter is in cell C5; 0.7854 is $\pi/4$, * means multiply, and C5^2 is the square of the value in cell C5, i.e., the diameter.

A:A1:	'STRESS-STRAIN DIAGRAM FOR		A:C15:	0.0026
A:D1:	'YELLOW BRASS		A:E15:	+B15/$H$3
A:A3:	'Original Length (in) =		A:F15:	+C15/$C$3
A:C3:	2		A:B16:	4500
A:E3:	'Cross sectional Area (in ^2) =		A:C16:	0.003
A:H3:	0.7854*C5^2		A:E16:	+B16/$H$3
A:A4:	'Final length (in) =		A:F16:	+C16/$C$3
A:C4:	2.438		A:B17:	5000
A:E4:	'Elongation (%) =		A:C17:	0.0034
A:H4:	100*(C4–C3)/C3		A:E17:	+B17/$H$3
A:A5:	'Original Diameter (in) =		A:F17:	+C17/$C$3
A:B5:	'		A:B18:	5500
A:C5:	0.5048		A:C18:	0.0039
A:B7:	'Load (lb)		A:E18:	+B18/$H$3
A:C7:	'Delta (in)		A:F18:	+C18/$C$3
A:E7:	'Stress (psi)		A:B19:	6000
A:F7:	'Strain (in/in)		A:C19:	0.0044
A:B8:	500		A:E19:	+B19/$H$3
A:C8:	0		A:F19:	+C19/$C$3
A:E8:	+B8/$H$3		A:B20:	6500
A:F8:	+C8/$C$3		A:C20:	0.0049
A:B9:	1000		A:E20:	+B20/$H$3
A:C9:	0.0004		A:F20:	+C20/$C$3
A:E9:	+B9/$H$3		A:B21:	7000
A:F9:	+C9/$C$3		A:C21:	0.0055
A:B10:	1500		A:E21:	+B21/$H$3
A:C10:	0.0007		A:F21:	+C21/$C$3
A:E10:	+B10/$H$3		A:B22:	7500
A:F10:	+C10/$C$3		A:C22:	0.0063
A:B11:	2000		A:E22:	+B22/$H$3
A:C11:	0.0011		A:F22:	+C22/$C$3
A:E11:	+B11/$H$3		A:B23:	8000
A:F11:	+C11/$C$3		A:C23:	0.0072
A:B12:	2500		A:E23:	+B23/$H$3
A:C12:	0.0015		A:F23:	+C23/$C$3
A:E12:	+B12/$H$3		A:B24:	8500
A:F12:	+C12/$C$3		A:C24:	0.0081
A:B13:	3000		A:E24:	+B24/$H$3
A:C13:	0.0018		A:F24:	+C24/$C$3
A:E13:	+B13/$H$3		A:B25:	9000
A:F13:	+C13/$C$3		A:C25:	0.0092
A:B14:	3500		A:E25:	+B25/$H$3
A:C14:	0.0022		A:F25:	+C25/$C$3
A:E14:	+B14/$H$3		A:B26:	9500
A:F14:	+C14/$C$3		A:C26:	0.014
A:B15:	4000		A:E26:	+B26/$H$3
			A:F26:	+C26/$C$3

(a) Quattro Pro cell entries

Figure 7.6

STRESS-STRAIN DIAGRAM FOR YELLOW BRASS						
Original Length (in.) =		2		Cross sectional Area (in. ^2) =	0.200138	
Final Length (in.) =		2.438		Elongation (%) =	21.9	
Original Diameter (in.) =		0.5048				
	Load (lb)	Delta (in.)		Stress (psi)	Strain (in./in.)	
	500	0		2498	0	
	1000	0.0004		4997	0.0002	
	1500	0.0007		7495	0.00035	
	2000	0.0011		9993	0.00055	
	2500	0.0015		12491	0.00075	
	3000	0.0018		14990	0.0009	
	3500	0.0022		17488	0.0011	
	4000	0.0026		19986	0.0013	
	4500	0.003		22484	0.0015	
	5000	0.0034		24983	0.0017	
	5500	0.0039		27481	0.00195	
	6000	0.0044		29979	0.0022	
	6500	0.0049		32478	0.00245	
	7000	0.0055		34976	0.00275	
	7500	0.0063		37474	0.00315	
	8000	0.0072		39972	0.0036	
	8500	0.0081		42471	0.00405	
	9000	0.0092		44969	0.0046	
	9500	0.014		47467	0.007	

(b) Quattro Pro Spreadsheet

Figure 7.6 (*Continued*)

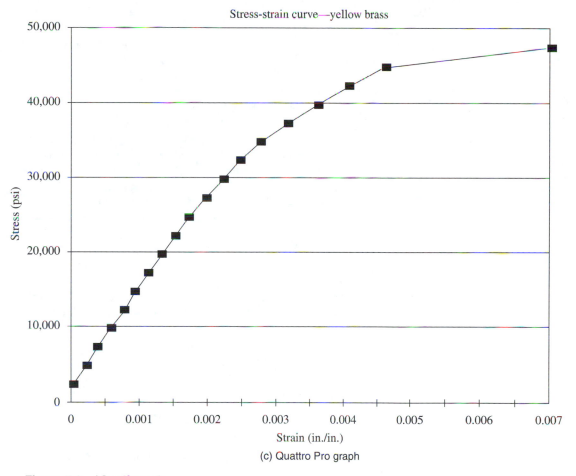

(c) Quattro Pro graph

Figure 7.6 (*Continued*)

H4 is 100* (C4-C3)/C3. Cell C3 contains the specimen's original length, and C4 contains its final length. Remembering that % elongation = [(final length − original length) × 100]/(original length), cell H4 calculates the % elongation.

5. Now enter the data: the loads into cells B8 through B26 and the extensions, Delta, into cells C8 through C26. Quattro Pro has a command FILL that you can use to enter the loads. To use FILL, the entries have to increase or decrease by a constant amount from cell to cell. This is called the STEP, which in the present case is 500.

Highlight cell B8, and holding down the mouse button, drag down over cells B9, B10, . . . , B26. All the cells from B8 to B26 should now be highlighted.

Click on BLOCK in the menu bar and then on Fill. A dialog box will appear, into which you can enter the START value, 500, and the STEP value, 500. Ignore the STOP value because it will stop at cell B26 anyway. Click on OK, and the load values will be entered. You cannot use FILL to enter the extension values. They will have to be entered one by one, but press the Down Arrow key instead of the Enter key after each entry.

6. Now look at cell E8 (Figure 7.6(a)). This is where you calculate the first stress value. Stress = load/area, which for the first load of 500 lb, in cell B8, is B8/H3 (the cross-sectional area is in cell H3). First notice that there is a plus sign in front of B. Quattro Pro *does not put this in for you.* As a matter of fact, if you do not type the + sign, Quattro Pro will assume you are entering text. By including the + you are saying that you want the contents of cell B8, not the letter B and the number 8. If we had entered +B8/H3, the correct value of 2498 psi would have resulted, but then we would have had to enter +B9/H3 in cell E9, +B10/H3 in cell E10, and so on.

A more convenient way to do this is to use the COPY command. This command copies the formula into the selected cells. It is very important to understand how COPY works. If you enter +B8/H3 into cell E8 and then use the COPY command to enter the formula into cells E9, E10, E11, . . . , E26, those cells would contain B9/H4, B10/H5, B11/H6, . . . , B26/H21, which you do not want. What you do want is B9/H3, B10/H3, B11/H3, . . . , B26/H3, because although you need to pick up the *next load value* each time, you have to use the *same area value* throughout. You do this by locking the H3 cell into the formula as $H$3.

To use the COPY command, highlight cell E8 and enter +B8/$H$3. Then click on BLOCK in the main menu and then on COPY. A dialog box will appear with

> From   A:E8..E8
> TO   A:E8..E8

Change the second line to

> TO A:E8..E26

and click on OK. The stress values shown in Figure 7.6(b) will be calculated and entered instantly.

7. You can calculate and enter the strain values in the same way, but this time you need to divide the extension values in column C by the fixed value of original specimen length in cell C3. The formula to enter into cell F8 is +C8/$C$3. Use the COPY command as before to enter the strain values into column F.

8. Before leaving the worksheet, we should tidy it up a little.
   a. Make the headings bold by highlighting the cells and then clicking on b, located below the main menu bar.
   b. Some of the cells are not wide enough to display all the text. We simply move the cursor (the arrow) up to the cell-identification bar with the A, B, C, . . . and slide it to the right edge of the column you want to widen. The arrow will change into a cross with arrows pointing left and right. Hold down the mouse button and drag the column to whatever width you need.
   c. The numbers after the decimal points in the stress values are meaningless. Eliminate them by first clicking on EDIT in the main menu bar, and then on DEFINE STYLE. Choose FORMAT in the dialog box and then FIXED. You are asked to enter the number of decimal places. Make it 0 and then exit the command by clicking on OK, then on OK again. Now highlight the column B8 . . . B26. You will see FIXED on the bar below the main menu,

with an arrow to its right pointing down. Click on the arrow and then select NORMAL. The numbers in the selected column will lose their decimals.
9. To plot the graph:
   a. Click on GRAPH in the main menu, then on NEW. Click on X-AXIS in the dialog box and you will be returned to the spreadsheet.
   b. Highlight the strain column numbers, cells F8 . . F26, by dragging the mouse; then click on the arrow head to the right of GRAPH NEW at the top of the screen. The dialog box will reappear. Click on 1st.
   c. Highlight the stress column numbers, cells E8 . . . E26, and again click on GRAPH NEW. Click OK. A graph will appear on the screen, but it will not have any titles, and it may not be an *xy* graph.
   d. Click on GRAPH in the main menu and then on TYPE. In the dialog box that appears, select the *xy* graph. This is probably the second icon in the second row. Click OK, and the graph will reappear on the screen.
   e. Go to GRAPH again in the main menu, select EDIT, and then select OK in the dialog box. Click on GRAPH again, and then on TITLES. Enter titles for the MAIN and SUBTITLES, the X-AXIS, and the Y-AXIS. Selecting VIEW from the GRAPH menu will give you a full-screen graph like the one in Figure 7.6(c).
   f. Press Esc to leave the full-screen view.

EXAMPLE PROBLEM 7.8

Write a program to calculate and plot a stress-strain curve using Excel, and apply it to the data from a tension test of yellow brass given in Example Problem 7.7.

Solution

1. Select the Microsoft Office icon in the Windows environment and then choose Excel. The spreadsheet is very similar to the Quattro Pro spreadsheet. (See Figure 7.7(a)).

2. Enter the text in the same way as in Example Problem 7.7 *except* enter Stress (psi) into cell F7 and Strain (in/in) into cell E7. You will see the reason for this when you come to plot the graph.

3. In Excel, a formula, such as the one in cell H3, must start with an $=$ sign, i.e., $= 0.7854{*}C5{^{\wedge}}2$.

4. Enter the Load values into cells B8 . . . B26 by first entering 500 into cell B8 and then entering the formula $= 500{+}B8$ INTO cell B9. (Or, you can use FILL under EDIT. Pick SERIES in COLUMNS, LINEAR, STEP $= 500$, STOP $= 9500$.) Highlight cells B10 . . . B26 by dragging; then click on EDIT in the main menu and then on COPY. Press Enter, and the Load values will be entered.

5. The extension values have to be entered one by one.

6. Because the stress and strain columns have been switched, change the formulas in cells E8 and F8. E8 will now be $=C8/\$C\$3$, and F8 will be $=B8/\$H\$3$.

STRESS-STRAIN DIAGRAM FOR			YELLOW BRASS			
Original Length (in.) =	2		Cross sectional Area (in. ^2) =	0.200138	
Final Length (in.) =	2.438		Elongation (%) =	21.9	
Original Diameter (in.) =	0.5048				
Load (lb)	Delta (in.)			Strain (in./in.)	Stress (psi)
500	0			0	2498
1000	0.0004			0.0002	4997
1500	0.0007			0.00035	7495
2000	0.0011			0.00055	9993
2500	0.0015			0.00075	12491
3000	0.0018			0.0009	14990
3500	0.0022			0.0011	17488
4000	0.0026			0.0013	19986
4500	0.003			0.0015	22484
5000	0.0034			0.0017	24983
5500	0.0039			0.00195	27481
6000	0.0044			0.0022	29979
6500	0.0049			0.00245	32478
7000	0.0055			0.00275	34976
7500	0.0063			0.00315	37474
8000	0.0072			0.0036	39972
8500	0.0081			0.00405	42471
9000	0.0092			0.0046	44969
9500	0.014			0.007	47467

(a) Excel spreadsheet

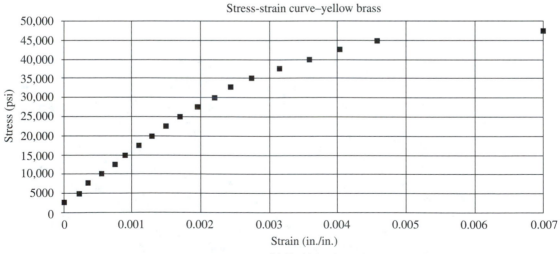

(b) Excel graph

Figure 7.7

7. COPY these into E9 . . . E26 and F9 . . . F26, respectively, in the same manner as for entering the Load values. (See Step 4.)

8. You can make the headings bold by highlighting the text; then click on the B in the main menu.

9. To control the number of decimal places displayed in the Stress column, highlight the column of numbers, click on FORMAT in the main menu, and choose CELLS and then NUMBER in the dialog box. Select 0; then choose OK.

10. Excel makes plotting the graph very easy. Click on CHART WIZARD and simply follow the instructions. For CHART TYPE select XY (Scatter) and the top right picture. Click on NEXT and make sure the correct data range is identified. Click on NEXT again and enter the titles. Click on FINISH. Your graph will look like Figure 7.7(b).

7.7.2 Program 'Proptab'

Excel program 'Proptab' quickly and conveniently gives you the average properties of many engineering materials. The following example demonstrates its use.

EXAMPLE PROBLEM 7.9

Use program 'Proptab' to find the following properties of AISI steel 1080 OQT 1100 in US units:

a. Ultimate strength
b. Yield strength
c. Elongation
d. Density
e. Elastic modulus

Solution
1. Open program 'Proptab' (enable the macro).
2. Enter US in cell C4.
3. Highlight number 14 in the selection column.

Answers
a. Ultimate strength = 145 ksi
b. Yield strength = 103 ksi
c. Elongation = 17%
d. Density = 0.283 lb/in^3
e. Elastic modulus = 30×10^6 psi

7.8.   PROBLEMS FOR CHAPTER 7

Hand Calculations

1. A $\frac{1}{10}$-in.-diameter, 5-ft-long steel rod is under tension.
   a. What is the tensile stress in the wire if it stretches 0.10 in.?
   b. What force is being applied to the rod? Young's modulus for steel is 30×10^6 psi.

2. A short, circular-cross-section support made of hard C17000 copper carries a varying compressive load of 10 kN. The ultimate compressive strength of hard C17000 copper is 1482 MPa, and its percent elongation is 2%. What diameter would you recommend for the support?

3. Poisson's ratio for brass is 0.34, and its elastic modulus is 16×10^6 psi. Calculate its modulus of rigidity in psi.

4. The temperature of a 3-m-long aluminum rod is increased by 300°C. By how much will it expand?

5. A tensile force of 50,000 lb is applied to a 100.000-in.-long, 2.000-in.-diameter rod. The rod stretches 5.305×10^{-2} in. If the material's modulus of rigidity is 11.5×10^6 psi, what is the resulting transverse strain in in./in.?

Computer Calculations

6. Using either Quattro Pro or Excel, plot a stress-strain diagram for 6061-T6 aluminum alloy. The test data are

$$\text{original gage length} = 2.000 \text{ in.}$$
$$\text{original specimen diameter} = 0.5012 \text{ in.}$$
$$\text{final gage length} = 2.417 \text{ in.}$$

Load (lb)	Elongation (mil)	Load (lb)	Elongation (mil)
500	0	5000	4.3
1000	0.4	5500	4.8
1500	0.9	6000	5.3
2000	1.3	6500	5.8
2500	1.8	7000	6.5
3000	2.2	7500	7.3
3500	2.7	8000	8.4
4000	3.3	8500	10.3
4500	3.8	9000	13.2

What would you estimate the yield strength to be from your graph? Check your answer by using program 'Proptab'.

7. Using program 'Proptab' find each of the following.
   a. Elastic modulus of gray iron ASTM A48 grade 20, in ksi
   b. Ultimate shear strength of 1100-H18 aluminum alloy, in MPa
   c. Allowable compressive stress of No. 2 Douglas fir parallel to the grain, in MPa

8. What force is required, in pounds, to punch a 2.5-in.-diameter hole in a 0.25 in thick AISI 4140 annealed steel plate? Use the program 'Proptab' to find the ultimate shear strength; then use the program 'Dstress' for the calculation.

9. An AISI 1141 OQT 900 steel plate, 10 cm wide and 4 mm thick, contains a 2.0-cm hole. What maximum tensile shock loading can be applied safely?

10. A stepped round bar whose diameters are 2.125 in. and 1.125 in., with a fillet radius of 0.125 in., is under a repeated tensile load of 1000 lb. The bar is made of 1100-H12 aluminum alloy. Is the allowable stress being exceeded?

11. A steady tensile force of 100 kN is applied to a 65-mm-diameter bar of AISI 430 annealed stainless steel that has a 45-mm diameter groove, with fillet radii of 6 mm. Is the allowable stress being exceeded?

12. Write a program to look up the
    a. Ultimate strength

b. Yield strength
c. Percent elongation for the following steels:
   1. AISI 1020 annealed
   2. AISI 1040 annealed
   3. AISI 1080 annealed
   4. AISI 1141 annealed
   5. AISI 4140 annealed
   6. AISI 5160 annealed

Use a *macro* for steel-type selection.

13. What is the weight of a rectangular bar of AISI 1020 that is 4.0 in. wide, 0.25 in. thick, and 10 in. long?

14. Using program 'Proptab' list the ultimate strength, yield strength, and percent elongation for AISI 1020 in the annealed condition, hot-rolled, and cold-drawn. What conclusions can you reach about the effects of hot-rolling and cold-drawing of AISI 1020?

15. What is the effect on AISI 1040 of water quenching at different temperatures? That is, what happens to the strengths and percent elongation as the quenching temperature is increased?

16. Two bars, one of cast zinc and one of cast magnesium, have identical dimensions. They are both subjected to the same tensile force. Which one stretches more?

17. What is the weight in newtons of a circular rod of titanium whose diameter is 5.0 cm and whose length is 1.0 m?

18. The density of concrete is approximately 150 lb/ft^3. Which is heavier, a block of concrete or a block of aluminum alloy of the same size?

19. Which material has the higher flexural strength; phenolic or molded epoxy?

20. Which stress-strain curve (stress as the ordinate) would have the steeper slope in the elastic region, that for C54400 soft bronze or that for C36000 soft brass?

8

Thermal Stress

■ Objectives

In this chapter you will learn how to calculate the stresses that develop when a body is heated or cooled but restrained from expanding. We will investigate the cases of

1. Single materials,
2. Materials in parallel,
3. Materials in series.

8.1 SINGLE MATERIAL

We saw in Chapter 7 that if the temperature of a bar of length l is changed by an amount ΔT, it will expand or contract by an amount δ, where

$$\delta = \alpha \cdot l \cdot \Delta T \qquad (8.1)$$

α is the coefficient of linear expansion of the material. In Chapter 7 we also defined *strain* as extension or compression divided by original length,

$$\varepsilon = \delta/l \qquad (8.2)$$

and the relationship between stress and strain as

$$s/\varepsilon = E, \qquad \text{the elastic modulus} \qquad (8.3)$$

So the question is, Have we stressed the bar by heating it? If it is free to expand, the answer is no. True, we have caused an increase in length, δ, and dividing that by the original length gives us δ/l, but this is not really a strain. Strain is caused by *external forces* that stretch or compress the material.

Imagine that we heat a bar that is free to expand, and its length increases by δ; then we compress it in the jaws of a vise back to its original length. Now we have caused a strain, because an external force has compressed the bar, and there will be a corresponding stress. This situation is exactly the same as if we had clamped the bar to prevent it from expanding and then heated it. So, if the material is restrained so that it cannot expand or contract then stresses will arise.

Starting with equation (8.3),

$$s/\varepsilon = E$$

or

$$s = \varepsilon E$$

Then from equation (8.2),

$$\varepsilon = \delta/l$$

so

$$s = (\delta/l)\cdot E$$

but from equation (8.1) we have

$$\delta/l = \alpha\cdot\Delta T$$

Finally, then,

$$s = \alpha\cdot\Delta T\cdot E \tag{8.4}$$

In words, if we clamp a body in order to prevent it from expanding and then heat or cool it the body will be subjected to a stress whose magnitude is $(\alpha\cdot\Delta T\cdot E)$, where α is the coefficient of linear expansion of the material, ΔT is the temperature change, and E is the elastic modulus of the material.

EXAMPLE PROBLEM 8.1

A 2-m-long, 35-mm-diameter aluminum bar is rigidly fixed between two walls at a temperature of 15°C. It is then heated to a uniform temperature of 100°C. $E = 69 \times 10^3$ MPa, and $\alpha = 23 \times 10^{-6}$ per °C.

  a. What will be the stress in the bar?
  b. What force will the bar exert on the walls?

Solution
  a.
$$s = \alpha\cdot\Delta T\cdot E$$
$$\alpha = (23 \times 10^{-6})°C^{-1}$$
$$\Delta T = 100°C - 15°C = 85°C$$
$$E = 69 \times 10^3 \text{ MPa}$$
$$s = (23 \times 10^{-6}°C^{-1})(85°C)(69 \times 10^3 \text{ MPa})$$
$$= 135 \text{ MPa}$$

  b. The cross-sectional area is $A = \pi(\text{diameter})^2/4$, so
$$A = 0.7854(35 \text{ mm})^2 = 962 \text{ mm}^2$$
$$\text{force} = \text{area} \times \text{stress} = (962 \text{ mm}^2)(135 \text{ MPa}) = 129.9 \times 10^3 \text{ N}$$

(because 1 MPa = 1 N/mm^2). So, the force is 129.9 kN (1kN = 10^3 N).

Answer
The stress in the bar is 135 MPa. The force exerted by the bar on the wall (and, of course, the wall on the bar) is 130 kN.

EXAMPLE PROBLEM 8.2

A 3-ft-long, 2-in.-diameter schedule 40 steel pipe is compressed end to end by a 2000-lb force. By how much would it have to be cooled to reduce the stress to zero?

$$E = 30 \times 10^6 \text{ psi}, \quad \alpha = 6.5 \times 10^{-6} \text{ per } °F$$

A nominal 2-in.-diameter, schedule 40 pipe has an outside diameter of 2.375 in. and an inside diameter of 2.067 in.

Solution

The cross-sectional area is $A = \pi \, [(\text{outside diameter})^2 - (\text{inside diameter})^2]/4$, so

$$A = 0.7854[(2.375 \text{ in.})^2 - (2.067 \text{ in.})^2] = 1.075 \text{ in.}^2$$

The initial stress is $s = F/A = 2000 \text{ lb}/1.075 \text{ in.}^2 = 1860$ psi.

$$s = \alpha \cdot \Delta T \cdot E$$

Therefore, $\Delta T = s/(\alpha E)$, with

$$s = 1860 \text{ psi}$$
$$\alpha = (6.5 \times 10^{-6})°F^{-1}$$
$$E = 30 \times 10^6 \text{ psi}$$

So

$$\Delta T = \frac{1860 \text{ psi}}{[(6.5 \times 10^{-6})°F^{-1}](30 \times 10^6 \text{ psi})}$$

$$= 9.6°F$$

Answer

The pipe must be cooled by 9.6°F to reduce the stress to zero.

8.2 MATERIALS IN PARALLEL

Figure 8.1 shows two different materials in parallel, such as a core of brass melted into an iron pipe, clamped between two solid walls. How will the stresses be distributed if we heat or cool the bar to a uniform temperature throughout, and what force will be exerted on the walls?

Let subscript 1 denote conditions for one of the materials and subscript 2, the other. The lengths are the same and stay the same:

$$l_1 = l_2 = l$$

Also, the temperature change is the same for both bodies, so

$$\Delta T_1 = \Delta T_2 = \Delta T$$

Applying equation (8.4) to each:

$$s_1 = \alpha_1 \cdot \Delta T \cdot E_1 \quad \text{and} \quad s_2 = \alpha_2 \cdot \Delta T \cdot E_2$$

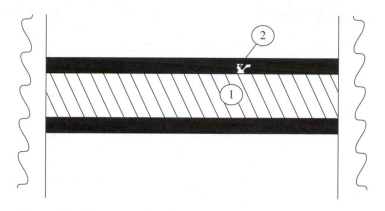

Figure 8.1   Two materials in parallel

Because force = stress × area, the forces will be

$$F_1 = \alpha_1 \cdot \Delta T \cdot E_1 \cdot A_1 \quad \text{and} \quad F_2 = \alpha_2 \cdot \Delta T \cdot E_2 \cdot A_2$$

The total force is the sum of the two forces:

$$F = \alpha_1 \cdot \Delta T \cdot E_1 \cdot A_1 + \alpha_2 \cdot \Delta T \cdot E_2 \cdot A_2$$

or
$$F = (\alpha_1 \cdot E_1 \cdot A_1 + \alpha_2 \cdot E_2 \cdot A_2) \Delta T \tag{8.5}$$

EXAMPLE PROBLEM 8.3

A 3.5-in.-diameter schedule 10S pipe is filled with concrete. It is rigidly fixed between two solid walls and heated from 45°F to 170°F. E for steel is 30×10^6 psi, E for concrete is 3×10^6 psi, α for steel is 6.5×10^{-6} per °F, and α for concrete is 5.5×10^{-6} per °F.

A nominal 3.5-in.-diameter schedule 10S pipe has an outside diameter of 4.000 in. and an inside diameter of 3.760 in.

a. What is the stress in each material?
b. What force is exerted on the walls?

Solution

a. Let the concrete be material *1* and the steel be material 2. The cross-sectional area of the concrete is

$$A_1 = \pi(\text{diameter})^2/4$$

The diameter of the concrete is the inside diameter of the pipe, 3.760 in.

$$A_1 = 0.7854(3.760 \text{ in.})^2 = 11.10 \text{ in.}^2$$

The cross-sectional area of the steel is

$$A_2 = \pi[(\text{outside diameter})^2 - (\text{inside diameter})^2]/4$$
$$= 0.7854[(4.000 \text{ in.})^2 - (3.760 \text{ in.})^2] = 1.463 \text{ in.}^2$$

The stress in the concrete is

$$s_1 = \alpha_1 \cdot \Delta T \cdot E_1$$

$$\alpha_1 = (5.5 \times 10^{-6})°F^{-1}$$

$$\Delta T = (170°F - 45°F) = 125°F$$

$$E = 3 \times 10^6 \text{ psi}$$

Therefore,

$$s = (5.5 \times 10^{-6}°F^{-1})(125°F)(3 \times 10^6 \text{ psi})$$

$$s_1 = 2063 \text{ psi}$$

The stress in the steel is

$$s_2 = \alpha_2 \cdot \Delta T \cdot E_2$$

$$\alpha_2 = (6.5 \times 10^{-6})°F^{-1}$$

$$\Delta T = (170°F - 45°F) = 125°F$$

$$E = 30 \times 10^6 \text{ psi}$$

Therefore,

$$s_2 = [(6.5 \times 10^{-6})°F^{-1}](125°F)(30 \times 10^6 \text{ psi})$$

$$= 24{,}380 \text{ psi}$$

The stress in the steel is 12 times that in the concrete.

b. The force exerted by the concrete is

$$F_1 = s_1 \cdot A_1$$

$$s_1 = 2063 \text{ psi}$$

$$A_1 = 11.10 \text{ in.}^2$$

So, $F_1 = (2063 \text{ psi})(11.10 \text{ in.}^2) = 22{,}900 \text{ lb}$

The force exerted by the steel is

$$F_2 = s_2 \cdot A_2$$

$$s_2 = 24{,}380 \text{ psi}$$

$$A_2 = 1.463 \text{ in.}^2$$

So,

$$F_2 = (24{,}380 \text{ psi})(1.463 \text{ in.}^2) = 35{,}700 \text{ lb}$$

So the total force exerted on the walls is

$$F = F_1 + F_2 = 22{,}900 \text{ lb} + 35{,}700 \text{ lb} = 58{,}600 \text{ lb}$$

Answer

stress in the concrete = 2,060 psi

stress in the steel = 24,380 psi

force exerted by the concrete = 22,900 lb

force exerted by the steel = 35,700 lb

total force on the walls = 58,600 lb

8.2.1 Materials in Parallel: General Equations

In preparation for writing a general computer program for materials in parallel, we now extend the preceding concepts to more than two materials.

Suppose there are three, four, or any number of materials in parallel. Let the number of different materials be N. Then we can write a general equation for the total force exerted:

$$F = \Delta T \sum_{n=1}^{n=N} (\alpha_n E_n A_n) \tag{8.6}$$

See Appendix 1 for an explanation of this type of notation.

To see how this compares with equation (8.5), note there were two materials, so $N = 2$. For that case, then,

$$F = \Delta T \sum_{n=1}^{n=2} (\alpha_n E_n A_n) \tag{8.7}$$

These materials were denoted as 1 and 2, so $n = 1$ and 2. The equation tells us to add $(\alpha_1 E_1 A_1)$ and $(\alpha_2 E_2 A_2)$ and then multiply by ΔT; i.e., $F = (\alpha_1 E_1 A_1 + \alpha_2 E_2 A_2)\,\Delta T$, which agrees with equation (8.5).

8.3 MATERIALS IN SERIES

An interesting question arises if two different materials are placed end to end, restrained from expanding, and then heated. Obviously, if one expands, the other must contract. Which one will expand, and which will contract?

We show two different materials placed end to end and rigidly fixed between two solid walls in Figure 8.2.

If materials 1 and 2 were free to expand, the total expansion would be

$$\begin{aligned}
\delta &= \delta_1 + \delta_2 \\
&= (\alpha_1 \cdot l_1 \cdot \Delta T) + (\alpha_2 \cdot l_2 \cdot \Delta T) \\
&= \Delta T(\alpha_1 l_1 + \alpha_2 l_2)
\end{aligned} \tag{8.8}$$

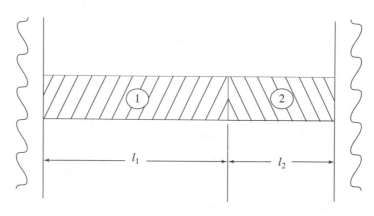

Figure 8.2 Two materials in series

If they are constrained so that $(l_1 + l_2)$ remains constant, then stresses will be set up in both materials. Again, we can determine these stresses by assuming that the bodies are free to expand or contract and then calculating the force necessary to return the combined body to its original length.

$$\text{strain} = \text{stress/elastic modulus}$$
$$\varepsilon = s/E$$

But strain = extension/original length

$$\varepsilon = \delta/l$$

so we can write

$$\delta/l = s/E$$

and

$$\text{stress} = \text{force/area}$$
$$s = F/A$$

So

$$\delta/l = F/AE$$

or

$$\delta = Fl/AE \qquad (8.9)$$

Rearranging this equation gives

$$F = \delta AE/l$$

This tells us the force needed to compress the body back to its original length (if it has expanded an amount δ) or to stretch it back (if it has contracted by an amount δ). F is, therefore, the force required to *prevent* the body from expanding or contracting. Applying this equation (8.9) to each material,

$$\delta_1 = F_1 l_1 / A_1 E_1$$
$$\delta_2 = F_2 l_2 / A_2 E_2$$

Therefore, the total expansion or contraction, δ, which is the sum of δ_1 and δ_2, is

$$\delta = (F_1 l_1 / A_1 E_1) + (F_2 l_2 / A_2 E_2)$$

But the same force must be acting on both pieces, i.e., $F_1 = F_2$. So,

$$\delta = F[l_1/(A_1 E_1) + l_2/(A_2 E_2)] \qquad (8.10)$$

Previously (equation (8.8)) we found that if the body were free to expand or contract, the amount that its length would change is

$$\delta = \Delta T(\alpha_1 l_1 + \alpha_2 l_2)$$

Substituting for δ from equation (8.8) into equation (8.10):

$$\Delta T(\alpha_1 l_1 + \alpha_2 l_2) = F[l_1/(A_1 E_1) + l_2/(A_2 E_2)]$$

or the force F may be expressed as

$$F = \Delta T(\alpha_1 l_1 + \alpha_2 l_2)/[l_1/(A_1 E_1) + l_2/(A_2 E_2)] \qquad (8.11)$$

8.3.1   Materials in Series: General Equations

Again, if we have N different materials, we can write an equation for the force in a compact form:

$$F = \frac{\Delta T \sum_{n=1}^{n=N} (\alpha_n l_n)}{\sum_{n=1}^{n=N} [l_n/(A_n E_n)]} \tag{8.12}$$

If all the pieces have the same cross-sectional area, they will have the same stress, because the same force is acting on each. Let the cross-sectional area be A; then

$$s_1 = s_2 = \cdots = s_n = F/A$$

The strains are, therefore,

$$\varepsilon_1 = s/E_1 = F/(AE_1), \qquad \varepsilon_2 = s/E_2 = F/(AE_2)$$

In general,

$$\varepsilon_n = F/(AE_n) \tag{8.13}$$

Suppose we have two bars of different materials set end to end between two solid walls. They have the same cross-sectional shape and size but are of different lengths. We then uniformly heat the bars. They will both try to expand, but since the total length of the two must remain constant, the only possibilities are (1) neither will expand, or (2) one will expand and the other will compress. Although (a) is possible, it would happen only in an extremely rare case, as we will discuss later.

Identifying the bars as 1 and 2, we have the following relationships:

$$F = \Delta T(\alpha_1 l_1 + \alpha_2 l_2)/[l_1/(A_1 E_1) + l_2/(A_2 E_2)] \tag{8.11}$$

$$s = F/A$$

$$\varepsilon_1 = s/E_1 = F/AE_1, \qquad \varepsilon_2 = s/E_2 = F/AE_2$$

Writing $\varepsilon = \delta/l$, we have $\delta_1/l_1 = F/(AE_1)$, and $\delta_2/l_2 = F/(AE_2)$.

If they had been free to expand, then from equation (8.1):

$$\delta = \alpha \cdot l \cdot \Delta T \qquad \text{or}$$

$$\delta_1/l_1 = \alpha_1 \cdot \Delta T \quad \text{and} \quad \delta_2/l_2 = \alpha_2 \cdot \Delta T$$

If the strain resulting from the combined restricting and heating, i.e., F/AE, is greater than this, then the bar has been compressed in addition to being prevented from expanding. On the other hand, if F/AE is less than $\alpha \cdot \Delta T$, the bar will expand at the expense of the other.

Let the amount of compression be δ_c. Then

$$\delta_{c1} = [F/(AE_1) - \alpha_1 \cdot \Delta T]l_1 \qquad \text{and} \tag{8.14(a)}$$

$$\delta_{c2} = [F/(AE_2) - \alpha_2 \cdot \Delta T]l_2 \tag{8.14(b)}$$

Of course, if δ_{c1} is positive—i.e., if bar 1 is compressed—then δ_{c2} will be negative, bar 2 will expand, and the sum of the two must be zero.

The only way the bars could both remain at their original lengths is if $\delta_c = 0$ for one of them. (δ_c would automatically be zero for the other one also.) So it would have to be that $F/AE = \alpha \cdot \Delta T$ for one of the bars. Realizing that F is given by equation (8.11), you can see that this is quite a complicated relationship.

EXAMPLE PROBLEM 8.4

A 2-in. × 2-in., square cast-iron class 40 bar and a 2-in. × 2-in. AISI 1020 steel bar are set end to end between two solid walls. The iron bar is 3 in. long, and the steel bar is 9 in. long. They are then heated uniformly from 60°F to 160°F.

$$\alpha \text{ for iron} = 6.3 \times 10^{-6} \text{ per °F, and } 6.5 \times 10^{-6} \text{ per °F for steel}$$

$$E \text{ for iron} = 19 \times 10^6 \text{ psi, and } 30 \times 10^6 \text{ psi for steel}$$

a. What force is exerted on the walls?
b. What is the stress in each piece?
c. By how much does each piece expand or contract?

Solution

Let the iron be 1 and the steel be 2.

a. Using equation (8.11) yields

$$F = \Delta T(\alpha_1 l_1 + \alpha_2 l_2)/[l_1/(A_1 E_1) + l_2/(A_2 E_2)]$$

$$\Delta T = (160°F - 60°F) = 100°F$$

$$\alpha_1 = (6.3 \times 10^{-6})°F^{-1}$$

$$l_1 = 3 \text{ in.}$$

$$\alpha_2 = (6.5 \times 10^{-6})°F^{-1}$$

$$l_2 = 9 \text{ in.}$$

$$A_1 = A_2 = 2 \text{ in.} \times 2 \text{ in.} = 4 \text{ in.}^2$$

$$E_1 = 19 \times 10^6 \text{ psi}$$

$$E_2 = 30 \times 10^6 \text{ psi}$$

Therefore,

$$F = (100°F)[(6.3 \times 10^{-6})°F^{-1} \times 3 \text{ in.} + (6.5 \times 10^{-6})°F^{-1}$$
$$\times 9 \text{ in.}]/[3 \text{ in.}/(4 \text{ in.}^2 \times 19 \times 10^6 \text{ psi}) + 9 \text{ in.}/(4 \text{ in.}^2 \times 30 \times 10^6 \text{ psi})]$$
$$= (100°F)(18.9 \times 10^{-6}°F^{-1} \text{ in.} + 58.5 \times 10^{-6}°F^{-1} \text{ in.})$$
$$\div (3.95 \times 10^{-8} \text{ in./lb} + 7.5 \times 10^{-8} \text{ in./lb})$$
$$= (100°F)(77.4 \times 10^{-6}°F^{-1} \text{ in.})/(11.45 \times 10^{-8} \text{ in./lb})$$

Therefore, $F = 67,600$ lb.

b. Stress = force/area = 67,600 lb/4 in.^{2} = 16,900 psi.

c. We calculate the contraction and expansion of each piece by applying equations (8.14(a)) and (8.14(b)).

$$\delta_{c1} = (F/AE_1 - \alpha_1 \cdot \Delta T) l_1$$

$$\delta_{c2} = (F/AE_2 - \alpha_2 \cdot \Delta T) l_2$$

$$\delta_{c1} = [67,600 \text{ lb}/(4 \text{ in.}^2 \times 19 \times 10^6 \text{ psi}) - (6.3 \times 10^{-6})°F^{-1} \times 100°F] \times 3 \text{ in.}$$

$$= (8.89 \times 10^{-4} - 6.3 \times 10^{-4}) \times 3 \text{ in.}$$

$$= +7.8 \times 10^{-4} \text{ in.} \quad \textbf{compression}$$

$$\delta_{c2} = [67,600 \text{ lb}/(4 \text{ in.}^2 \times 30 \times 10^6 \text{ psi}) - (6.5 \times 10^{-6})°F^{-1} \times 100°F] \times 9 \text{ in.}$$

$$= (5.63 \times 10^{-4} - 6.5 \times 10^{-4}) \times 9 \text{ in.}$$

$$= -7.8 \times 10^{-4} \text{ in.} \quad \textbf{expansion}$$

So the steel expands an amount 0.00078 in. and the iron contracts an equal amount, as it must.

Answer

The force exerted on the walls is 67,600 lb. The stress in both pieces is 16,900 psi. The iron compresses an amount 7.8×10^{-4} in. The steel expands an amount 7.8×10^{-4} in.

8.4   COMPUTER ANALYSES

One of the powerful and attractive characteristics of spreadsheet programs is their What if? capability: What if we change the area, the length, the temperature rise, etc.? Program 'Tstress', which stands for *thermal stress*, is an Excel program that demonstrates this What if? characteristic.

8.4.1   Program 'Tstress'

Program 'Tstress' may be used to determine the stresses, force, and length changes for constrained heated or cooled materials in series or the stresses and forces for constrained, heated or cooled materials in parallel. It also gives the strains (from which length changes can be calculated) for unrestrained heated or cooled members. Up to five materials may be combined from a selection of 16. Either US or SI units may be used. The following examples demonstrate its use.

EXAMPLE PROBLEM 8.5

A 2-ft-long, 3-in. square-cross-section bar of cast-iron class 40 is clamped between two walls in order to prevent it from expanding. Its temperature is raised from 50°F to 200°F. Calculate (a) the stress in the bar, (b) the force exerted on the walls, and (c) the expansion if the bar were free to expand.

Solution

1. Open program 'Tstress'.
2. Enter *either* S or P in cell C5, because there is only one member.
3. Enter US in cell C7.
4. Enter 1 in cell F5.
5. Note that cast-iron class 40 is number 5 in the selection list in column K.
6. Enter 5 in cell A10.
7. Enter 24 for the length in cell C10.
8. Enter 9 for the area in cell D10.
9. Enter 150 for the temperature rise in cell E10.

Answers

The following output is shown:

a. The stress is 18,300 psi.
b. The force is 165,000 lb.
c. The strain would be 0.00095 in./in. if the bar could expand, so $\delta = \varepsilon l = 0.00095$ in./in. $\times$ 24 in. = 0.023 in.

EXAMPLE PROBLEM 8.6

A 2-in.-diameter hole is drilled along the axis of the bar in Example Problem 8.5 and is filled with copper. It is then constrained between two walls and heated from 50°F to 200°F. Calculate (a) the stress in each material and (b) the force exerted on the walls.

Solution

1. The cross-sectional area of the copper is

$$A = \pi(2 \text{ in.})^2/4 = 3.142 \text{ in.}^2$$

2. The cross-sectional area of the iron is

$$A = (9 \text{ in.}^2 - 3.142 \text{ in.}^2) = 5.858 \text{ in.}^2$$

3. The material ID number for copper is 8.
4. Change cell F5 to 2.
5. Make sure cell C5 is P

Answers

The results are as follows.

a. The stress in the iron is 18,300 psi. Notice that this has not changed because it is independent of the area ($s = \alpha \cdot E \cdot \Delta T$). The stress in the copper is 24,840 psi.
b. The force on the iron is 107,400 lb, the force on the copper is 78,050 lb, and the force on the walls is 185,400 lb.

EXAMPLE PROBLEM 8.7

A 100-mm-long bar comprises a center core of 5-mm-diameter monel, surrounded by a 1.5-mm-thick layer of aluminum, inside a 10-mm-diameter bar of steel. The composite bar is prevented from expanding as its temperature is increased by 90°C. Calculate the stress in each piece and the force on the walls.

Solution

1. Enter P for a parallel member calculation.
2. Enter SI for the units.
3. Enter 3 in cell F5.
4. The material ID numbers are

    monel = 9,   aluminum = 1,   steel = 15

    Enter these numbers in cells A10, A11, and A12, and clear the remaining A13 and A14 cells.

5. We need to know the cross-sectional areas of each piece.

Material	Diameters (OD/ID)	Area
Monel	5 mm/0	19.635 mm^2
Aluminum	8 mm/5 mm	30.630 mm^2
Steel	10 mm/8 mm	28.275 mm^2

6. Enter the lengths, areas, and temperature rise.

Answer

The results are

$$\text{stress in the steel} = 218 \text{ MPa}$$
$$\text{stress in the aluminum} = 143 \text{ MPa}$$
$$\text{stress in the monel} = 227 \text{ MPa}$$
$$\text{force on the walls} = 15,000 \text{ N}$$

EXAMPLE PROBLEM 8.8

Three 1-in-diameter rods are placed end to end, prevented from expanding, and uniformly heated through 60°F. The rods are brass, 1 in. long; nylon 6/6, 1.5 in. long; and polyethylene, 2 in. long. By how much does each expand or compress?

Solution

Each piece has a cross-sectional area of 0.7854 in^2. Try to do this problem without help.

> **Answer**
> The results are
>
> | | |
> |---|---|
> | brass | 0.00057 in. expansion |
> | nylon 6/6 | 0.00259 in. expansion |
> | polyethylene | 0.00316 in. compression |

8.5   PROBLEMS FOR CHAPTER 8

Hand Calculations

1. A 2.5-m-long, 35-mm-diameter brass rod is subjected to an 80°C temperature rise. Calculate each of the following.
   a. Total elongation if the end is free to expand
   b. Stress if the ends are rigidly fixed
   c. Force if the ends are rigidly fixed

2. A concrete pavement is to be laid at a temperature of 60°F, but it is expected that the temperature could rise to 95°F during the summer. The pavement is made of concrete slabs, each 40 ft long.
   a. How wide a gap should be left between the slabs?
   b. What would be the stress in the slabs if the temperature rose to 115°F?

3. A steel measuring tape measures exactly at 15°C under a pull of 20 N. The tape is 1 mm thick and 8 mm wide. Using this tape, a building was measured to be 45.750 m long, but the temperature was 35°C and the pull on the tape during that measurement was 85 N. What was the correct length of the building?

4. A 3-ft-long, round steel pipe has an inside diameter of 4 in. and an outside diameter of 6 in. It is filled with concrete and then rigidly fixed between two walls. What are the stresses in the materials and the force on the walls for

   a. A 50°F temperature rise,
   b. A 200°F temperature rise.

5. A 5-mm-diameter rod inside a refrigerator consists of a 50-mm brass piece and a 30-mm polyethylene piece attached end to end. It is installed at 20°C with its ends rigidly fixed. During operation its temperature drops to −18°C.
   a. What force will be developed on the rod?
   b. Will it be tensile or compressive?
   c. By how much will each piece expand or compress?

6. A 10-in.-long aluminum rod is supported at one end. Its free end provides a gap of 0.05 in. at 50°F with the free end of a 7-in. bronze rod supported in a like manner and in line with it. Both rods are 1.5 in. in diameter.
   a. If both rods are heated to the same temperature, at what temperature will they expand to just touch each other?
   b. If they are heated in such a way that the aluminum rod is always 100°F greater than the bronze rod, what will be the temperature of the aluminum rod when they just touch?
   c. If they are both heated to 500°F, what will be the stress in each rod, and what will be the force on the walls?

Computer Calculations

7. Repeat Problem 4.
8. Repeat Problem 5.
9. Repeat Problem 6(c).

10. Use program 'Tstress' to solve Example Problem 8.1.
11. Use program 'Tstress' to solve Example Problem 8.2.

12. Use program 'Tstress' to solve Example Problem 8.3.

13. Use program 'Tstress' to solve Example Problem 8.4.

14. An aluminum plate and a bronze plate, both 4.0 in. wide and 0.25 in. thick, are fastened together face to face by two 0.5-in. rivets located 5.0 in. apart on the center line. The plates are then heated uniformly through 100°F. The allowable shear stress for the rivets is 17,500 psi. Will the unit fail?

15. If you found that the rivets failed in the previous problem, what temperature rise would be safe? If you found that the rivets did not fail, what is the maximum temperature that would be safe?

16. Five 1.0-in.-long, 0.50-in.-diameter pieces of metal are placed in series between two solid walls. The combined 5.0-in.-long piece just touches the walls when the temperature is 15°C. The pieces are made of aluminum, brass, bronze, monel, and steel. Do any of the pieces contract when they are heated uniformly? If so, which? Which one expands the most?

17. Write a computer program to calculate the thermal stresses in two materials in series that are prevented from expanding. Use the SI system of units. Write the program so that any two of the following five materials may be selected. α is the coefficient of linear expansion, and E is the modulus of elasticity.

Material	$\alpha = \delta/(l \cdot \Delta T)$ (mm/mm/°C)	E (MPa)
Brass	18.7×10^{-6}	103×10^3
Concrete	9.9×10^{-6}	20.7×10^3
Monel	14.0×10^{-6}	179×10^3
Steel	11.7×10^{-6}	207×10^3
Wrought iron	11.5×10^{-6}	193×10^3

18. Write a computer program to calculate the thermal stresses in two materials in parallel that are prevented from expanding. Use the SI system of units. Write the program so that any two of the following five materials may be selected. Use a macro to automatically enter the material and its properties by simply selecting the material number.

α is the coefficient of linear expansion, and E is the modulus of elasticity.

Material	$\alpha = \delta/(l \cdot \Delta T)$ (mm/mm/°C)	E (MPa)
Brass	18.7×10^{-6}	103×10^3
Concrete	9.9×10^{-6}	20.7×10^3
Monel	14.0×10^{-6}	179×10^3
Steel	11.7×10^{-6}	207×10^3
Wrought iron	11.5×10^{-6}	193×10^3

9

Connections

■ *Objectives*

In this chapter you will learn about different ways of joining materials and how to calculate the strengths and efficiencies of those joints.

9.1 TYPES OF CONNECTIONS

Pieces may be joined together either permanently or with the ability to be disassembled. Welding, brazing, soldering, adhesive bonding, riveting, and nailing are all examples of permanent connections. Bolted, screwed, and clamped connections may be taken apart fairly readily with little or no damage to the parts. In most cases a joint is not as strong as the parent material, so we must be able to determine its strength. The ratio of joint strength to the strength it would be if it were one continuous piece is called the **joint efficiency**. It is usually expressed as a percent.

In this chapter we learn how to calculate the strengths and efficiencies of the most commonly used connections in engineering: riveted, bolted, and welded joints.

9.2 WELDED CONNECTIONS

The parent organization for welders in the United States is the *American Welding Society (AWS)*. That organization defines a weld as "a localized coalescence of metals or nonmetals produced either by heating the materials to suitable temperatures, with or without the application of pressure, or by the application of pressure alone and with or without the use of filler material."

9.2.1 Types of Welds

All welding processes can be divided into three major categories: *fusion welding, soldering* and *brazing*, and *solid-state welding*.

Fusion welding, which is the most common type of welding, uses heat to melt the base metals (the metals being welded together) and the filler, if one is being used. In the molten state, the metals flow together and, as they solidify, form one continuous body. The major fusion welding processes are:

1. Oxygas welding, usually oxyacetylene (OAW),
2. Shielded-metal arc welding (SMAW),
3. Gas-metal arc welding (GMAW), also known as metal inert gas welding (MIG),

4. Gas-tungsten arc welding (GTAW), also known as tungsten inert gas welding (TIG),
5. Plasma arc welding (PAW),
6. Flux-cored arc welding (FCAW).

Oxyacetylene welding equipment is relatively inexpensive and easily portable, but it is slow and limited to thin base metals. The reason for its limited capability is its flame temperature, which is only 6000°F, compared with 10,000°F for arc welding processes. The filler rods used with this type of welding are designated RGxx, where *xx* is a two-digit number equal to the rod material's tensile strength in kpsi. The most commonly used are

RG45 with a tensile strength of 45–50 kpsi,
RG60 with a tensile strength of 50–65 kpsi,
RG65 with a tensile strength of 65–75 kpsi.

Shielded-metal arc welding electrodes are designated Exxyz or Exxxyz, where *xx* or *xxx* is a two- or three-digit number equal to its tensile strength in kpsi. This number can range from 60 to 140 kpsi. The digit *y* indicates the welding positions possible with the electrode, which can be 1, 2, or 4 (3 is not used). The *z* indicates the type of coating on the electrode and can be a number from 0 to 8. For example, E6010 is a high-cellulose-sodium coated electrode (indicated by the last 0) having a tensile strength of 60 kpsi (the first two digits) and suitable for welding in any position (indicated by the 1). We concern ourselves only with the strengths of welds, so the important part of the electrode designation is the E60, E70, . . . , E140.

Gas-metal arc welding filler wires have two types of designation. The AWS classification for low-carbon steel electrode wires is ERxxS-z, where the *R* indicates that the wire can be used as a filler rod; *xx* is the tensile strength in kpis; *S* stands for *solid core*; and *z* is a number indicating the chemical composition of the wire. Again we are concerned only with the xx. For example, the tensile strength of ER70S-5 is 70 kpsi. Wires classified for use on aluminum, magnesium, copper, nickel, titanium, and their alloys and also for stainless steels are designated according to alloy composition. For example, a wire suitable for welding stainless steel 301, 302, 304, or 308 is ER308, and one for monel 400 (which is a nickel- (Ni-) copper (Cu) alloy) is ERNiCu-7. The tensile strengths of these metals can be found in Excel program 'Proptab'.

Gas-tungsten arc welding fillers often contain deoxidizers to reduce the possibility of oxide contamination in the weld. Carbon-steel filler rods for oxyacetylene welding cannot be used for GTAW. As in all fusion welding, the filler material must be compatible with the base metal.

In **brazing and soldering**, the base metals are heated, but not to their melting points. A third metal is melted between them and forms a bond.

In **solid-state welding**, coalescence is produced at temperatures below the melting point of the base metals. No filler is used, but pressure is required for most processes.

9.2.2 Welded Joint Strength

In **butt welding**, the pieces to be joined are placed end to end, usually with a small gap between them (root opening). If the plates are 0.25 in. or less in thickness, the ends are left square and the plates are welded from either one side, forming a *single square-butt weld*, or from both sides, for a *double-square butt weld*. See Figure 9.1(a). Thicker pieces require some kind of preparation to provide a space for the welding tip. Depending on the thickness, this can be one of many configurations. There are bevel-groove, J-groove, V-groove, and U-groove welds in both single and double types. We show these in Figure 9.1(b)–(e).

Joint efficiency is defined as the ratio of joint strength to plate tensile strength (the weaker one if they are different) expressed as a percent.

For complete penetration of the weld through the material, a butt weld is as strong as the base material in tension, compression, and bending, i.e., *a complete penetration*

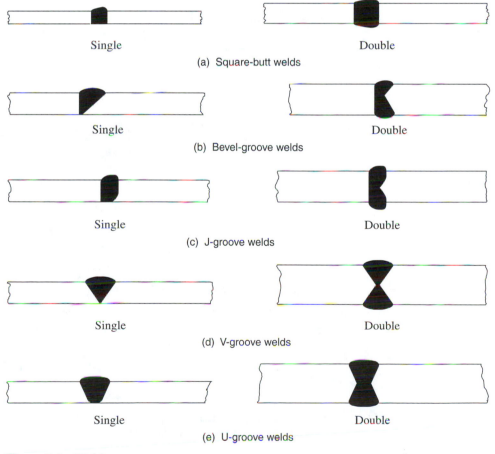

Figure 9.1   Welds.

butt weld has an efficiency of 100% in all but shear. The shear strength is the same as for fillet welds, which we consider next.

We show two types of **fillet welds** in Figure 9.2(a) and (b). They are *side fillet welds with corner return* and *end fillet welds with corner return*. Side or end fillet welds should be continued around the corners for a distance *not less than twice the nominal size of the weld*. For example, a 0.5-in. weld should continue around both corners a distance of at least 1 in. The size of a fillet weld is designated by the length of its *leg*, as shown in Figure 9.2(c). We assume that both legs are of equal length, so that the angle is 45°. Tests have shown that fillet welds usually fail by shear through the throat, regardless of the direction of the applied load.

Note: The maximum size of a fillet weld applied to the square edge of a plate should be $\frac{1}{16}$ in. less than the plate thickness for plate thicknesses of $\frac{1}{4}$ in. or greater and equal to the plate thickness for plate thicknesses less than $\frac{1}{4}$ in.

Because force = stress × area, we need to know the allowable stress and the cross-sectional area through the throat. According to the AISC (American Institute of Steel Construction) code, the allowable stress is equal to 30% of the electrode tensile strength. We assume also that in the case of oxyacetylene welding, it is equal to 30% of the welding rod's tensile strength. The cross-sectional area through the throat is equal to the throat width times the weld length. Referring to Figure 9.2(d) we see that the throat width is equal to (leg × sin 45°). If we call the weld length l, then the area is (l × leg × sin 45°):

$$A = l \times \text{leg} \times \sin 45° = (l)(\text{leg})(0.707)$$

The allowable stress is $s = 0.30 \times$ (electrode tensile stress), or

$$s = 0.30 \ (E\#)$$

where E# stands for the electrode's two- or three-digit number. So, the force is $F = sA = [0.30 \ (E\#)] \ [(l) \ (\text{leg}) \ (0.707)]$, or

$$F = 0.212 \ (E\#)(l)(\text{leg}) \tag{9.1}$$

The E# is the electrode tensile strength in kpsi. If the weld length, l, is in inches and the leg is in inches, then the force will be in kilopounds (thousands of pounds). Thus, in pounds,

$$F \ (\text{lb}) = 212 \ (E\# \ \text{kpsi})(l \ \text{in.})(\text{leg in.}) \tag{9.1(a)}$$

In SI units, the weld length and the leg are in mm, so we must convert the E# to MPa to get the force in newtons. We know 1 kpsi × 6.895 = 1 MPa, so equation (9.1) becomes

$$F = 0.212[(E\# \ \text{kpsi} \times 6.895) \ \text{MPa}](l \ \text{mm})(\text{leg mm})$$
$$F \ (\text{N}) = 1.462 \ (E\# \ \text{kpsi})(l \ \text{mm})(\text{leg mm}) \tag{9.1(b)}$$

Note: Observe that the electrode designation is still in kpsi in this equation, although the weld length and size are in mm.

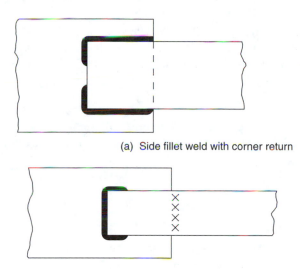

(a) Side fillet weld with corner return

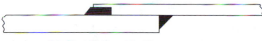

(b) End fillet weld with corner return

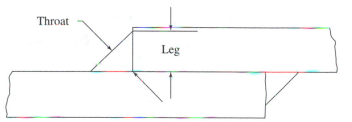

(c) Leg height and throat width of fillet weld

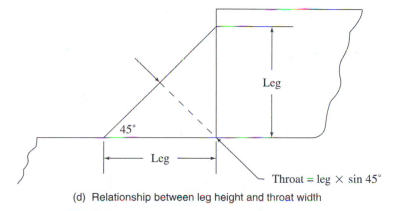

(d) Relationship between leg height and throat width

Figure 9.2

EXAMPLE PROBLEM 9.1

The A36 plates shown in Figure 9.2(b) are 5 in. wide by 0.375 in. thick and 4 in. wide by 0.25 in. thick. A 0.19-in. weld is made on both ends, with corner returns. E70 electrodes were used in the welding process. There is a 25,000-lb tensile force on the plates. Is it a safe design? If not, why?

Solution

The total weld length, l, is 4 in. + 4 in. + 4(corner returns). One corner return is twice the weld size; i.e., 2(0.19 in.) = 0.38 in. Therefore, l = 8 in. + 4(0.38 in.),

or
$$l = 9.52 \text{ in.}$$
$$E\# = 70 \text{ kpsi}$$
$$\text{leg} = 0.19 \text{ in.}$$

Using equation (9.1(a)):
$$F = 212(E\# \text{ kpsi})(l \text{ in.})(\text{leg in.})$$
$$= 212(70 \text{ klb})(9.52 \text{ in.})(0.19 \text{ in.}) = 26,840 \text{ lb}$$

So it would appear to be safe. *But, is the thinner plate strong enough in tension?*

According to the AISC code, *the allowable tensile stress in the plate is* 60% *of the material yield strength.* The yield strength of A36 steel is 36,000 psi, so the allowable stress is 0.60 (36,000 psi) = 21,600 psi. The cross-sectional area of the thinner plate is (4 in.)(0.25 in.) = 1 in^2.

$$F = sA = (21,600 \text{ psi})(1 \text{ in.}^2) = 21,600 \text{ lb}$$

Therefore, the conclusion is that although the joint is strong enough, the *thinner plate will fail in tension.*

It is important always to check the plate strength as well as the weld strength.

Answer

The weld is acceptable, but the plate will fail in tension.

EXAMPLE PROBLEM 9.2

The plates shown in Figure 9.2(a) are both 10-mm-thick A441 steel. One is 10 cm wide and the other is 8 cm wide. The total weld length is 12 cm, and the weld size is 8 mm. E70 electrodes were used in the welding process. What is the joint efficiency? The yield strength of A441 steel is 317 MPa.

Solution

The total weld length is 12 cm = 120 mm, so
$$l = 120 \text{ mm}$$
$$E\# = 70 \text{ kpsi}$$
$$\text{leg} = 8 \text{ mm}$$

Using equation (9.1 (b)):

$$F = 1.462 \ (E\# \ klb)(l \ mm)(leg \ mm)$$
$$= 1.462(70 \ klb)(120 \ mm)(8 \ mm) = 98{,}200 \ N = 98.2 \ kN$$

The smaller plate's cross-sectional area is (80 mm) (10 mm) = 800 mm². The allowable plate stress is 0.60 (317 MPa) = 190 MPa.

$$F = sA = (190 \ MPa)(800 \ mm^2) = 15{,}220 \ N = 152 \ kN$$

The joint efficiency is the joint strength divided by the plate tensile strength (the thinner, if they are of different sizes), expressed as a percent:

$$\eta = (98.2 \ kN/152 \ kN) \times 100\% = 64.6\%$$

Answer
The joint efficiency is 64.6%.

9.3 RIVETED AND BOLTED CONNECTIONS

Figure 9.3(a) shows two plates riveted or bolted together to form a *lap joint*. There is one shear plane per rivet (or bolt), so this is called *single shear*. As we discussed in Chapter 6, some configurations result in more than one shear plane, such as the butt joint in Figure 9.3(b).

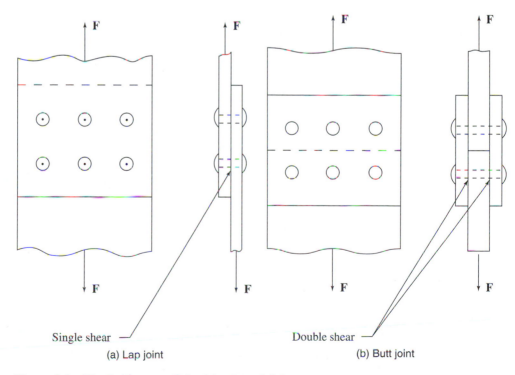

Single shear

Double shear

(a) Lap joint

(b) Butt joint

Figure 9.3   Single shear and double shear joints.

For lap joints, the number of *rows* of rivets or bolts determines the type of joint, regardless of the number of columns. For example, both configurations on the left in Figure 9.4(a) are *single-bolted* connections, and both on the right are *double-bolted*.

Butt joints are made by *butting* the two plates end to end, placing a cover plate on the top and another on the bottom, and then fastening them together with rivets or bolts that pass right through all three plates. The type of joint, single-bolted, double-bolted, etc., depends on the number of rows of bolts *on one side of the joint*—i.e., one-half of the total number of rows. Both configurations on the left in Figure 9.3(b) are single-bolted, and both on the right are double-bolted. The reason for this is due to the way we analyze the joint strength, which we show in Figure 9.5. Both sides of the connection are being subjected to the same force, and with regard to one side, it is exactly the same as a double-shear lap joint, with the force being shared equally between the two cover plates. The configuration on the right in Figure 9.5 has exactly the same strength as the one on the left.

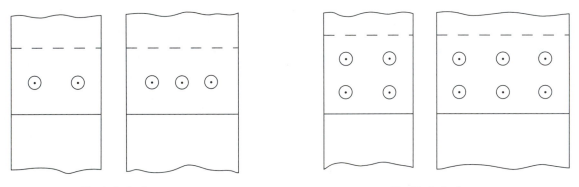

Single-bolted   Double-bolted

(a) Examples of single-bolted and double-bolted lap joints

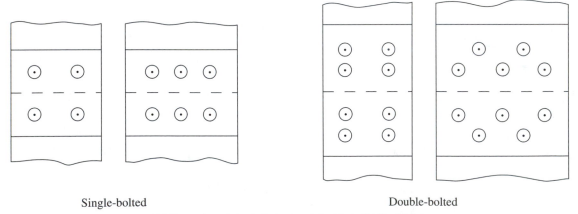

Single-bolted   Double-bolted

(b) Examples of single-bolted and double-bolted butt joints

Figure 9.4

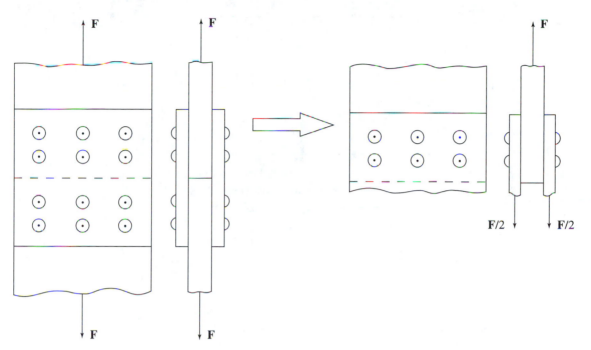

Figure 9.5   Analysis of butt joints.

9.3.1   Possible Modes of Failure

Generally there are three ways in which a riveted or bolted connection can fail.

1. Shear of the rivets or bolts. Figure 9.6(a) and (b) show examples of single- and double-shear failure.

2. Failure of the plates in tension. Figure 9.7(a) shows plate failure due to tension, and Figure 9.7(b) shows the area under stress.

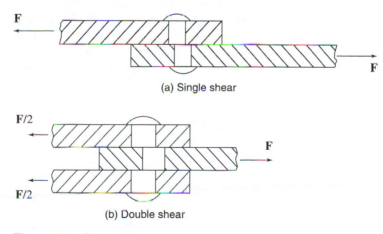

Figure 9.6   Shear failure

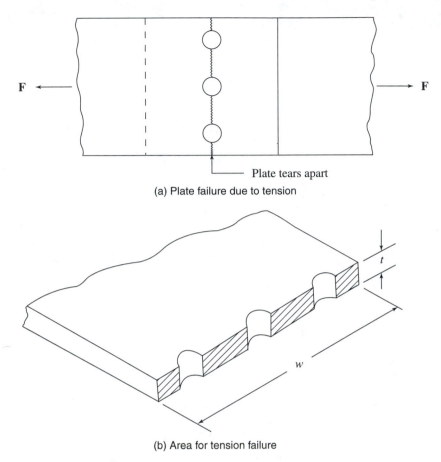

(a) Plate failure due to tension

(b) Area for tension failure

Figure 9.7   Tension failure.

3. Failure of the plates in bearing. This can occur only if the connected plates can move relative to each other. This failure is the local crushing of the plate in the region of the rivets or bolts. See Figure 9.8. Rivets and low-carbon-steel bolts cannot be made tight enough to prevent the possibility of slippage between the plates. These result in what are called *bearing-type* connections. High-strength bolts, if tightened to 70% of their tensile strength, can prevent slippage, resulting in *friction-type* connections.

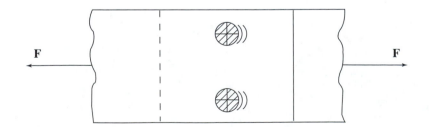

Figure 9.8   Failure due to bearing.

Shear Failure

As we discussed previously, shear may be single or multiple. Generally, lap joints are subjected to single-shear and butt joints, to double-shear. The allowable force is equal to the allowable shear stress times the total cross-sectional area of the rivets or bolts in the shear planes.

1. For a lap joint with single shear, the shear area is

$$A_s = 0.7854(d^2)(N) \qquad (9.2(a))$$

where d is the rivet or bolt diameter and N is the total number of rivets or bolts. The constant, 0.7854, is $\pi/4$.

Note: For bolts, if the threads lie in the shear plane, d is the *root* diameter.

2. For a lap joint with double shear, there are two shear areas per bolt, so the shear area is

$$A_s = 0.7854(d^2)(2N) = 1.5708(d^2)(N) \qquad (9.2(b))$$

3. For a butt joint, the shear area is the total of the shear areas on one side of the joint, and each bolt is in double shear, so

$$A_s = 0.7854(d^2)(2n) = 1.5708(d^2)(n) \qquad (9.2(c))$$

where n is the number of rivets or bolts *on one side of the connection*; i.e., $n = N/2$.

The following allowable shear stresses are specified by the AISC and by the Aluminum Association:

For rivets (steel):

$$s_s = 17.5 \text{ kpsi} \quad (121 \text{ MPa}) \qquad \text{for A502-1}$$
$$s_s = 22 \text{ kpsi} \quad (152 \text{ MPa}) \qquad \text{for A502-2}$$

For rivets (aluminum alloy):

$$s_s = 4 \text{ kpsi} \quad (27 \text{ MPa}) \qquad \text{for 1100-H14}$$
$$s_s = 14.5 \text{ kpsi} \quad (100 \text{ MPa}) \qquad \text{for 2017-T4}$$
$$s_s = 8.5 \text{ kpsi} \quad (58 \text{ MPa}) \qquad \text{for 6053-T61}$$
$$s_s = 11 \text{ kpsi} \quad (76 \text{ MPa}) \qquad \text{for 6061-T6}$$

For bolts (steel):

$$s_s = 10 \text{ kpsi} \quad (69 \text{ MPa}) \qquad \text{for A307 low-carbon bolts}$$
$$s_s = 17.5 \text{ kpsi} \quad (121 \text{ MPa}), \qquad \text{friction-type connection for A325}$$
$$s_s = 30 \text{ kpsi} \quad (207 \text{ MPa}), \qquad \text{bearing-type connection for A325}$$
$$s_s = 22 \text{ kpsi} \quad (152 \text{ MPa}), \qquad \text{friction-type connection for A490}$$
$$s_s = 40 \text{ kpsi} \quad (276 \text{ MPa}), \qquad \text{bearing-type connection for A490}$$

For steel bolts not listed, *use 22% of the ultimate tensile strength*.

For bolts (aluminum alloy):

$$s_s = 16 \text{ kpsi} \quad (110 \text{ MPa}) \qquad \text{for 2024-T4}$$
$$s_s = 12 \text{ kpsi} \quad (83 \text{ MPa}) \qquad \text{for 6061-T6}$$
$$s_s = 17 \text{ kpsi} \quad (117 \text{ MPa}) \qquad \text{for 7075-T73}$$

The allowable shear force is

$$F_s = A_s s_s \tag{9.3}$$

Tension Failure

The area for tension failure is the sum of the shaded areas in Figure 9.7(b). It consists of the cross-sectional area of the plate minus the areas taken out by the holes. We usually allow a clearance of 0.0625 in., or 1.5 mm, for the holes, and an additional 0.0625 in., or 1.5 mm, for possible damage to the rim of the holes caused by stamping or drilling. For calculation purposes, then, we assume that the hole diameters are (d in. + 0.125 in.) or (d mm + 3 mm). If there are n_r bolts or rivets *in one row*, each of diameter d, and the plate thickness is t, then the tension area will be reduced by $[n_r(d + .125)t]$ in.^{2} or by $[n_r(d + 3) t]$ mm.^{2} If there were no rivets or bolts, the plate cross-sectional area would be (wt), where w is the plate width. The tension area is, therefore,

$$A_t = (wt) \text{ in.}^2 - [n_r(d + 0.125)t] \text{ in.}^2 \tag{9.4(a)}$$
or
$$A_t = (wt) \text{ mm}^2 - [n_r(d + 3)t] \text{ mm}^2 \tag{9.4(b)}$$

According to the AISC code, the allowable tensile stress is equal to 60% of the yield stress.

$$s_t = 0.60 s_y \tag{9.5}$$

The allowable tension force is

$$F_t = A_t s_t \tag{9.6}$$

Bearing Failure

If the joint is *not* a friction type, it may fail by the rivets or bolts crushing the plates, as we show in Figure 9.8. Again, the allowable force equals the allowable stress times the area, but what area? The magnitude of the bearing stress varies around the rivets (or bolts), but it is customary to take the bearing area for each as a rectangle of width d and height t. So the total bearing area is

For lap joints:

$$A_b = (d)(t)(N) \tag{9.7(a)}$$

For butt joints:

$$A_b = (d)(t)(n) \tag{9.7(b)}$$

The AISC specification (for steel) gives the allowable bearing stress on the projected area as 120% of the plate material's ultimate stress.

$$s_b = 1.20 s_u \tag{9.8}$$

For aluminum alloys, the allowable bearing stresses are

$$s_b = 12.5 \text{ kpsi} \quad (86 \text{ MPa}) \qquad \text{for 1100-H14}$$
$$s_b = 49 \text{ kpsi} \quad (338 \text{ MPa}) \qquad \text{for 2014-T6}$$
$$s_b = 15 \text{ kpsi} \quad (103 \text{ MPa}) \qquad \text{for 3003-H14}$$
$$s_b = 34 \text{ kpsi} \quad (234 \text{ MPa}) \qquad \text{for 6061-T6}$$
$$s_b = 24 \text{ kpsi} \quad (165 \text{ MPa}) \qquad \text{for 6063-T6}$$

The allowable bearing force is

$$F_b = A_b s_b \tag{9.9}$$

9.3.2   Joint Strength and Efficiency

The joint strength is the smallest of F_s, F_t, and F_b. The joint efficiency is the ratio of the joint strength to the tensile strength of the solid plate, expressed as a percent. The tensile strength of the solid plate is

$$F_p = wts_t \tag{9.10}$$

so the joint efficiency is

$$\eta = 100 \times (\text{the smallest of } F_s, F_t, F_b)/F_p \tag{9.11}$$

EXAMPLE PROBLEM 9.3

Calculate the strength and the efficiency of the lap joint shown in Figure 9.9. The 10-mm rivets are made of A502-2 steel and the plates are made of A441 structural steel. The plates are 1.5 cm thick and 7 cm wide.

Solution

1. Allowable shear force.

This connection is a single-shear lap joint, so we will use equation (9.2 (a)) for the total shear area:

$$d = 10 \text{ mm}$$
$$N = 6$$

so $\qquad A_s = 0.7854(d^2)(N) = 0.7854(10 \text{ mm})^2(6) = 471 \text{ mm}^2$

The allowable shear stress for A502-2 rivets is $s_s = 152$ MPa. Therefore, $F_s = (471 \text{ mm}^2)$ (152 MPa) = 71,592 N = 71.6 kN.

2. Allowable tensile force.

$$w = 70 \text{ mm}$$
$$t = 15 \text{ mm}$$
$$d = 10 \text{ mm}$$

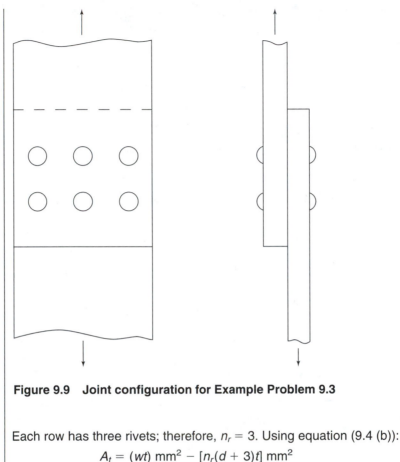

Figure 9.9   Joint configuration for Example Problem 9.3

Each row has three rivets; therefore, $n_r = 3$. Using equation (9.4 (b)):

$$A_t = (wt) \text{ mm}^2 - [n_r(d + 3)t] \text{ mm}^2$$
$$= (70 \text{ mm} \times 15 \text{ mm}) - [3(10 \text{ mm} + 3)15 \text{ mm}]$$
$$= 1050 \text{ mm}^2 - 585 \text{ mm}^2 = 465 \text{ mm}^2$$

The yield strength of A441 steel is 317 Mpa, and its ultimate strength is 463 Mpa. Using equation (9.5), $s_t = 0.60s_y = 0.60 \times 317$ MPa = 190 MPa. Therefore, $F_t =$ (465 mm^2)(190 MPa) = 88.4 kN.

   3. Allowable bearing force.

   The bearing area is, from equation (9.7 (a)),

$$A_b = (d)(t)(N) = (10 \text{ mm})(15 \text{ mm})(6) = 900 \text{ mm}^2$$

The allowable bearing stress, from equation (9.8), is

$$s_b = 1.20s_u = 1.20(463 \text{ MPa}) = 556 \text{ MPa}$$

Therefore, $F_b = (900 \text{ mm}^2)(556 \text{ MPa}) = 500,400 \text{ N} = 500 \text{ kN}$.

   4. Summary.

$$F_s = 71.6 \text{ kN}, \qquad F_t = 88.4 \text{ kN}, \qquad F_b = 500 \text{ kN}$$

The joint strength is the smallest of these, which is 71.6 kN. The solid-plate strength, from equation (9.10) is

$$F_p = wts_t = (70 \text{ mm})(15 \text{ mm})(190 \text{ MPa}) = 199.5 \text{ kN}$$

The joint efficiency, from equation (9.11) is

$$\eta = 100 \times (\text{the smallest of } F_s, F_t, F_b)/F_p$$

$$= (71.6 \text{ kN}/199.5 \text{ kN}) \times 100\% = 35.9\%$$

Answer

The joint strength is 72 kN, and the joint efficiency is 36%.

EXAMPLE PROBLEM 9.4

Calculate the strength and the efficiency of the butt joint shown in Figure 9.10. The UNC $\frac{7}{8}$-in. bolts are made of ASTM A490 high-strength steel, and the plates are made of A441 structural steel. The main plates are $\frac{3}{4}$ in. thick and 8 in. wide, and the cover plates are $\frac{1}{2}$ in. thick by 8 in. wide. It is a friction-type connection, with the bolt threads occurring in the shear planes.

Note: From standard UNC bolt-size tables, the root diameter of a $\frac{7}{8}$-in. bolt is 0.766 in.

Solution

1. Because this is a friction-type connection, we need to determine only the allowable shear and tensile forces because it can not fail in bearing.

2. The total thickness of the two cover plates is 1 in., which is greater than the main plate thickness (as it should be for good design), so there is no need to calculate the allowable tensile force on the cover plates. They will be stronger than the main plates.

3. To find allowable shear force, remember, we only use one side of the joint. There are 9 bolts per side, so $n = 9$.

The threads are in the shear planes, so we must use the bolt root diameter, which is 0.766 in. Using equation (9.2(c)),

$$A_s = 0.7854(d^2)(2n) = 0.7854(0.766 \text{ in.})^2 \, (2 \times 9)$$

$$= 8.295 \text{ in.}^2$$

The allowable shear stress for A490 bolts, being used in a friction-type connection, is 22 kpsi, so the allowable shear force is $(8.295 \text{ in.}^2)(22 \text{ kpsi}) = 182.5$ klb; i.e., $F_s = 182.5$ klb.

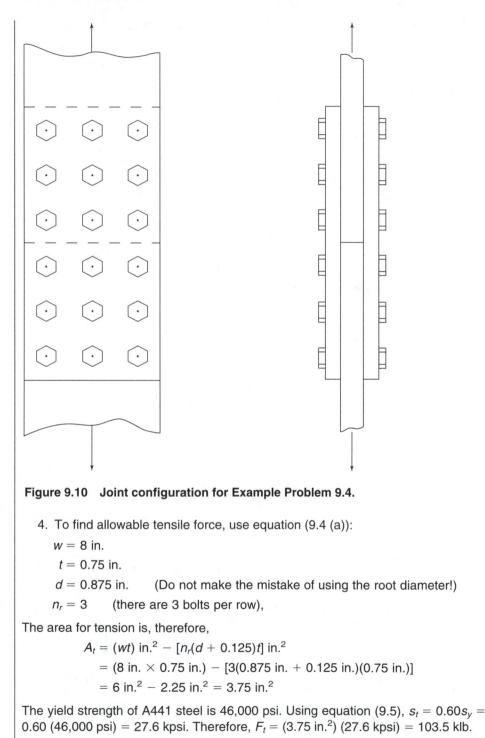

Figure 9.10   Joint configuration for Example Problem 9.4.

4. To find allowable tensile force, use equation (9.4 (a)):

   $w = 8$ in.

   $t = 0.75$ in.

   $d = 0.875$ in.   (Do not make the mistake of using the root diameter!)

   $n_r = 3$ (there are 3 bolts per row),

The area for tension is, therefore,

$$A_t = (wt) \text{ in.}^2 - [n_r(d + 0.125)t] \text{ in.}^2$$
$$= (8 \text{ in.} \times 0.75 \text{ in.}) - [3(0.875 \text{ in.} + 0.125 \text{ in.})(0.75 \text{ in.})]$$
$$= 6 \text{ in.}^2 - 2.25 \text{ in.}^2 = 3.75 \text{ in.}^2$$

The yield strength of A441 steel is 46,000 psi. Using equation (9.5), $s_t = 0.60s_y = 0.60$ (46,000 psi) = 27.6 kpsi. Therefore, $F_t = (3.75 \text{ in.}^2)$ (27.6 kpsi) = 103.5 klb.

5. In summary,

$$F_s = 182.5 \text{ klb}, \qquad F_t = 103.5 \text{ klb}$$

So the joint strength is 103.5 klb. The solid-plate strength is $F_p = wts_t = $ (8 in.) (0.75 in.) (27.6 kpsi).

i.e.

$$F_p = 165.6 \text{ klb, so the joint efficiency is}$$

$$\eta = (103.5 \text{ klb/165.6 klb}) \times 100\% = 62.5\%$$

Answer

The joint strength is 103,500 lb, and the joint efficiency is 62.5%.

9.3.3   Eccentrically Loaded Riveted and Bolted Connections

In the previous sections we assumed (but did not state) that the force direction passed through the center of the rivet or bolt pattern. This resulted in all the rivets or bolts sharing the load equally. If the line of action of the force—i.e., the force direction—does not pass through the center of the pattern, there will be a moment that will increase or decrease the shear forces. Referring to Figure 9.11, we see that the combined shear forces on the rivets or bolts due to the direct shear and the moment act in different directions to counteract both the "pulling" and the "twisting."

We revisit this in Chapter 18, but we mention it here because the Excel program 'Connect,' to be introduced in Section 9.4, can take account of eccentric loadings.

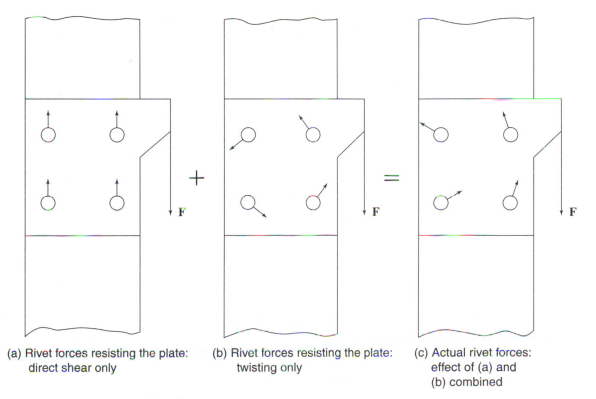

(a) Rivet forces resisting the plate: direct shear only

(b) Rivet forces resisting the plate: twisting only

(c) Actual rivet forces: effect of (a) and (b) combined

Figure 9.11   Eccentrically loaded connection.

9.4 COMPUTER PROGRAMS FOR CONNECTIONS

Spreadsheet programs are ideal for these types of calculations because you can change parameters and see what happens. For example, if you find that your design is too weak in shear, simply increase the rivet or bolt diameter and the new allowable forces will be calculated immediately.

9.4.1 Program 'Connect'

This program has the capability of analyzing welded connections, bolted and riveted connections, and bolted and riveted connections with eccentric loadings. When a joint is subjected to an eccentric load, each rivet or bolt is acted upon by different

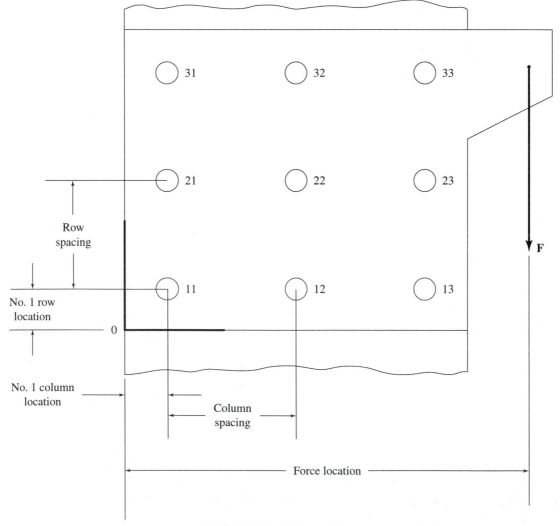

Eccentrically loaded connection.

Figure 9.12   Eccentrically loaded connection.

forces. It is necessary, therefore, to identify the location of each connector and the line of action of the load with reference to some origin. We select the bottom left corner of the connected plate as this origin. See Figure 9.12. Defining the x-location of the first (left-hand) column, the column spacing, the y-location of the first row (bottom row), and the row spacing defines the positions of all the connectors. The x-location of the load completes the additional inputs required for eccentric loadings. For identification, we label each connector with a subscript xy, where x refers to the row and y refers to the column. For example, r_{23} is the rivet in the second row of the third column. We include the following examples to demonstrate the use and flexibility of the program 'Connect.'

EXAMPLE PROBLEM 9.5

The lap joint shown in Figure 9.13 consists of two $\frac{3}{16}$-in.-thick by 7-in.-wide A242 steel plates connected by two rows of three rivets. The $\frac{7}{16}$-in.-diameter rivets are made of A502 grade 1 steel.

    a. Calculate the joint strength and efficiency.

    b. Change the rivets to $\frac{1}{2}$-in. A490 bolts, bearing-type connection, having the threads in the shear plane.

    c. Repeat with the threads not in the shear plane.

    d. Repeat (c) with the bolts tightened down to form a friction-type connection.

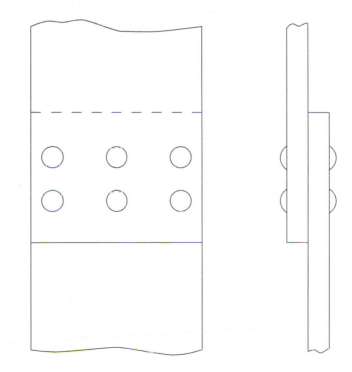

Figure 9.13   Connection for Example Problem 9.5.

Solution

a. Open program 'Connect'.

> The units are US (cell C4).
> It is a lap joint: L (cell C8).
> It is not a friction type connection: N (cell C10).
> The plate width is 7 (cell D11).
> The plate thickness is 0.1875 (cell D12).
> The rivet diameter is 0.4375 (cell D13).
> There are no threads in the shear plane N (cell C15).
> The number of rows is 2 (cell D17).
> The number of columns is 3 (cell D18).
> Eccentric loading? N (cell D20).

Go to line 28:

> "Highlight the plate material in column O; then press Ctrl-p."
> Go to column O and highlight 2; then, holding down the Ctrl key, press p.
> "Highlight the bolt or rivet material in column U; then press Ctrl-b."
> Go to column U and highlight 1; then holding down the Ctrl key, press b.

The results (under columns G to J) show the allowable maximum forces, the least being 15,785 lb due to rivet shear stress. The joint efficiency is 40.09%.

Answers for part (a)

The maximum acceptable shear force is 15,800 lb, maximum acceptable tensile force is 29,900 lb, maximum acceptable bearing force is 33,200 lb, and joint efficiency is 40%.

b. The only entries we have to change are the bolt diameter, the bolt ID number, and Y for threads in the shear plane.

> Bolt diameter: 0.5
> Bolt ID#: 8

Answers for part (b)

The joint strength is 23,500 lb and the joint efficiency is 65%. Plate tensile strength is the limiting factor.

c. The only change this time is that the threads are not in the shear plane.

> Are the threads in the shear plane? N

Answers for part (c)

The joint strength is 26,500 lb and joint efficiency is 73%. As before, the joint strength is determined by the plate strength.

d. Change to a friction-type joint.

> Is this a friction-type connection? Y

Answers for part (d)

The joint strength is 25,900 lb, determined by bolt shear strength. The joint efficiency is 72%.

EXAMPLE PROBLEM 9.6

Figure 9.14 shows a double-bolted butt joint connecting two 6061-T6 aluminum alloy plates with ten 8-mm-diameter, 6061-T6 bolts on each side of the joint. The plates are 120 mm wide and 10 mm thick, and the cover plates are each 3.97 mm thick. The bolt threads are not in the shear planes, and the bolts do not have sufficient strength to form a friction-type connection. The load, which is not eccentric, is 64 kN.

    a. Is this an acceptable design?

    b. Would it be better if the 8-mm bolts were replaced with 10-mm bolts of the same material?

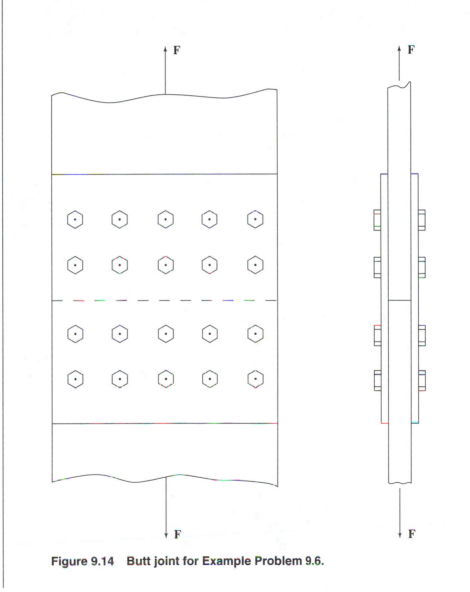

Figure 9.14   Butt joint for Example Problem 9.6.

Solution

The first thing to notice is that this is not a very good design because the thickness of the two cover plates together is less than the base plate thickness. If the joint were to fail in tension, it would be in the cover plates.

a. Open program 'Connect' and enter the data.

  The units are SI: SI.
  It is a butt joint: B.
  It is not a friction-type connection: N.
  Plate width: 120.
  Plate thickness: 7.94 (two times the cover-plate thickness).

Note: We are using the cover-plate thicknesses rather than the main-plate thickness, because their combined thickness is less than that of the main plate.

  Bolt diameter is 8 mm: 8
  The threads are not in the shear plane: N
  Number of columns: 5
  Is the pattern eccentrically loaded? N

Skip down to line 28:

  The plate material (6061-T6) ID# is 27 in the selection column O.
  The bolt material ID# is 10 in the selection column U.

Answer

The results show that the joint strength is 81,668 N, which is greater than required.

b. Change the bolt diameter to 10 mm. The joint strength is now 69,104 N, which is acceptable but less than that obtained with 8-mm bolts. *Why?*

EXAMPLE PROBLEM 9.7

Design a fillet weld for the lap joint shown in Figure 9.15. One plate is 10 in. wide by $\frac{3}{4}$ in. thick, and the other is 7 in. wide by $\frac{5}{16}$ in. thick. They are both made of A36 structural steel, and E60 electrodes are to be used. The load on the joint is 19,000 lb.

Solution

Open the program 'Connect.'

  Enter US in cell C4.
  Move right on the screen to column AJ.
  Enter one plate width, 10 (cell AN3).
  Enter its thickness, 0.75 (cell AN4).
  Enter the other plate width, 7 (cell AN5).
  Enter its thickness, 0.3125 (cell AN6).
  Enter the weld size.

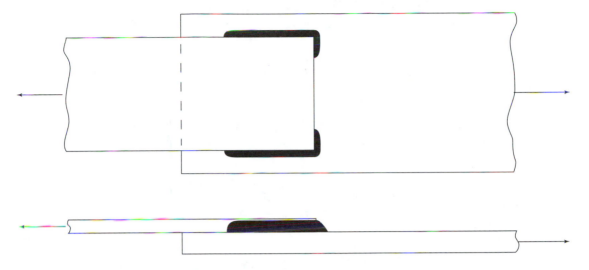

Figure 9.15   Welded lap joint for Example Problem 9.7.

Note: Weld size is based on the *thinner plate* and should be equal to the plate thickness for thicknesses less than $\frac{1}{4}$ in. and $\frac{1}{16}$ in. less than the thickness for $\frac{1}{4}$-in. and larger plates. So we should use a $(\frac{5}{16}\text{-in.} - \frac{1}{16}\text{-in.}) = \frac{1}{4}\text{-in.}$ weld.

> Enter 0.25
> Enter the electrode E#, 60.
> Enter the total weld length. This is what we are trying to determine, so try 4.
> Highlight the plate material ID#, 1, in selection column O and press Ctrl-w.

The results show a joint strength of 12,720 lb, which is too low. Change the weld length to 5 in. Again, the joint strength of 15,900 lb is too low. Change the weld length to 6 in. The joint strength is now acceptable at 19,080 lb. The efficiency is 40.4%. Corner returns should be at least twice the weld size, i.e., $\frac{1}{2}$ in. Therefore, the weld distribution should be 2.5 in., with 0.5-in. Corner return on each side.

Answer
The required weld has a $\frac{1}{4}$-in. leg, 2 in. long on each side, with $\frac{1}{2}$-in. returns.

EXAMPLE PROBLEM 9.8

Calculate the strength and efficiency of the eccentrically loaded joint shown in Figure 9.16. The plate material is A588 structural steel, and the $\frac{3}{16}$-in. rivets are made of A502 class 2 steel. Determine which rivet(s) limit the joint strength.

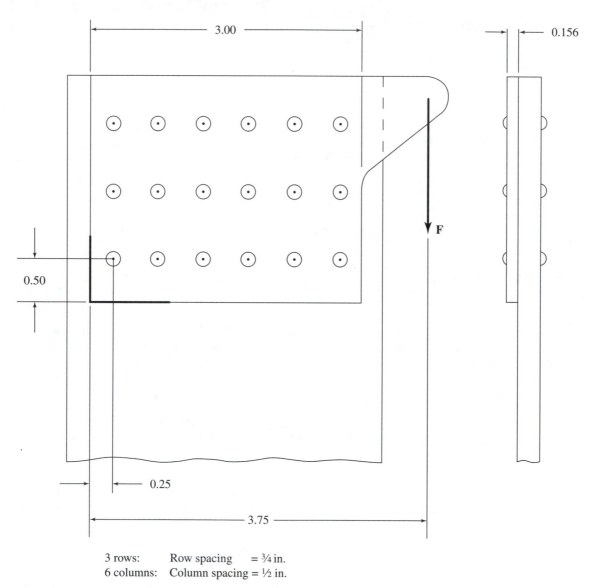

3 rows: Row spacing   = ¾ in.
6 columns: Column spacing = ½ in.

Figure 9.16   Eccentrically loaded joint for Example Problem 9.8.

Solution

Figure 9.17 shows the input/output sheet for this problem. You will notice that, because the loading is eccentric, we have to enter additional data concerning the locations of the rivets and the line of action of the force.

The results show that the joint strength is 2831 lb, the limit being due to rivet shear. The rivet shear force distribution table gives the shear force on each rivet

RIVETED, BOLTED & WELDED CONNECTIONS 'Connect'

Bolt or Rivet Shear Force Distribution

Column	1	2	3	4	5	6
Row 6	0	0	0	0	0	0
Row 5	0	0	0	0	0	0
Row 4	0	0	0	0	0	0
Row 3	56.30512	41.866145	41.55721	55.61428	76.48931	100
Row 2	40.05723	13.678079	12.70107	39.08023	65.45938	91.83853
Row 1	56.30512	41.866145	41.55721	55.61428	76.48931	100

Max Force (Shear)   2831 lb
Max Force (Tension)   4423 lb
Max Force (Bearing)   29,853 lb

MAX. STRENGTH =   2831 lb

JOINT EFFICIENCY = 24.00%

Are you using US or SI units?

Enter US or SI   US

If this is a welded j joint go to column AJ

Riveted and bolted connections

Is this a Lap or a Butt joint?

Enter L or B   L

Is this a friction-type connection?

Enter Y or N   N

Enter the plate width in inches   3

Enter the plate thickness in inches   0.156

Enter the bolt or rivet dia. in inches   0.1875

Are the threads in the shear plane?

Enter Y or N   N

Bolt or rivet pattern

Enter the number of rows   3

Enter the number of columns   6

Is the pattern eccentrically loaded?

Enter Y or N   Y

If line 20 is N, skip down to line 28

Enter the x-location of the force in inches   3.75

Enter the x-location of column #1 in inches   0.25

Enter the y-location of row #1 in inches   0.5

Enter the spacing between columns in inches   0.5

Enter the spacing between rows in inches   0.75

Highlight the plate material in column 0, then press Ctrl-p

Plate material ID #   5

Highlight the bolt or rivet material from column U, then press Ctrl-b

Bolt or rivet ID #   2

Figure 9.17 Input/output sheet for Example Problem 9.8.

227

compared to the joint strength, expressed as a percent. We see that the rivets r_{16} and r_{36} are both 100% loaded and are, therefore, the ones limiting the joint strength. Notice that r_{23} is loaded to only 12.7% of its maximum capability. There are ways to design connections having maximum efficiency, but these are beyond the scope of this text.

9.5   PROBLEMS FOR CHAPTER 9

Hand Calculations

1. A lap joint is to be designed for a load of 40,000 lb. It is to be end-fillet welded with corner returns. The A36 steel plates are $\frac{3}{8}$ in. thick. One is 10 in. wide, and the other is 12 in. wide. Class E60 electrodes are to be used. Determine the weld size, weld length, and weld distribution.

   Note: You should weld both ends for good design.

2. Determine the joint strength and efficiency of a lap-welded connection of two A588 steel plates. One is 6 in. wide by $\frac{3}{8}$ in. thick, and the other is 3 in. wide by $\frac{1}{4}$ in. thick. The 0.19-in. fillet welds were made using Class E80 electrodes. The total weld length is 9 in. How would this piece fail if the load exceeded the maximum?

3. Two 6-in.-wide, $\frac{1}{4}$-in.-thick plates are joined by *top and bottom* $\frac{3}{16}$-in. fillet welds. The plate material is A242 steel, and Class E70 electrodes are used.
   a. What safe load can be carried?
   b. What is the joint efficiency?
   c. What safe load could the joint carry if the plates were butt-welded?

4. In a lap weld, the narrower plate is 7 in. wide and $\frac{7}{16}$ in. thick. It is subjected to a pull of 55,000 lb. The plate material is A441 steel, and Class E80 electrodes are used. What total length of $\frac{3}{8}$-in. weld should be used?

5. Two A572 steel plates 90 mm wide are to carry a load of 110 kN. They are to be top and bottom end-fillet welded with corner returns, the weld size being 1.5 mm less than the plate thickness and the corner returns being twice the weld size. Class E70 electrodes are to be used.
   a. What is the required plate thickness?
   b. What would be the required plate thickness if a complete penetration butt-weld were used?
   c. Using the plate thickness determined in (a), could 24-mm-diameter A325 steel bolts be used instead of a weld?

6. A double-bolted butt joint, with four 20-mm-diameter A490 bolts on each side and two bolts per row, connects two 25-cm by 2-cm A441 steel plates. Each cover plate is 12 mm thick. Calculate the strength and efficiency of the connection for the following conditions:
   a. Bearing-type connection, threads occur in the shear planes.
   b. Bearing-type connection, threads not in the shear planes.
   c. Friction-type connection, threads not in the shear planes.
   d. Bolts replaced with 16-mm A502 grade 2 steel rivets.

7. A single-bolted lap joint with 10-in. by $\frac{1}{2}$-in. A36 structural steel plates contains three $\frac{7}{8}$-in.-diameter A325 bolts. The threads do not occur in the shear plane, and it is a bearing-type connection. The load on the joint is 45,000 lb. Is it safe? If not, propose a modification that would result in a safe design.

Computer Calculations

8. Repeat Problem 1.
9. Repeat Problem 2.
10. Repeat Problem 3.
11. Repeat Problem 4.

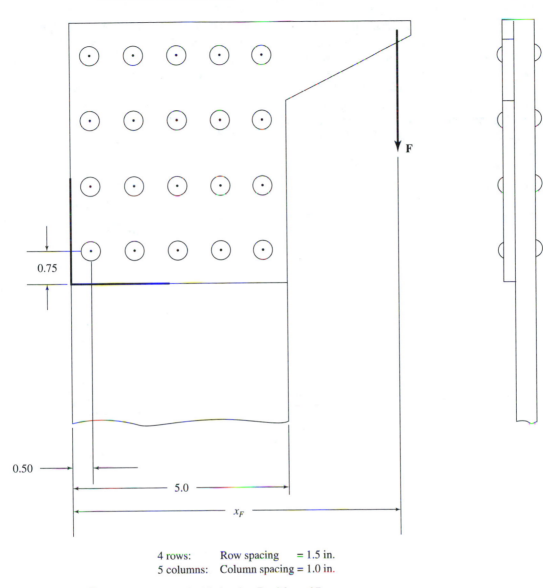

4 rows: Row spacing   = 1.5 in.
5 columns: Column spacing = 1.0 in.

Figure 9.18   Eccentrically loaded joint for Problem 15.

12. Repeat Problem 5.

13. Repeat Problem 6.

14. Repeat Problem 7.

15. Use Program 'Connect' to solve Example Problem 9.1.

16. Use Program 'Connect' to solve Example Problem 9.2.

17. Use Program 'Connect' to solve Example Problem 9.3.

18. Use Program 'Connect' to solve Example Problem 9.4.

19. Figure 9.18 shows an eccentrically loaded joint with four rows and five columns of $\frac{3}{16}$-in.-diameter A502 class 1 steel rivets. The

plates are $\frac{3}{16}$ in. thick and are made of A36 steel. What is the maximum safe distance, x_F (to the nearest 0.1 in.), that the 2000-lb load can be from the left edge of the plate? Which rivet(s) would limit this distance?

20. Referring to Figure 9.18, remove the first row and the first column of rivets (the bottom row and the left-hand column). Using the x_F-value determined in Problem 19, what maximum force F can be safely applied?

21. Referring again to Figure 9.18, remove the fourth row and the fifth column of rivets (the top row and the right-hand column). Using the x_F-value determined in Problem 19, what maximum force F can be safely applied?

10

Pressure Vessels

■ Objectives

In this chapter you will learn about stresses in cylindrical and spherical pressure vessels. These stresses arise because the internal pressure is greater than the external pressure; of course, if the pressure difference is too great, the vessel will burst.

10.1   INTERNAL PRESSURE

A **pressure vessel** is a container that holds a fluid that is at a higher pressure than the outside atmospheric pressure. In scientific terms the word *fluid* means liquid or gas (including air). The average atmospheric pressure at sea level is 14.7 psi (0.10 MPa). The difference between absolute pressure and atmospheric pressure is called **gauge pressure.** We use gauge pressure in vessel calculations because it is the difference between the internal pressure and the atmospheric pressure that creates the force, which in turn creates the stresses.

The most common forms of pressure vessels are closed-end cylinders and spheres, and our first task is to determine how the pressure will act on the vessel.

10.2   FORCES INSIDE A CYLINDRICAL PRESSURE VESSEL

The pressure creates forces *radially* on the cylinder and *longitudinally* on the ends.

10.2.1   Radial Forces

Figure 10.1(a) shows the internal gauge pressure, p, acting on the cylindrical wall of a vessel whose inside diameter is D_i. The internal force created by the pressure is trying to expand the vessel, but this is being resisted by forces in the vessel wall. We make a free-body diagram by taking a piece of length L of the upper half of the cylinder. See Figure 10.1(b). For this free body to be in equilibrium,

1. the algebraic sum of the horizontal forces must be zero, and
2. the algebraic sum of the vertical forces must be zero.

The pressure acts normal to the surface everywhere, so for every horizontal component to the right there is an equal and opposite component to the left. Condition (1) is automatically satisfied.

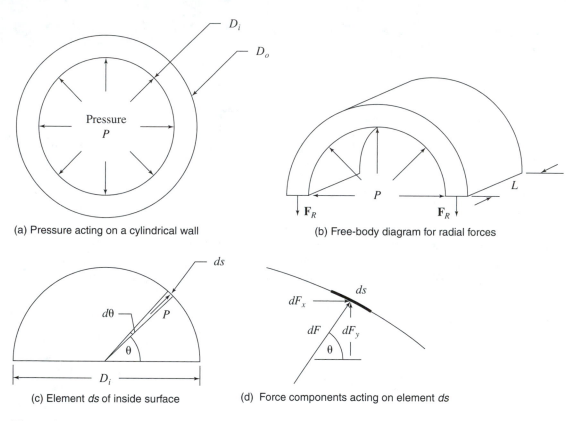

(a) Pressure acting on a cylindrical wall

(b) Free-body diagram for radial forces

(c) Element *ds* of inside surface

(d) Force components acting on element *ds*

Figure 10.1

With regard to the vertical components, however, there is a net upward force that must be balanced by internal forces in the shell of the vessel. We show these in Figure 10.1(b) as F_R. We can show (see Using Calculus) that the force acting on the inside of the half cylinder is equal to the *pressure times the projected area* of the cylinder on the plane of its diameter. That is,

$$\text{vertical component of force} = pD_iL$$

Because this is balanced by $2F_R$, we have:

$$F_R = pD_iL/2 \tag{10.1}$$

Now, of course, if we had chosen our free body as the left half of the cylinder by slicing vertically, then we would have forces F_R acting to the right. Similarly, if we had taken a 45° slice, the forces F_R would have been acting at 45°. This tells us that the forces in the shell are acting circumferentially. These forces create stresses known as *hoop stresses* that tend to rupture any *longitudinal* seams.

USING CALCULUS

Consider an element of longitudinal length L and circumferential length ds, as shown in Figure 10.1(c). The element is at angle θ from the horizontal diameter, and ds subtends angle $d\theta$ at the center of the half-cylinder. The force acting on the elemental area is

$$dF = p\,ds\,L \qquad \text{(force = pressure} \times \text{area)}$$

But
$$ds = (D_i/2)d\theta$$

Therefore,
$$dF = pD_i\,d\theta\,L/2$$

This force is resolved into its horizontal and vertical components in Figure 10.1(d), where we see that the vertical component is $dF_y = dF \sin\theta$. So,

$$dF_y = pD_iL \sin\theta\,d\theta/2$$

Integrating between the limits $\theta = 0$ and $\theta = \pi$ yields

$$F_y = pD_iL \int_0^\pi \sin\theta\,d\theta/2$$
$$F_y = pD_iL(\cos 0 - \cos \pi)/2$$

Because $\cos 0 = 1$ and $\cos \pi = -1$,

$$F_y = pD_iL$$

But $F_y = 2F_R$, therefore

$$F_R = pD_iL/2$$

10.2.2 Longitudinal Forces

Figure 10.2 shows the internal pressure acting on the hemispherical end of a cylindrical vessel. Using a similar analysis to that just shown, we arrive at the same conclusion: the force acting on the end of the cylinder is equal to the *pressure times the projected area* of the hemisphere on the plane of its diameter. That is,

$$\text{horizontal component of force} = p\pi D_i^2/4$$

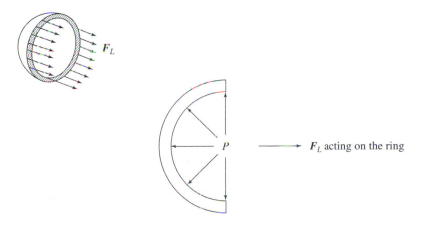

Figure 10.2 Free-body diagram for longitudinal forces.

This is balanced by the longitudinal force F_L, therefore,

$$F_L = p\pi D_i^2/4 \qquad (10.2)$$

This force creates *longitudinal stresses* that tend to rupture any *circumferential* seams.

10.3   FORCES INSIDE A SPHERICAL PRESSURE VESSEL

The pressure creates radial forces only, and if we take a hemisphere as our free body, the result will be the same as equation (10.2).

$$F_{sphere} = p\pi D_i^2/4 \qquad (10.3)$$

10.4   STRESSES IN PRESSURE VESSELS

The pressure difference between the inside and the outside of a vessel creates three types of stresses in the vessel walls:

1. **Hoop stresses** are tangential stresses tending to split any longitudinal seams. These stresses vary throughout the thickness of the vessel wall, decreasing from a maximum at the inner surface to a minimum at the outer surface.
2. **Longitudinal stresses** tend to split any circumferential seams. These stresses are constant throughout the wall thickness.
3. **Radial stresses** are compressive stresses that act radially through the vessel wall, increasing from zero at the outer surface to a maximum value equal to the internal pressure at the inner surface.

10.4.1   Thin-Walled Pressure Vessels

*A **thin-walled pressure vessel** is defined as a pressure vessel whose mean diameter is 20 times or more greater than the wall thickness.*

If the wall thickness is t, the inside diameter is D_i, and the outside diameter is D_o, then for a pressure vessel to be considered thin-walled,

$$(D_i + D_o)/2 \geq 20t \qquad (10.4)$$

Many pressure vessels satisfy this criterion, which leads to simplified expressions for the stresses. In a thin-walled vessel the radial stress variations through the wall are negligible, and the longitudinal and hoop stress equations are easily derived.

10.4.2   Hoop Stress in a Thin-Walled Cylindrical Vessel

We showed (equation (10.1)) that the force, F_R, on each side of the half-cylinder of Figure 10.1(b) is

$$F_R = pD_iL/2$$

The area over which F_R is distributed is

$$A_R = tL$$

Therefore, the hoop stress is

$$s_H = F_R/A_R = (pD_iL/2)/(tL)$$

or

$$s_H = pD_i/2t \tag{10.5}$$

10.4.3 Longitudinal Stress in a Thin-Walled Cylindrical Vessel

The longitudinal force F_L (equation (10.2)) acts over an area A_L (see Figure 10.2), where

$$F_L = p\pi D_i^2/4 \quad \text{and} \quad A_L = \pi D_i t$$

Note: For thin-walled vessels, the error incurred by using the inside diameter instead of the mean diameter results in errors in the stresses that are less than 5%.

The longitudinal stress is

$$s_L = F_L/A_L = (p\pi D_i^2/4)/(\pi D_i t)$$

or

$$s_L = pD_i/4t \tag{10.6}$$

Notice that the hoop stress is two times the longitudinal stress.

10.4.4 Stress in a Thin-Walled Spherical Vessel

From equation (10.3),

$$F_{\text{sphere}} = p\pi D_i^2/4$$

The area over which this force acts is A_{sphere}, where

$$A_{\text{sphere}} = \pi D_i t$$

So the stress is

$$s_{\text{sphere}} = pD_i/4t \tag{10.7}$$

which is the same as equation (10.6). Note that S_{sphere} is a tangential stress.

EXAMPLE PROBLEM 10.1

A cylindrical pressure vessel is 6 ft in diameter and has a wall thickness of $\frac{1}{2}$ in. The design stress is limited to 5000 psi. What maximum internal pressure may be used?

Solution
The vessel diameter is 144 times the wall thickness, so we can use the thin-wall equations, and we know that the greater stress will be the hoop stress:

$$s_H = pD_i/2t$$

$$s_H = 5000 \text{ psi}, \qquad D_i = 72 \text{ in.}, \qquad t = 0.5 \text{ in.}$$

Solving for the pressure,

$$p = 2ts_H/D_i = 2(0.5 \text{ in.})(5000 \text{ psi})/(72 \text{ in.})$$
$$= 69 \text{ psi}$$

Answer
The maximum allowable internal pressure is 69 psi.

EXAMPLE PROBLEM 10.2

A spherical pressure vessel is 5 m in diameter and has a wall thickness of 2.5 cm. Can it be safely used to hold nitrogen at 7 MPa if its maximum allowable stress is 310 MPa?

Solution
This is classified as a thin-walled vessel.

$$p = 7 \text{ MPa}, \qquad D_i = 5 \text{ m} = 5000 \text{ mm}, \qquad t = 2.5 \text{ cm} = 25 \text{ mm}$$
$$s_{\text{sphere}} = pD_i/4t = (7 \text{ MPa})(5000 \text{ mm})/(4 \times 25 \text{ mm})$$
$$= 350 \text{ MPa}$$

Answer
No, the vessel cannot be used for this application.

10.4.5 Thick-Walled Pressure Vessels

Although many pressure vessels satisfy the thin-wall criterion of equation (10.4), there are also many that do not, such as cylinders used for hydraulic and pneumatic actuation, water lines, and gas lines. In these cases we must use the thick-wall equations tabulated in the following table.

Vessel Shape	Type of Stress	Maximum Stress	
Cylindrical	Hoop, s_H	$p(D_o^2 + D_i^2)/(D_o^2 - D_i^2)$	(*)
Cylindrical	Longitudinal, s_L	$pD_i^2/(D_o^2 - D_i^2)$	(**)
Cylindrical	Radial, s_R	$-p$	(*)
Spherical	Tangential, s_{sphere}	$p(D_o^3 + 2D_i^3)/2(D_o^3 - D_i^3)$	(*)
Spherical	Radial, s_R	$-p$	(*)

* At the inside surface.
** Uniform throughout the wall.

EXAMPLE PROBLEM 10.3

Repeat Example Problem 10.2 using the thick-wall equation.

Solution

$$p = 7 \text{ MPa}, \qquad D_i = 5 \text{ m} = 5000 \text{ mm}, \qquad t = 2.5 \text{ cm} = 25 \text{ mm}$$

$$D_o = D_i + 2t = 5050 \text{ mm}$$

$$
\begin{aligned}
s_{\text{sphere}} &= \frac{p(D_o^3 + 2D_i^3)}{2(D_o^3 - D_i^3)} \\[2mm]
&= \frac{(7 \text{ MPa})[(5050 \text{ mm})^3 + 2(5000 \text{ mm})^3]}{2[(5050 \text{ mm})^3 - (5000 \text{ mm})^3]} \\[2mm]
&= \frac{(7 \text{ MPa})(1.288 \times 10^{11} \text{ mm}^3 + 2.5 \times 10^{11} \text{ mm}^3)}{2(1.288 \times 10^{11} \text{ mm}^3 - 1.25 \times 10^{11} \text{ mm}^3)} \\[2mm]
&= (7 \text{ MPa})(3.788 \times 10^{11} \text{ mm}^3)/(0.076 \times 10^{11} \text{ mm}^3) \\[2mm]
&= 349 \text{ MPa}
\end{aligned}
$$

Answer
This compares with 350 MPa of Example Problem 10.2.

EXAMPLE PROBLEM 10.4

You are designing a hydraulic cylinder to actuate a snowplow. Its inside diameter is 5.0 in., and the maximum operating pressure is 3000 psi. The allowable cylinder wall stress is 11,700 psi. What wall thickness would you select?

Solution
We do not know whether this will be classified as a thin-wall or a thick-wall vessel, but for it to be a thin-walled vessel the wall thickness will have to be less than (5.0 in.)/20 = $\frac{1}{4}$ in. This seems small. Using the thick-wall equation for hoop stress,

$$s_H = p(D_o^2 + D_i^2)/(D_o^2 - D_i^2)$$

$$p = 3000 \text{ psi}, \qquad D_i = 5.0 \text{ in.}, \qquad s_H = 11{,}700 \text{ psi.}$$

Therefore, $(11{,}700 \text{ psi}) = (3000 \text{ psi})(D_o^2 + 25 \text{ in.}^2)/(D_o^2 - 25 \text{ in.}^2)$

Rearranging $(D_o^2 + 25 \text{ in.}^2)/(D_o^2 - 25 \text{ in.}^2) = (11{,}700 \text{ psi})/(3000 \text{ psi}) = 3.9$

So

$$D_o^2 + 25 \text{ in.}^2 = 3.9D_o^2 - 97.5 \text{ in.}^2$$

$$2.9D_o^2 = 122.5 \text{ in.}^2$$

$$D_o^2 = 42.24 \text{ in.}^2$$

$$D_o = 6.5 \text{ in.}$$

The wall thickness, t, is

$$t = (D_o - D_i)/2$$
$$= (6.5 \text{ in.} - 5.0 \text{ in.})/2$$
$$= 0.75 \text{ in.}$$

Answer

The required wall thickness is $\frac{3}{4}$ in.

Note: If we had used the thin-wall equation (10.5):

$$s_H = pD_i/2t$$
$$t = pD_i/2s_H$$
$$= (3000 \text{ psi})(5.0 \text{ in.})/(2 \times 11{,}700 \text{ psi})$$
$$= 0.64 \text{ in.}$$

This is 15% undersize. If in doubt, always use the thick-wall equations.

EXAMPLE PROBLEM 10.5

A 3-m-long, 1-m-inside-diameter cylindrical tank is to hold compressed air. The tank is made by rolling 20-mm-thick A36 steel plate and welding a seam, as shown in Figure 10.3. Class E 70 series electrodes are used. What is the maximum safe tank pressure?

Solution

The ratio of vessel diameter to wall thickness is 1000 mm/20 mm = 50. We may use the thin-wall equations. It is the hoop stress that will tend to rupture the welded seam.

$$s_H = pD_i/2t$$
$$D_i = 1000 \text{ mm}, \qquad t = 20 \text{ mm}$$

So,
$$s_H = 25p \tag{1}$$

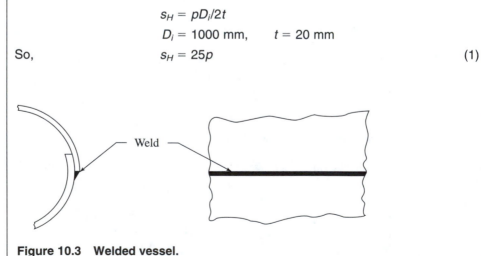

Weld

Figure 10.3   Welded vessel.

We saw in equation (9.1(b)) that the allowable force on a fillet weld may be expressed as

$$F \text{ (N)} = 1.462(E\text{\# kpsi})(l\text{ mm})(\text{leg mm}) \text{ N}$$

The weld leg should be 1.5 mm less than the plate thickness, so

$$\text{leg} = 18.5 \text{ mm}$$

The weld length is 3 m = 3000 mm, so

$$l = 3000 \text{ mm}$$

Therefore, $$F = (1.462)(70)(3000)(18.5) \text{ N}$$

$$= 5680 \text{ kN}$$

The force exerted by the hoop stress is the stress times the area:

$$F_H = (tl)s_H$$

so $$5680 \text{ kN} = (20 \text{ mm})(3000 \text{ mm})s_H$$

or $$s_H = 95 \text{ N/mm}, \quad \text{i.e.,} \quad 95 \text{ MPa}$$

Inserting this stress into equation (1) gives

$$p = 95 \text{ MPa}/25$$

$$= 3.8 \text{ MPa}, \quad \text{which is approximately 550 psi}$$

We must check that the vessel wall is strong enough.

According to the AISC code, the allowable tensile stress in the plate is 60% of the material yield strength. The yield strength of A36 steel is 248 MPa, so the allowable stress is 0.60(248 MPa) = 149 MPa. This shows that the seam is limiting the maximum allowable pressure.

Answer
The maximum allowable pressure is 3.8 MPa.

10.5   COMPUTER PROGRAM FOR VESSELS

Spreadsheet program 'Vessel' uses the thick-wall equations for cylindrical and for spherical vessels. The required inputs are as follows:

1. US or SI for units
2. C or S for cylindrical or spherical vessel
3. Vessel inside diameter
4. Wall thickness
5. Internal gauge pressure

The values must be entered in the units specified. The output values are

Maximum hoop stress and maximum longitudinal stress for cylindrical vessels, or
Maximum tangential stress for spherical vessels.

EXAMPLE PROBLEM 10.6

Repeat Example Problem 10.4.

Solution

> Call up program 'Vessel'.
> Enter US for units.
> Enter C for cylindrical.
> The vessel inside diameter is 5.
> For wall thickness, try any reasonable value, say, 0.5
> Internal gauge pressure is 3000.

The output shows that the maximum hoop stress exceeds 11,700 psi, so increase the wall thickness. One or two tries show that a wall thickness of 0.75 in. results in a maximum hoop stress of 11,696 psi. (See Figure 10.4.)

Answer

The required wall thickness is $\frac{3}{4}$ in.

Program 'Pressure Vessel'

Input

Enter US or SI for units	US
Enter C or S for cyl. or spherical vessel.	C
Vessel inside diameter	5	in.
Wall thickness	0.75	in.
Internal gauge pressure	3000	psi.

Output
Cylindrical Vessel:

Maximum hoop stress	11696	psi
Maximum longitudinal stress	4348	psi

Spherical Vessel:

Maximum tangential stress		psi

Input/output sheet for Example Problem 10.6.

Figure 10.4   Input/output sheet for Example Problem 10.6.

EXAMPLE PROBLEM 10.7

Repeat Example Problem 10.5.

Solution

1. Call up program 'Connect', enter SI for units, and then move across the screen to *Welded Connections.* Make the entries shown in Figure 10.5(a). We see that the joint strength is 5680 kN and that it is lower than the plate strength.

$$\text{stress} = \text{force/area} = (5680 \text{ kN})/(60{,}000 \text{ mm}^2) = 95 \text{ MPa}$$

Welded Connections

Enter the width of one plate in mm	3000
Enter the thickness of that plate in mm	20
Enter the width of the other plate in mm	3000
Enter the thickness of that plate in mm	20
Enter the weld size (leg) in mm	18.5
Enter the electrode designation (E#) in kpsi	70
Enter the total weld length in mm	3000
Highlight the plate material ID# in column O, then press Ctrl-w	1

Joint Strength =	5679870 N

Plate strength =	8935920 N

Joint Efficiency =	63.56 %

(a) Input/output sheet for Part 1

Program 'Pressure Vessel'

Input

Enter US or SI for units	SI	
Enter C or S for cyl. or spherical vessel.	C	
Vessel inside diameter	1000	mm
Wall thickness	20	mm
Internal gauge pressure	3.73	MPa

Output

Cylindrical Vessel:

Maximum hoop stress	95	MPa
Maximum longitudinal stress	46	MPa

Spherical Vessel:

Maximum tangential stress		MPa

(b) Input/output sheet for Part 2

Figure 10.5   Example Problem 10.7.

2. Call up program 'Vessel'. Make the entries shown in Figure 10.5(b) except for the pressure. Try various pressures until the maximum hoop stress equals 95 MPa. The required pressure is found to be 3.73 MPa.

Answer

The maximum allowable pressure is 3.7 MPa.
Compare this with 3.8 MPa found by using the thin-wall equation.

10.6   PROBLEMS FOR CHAPTER 10

Hand Calculations

1. A spherical oxygen tank for a spacecraft contains oxygen at 10,000 psi. Its inside diameter is 1 ft, and its wall thickness is $\frac{3}{4}$ in. Compute the maximum tangential stress.

2. A spherical gas tank is subjected to an internal pressure of 250 kPa. Its inside diameter is 3.5 m, and the tangential stress is limited to 10 MPa. What minimum wall thickness is required?

3. A 1.5-ft-diameter rocket combustion chamber is subjected to a gas pressure of 2500 psi. If the chamber is made of titanium alloy with a yield strength of 155 ksi and a safety factor of 4 based on yield is used, what minimum thickness should the wall be?

4. Hemispherical caps are bolted on to the ends of a cylindrical tank. See Figure 10.6. The tank's inside diameter is 4.5 ft, and the wall thickness is $\frac{3}{4}$ in. The wall stress is limited to 20,000 psi. What maximum pressure can the tank sustain? How many $\frac{3}{4}$ in.-diameter bolts per end cap are required if the allowable stress for the bolt is 26,000 psi?

5. A cylindrical hydraulic car jack has an inside diameter of 4.5 in. and a wall thickness of $\frac{1}{4}$ in. The maximum allowable hoop stress is 6000 psi. What maximum allowable load can it lift? Use the thick-wall equation.

6. Repeat the previous problem using the thin-wall equation. What percent error is incurred by using the thin-wall equation?

7. An octagonal rifle barrel, $\frac{15}{16}$ in. across the flats, has a 0.450-in. bore. The peak pressure during firing is 26,000 psi. The barrel is

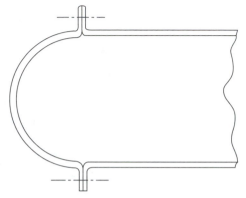

Figure 10.6   Problem 4.

made of AISI 4140 OQT 700 alloy steel that has a yield strength of 212,000 psi. What factor of safety, based on yield strength, was used in the barrel design?

8. Repeat Problem 7 using the thin-wall equation. What percent error is incurred by using the thin-wall equation?

9. A 2-m-long, 25-cm-inside-diameter cylindrical tank is to hold compressed air at 3.0 MPa. The tank is made by rolling A36 steel plate and welding a seam, as shown in Figure 10.3. Class E 60 series electrodes are used. What minimum wall thickness is required?

10. A nominal $\frac{3}{4}$-in. schedule 40 water pipe has an outside diameter of 1.050 in. and a wall thickness of 0.113 in. If its ultimate strength is 60 ksi and a safety factor of 12 based on the ultimate strength is required for protection against water hammer, what maximum safe water pressure may be used?

Computer Calculations

11. Repeat Problem 1.
12. Repeat Problem 2.
13. Repeat Problem 3.
14. Repeat Problem 4.
15. Repeat Problem 5.
16. Repeat Problem 7.

17. Repeat Problem 9.
18. Repeat Problem 10.
19. Design a hydraulic jack (diameter and wall thickness) to lift a load of 10 tons. Use AISI 1040 WQT 700 alloy steel (ultimate strength = 127 ksi) and a safety factor of 8.

20. Show that the hoop stress equation, $S_H = p(D_o^2 + D_i^2)/(D_o^2 - D_i^2)$, can be rearranged to give the wall thickness as

$$t = (D_i/2)[\sqrt{1 + 2p/(s_H - p)} - 1]$$

Hint: Write $D_o = D_i + 2t$. Obtain a quadratic equation in t. Solve the quadratic.

21. Using the result of Problem 20, write a computer program to solve for the minimum wall thickness for a cylindrical pressure vessel, given the maximum allowable hoop stress, the inside diameter, and the internal gauge pressure. Write the program so that either US or SI units may be used.

11

Centroids and Moments of Inertia

■ *Objectives*

In this chapter you will learn about centers of gravity (centers of mass), centroids (centers of areas), and moments of inertia.

11.1 CENTER OF GRAVITY

Figure 11.1 shows a beam, with a heavy weight on one end and a lighter weight on the other, balancing on a knife-edge support. It is intuitively obvious that we had to place the support closer to the heavier end for the system to balance. The support must be pushing up with a force equal to the total weight of the system. This is called the **reaction.** If the reaction were greater than the total weight of the beam plus end weights, the whole system would rise off the ground. If it were less, then the support would collapse. In reality, the force pushing down against the support is the sum of the weights of all the molecules in the beam and the weights. Because it is balanced, though, it is as if the total weight were concentrated in one point directly above the support. This point is called the **center of gravity.** The concept of center of gravity is used throughout all engineering and physics. We imagine that the system, in this case the beam and the weights, is weightless except for a tiny spot where all the mass is concentrated.

It is easy to guess where the center of gravity is located for some shapes. For a ball, for example, it is at its center, and for a rectangular box it is where the diagonals cross. More complicated shapes have to be analyzed to find their centers of gravity.

Figure 11.1   Center of gravity.

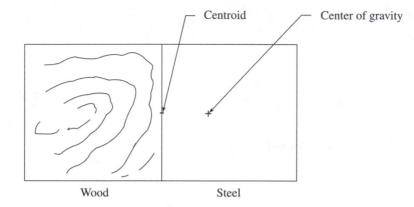

Figure 11.2   Centroid and center of gravity.

We generally refer to the center of gravity of a two-dimensional object, such as a piece of sheet metal or cardboard, as the center of area, or the **centroid.** Actually, there *is* a difference between the center of gravity and the centroid if the body is not homogeneous. The rectangular plate in Figure 11.2 consists of a square of plywood attached to a square of steel, both squares being the same size. Obviously, it would not balance along the joint because the steel is heavier than the wood. Its center of gravity is somewhere to the right of the joint, but its center of area, its *centroid,* is right in the middle of the joint.

11.2   MOMENT OF AREA

The **moment of an area** about an axis is similar to the moment of a force, except that we use the area (square inches or square millimeters) instead of the force (lb or N), and the distance is the perpendicular distance from the axis to the *centroid* of the area. The centroid of an area is its geometric center. Take a rectangular piece of cardboard, such as the back of a pad of paper, and balance it on a pencil point. You would have to place the point at the exact center of the rectangle for it to balance. This point is its centroid. For example, the centroid of the area shown in Figure 11.3(a) is at the center of the rectangle, which is 2 in. up from the base. Let us draw any axis, x_1x_1, parallel to the edge of the rectangle. For demonstration purposes place x_1x_1 at a perpendicular distance of 3 in. from the bottom edge. The distance from the axis to the centroid is $\bar{y} =$ (3 in. + 2 in.) = 5 in. The rectangular area is (3 in.)(4 in.) = 12 in.^{2}, so the moment of area about the x_1x_1 axis is

$$A\bar{y} = (5\ \text{in.})(12\ \text{in.}^2) = 60\ \text{in.}^3$$

Let us look at this another way. Divide the rectangle into 12 one-inch squares and number them as in Figure 11.3(b). The centroid of each square is at its center, so the perpendicular distance from the x_1x_1 axis to the centroid of square 1 is 3.5 in.

The area of square 1 is 1 in.^{2}, so its moment of area about the axis is

$$Ay = (3.5\ \text{in.})(1\ \text{in.}^2) = 3.5\ \text{in.}^3$$

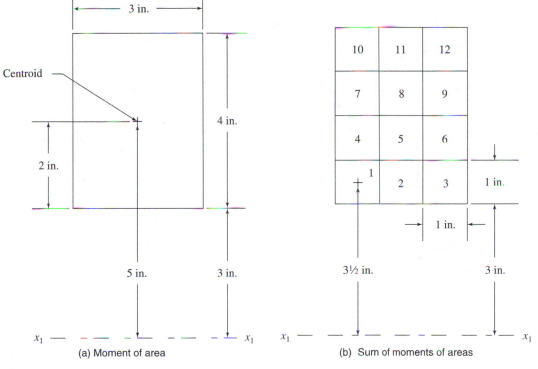

(a) Moment of area   (b) Sum of moments of areas

Figure 11.3

The centroidal distance of square 2 from the axis is also 3.5 in. and its area is 1 in.^{2}, making its moment of area about the axis

$$Ay = 3.5 \text{ in.}^3$$

The same is true for square 3, so the total moment of area of the first row about x_1x_1 is

$$\Sigma Ay = 3.5 \text{ in.}^3 + 3.5 \text{ in.}^3 + 3.5 \text{ in.}^3 = 10.5 \text{ in.}^3$$

The centroidal distances of the next row of squares, 4, 5, and 6, are all the same at 4.5 in. Each square is again 1 in.^{2}, so the sum of the second-row moments of area about $x_1x_1 = 3 \times 4.5 \text{ in.}^3 = 13.5 \text{ in}^3$. Continuing this for the next row gives $3 \times 5.5 \text{ in.}^3 = 16.5 \text{ in.}^3$, and for the last row, $3 \times 6.5 \text{ in.}^3 = 19.5 \text{ in}^3$. The total moment of area about x_1x_1 is the sum of these:

$$\Sigma Ay = 10.5 \text{ in.}^3 + 13.5 \text{ in.}^3 + 16.5 \text{ in.}^3 + 19.5 \text{ in.}^3 = 60 \text{ in.}^3$$

which is the same result we got by taking the total area times the centroidal distance of that area. The sum of all the little areas (12 one-inch squares) is equal to the total area, 12 in^2. In symbols, then,

$$\bar{y}\Sigma A = \Sigma Ay \tag{11.1}$$

The total area of a body times the perpendicular distance from any axis to the centroid of that area is equal to the sum of partial areas of that body times the perpendicular distances from the same axis to their centroids.

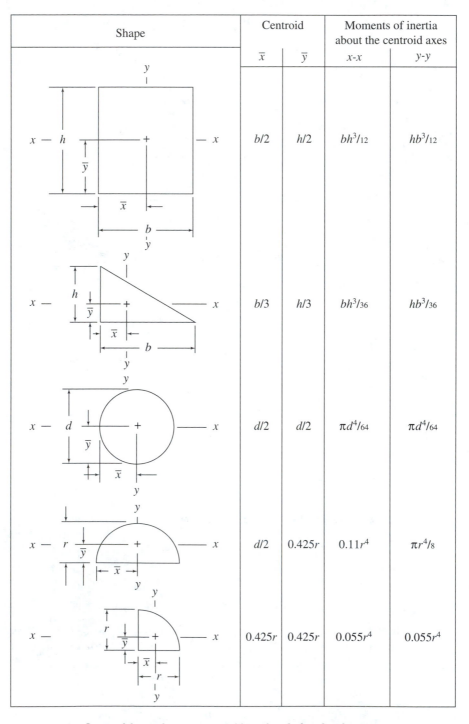

Shape	Centroid		Moments of inertia about the centroid axes	
	$\overline{x}$	$\overline{y}$	x-x	y-y
	$b/2$	$h/2$	$bh^3/12$	$hb^3/12$
	$b/3$	$h/3$	$bh^3/36$	$hb^3/36$
	$d/2$	$d/2$	$\pi d^4/64$	$\pi d^4/64$
	$d/2$	$0.425r$	$0.11r^4$	$\pi r^4/8$
	$0.425r$	$0.425r$	$0.055r^4$	$0.055r^4$

Figure 11.4   Centroids and moments of inertia of simple shapes.

Rearranging equation (11.1):

$$\bar{y} = \Sigma Ay / \Sigma A \qquad (11.2)$$

This allows us to determine the location of the centroid of a shape if we can divide its area into areas whose centroidal locations are known. For example, a tee consists of two rectangles. We know where the centroid of each rectangle is located, so we can find the centroid of the tee. But before we do that, why do we want to know the location of a centroid? In Chapter 13, we will learn that the stress at any point in a loaded beam is directly proportional to the distance from that point to the centroidal axis of the beam, the centroidal axis being the axis that runs through the centroids of the beam's cross sections. Also, in Chapter 9, where we discussed eccentric loadings on riveted and bolted connections, we referred to the center of the rivet or bolt pattern. What we were really referring to was the centroid of the pattern, as we shall see in Chapter 18.

11.3 CENTROIDS OF SIMPLE SHAPES

We can determine the location of the centroid of a composite shape by using equation (11.2) if we can divide the shape into simple shapes whose centroid locations are known, such as, squares, rectangles, or circles. Figure 11.4 gives the centroidal locations of some simple shapes.

Try This: Cut a triangle of any shape (it doesn't have to be a right-angled triangle or an equilateral triangle) from a piece of cardboard. Draw a line, parallel to one edge, at a perpendicular distance to that edge equal to one-third of the height. Now do the same from another edge. Where the lines cross is the centroid of the triangle. See if it will balance on a pencil point at that location.

11.4 CENTROIDS OF COMPOSITE SHAPES

Unless an area is totally irregular, it can usually be divided into a group of connected shapes whose centroid locations are known. The following examples show you how to determine the centroid locations of composite shapes.

EXAMPLE PROBLEM 11.1

Find the location of the centroid of the tee shown in Figure 11.5.

Solution
First we need to define an *xx*-axis from which to take measurements. We usually choose this as the bottom edge of the figure, although any axis may be selected. We are going to make use of equation (11.2)

$$\bar{y} = \Sigma Ay / \Sigma A$$

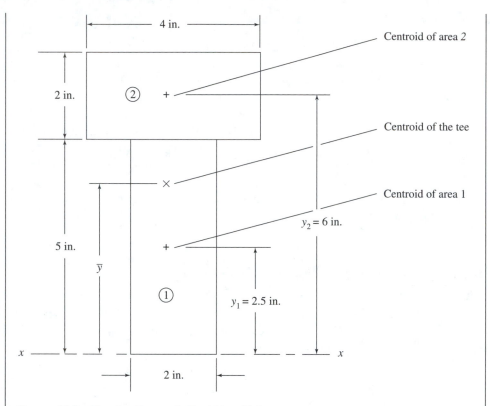

Figure 11.5   Tee for Example Problem 11.1

so we need to divide the tee into convenient areas. Call them areas 1 and 2, as shown in Figure 11.5. There are only two areas, so we can write equation (11.2) as

$$\overline{y} = (A_1y_1 + A_2y_2)/(A_1 + A_2)$$
$$A_1 = (2 \text{ in.})(5 \text{ in.}) = 10 \text{ in.}^2$$
$$A_2 = (2 \text{ in.})(4 \text{ in.}) = 8 \text{ in.}^2$$

y_1 is the perpendicular distance from the axis to the centroid of area 1, which is located at the center of the rectangle. Therefore, $y_1 = 2.5$ in.

y_2 is the perpendicular distance from the axis to the centroid of area 2; therefore, $y_2 = (5 \text{ in.} + 1 \text{ in.}) = 6$ in.

$$\overline{y} = \frac{A_1y_1 + A_2y_2}{A_1 + A_2}$$

$$= \frac{(10 \text{ in.}^2)(2.5 \text{ in.}) + (8 \text{ in.}^2)(6 \text{ in.})}{10 \text{ in.}^2 + 8 \text{ in.}^2}$$

$$= \frac{25 \text{ in.}^3 + 48 \text{ in.}^3}{18 \text{ in.}^2}$$

$$= 4.056 \text{ in.}$$

Since the tee is symmetrical about a vertical centerline, the centroid will lie on that line. (It would balance on a knife edge along that line).

Answer
The centroid is on the center line at $\overline{y} = 4.06$ in.

EXAMPLE PROBLEM 11.2

Repeat Problem 11.1 with the axis as shown in Figure 11.6.

Solution
The axis location should have no effect on the result. This time, $y_1 = -2.5$ in. and $y_2 = 1.0$ in.

So,
$$\overline{y} = \frac{A_1 y_1 + A_2 y_2}{A_1 + A_2}$$

$$= \frac{(10 \text{ in.}^2)(-2.5 \text{ in.}) + (8 \text{ in.}^2)(1.0 \text{ in.})}{18 \text{ in.}^2}$$

$$= \frac{-25 \text{ in.}^3 + 8 \text{ in.}^3}{18 \text{ in.}^2}$$

$$= -0.944 \text{ in.}$$

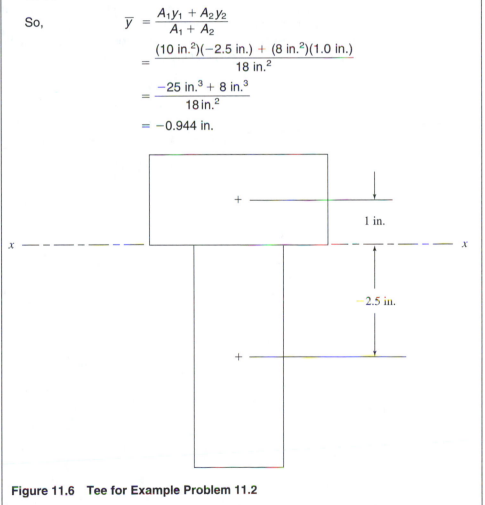

Figure 11.6 Tee for Example Problem 11.2

This tells us that the centroid of the tee is located 0.944 in. below the joint, which is (5 in. − 0.944 in.) = 4.056 in. from the bottom edge of the figure, which agrees with the previous result.

Answer

The position of the centroid does not depend on the location of the axis chosen for its calculation.

EXAMPLE PROBLEM 11.3

Determine the location of the centroid of the area shown in Figure 11.7.

Solution

The area consists of a rectangle with a rectangular cutout. Notice that the combined figure is not symmetrical about any axis. So we will have to determine both the x- and the y-positions of the centroid.

First the y location, $\overline{y}$: Set up an xx-axis along the bottom edge of the figure. Call the large (total) area, area 1, and the cutout area, area 2.

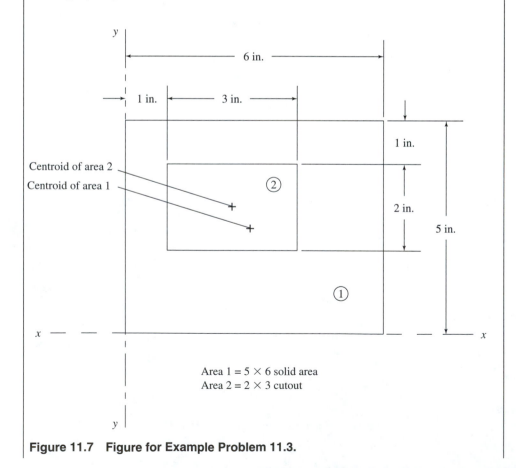

Area 1 = 5 × 6 solid area
Area 2 = 2 × 3 cutout

Figure 11.7   Figure for Example Problem 11.3.

$$A_1 = (5 \text{ in.})(6 \text{ in.}) = 30 \text{ in.}^2$$

$$A_2 = -(2 \text{ in.})(3 \text{ in.}) = -6 \text{ in.}^2$$

Note that this area is negative.

y_1 is the perpendicular distance from the *xx*-axis to the centroid of area 1 *assuming no cutout.*

$$y_1 = 2.5 \text{ in.}$$

$$y_2 = 2 \text{ in.} \quad \text{(from the } xx\text{-axis to the bottom edge of area 2)}$$

$$+ 1 \text{ in.} \quad \text{(from the bottom edge of area 2 to its centroid)}$$

$$= 3 \text{ in.}$$

$$\overline{y} = \frac{A_1 y_1 + A_2 y_2}{A_1 + A_2}$$

$$= \frac{(30 \text{ in.}^2)(2.5 \text{ in.}) + (-6 \text{ in.}^2)(3 \text{ in.})}{30 \text{ in.}^2 + (-6 \text{ in.}^2)}$$

$$= \frac{75 \text{ in.}^3 - 18 \text{ in.}^3}{24 \text{ in.}^2}$$

$$= 2.375 \text{ in.}$$

Now with regard to the *yy*-axis, which we select as the left edge of the object (see Figure 11.7), we have to determine $\overline{x}$, which is the location of the centroid from the *yy*-axis. In order to do this we need to know x_1 and x_2, which are the *x*-locations of the individual centroids.

$$x_1 = 3 \text{ in. to the center of the large rectangle}$$

$$x_2 = 1 \text{ in.} \quad \text{(to the left edge of the small rectangle)}$$

$$+ 1.5 \text{ in.} \quad \text{(to the centroid of that rectangle)}$$

$$x_2 = 2.5 \text{ in.}$$

Equation (11.2), $\overline{y} = \Sigma Ay/\Sigma A$, written for the centroid *x*-location, becomes $\overline{x} = \Sigma Ax/\Sigma A$, which for two areas becomes

$$\overline{x} = \frac{A_1 x_1 + A_2 x_2}{A_1 + A_2}$$

$$= \frac{(30 \text{ in.}^2)(3 \text{ in.}) + (-6 \text{ in.}^2)(2.5 \text{ in.})}{24 \text{ in.}^2}$$

$$= \frac{90 \text{ in.}^3 - 15 \text{ in.}^3}{24 \text{ in.}^2}$$

$$= 3.125 \text{ in.}$$

Answer

The centroid is located at (3.125 in., 2.375 in.).

Note: Always check that your answer makes sense. In this case, if there were no cutout, the centroid would be at (3, 2.5). The cutout is up and to the left, leaving more material down and to the right, so you would expect the centroid to be lower than 2.5, which it is (2.375), and bigger than 3, which it is (3.125).

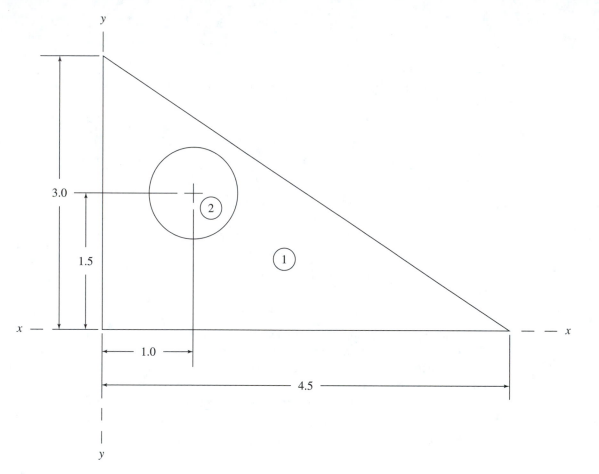

Figure 11.8   Example Problem 11.4.

EXAMPLE PROBLEM 11.4

The triangle shown in Figure 11.8 has a 1-in.-diameter circular cutout. Find the location of its centroid.

Solution

Choose the axes *xx* and *yy* as shown. Call the complete triangle, with no cutout, area 1 and the circular cutout area 2. The area of a triangle is equal to one-half the base times its height, so

$$A_1 = (4.5 \text{ in.})(3 \text{ in.})/2 = 6.75 \text{ in.}^2$$

The area of a circle is $(\pi/4)(\text{diameter})^2$, so

$$A_2 = -0.7854(1 \text{ in.})^2 = -0.7854 \text{ in.}^2 \qquad \text{(negative area because it is a cutout)}$$

First, with regard to axis *xx:*

$$y_1 = \frac{h}{3} \quad \text{(see Figure 11.4)}$$

$$y_1 = \frac{3 \text{ in.}}{3} = 1.0 \text{ in.} \quad (\tfrac{1}{3} \text{ of the height})$$

$$y_2 = 1.5 \text{ in.} \quad \text{(to the center of the circle)}$$

$$\bar{y} = \frac{A_1 y_1 + A_2 y_2}{A_1 + A_2}$$

$$= \frac{(6.75 \text{ in.}^2)(1.0 \text{ in.}) + (-0.7854 \text{ in.}^2)(1.5 \text{ in.})}{6.75 \text{ in.}^2 - 0.7854 \text{ in.}^2}$$

$$= \frac{6.75 \text{ in.}^3 - 1.178 \text{ in.}^3}{5.965 \text{ in.}^2}$$

So, $\bar{y} = 0.934$ in.

Now with regard to the *yy*-axis:

$$x_1 = \frac{b}{3} = \frac{4.5 \text{ in.}}{3} = 1.5 \text{ in.}$$

$$x_2 = 1 \text{ in.} \quad \text{(distance to the center of the circle)}$$

$$\bar{x} = \frac{(A_1 x_1 + A_2 x_2)}{A_1 + A_2}$$

$$= \frac{(6.75 \text{ in.}^2)(1.5 \text{ in.}) + (-0.7854 \text{ in.}^2)(1 \text{ in.})}{5.965 \text{ in.}^2}$$

So, $\bar{x} = 1.566$ in.

Answer

The centroid is located at (1.57 in., 0.93 in.) Does this make sense?

11.5   CENTROIDS OF ANY SHAPE

The centroids of areas whose boundaries can be expressed as mathematical equations can be found by calculus. (See Using Calculus). However, if the shape can be drawn on **AutoCad,** its centroid can be located, and its *moments of inertia* (to be discussed next) can be calculated by a special AutoCad routine.

USING CALCULUS

To find the centroid of an area bounded by the curves $y_1 = f_1(x)$ and $y_2 = f_2(x)$ and the lines $x = x_1$ and $x = x_2$ (see Figure 11.9), we draw an infinitesimally thin element of area $dA = (y_2 - y_1)\, dx$ and take its moment about the y-axis.

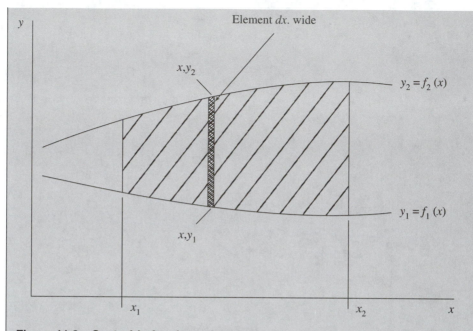

Figure 11.9   Centroid of an irregular area.

The moment of the element about the y-axis is

$$dM_y = xdA = x(y_2 - y_1) \, dx$$

The total moment of the area about the y-axis is

$$M_y = \int_{x_1}^{x_2} x(y_2 - y_1) \, dx$$

But M_y is equal to the area times the distance $\bar{x}$ from the axis to the centroid:

$$\bar{x} = M_y/A = (1/A) \int_{x_1}^{x_2} x(y_2 - y_1) \, dx$$

To find the moment about the x-axis, we know that the centroid of the element is at its midpoint, which is at a distance $(y_1 + y_2)/2$ from the x-axis. The moment of the element about the x-axis is, therefore,

$$dM_x = dA(y_1 + y_2)/2 = (y_2 - y_1)dx(y_1 + y_2)/2$$

So the total moment is

$$M_x = \int_{x_1}^{x_2} [(y_2 - y_1)(y_1 + y_2)/2] \, dx = \int_{x_1}^{x_2} [(y_2^2 - y_1^2)/2] \, dx$$

Therefore,

$$\bar{y} = (1/2A) \int_{x_1}^{x_2} (y_2^2 - y_1^2) \, dx$$

EXAMPLE

Find the centroid of the area bounded by the parabola $y^2 = 9x$, the x-axis, and the line $x = 2$.

Solution

First we need to determine the area, A.

$$A = \int_{x_1}^{x_2} (y_2 - y_1)\, dx$$

Here $y_2 = 3x^{1/2}$, $y_1 = 0$, $x_1 = 0$, and $x_2 = 2$; therefore,

$$A = \int_0^2 3x^{1/2}\, dx = (3)(\tfrac{2}{3})x^{3/2}\, \big|_0^2 = 5.657 \text{ square units}$$

To determine $\bar{x}$:

$$\bar{x} = \int_{x_1}^{x_2} (1/A)x(y_2 - y_1)\, dx$$

$$\bar{x} = (1/5.657) \int_0^2 (x)(3x^{1/2})\, dx$$

$$= (3/5.657) \int_0^2 x^{3/2}\, dx$$

$$= (3/5.657)(\tfrac{2}{5})x^{5/2}\, \big|_0^2 = (3/5.657)(\tfrac{2}{5})(5.657)$$

$$= \tfrac{6}{5} \text{ units}$$

To determine $\bar{y}$:

$$\bar{y} = (1/2A) \int_{x_1}^{x_2} (y_2^2 - y_1^2)\, dx$$

$$= (1/11.314) \int_0^2 (3x^{1/2})^2\, dx$$

$$= (1/11.314)(9) \int_0^2 x\, dx$$

$$= (1/11.314)(9)(\tfrac{1}{2}) x^2\, \big|_0^2 = (1/11.314)(9)(\tfrac{1}{2})(4)$$

$$= 1.591 \text{ units}$$

Answer

The centroid is located at (1.20, 1.591).

11.6 MOMENT OF INERTIA

You will understand the significance of the moment of inertia after reading Chapter 13, but for now it is sufficient to know that the larger the moment of inertia of a beam, the greater resistance it has to bending. Take a wooden or plastic ruler and bend it. It's easy. Now turn it through 90° so that you are holding the edges. Bend it. You can't, because the moment of inertia in the second position is about 100 times that in the first. It is important to realize that you cannot say that a given shape has a certain moment of inertia. It is a *moment* and *must, therefore, be referred to some axis*. We refer it to an axis through the *centroid* of the area, for reasons you will understand in Chapter 13.

The moment of inertia of an area (also called the *second moment of area*) about an axis is calculated by dividing the area into infinitesimal areas, multiplying the area of each by the *square* of its perpendicular distance from the axis and adding all the products. We denote the moment of inertia of an area about axis *xx* as I_x and about axis *yy* as I_y. Expressed mathematically,

$$I_x = \Sigma A y^2 \qquad\qquad (11.3(a))$$

and

$$I_y = \Sigma A x^2 \qquad\qquad (11.3(b))$$

11.7   MOMENT OF INERTIA OF SIMPLE SHAPES

Figure 11.4 shows the moments of inertia about the centroidal axes for some simple shapes.

EXAMPLE PROBLEM 11.5

Determine the moment of inertia about the centroidal axis for the rectangle shown in Figure 11.10(a) and for the same rectangle turned through 90°, as in Figure 11.10(b).

Solution
From Figure 11.4, the moment of inertia of a rectangle about its centroidal axis is $I = bd^3/12$.

In Figure 11.10(a), $b = 40$ mm and $d = 60$ mm; therefore,

$$I_x = (40 \text{ mm})(60 \text{ mm})^3/12 = 7.2 \times 10^5 \text{ mm}^4$$

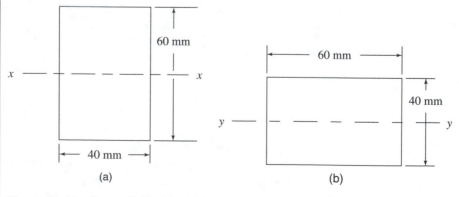

(a)   (b)

Figure 11.10   Example Problem 11.5.

In Figure 11.10(b), $b = 60$ mm and $d = 40$ mm; therefore,
$$I_y = (60 \text{ mm})(40 \text{ mm})^3/12 = 3.2 \times 10^5 \text{ mm}^4$$

Answer
The moments of inertia are
$$I_x = 7.2 \times 10^5 \text{ mm}^4$$
$$I_y = 3.2 \times 10^5 \text{ mm}^4$$

EXAMPLE PROBLEM 11.6

Compare the moments of inertia of a square and an equilateral triangle of the same area about their centroidal axes.

Solution
An equilateral triangle is one with all sides equal in length (see Figure 11.11). Suppose one side of the triangle is of unit length. This could be 1 in., 1 ft, 1 cm—any unit length. Divide the triangle into two parts by dropping a perpendicular from the apex to the base. It will divide the base into two equal parts, so the right-angled triangles have a base length of $\frac{1}{2}$, a hypotenuse of 1, and a third side of length h (see Figure 11.11). Using the Pythagorean theorem,
$$1^2 = (\tfrac{1}{2})^2 + h^2$$
$$1 = \tfrac{1}{4} + h^2$$
$$h^2 = \tfrac{3}{4}$$
$$h = \sqrt{\tfrac{3}{4}} = \sqrt{3}/2$$

So, the area of the equilateral triangle is
$$A = (1)(\sqrt{3}/2)/2 = \sqrt{3}/(2 \times 2) = \sqrt{3}/4$$

The moment of inertia of a triangle about its centroidal axis is $I_x = bh^3/36$ (see Figure 11.4). Substituting $b = 1$ and $h = \sqrt{3}/2$ gives
$$I_x = (1)(\sqrt{3}/2)^3/36 = (1)(\sqrt{3})^3/[(2)^3(36)] = \sqrt{3}/96$$

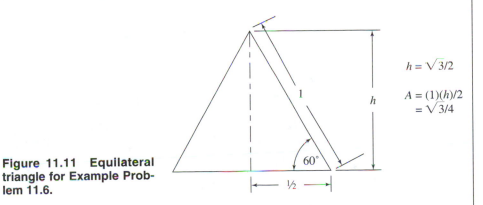

$$h = \sqrt{3}/2$$
$$A = (1)(h)/2$$
$$= \sqrt{3}/4$$

Figure 11.11 Equilateral triangle for Example Problem 11.6.

The moment of inertia of a square about its centroidal axis is $I_x = b^4/12$. But, since the area of a square is b^2, we can write this as

$$I_x = A^2/12$$

If the area of the square is the same as the area of the triangle,

$$A = \sqrt{3}/4$$

so $I_x = (\sqrt{3}/4)^2/12 = \frac{3}{192}$ The ratio of moments of inertia, triangle to square, is

$$(\sqrt{3}/96)/(\tfrac{3}{192}) = 0.01804/0.015625 = 1.1546$$

This shows that the *shape* of an area, in addition to the actual area, has an effect on the moment of inertia and therefore on the resistance to bending.

Answer
The ratio of moments of inertia, triangle to square, is 1.1546.

11.8   TRANSFER-OF-AXIS THEOREM

We often need to calculate the moment of inertia of an area about an axis parallel to the centroidal axis—one that does not pass through the centroid. We do this by calculating the moment of inertia about the centroidal axis and adding to this value the product of the area of the figure and the square of the distance between the centroidal axis and the new axis. This is called the *transfer-of-axis theorem*, or simply the *transfer theorem*.

If we shift the axis from the centroidal axis, xx, to a parallel axis, aa, the distance between the two axes being d, then

$$I_a = I_x + Ad^2 \tag{11.4}$$

EXAMPLE PROBLEM 11.7

What is the moment of inertia of the rectangle shown in Figure 11.12 about its bottom edge?

Solution

$$b = 5 \text{ in.,} \qquad h = 8 \text{ in.}$$
$$I_x = bd^3/12 = (5 \text{ in.})(8 \text{ in.})^3/12 = 213.33 \text{ in.}^4$$

Its centroidal axis, xx, is halfway up, so the distance between the two axes is

$$d = 4 \text{ in.}$$

and the area of the rectangle is

$$A = (5 \text{ in.})(8 \text{ in.}) = 40 \text{ in.}^2$$

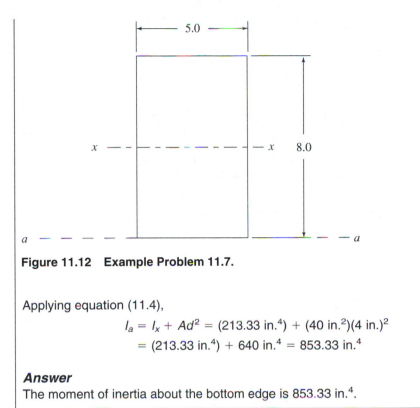

Figure 11.12   Example Problem 11.7.

Applying equation (11.4),

$$I_a = I_x + Ad^2 = (213.33 \text{ in.}^4) + (40 \text{ in.}^2)(4 \text{ in.})^2$$
$$= (213.33 \text{ in.}^4) + 640 \text{ in.}^4 = 853.33 \text{ in.}^4$$

Answer
The moment of inertia about the bottom edge is 853.33 in.^{4}.

We now derive a general formula for the moment of inertia of a rectangle about its bottom edge. The result will be used in the section Using Calculus.

The moment of inertia of a rectangle about its centroidal axis is $I_x = bd^3/12$. We are going to shift the axis a distance $d = d/2$, so the distance squared is $d^2/4$.

The area of the rectangle is $A = bd$.

Applying equation (11.4),

$$I_a = I_x + Ad^2 = bd^3/12 + (bd)(d^2/4) = bd^3/12 + bd^3/4 = bd^3/3$$

Answer
The moment of inertia of a rectangle about its bottom edge is $I_a = bd^3/3$.

11.9   MOMENT OF INERTIA OF COMPOSITE SHAPES

Calculating the moment of inertia of a composite shape about its centroidal axis is a five-step process.

1. Divide the area into simple areas.
2. Locate the centroid.
3. Determine the moment of inertia of each simple area *about its own centroidal axis*.
4. Use the transfer theorem.
5. Add up the contributions to the moment of inertia.

EXAMPLE PROBLEM 11.8

Calculate the moment of inertia of the area shown in Figure 11.13 about its centroidal axis.

Solution

1. Let the square be area 1 and the triangle be area 2.

$$A_1 = (8 \text{ in.})(8 \text{ in.}) = 64 \text{ in.}^2 \qquad A_2 = (8 \text{ in.})(6 \text{ in.})/2 = 24 \text{ in.}^2$$

2. Select the bottom edge of the square as the xx-axis.
The distance from the axis to the centroid of the square is

$$y_1 = 4 \text{ in.}$$

The centroid of the triangle is one-third of the way up from its base—i.e., 2 in. from its base—so the distance from the axis to the centroid of the triangle is

$$y_2 = (8 \text{ in.} + 2 \text{ in.}) = 10 \text{ in.}$$

We find the location of the centroid of the combined figure by using equation (11.2):

$$\bar{y} = \Sigma Ay/\Sigma A$$

which for two areas becomes

$$\bar{y} = \frac{A_1 y_1 + A_2 y_2}{A_1 + A_2}$$

$$= \frac{(64 \text{ in.}^2)(4 \text{ in.}) + (24 \text{ in.}^2)(10 \text{ in.})}{64 \text{ in.}^2 + 24 \text{ in.}^2}$$

$$= \frac{(256 \text{ in.}^3) + (240 \text{ in.}^3)}{88 \text{ in.}^2}$$

$$= 5.636 \text{ in.}$$

We know by symmetry that the centroid lies on the vertical center line.

3. The moment of inertia of the square about its own centroidal axis is

$$I_x = b^4/12 = (8 \text{ in.})^4/12 = 341.33 \text{ in.}^4$$

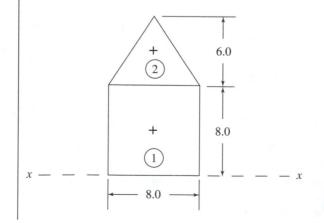

Figure 11.13   Example Problem 11.8.

The moment of inertia of the triangle about its own centroidal axis is

$$I_x = bh^3/36 = (8 \text{ in.})(6 \text{ in.})^3/36 = 48 \text{ in.}^4$$

4. The distance from the centroid of the combined figure to the centroid of the square is

$$d_1 = 5.636 \text{ in.} - 4 \text{ in.} = 1.636 \text{ in.}$$

The distance from the centroid of the combined figure to the centroid of the triangle is

$$d_2 = 10 \text{ in.} - 5.636 \text{ in.} = 4.364 \text{ in.}$$

5. Using the transfer theorem, the moment of inertia of the square about the centroid of the combined figure is

$$I_c = I_x + Ad^2 = (341.33 \text{ in.}^4) + (64 \text{ in.}^2)(1.636 \text{ in.})^2 = 512.63 \text{ in.}^4$$

and for the triangle, it is

$$I_c = I_x + Ad^2 = (48 \text{ in.}^4) + (24 \text{ in.}^2)(4.364 \text{ in.})^2 = 505.07 \text{ in.}^4$$

The total moment of inertia is the sum of these:

$$I_c = (512.63 \text{ in.}^4) + (505.07 \text{ in.}^4) = 1017.70 \text{ in.}^4$$

Answer
The moment of inertia about the centroidal axis is 1018 in^4.

EXAMPLE PROBLEM 11.9

a. Locate the centroid of the area shown in Figure 11.14.
b. Determine its moment of inertia about its centroidal axis, *xx*.
c. Determine its moment of inertia about axis *aa*, 1 in. below and parallel to the bottom edge of the figure.

Solution
a. *Centroid location*   There are four areas:

1. The rectangle, 2 in. × 4 in. ($A = 8 \text{ in.}^2$).
2. The triangle, 2-in. base, 3-in. height ($A = 3 \text{ in.}^2$).
3. The semicircle, 2-in. radius ($A = \pi (2 \text{ in.})^2/2 = 6.283 \text{ in.}^2$).
4. The circle, 2-in. diameter ($A = -\pi(2 \text{ in.})^2/4 = -3.142 \text{ in.}^2$). *This is a negative area.*

Because the shape is not symmetrical, we will have to find both $\bar{x}$ and $\bar{y}$. Use the reference axes shown in Figure 11.14. When there are more than two areas, it is probably easier to set up a table of calculations:

Note: $\bar{y}$ for the triangle is (4 in.) + ($\frac{1}{3}$) (3 in.) = 5.0 in. $\bar{x}$ for the triangle is ($\frac{1}{3}$) (2 in.) = 0.667 in.

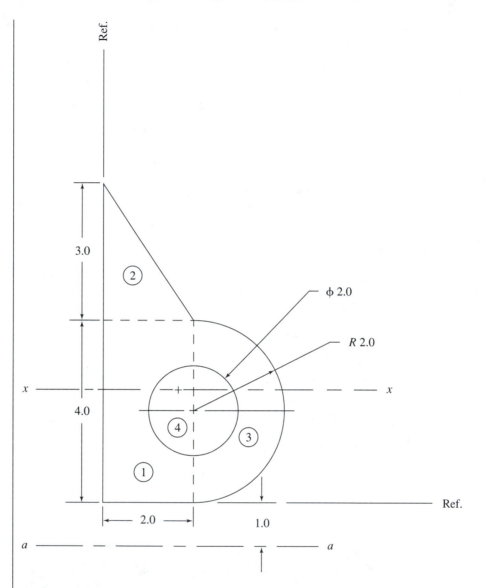

Figure 11.14   Example Problem 11.9.

From Figure 11.4:

$\bar{x}$ for the semicircle = (2 in.) + (0.425)(2 in.) = 2.850 in.

No.	A (in.²)	$\bar{y}$ (in.)	A$\bar{y}$ (in.³)	$\bar{x}$ (in.)	A$\bar{x}$ (in.³)
1	8.0	2.0	16.0	1.0	8.0
2	3.0	5.0	15.0	0.667	2.0
3	6.283	2.0	12.566	2.850	17.907
4	−3.142	2.0	−6.284	2.0	−6.284

$\Sigma A = 14.141$ in.^{2}, $\Sigma Ay = 37.282$ in.^{3}, $\Sigma Ax = 21.623$ in.^{3}

$\bar{y} = \Sigma Ay/\Sigma A$ $\bar{x} = \Sigma Ax/\Sigma A$

$= 37.282$ in.^{3}/14.141 in.^{2} $= 21.623$ in.^{3}/14.141 in.^{2}

$= 2.636$ in. $= 1.529$ in.

b. *Moment of inertia about the centroidal axis xx*

Note: I_x is the moment of inertia of the area about its own centroidal axis.

1. The rectangle, $I_x = bd^3/12 = (2$ in.$)(4$ in.$)^3/12 = 10.667$ in.^{4}
2. The triangle, $I_x = bh^3/36 = (2$ in.$)(3$ in.$)^3/36 = 1.5$ in.^{4}
3. The semicircle, $I_x = \pi d^4/128 = \pi(4$ in.$)^4/128 = 6.283$ in.^{4}
4. The hole, $I_x = -\pi d^4/64 = -\pi(2$ in.$)^4/64 = -0.785$ in.^{4}

No.	I_x (in.^{4})	d (in.)	Ad^2 (in.^{4})	(I_x + Ad^2)(in.^{4})
1	10.667	−0.636	3.236	13.903
2	1.5	2.364	16.765	18.265
3	6.283	−0.636	2.541	8.824
4	−0.785	−0.636	−1.271	−2.056

$$I_c = \Sigma(I_x + Ad^2) = 38.936 \text{ in.}^4$$

c. *Moment of inertia about the axis aa.* The distance between the centroidal axis *xx* and the parallel axis *aa* is (1 in.) + (2.636 in.) = 3.636 in. The total area of the figure is $A = 14.141$ in^2. Using the transfer formula,

$$I_a = I_c + Ad^2 = (38.936 \text{ in.}^4) + (14.141 \text{ in.}^2)(3.636 \text{ in.})^2$$

$$= 225.89 \text{ in.}^4$$

Answers
   a. The centroid is at $\bar{x} = 1.529$ in., $\bar{y} = 2.636$ in.
   b. $I_c = 38.9$ in^4.
   c. $I_a = 225.9$ in^4.

11.10   MOMENT OF INERTIA OF ANY SHAPE

Determining the moment of inertia of an irregular shape requires the use of calculus (see Using Calculus) unless the shape can be drawn accurately on AutoCad. Auto-Cad's Advanced Modeling Extension will determine the location of the centroid and calculate the moments of inertia of any shape that comprises *entities*. Entities are shapes created by rectangles, polygons, or circles or any shape drawn with continuous polylines. The polylines may be modified by the *spline* routine, the *fit* routine, etc.

AutoCAD also lists the *product of inertia*, the *radii of gyration*, and the *principal moments*. (See Sections 11.10.1, 11.10.2, and 11.10.3).

USING CALCULUS

We derived expressions for the moments of inertia about the centroidal axes in Section 11.6. They were

$$I_x = \Sigma A y^2 \tag{11.3(a)}$$

and
$$I_y = \Sigma A x^2 \tag{11.3(b)}$$

These were the results of *summation,* but the true mathematical expressions are the results of *integration:*

$$I_x = \int y^2 \, dA$$

and
$$I_y = \int x^2 \, dA$$

To find the moments of inertia of an area bounded by a curve $y = f(x)$, the two lines $x = x_1$ and $x = x_2$, and the x-axis, as shown in Figure 11.15, we first draw an infinitesimally narrow element whose area $dA = y \, dx$.

The moment of inertia of the element about the y-axis is

$$dI_y = x^2 \, dA = x^2 y \, dx$$

Therefore, the moment of inertia of the entire area about the y-axis is

$$I_y = \int_{x_1}^{x_2} x^2 y \, dx$$

To find the moment of inertia about the x-axis, we use the expression for the moment of inertia of a rectangle about its bottom edge found in Example Problem 11.7:

$$I_a = bd^3/3$$

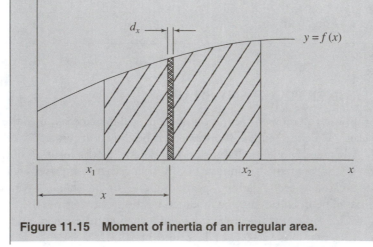

Figure 11.15   Moment of inertia of an irregular area.

Applying this to the element in Figure 11.15,

$$dI_x = dx\, y^3/3$$

So,

$$I_x = \int_{x_1}^{x_2} y^3/3\ dx$$

EXAMPLE

Find the moments of inertia about the x- and y-axes of an area bounded by the curve $y = 2x^3$, the lines $x = 2$ and $x = 3$, and the x-axis.

Solution

Moment of inertia about the y-axis:

$$I_y = \int_{x_1}^{x_2} x^2 y\ dx$$

Here $y = 2x^3$, $x_1 = 2$, and $x_2 = 3$.

$$I_y = \int_2^3 (x^2)(2x^3)\ dx = 2\int_2^3 x^5\ dx$$

$$= (2)(\tfrac{1}{6})\ x^6\ \big|_2^3 = (\tfrac{1}{3})(3^6 - 2^6)$$

$$= (\tfrac{1}{3})(729 - 64)$$

$$= 221.67\ (\text{units})^4$$

Moment of inertia about the x-axis:

$$I_x = \int_{x_1}^{x_2} y^3/3\ dx$$

$$= \int_2^3 (2x^3)^3/3\ dx = (\tfrac{8}{3})\int_2^3 x^9\ dx$$

$$= (\tfrac{8}{3})(\tfrac{1}{10})\ x^{10}\ \big|_2^3 = (\tfrac{8}{30})(3^{10} - 2^{10})$$

$$= (\tfrac{8}{30})\ (59{,}049 - 1024)$$

$$= 15{,}473\ (\text{units})^4$$

Answer

The moment of inertia about the x-axis is $15{,}473\ (\text{units})^4$; the moment of inertia about the y-axis is $221.67\ (\text{units})^4$.

11.10.1 Product of Inertia

As we discussed earlier, the moment of inertia about the xx-axis is $I_x = \Sigma Ay^2$; about the yy axis it is $I_y = \Sigma Ax^2$. The product of inertia is defined as $I_{xy} = \Sigma Axy$.

11.10.2   Radius of Gyration

We use the radius of gyration in Chapter 19, which deals with columns. It is defined by the equation

$$r = \sqrt{I/A}$$

where I is the moment of inertia and A is the cross-sectional area of the column.

11.10.3   Principal Moments

The **principal axes** of an area are the axes for maximum and minimum moments of inertia. They are perpendicular to each other, and the products of inertia corresponding to the principal axes are zero.

11.11   POLAR MOMENT OF INERTIA

The **polar moment of inertia** is the moment of inertia about an axis *perpendicular* to an area. If r is the radial distance of an infinitesimal area A from the axis (see Figure 11.16), then the polar moment of inertia is defined as

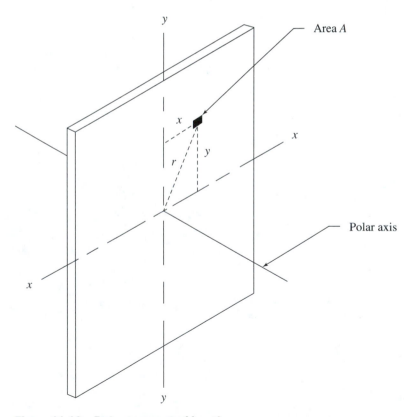

Figure11.16   Polar moment of inertia.

$$J = \Sigma A r^2 \tag{11.5}$$

From the Pythagorean theorem $r^2 = x^2 + y^2$, so

$$J = \Sigma A r^2 = \Sigma A(x^2 + y^2) = \Sigma A x^2 + \Sigma A y^2 = I_y + I_x$$

We see then that the polar moment of inertia is equal to the sum of the *x*- and *y*-moments of inertia. For a doubly symmetric area, such as a square or a circle, $I_x = I_y$ and $J = 2I_x$. We use the polar moment of inertia in Chapter 12, which deals with torsion.

EXAMPLE PROBLEM 11.10

Calculate the polar moment of inertia about an axis normal to the plane and through the centroid of a 20-mm by 60-mm rectangle.

Solution

$$I_x = bh^3/12 = (20mm)(60mm)^3/12 = 3.6 \times 10^5 \ mm^4$$

$$I_y = hb^3/12 = (60mm)(20mm)^3/12 = 0.4 \times 10^5 \ mm^4$$

$$J = I_x + I_y = 4.0 \times 10^5 \ mm^4$$

Answer

The polar moment of inertia is $4.0 \times 10^5 \ mm^4$.

USING CALCULUS

Figure 11.17 shows a hollow cylindrical bar of inside radius r_1 and outside radius r_2. The shaded part is a cylindrical ring of radius ρ and thickness $d\rho$.

Because the circumference of the ring is $2\pi\rho$ and its thickness is $d\rho$, its area is $dA = 2\pi\rho \ d\rho$. So, the polar moment of inertia is

$$J = \int \rho^2 \ dA = 2\pi \int \rho^3 \ d\rho$$

$$= 2(\tfrac{1}{4})\pi \ \rho^4 \ \big|_{r_1}^{r_2} = \pi(r_2^4 - r_1^4)/2$$

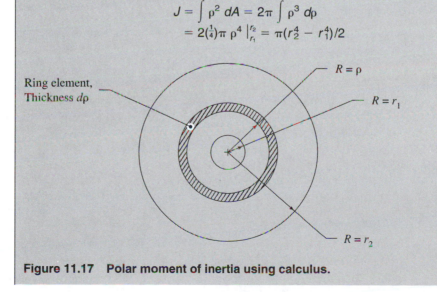

Ring element, Thickness $d\rho$

$R = \rho$

$R = r_1$

$R = r_2$

Figure 11.17   Polar moment of inertia using calculus.

In terms of the inside diameter, D_1, and the outside diameter, D_2:
$$J = \pi[(D_2/2)^4 - (D_1/2)^4]/2$$
$$= \pi(D_2^4 - D_1^4)/32$$

If the bar is solid with diameter D, then $D_1 = 0$ and
$$J = \pi D^4/32$$

11.12   COMPUTER ANALYSES

We discussed using AutoCad's Advanced Modeling Extension (AME) to determine centroids and moments of inertia in Section 11.10. If you do not have AutoCad or are not familiar with it, you can use the Excel program 'Centroid'.

11.12.1   Program 'Centroid'

The program 'Centroid' calculates the centroid location and the centroidal moments of inertia for any shape that can be divided into a group of simple shapes. The shapes include rectangles, triangles, circles, semicircles, and quarter circles.

Figure 11.18 shows the input/output sheet for the program. We have made provisions for three rectangles, three triangles, three circles, two semicircles, and two quarter-circles. Each has a cell labeled 'FACTOR'. If the area is positive, enter a 1 in the FACTOR cell. If the area is negative (for example, a hole), enter a -1 in the FACTOR cell. It is important to fully understand the inputs for this program.

First, you must decide on an origin. This is where $x = 0$ and $y = 0$ and is usually selected as the lower-left corner of the object or the center of a circle. It can be any convenient point, from which all other points can be located.

Rectangles

The x- and y-coordinates of the four corners are entered in a *counterclockwise* direction, starting at the *lower-left* corner. (See Figure 11.19.)

Triangles

The x- and y-coordinates are entered as shown in Figure 11.20. The numbering starts at the right-angled corner; then, depending on the orientation of the triangle, it goes *horizontally* right or left to the second point. The third point is either straight up or straight down from the first point.

Circles

You simply enter the radius and the x- and y-location of the center.

Semicircles

In addition to the radius and the x-, y-location of the radius center point, you have to define the semicircle's orientation. You do this by entering the two quadrants in which it lies, and these are expressed in a *counterclockwise* direction. Figure 11.21 should

Centroid

CENTROID and MOMENTS OF INERTIA CALCULATION PROGRAM 'Centroid'

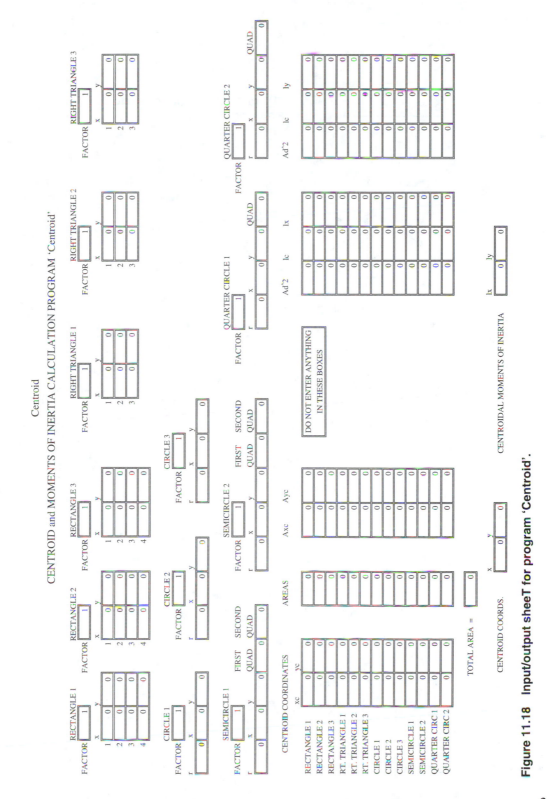

Figure 11.18   Input/output sheeT for program 'Centroid'.

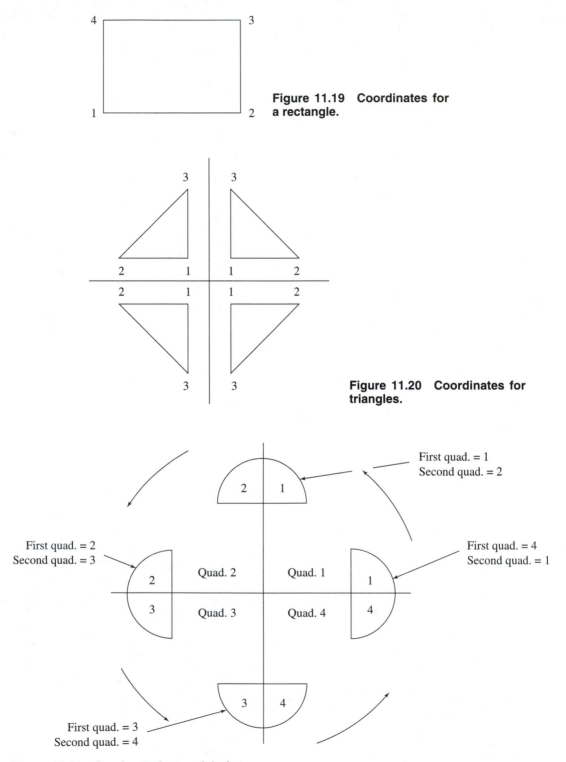

Figure 11.19   Coordinates for a rectangle.

Figure 11.20   Coordinates for triangles.

Figure 11.21   Quadrants for semicircles.

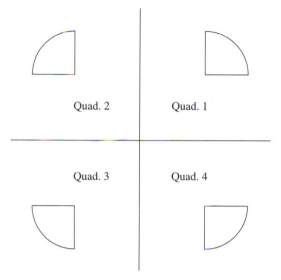

Figure 11.22 Quadrants for quarter-circles.

make this clear. For example, the one shaped like a *D* is in the *fourth* and the *first* quadrants. Suppose its center is at $x = 2$ in. and $y = 3$ in. and it has a 4-in. radius. The entries would be

4	2	3	4	1
radius	*x*-location	*y*-location	1st quad.	2nd quad.

Quarter-Circles

Quarter-circles need the radius, the radius center's *x*, *y* location, and the quadrant, as shown in Figure 11.22.

EXAMPLE PROBLEM 11.11

Find the centroid and the centroidal moments of inertia of the shape shown in Figure 11.23(a).

Solution

We first divide the shape into two rectangles, as shown in Figure 11.23(a). The coordinates of the first rectangle are (0, 0), (10, 0), (10, 1.2), (0, 1.2). Those of the second rectangle are (0, 1.2), (1.2, 1.2), (1.2, 15), (0, 15). We enter these values, both with FACTOR = 1, as shown in Figure 11.24(a).

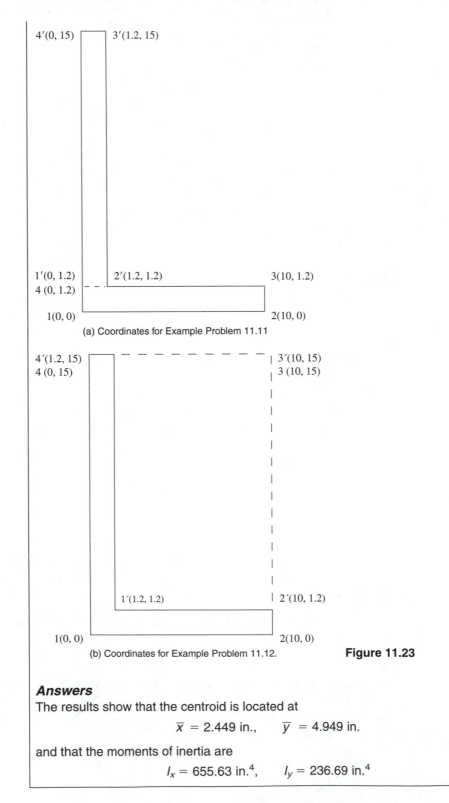

(a) Coordinates for Example Problem 11.11

(b) Coordinates for Example Problem 11.12.

Figure 11.23

Answers

The results show that the centroid is located at

$$\overline{x} = 2.449 \text{ in.,} \qquad \overline{y} = 4.949 \text{ in.}$$

and that the moments of inertia are

$$I_x = 655.63 \text{ in.}^4, \qquad I_y = 236.69 \text{ in.}^4$$

Centroid

CENTROID and MOMENTS OF INERTIA CALCULATION PROGRAM 'Centroid'

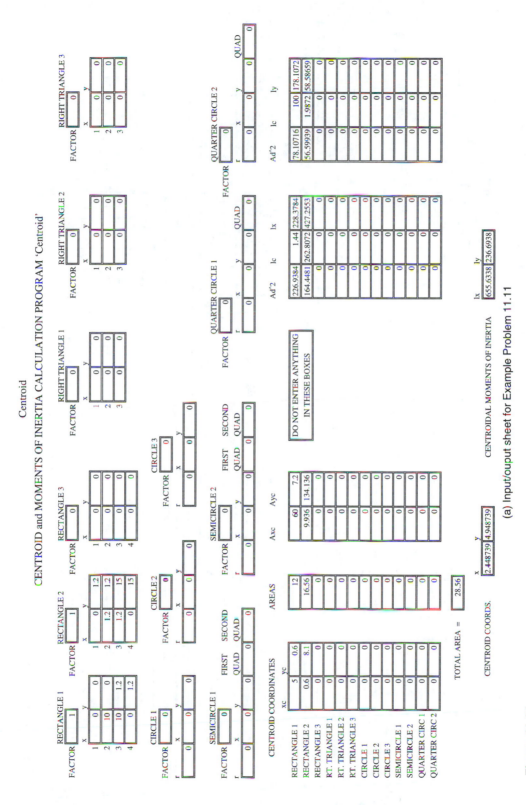

(a) Input/ouput sheet for Example Problem 11.11

Figure 11.24

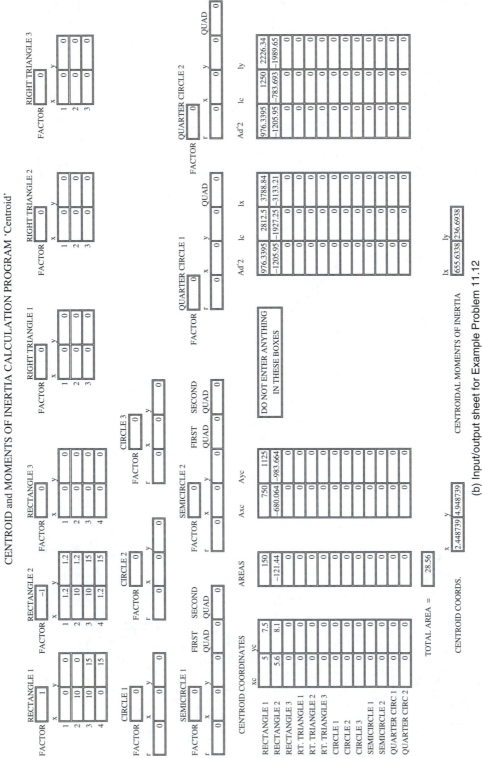

Centroid

CENTROID and MOMENTS OF INERTIA CALCULATION PROGRAM 'Centroid'

(b) Input/output sheet for Example Problem 11.12

Figure 11.24 *(Continued)*

EXAMPLE PROBLEM 11.12

Repeat Example Problem 11.11 by completing the rectangle as shown in Figure 11.23(b) and subtracting the smaller rectangle from the larger one.

Solution

The coordinates of the first rectangle are (0, 0), (10, 0), (10, 15), (0, 15). Those of the second rectangle are (1.2, 1.2), (10, 1.2), (10, 15), (1.2, 15). We enter the first with FACTOR = 1 and the second with FACTOR = −1, as shown in Figure 11.24(b). The results show that the centroid is located at

$$\bar{x} = 2.449 \text{ in.}, \qquad \bar{y} = 4.949 \text{ in.}$$

and that the moments of inertia are

$$I_x = 655.63 \text{ in.}^4, \qquad I_y = 236.69 \text{ in.}^4$$

Answer

The results are exactly the same as for Example Problem 11.11, as they should be.

EXAMPLE PROBLEM 11.13

To show the ease with which a more complicated problem can be solved, let us return to Example Problem 11.10 (Figure 11.14) and redo it using 'Centroid'.

Solution

1. Refer to Figure 11.14.

   Rectangle: FACTOR = 1, and the coordinates are (0, 0), (2, 0), (2, 4), (0, 4).
   Triangle: FACTOR = 1, and the coordinates are (0, 4), (2, 4), (0, 7).
   Circle: FACTOR = −1, radius = 1.0, x = 2.0, y = 2.0.
   Semicircle: FACTOR = 1, radius = 2.0, x = 2.0, y = 2.0, Quad 1 = 4, Quad 2 = 1.

2. Enter the data as shown in Figure 11.25.
3. The results show that the centroid is located at

$$\bar{x} = 1.529 \text{ in.}, \qquad \bar{y} = 2.636 \text{ in.}$$

Centroid

CENTROID and MOMENTS OF INERTIA CALCULATION PROGRAM 'Centroid'

	RECTANGLE 1		RECTANGLE 2		RECTANGLE 3	
FACTOR	1		0		0	
	x	y	x	y	x	y
1	0	0	0	0	0	0
2	2	0	0	0	0	0
3	2	4	0	0	0	0
4	0	4	0	0	0	0

	RIGHT TRIANGLE 1		RIGHT TRIANGLE 2		RIGHT TRIANGLE 3	
FACTOR	1		0		0	
	x	y	x	y	x	y
1	0	4	0	0	0	0
2	2	4	0	0	0	0
3	0	7	0	0	0	0

	FACTOR	r	x	y
CIRCLE 1	–1	1	2	2
CIRCLE 2	0	0	0	0
CIRCLE 3	0	0	0	0

	FACTOR	r	x	y	FIRST QUAD	SECOND QUAD
SEMICIRCLE 1	1	2	2	2	4	1
SEMICIRCLE 2	0	0	0	0	0	0

	FACTOR	r	x	y	QUAD
QUARTER CIRCLE 1	0	0	0	0	0
QUARTER CIRCLE 2	0	0	0	0	0

DO NOT ENTER ANYTHING IN THESE BOXES

	CENTROID COORDINATES xc	yc	AREAS	Axc	Ayc
RECTANGLE 1	1	2	8	8	16
RECTANGLE 2	0	0	0	0	0
RECTANGLE 3	0	0	0	0	0
RT. TRIANGLE 1	0.666667	5	3	2	15
RT. TRIANGLE 2	0	0	0	0	0
RT. TRIANGLE 3	0	0	0	0	0
CIRCLE 1	2	2	-3.14159	-6.28318	-6.28318
CIRCLE 2	0	0	0	0	0
CIRCLE 3	0	0	0	0	0
SEMICIRCLE 1	2.85	2	6.28318	17.90706	12.56636
SEMICIRCLE 2	0	0	0	0	0
QUARTER CIRC 1	0	0	0	0	0
QUARTER CIRC 2	0	0	0	0	0

	Ad^2	Ic	Ix	Ad^2	Ic	Iy
RECTANGLE 1	3.24025	10.66667	13.90692	2.239561	2.666667	4.906228
RECTANGLE 2	0	0	0	0	0	0
RECTANGLE 3	0	0	0	0	0	0
RT. TRIANGLE 1	16.75952	1.5	18.25952	2.231366	0.666667	2.898032
RT. TRIANGLE 2	0	0	0	0	0	0
RT. TRIANGLE 3	0	0	0	0	0	0
CIRCLE 1	–1.27244	–0.7854	–2.05784	–0.69664	–0.7854	–1.48204
CIRCLE 2	0	0	0	0	0	0
CIRCLE 3	0	0	0	0	0	0
SEMICIRCLE 1	2.544884	6.2832	8.828084	10.96277	1.76	12.72277
SEMICIRCLE 2	0	0	0	0	0	0
QUARTER CIRC 1	0	0	0	0	0	0
QUARTER CIRC 2	0	0	0	0	0	0

TOTAL AREA = 14.14159

CENTROID COORDS. x 1.529098 y 2.636421

CENTROIDAL MOMENTS OF INERTIA Ix 38.93668 Iy 19.04499

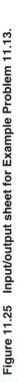

Figure 11.25 Input/output sheet for Example Problem 11.13.

and that the moments of inertia are

$$I_x = 38.936 \text{ in.}^4, \qquad I_y = 19.045 \text{ in.}^4$$

Answers

The centroid is located at

$$\overline{x} = 1.529 \text{ in.}, \qquad \overline{y} = 2.636 \text{ in.}$$

and the moments of inertia are

$$I_x = 38.936 \text{ in.}^4, \qquad I_y = 19.045 \text{ in.}^4$$

These are exactly the same as for Example Problem 11.10.

EXAMPLE PROBLEM 11.14

Find the centroid location and the centroidal moments of inertia of the area shown in Figure 11.26(a).

Solution

We show the FACTORS and the coordinates in Figure 11.26(b) and the input/output sheet in Figure 11.26(c). You should study these carefully to fully understand the input.

Answer

The results show that the centroid is located at

$$\overline{x} = 2.742 \text{ in.}, \qquad \overline{y} = 2.996 \text{ in.}$$

and that the moments of inertia are

$$I_x = 293.724 \text{ in.}^4, \qquad I_y = 146.721 \text{ in.}^4$$

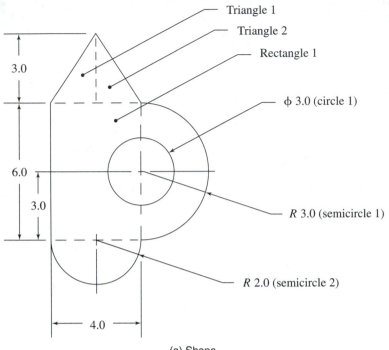

(a) Shape

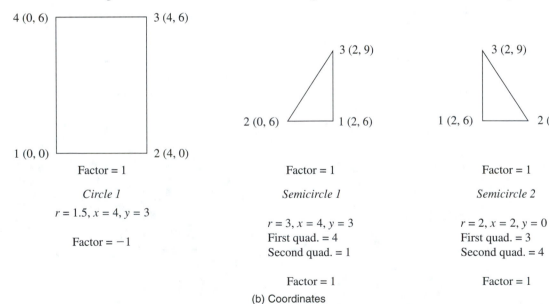

(b) Coordinates

Figure 11.26   Example Problem 11.14.

CENTROID and MOMENTS OF INERTIA CALCULATION PROGRAM 'Centroid'

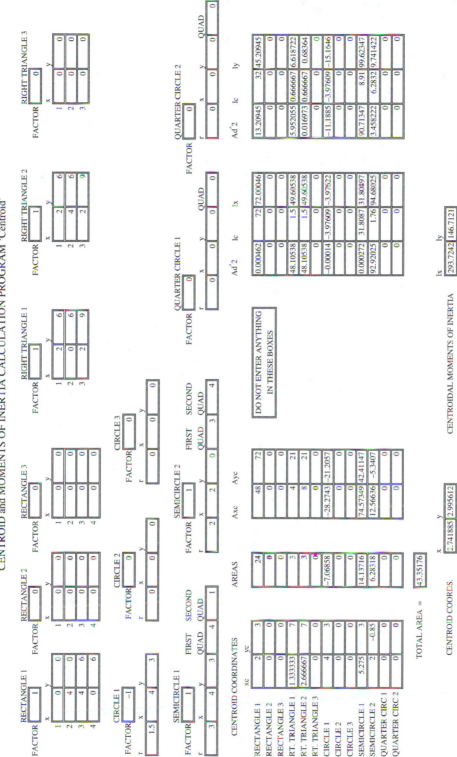

(c) Input/output sheet

Figure 11.26 *(Continued)*

11.13   PROBLEMS FOR CHAPTER 11

Hand Calculations

1. A 1.0-in. × 1.0-in. square plate of C14500 copper and a 1.0-in. × 1.0-in. square plate of 6061-T6 aluminum alloy are joined along one edge to form a 1.0-in. × 2.0-in. rectangle. Both plates are of the same thickness. The density of C14500 copper is 0.323 lb/in.^{3}, and that of 6061-T6 is 0.10 lb/in^3. Determine the position of the center of gravity of the rectangle.

2. Calculate the product of inertia of a 3-cm × 3-cm square.

3. Calculate the polar moment of inertia of a 3-cm × 3-cm square.

4. Find the centroid locations and the centroidal moments of inertia for the shape shown in Figure 11.27.

5. Find the centroid locations and the centroidal

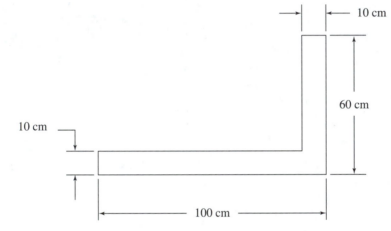

Figure 11.27   Problems 4, 9, 10, 11, and 12.

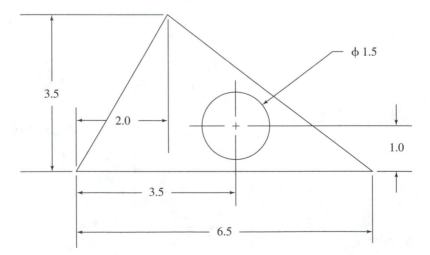

Figure 11.28   Problems 5, 13, 14, 15, and 16.

moments of inertia for the shape shown in Figure 11.28.

6. Show that the moment of inertia of an isosceles triangle about its axis of symmetry is $hb^3/48$, where b is the base and h is the height. (*Hint:* Divide it into two right triangles, and use the transfer theorem.)

7. Using Problem 6, calculate the moment of inertia of an isosceles triangle about its axis of symmetry if the base is 9 cm and the height is 15 cm.

8. Using calculus, determine the location of the centroid and the moments of inertia about the x- and y-axes of an area bounded by the parabola $y = 4x^2$, the lines $x = 1$ and $x = 3$, and the x-axis.

Computer Calculations

9. Using program 'Centroid,' find the centroid locations and the centroidal moments of inertia for the shape shown in Figure 11.27.

10. Referring to Figure 11.27, we want to move the position of the centroid to $\bar{x} = 70$ cm, $\bar{y} = 20$ cm by increasing the width of the vertical leg. What should be the new leg width?

11. If both legs in Figure 11.27 are increased in width to 20 cm, what will be the location of the centroid?

12. Using AutoCAD, find the centroid locations and the centroidal moments of inertia for the shape shown in Figure 11.27.

13. Using program 'Centroid,' find the centroid locations and the centroidal moments of inertia for the shape shown in Figure 11.28.

14. If the hole in Figure 11.28 is moved by 1.5 in. to the left and 0.50 in. up, where would the

centroid be located, and what would be the new centroidal moments of inertia?

15. If there were no hole in the piece shown in Figure 11.28, where would the centroid be located, and what would be the centroidal moments of inertia?

16. Using AutoCAD, find the centroid locations and the centroidal moments of inertia for the shape shown in Figure 11.28.

17. Find the location of the centroid and the centroidal moments of inertia of the area shown in Figure 11.29 using AutoCad.

18. Find the location of the centroid and the centroidal moments of inertia of the area shown in Figure 11.29 using program 'Centroid'.

19. If the hole diameter were increased to 30 cm in the piece shown in Figure 11.29, where would the centroid be located and what

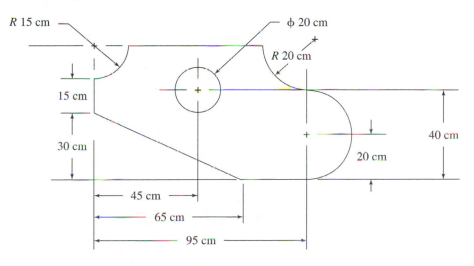

Figure 11.29 Problems 17, 18, 19, and 20.

would be the new centroidal moments of inertia?

20. If the radius of the 15-cm. quarter-circle in Figure 11.29 were doubled, where would the centroid be located and what would be the new centroidal moments of inertia?

21. Write a program to determine the location of the centroid and the moments of inertia about the x- and y-axes of an area bounded by the parabola $y = ax^2$, the lines $x = x_1$ and $x = x_2$, and the x-axis; a, x_1, and x_2 can have any values.

12

Torsion

■ **Objectives**

In this chapter you will learn about torque, work, power, and the stresses involved with shafts and couplings.

12.1   TORQUE

Torque is the twisting action applied in a plane perpendicular to the axis of an object (usually a shaft). We show a force acting at the end of an arm attached to a shaft in Figure 12.1(a) and the end view in Figure 12.1(b). The *external torque* acting on the shaft is equal to the force times the perpendicular distance from the line of action of the force to the center of the shaft. In Figure 12.1(a),

$$T = Fd \tag{12.1}$$

If the force does not act at a right angle to the center line of the arm (as in Figure 12.1(c)), then

$$T = Fd \sin \theta \tag{12.2}$$

You will notice that there is really no difference between torque and moment, as discussed in Section 3.2. We usually think of torque as something that is twisting a shaft or causing it to rotate, whereas a moment is something bending a beam. In general, we reserve the word *torque* for circular motion, or twisting.

Figure 12.2 shows an overhead shaft in a machine shop. It is driven by an electric motor through the pulley and belt A. If the bearings supporting the shaft were completely free of friction and if the shaft were not driving anything, then there would be no resistance to rotation and no net force ($F_1 - F_2$). However, the bearings are not frictionless, and the shaft is driving a lathe through the pulley and belt B. This creates tensions in the belts that are different on each side of the pulleys.

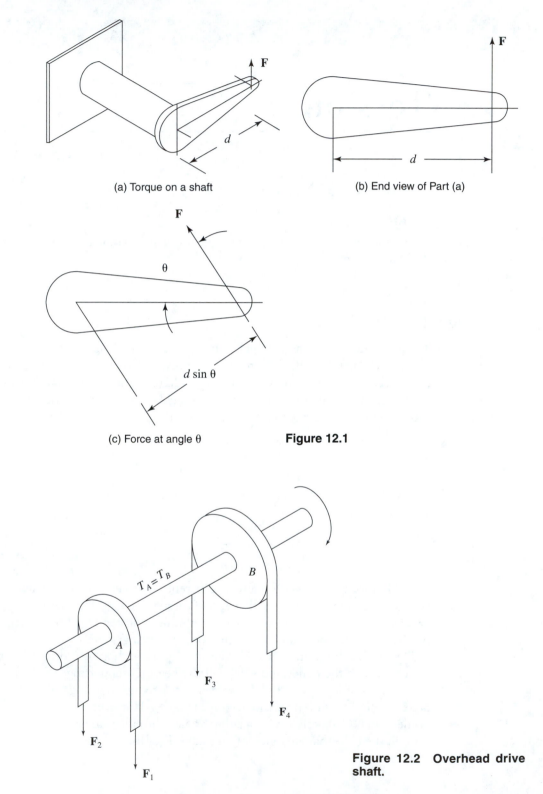

(a) Torque on a shaft

(b) End view of Part (a)

(c) Force at angle θ

Figure 12.1

Figure 12.2   Overhead drive shaft.

EXAMPLE PROBLEM 12.1

Suppose that in Figure 12.2 pulley A is 25 cm in diameter and pulley B is 60 cm. F_1 = 4 kN, F_2 = 1.5 kN, and F_3 = 1.75 kN. Calculate the force (tension) F_4. Assume that the bearing friction is negligibly small.

Solution

Because we are neglecting the bearing friction, all the torque going into pulley A must be transmitted to pulley B. The input torque is the *net* force acting on pulley A times the pulley radius. The net force is equal to F_1 on one side minus F_2 on the other.

$$T_A = (F_1 - F_2)r_A = (4 \text{ kN} - 1.5 \text{ kN})(12.5 \text{ cm}) = 31.25 \text{ kN·cm}$$

We usually work in N·m or N·mm, so T_A = 312.5 N·m (1 kN = 1000 N and 1 cm = 0.01 m). This is the torque on the shaft between the two pulleys, and that torque is being transmitted to the driven pulley B. The radius of pulley B is 30 cm (0.3 m), and F_3 = 1.75 kN (= 1750 N).

$$T_B = T_A = 312.5 \text{ N·m} = (F_3 - F_4)r_B = (1750 \text{ N} - F_4)(0.3 \text{ m})$$

Therefore,

$$1750 \text{ N} - F_4 = 312.5 \text{ N·m}/0.3 \text{ m} = 1042 \text{ N}$$
$$F_4 = 1750 \text{ N} - 1042 \text{ N} = 708 \text{ N}$$

Answer

The tension is F_4 = 708 N.

12.2   TORSIONAL DEFORMATION

Suppose we take a shaft, as in Figure 12.3, anchor one end so that it cannot rotate, and apply a torque to the shaft. Depending on the magnitude of the torque and the shaft material, there will be a certain amount of twisting. Obviously, for a given material, the greater the torque, the greater the twist. Before twisting, we draw a straight line AD parallel to the center line and on the surface of the shaft. As we apply the torque, point A, on the end face, will move through angle θ to point B. θ is the **angle of twist**. The amount of deformation, AB, on the surface is greater than that, $A'B'$, at an inner radius r. There is no deformation at the center. If we think of the shaft as consisting of millions of rings of material, each ring is sliding over the next. Starting at zero at the center, the strain increases linearly as the radius increases. So these rings are actually *shearing* over each other and shear stresses are being created, from zero at the center to a maximum at the surface. At the wall, however, there is no twisting, so the amount of deformation increases linearly as the shaft length, L, increases.

We introduced the *modulus of rigidity*, G, in Section 7.5. It is the *elastic modulus in shear*, like the elastic modulus, E, for tension and compression. G relates the shear stress to the shear strain, just as E relates direct stress to direct strain.

$$G = \frac{\text{shear stress}}{\text{shear strain}} = \frac{s_s}{\varepsilon_s} \qquad (12.3)$$

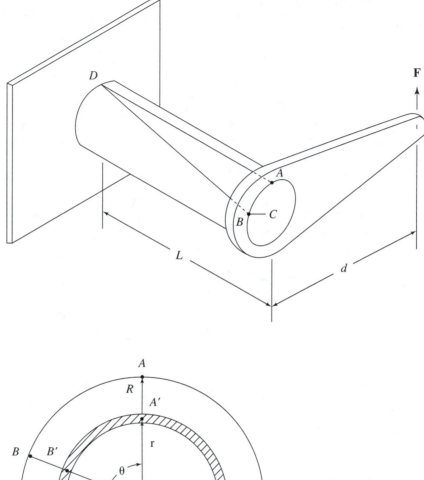

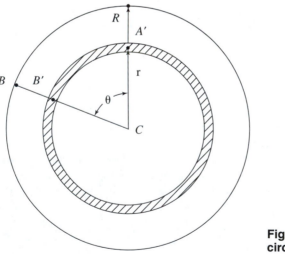

Figure 12.3   Deformation of a circular shaft.

The question is, what are shear stress and shear strain? Dealing with shear strain first, it is defined as the deformation divided by the shaft length.

$$\varepsilon_s = AB/L \qquad \text{(on the surface)}$$

It is important to have a physical understanding of this. For relatively small amounts of twist, the line AD (see Figure 12.3) will remain straight. So halfway from the wall, at $L/2$, the deformation AB will be half of that on the end face. The shear strain there will be $\varepsilon_s = (AB/2)/(L/2) = AB/L$. This shows that there is a *constant shear strain* all over the surface of the shaft. At some inner radius, such as the one shown in Figure 12.3, the deformation $A'B'$ on the end face will be less than that on the surface, but the distance from the wall is still L. So the strain on the shaded ring is less than that on the surface, but, again, it is constant at all points on that radius.

If we measure θ in radians and call the radius of the shaft R and the radius of the shaded ring r, then

$$AB = R\theta, \quad \text{and} \quad A'B' = r\theta$$

so

$$\varepsilon_s = AB/L \quad \text{(on the surface)} = R\theta/L \quad \text{and} \tag{12.4}$$
$$\varepsilon_s' = A'B'/L \quad \text{(at radius } r) = r\theta/L$$

Rearranging equation (12.4),

$$\theta = \varepsilon_s L/R \tag{12.5}$$

From equation (12.3), $G = s_s/\varepsilon_s$, or $\varepsilon_s = s_s/G$. Substituting for ε_s in equation (12.5):

$$\theta = (s_s L)/(GR) \text{ radians} \tag{12.6}$$

In the next section we relate shear stress to torque and return to equation (12.6)

12.3   TORSIONAL SHEAR STRESS

Referring again to Figure 12.3, the external torque on the shaft is $T = Fd$. This torque is resisted throughout the shaft, the sum of all the infinitesimal torques, ΔT, acting on all the shaft elements. The force, ΔF, acting on the shaded ring, whose area is A, is

$$\Delta F = \Delta T/r$$

Since stress = force/area,

$$s_s' = \Delta F/A = \Delta T/(Ar) \tag{12.7}$$

The shear stress increases linearly from zero at $r = 0$ to s_s at $r = R$. Expressed mathematically,

$$s_s'/r = s_s/R \quad \text{or} \quad s_s' = s_s r/R$$

Substituting for s_s' in equation (12.7),

$$s_s r/R = \Delta T/(Ar)$$

Therefore, $s_s = (R\Delta T)/(Ar^2)$.

Adding all the infinitesimal rings gives

$$s_s = RT/\Sigma(Ar^2)$$

$\Sigma(Ar^2)$ will be recognized as J, the *polar moment of inertia*, introduced in Section 11.11. Finally,

$$s_s = RT/J \tag{12.8}$$

J is a function only of the geometry. For example, for a solid circular shaft, $J = \pi D^4/32$. R is also a geometric value, so the two are often combined as J/R, which is called the

polar section modulus, Z_p.

$$Z_p = J/R \tag{12.9}$$

so

$$s_s = T/Z_p \tag{12.10}$$

For a *solid* circular shaft,

$$J = \pi D^4/32 \tag{12.11}$$
$$R = D/2, \text{ so}$$
$$Z_p = \pi D^3/16 \tag{12.12}$$

For a *hollow* circular shaft with an inside diameter of D_i and an outside diameter of D_o,

$$J = \pi(D_o^4 - D_i^4)/32 \tag{12.13}$$
$$Z_p = \pi(D_o^4 - D_i^4)/16D_o \tag{12.14}$$

Previously, we found an expression for the angle of twist, θ, which was (equation 12.6):

$$\theta = (s_s L)/(GR) \text{ radians}$$

Substituting for s_s from equation (12.8) gives

$$\theta = (RTL/J)/(GR)$$

or

$$\theta = (TL)/(JG) \text{ radians} \tag{12.15}$$

EXAMPLE PROBLEM 12.2

How much torque can be safely transmitted by a 10-cm-diameter solid shaft made of AISI 1141 OQT 700? Use a factor of safety of 8 based on the ultimate shear strength. The ultimate shear strength of AISI 1141 OQT 700 is 1089 MPa.

Solution
Using equation (12.10),

$$s_s = T/Z_p \quad \text{or} \quad T = s_s Z_p$$

From equation (12.12), with D = 10 cm = 100 mm:

$$Z_p = \pi D^3/16 = \pi(100 \text{ mm})^3/16 = 196.3 \times 10^3 \text{ mm}^3$$

Ultimate shear strength of AISI 1141 OQT 700 is 1089 MPa; therefore, s_s = 1089 MPa/8 = 136 MPa. So the torque is

$$T = s_s Z_p$$
$$= (136 \text{ MPa})(196.3 \times 10^3 \text{ mm}^3)$$
$$= 26.7 \times 10^6 \text{ MPa} \cdot \text{mm}^3$$

What are these units?

$$1 \text{ MPa} = 1 \text{ N/mm}^2$$

So

$$1 \text{ MPa} \cdot \text{mm}^3 = 1 \text{ N} \cdot \text{mm}$$

Thus, $T = 26.7 \times 10^6$ N mm; split this into

$$T = 26.7(1000 \text{ N})(1000 \text{ mm})$$
$$= 26.7 \text{ kN} \cdot \text{m}$$

Answer
The amount of torque that can be safely transmitted is 26.7 kN · m.

EXAMPLE PROBLEM 12.3

A torque of 400 N · m is applied to a 5-cm-diameter solid shaft made of AISI 4140 OQT 1100 steel. The modulus of elasticity in shear for steel is $G = 80 \times 10^3$ MPa. If the shaft is 1 m long, what is the twist angle in degrees?

Solution
From equation (12.15)

$$\theta = (TL)/(JG) \text{ radians}$$

$$T = 400 \text{ N·m}, \qquad L = 1 \text{ m}, \qquad J = \pi D^4/32 \qquad \text{from equation (12.11)}$$

$$G = 80 \times 10^3 \text{ MPa}, \quad \text{and} \quad D = 5 \text{ cm}$$

$$J = \pi(50 \text{ mm})^4/32 = 6.136 \times 10^5 \text{ mm}^4$$

So,

$$\theta = \frac{[(400 \times 10^3 \text{ N·mm}) (1000 \text{ mm})]}{(6.136 \times 10^5 \text{ mm}^4) (80 \times 10^3 \text{ MPa})}$$

$$= (4.0 \times 10^8 \text{ N·mm}^2)/(490.1 \times 10^8 \text{ MPa·mm}^4) \text{ radians}$$

Note: 1 MPa = 1 N/mm^2; therefore, 1 MPa·mm^4 = 1 N·mm^2. So the units on the top and bottom cancel out, as they should because the answer is in radians.

$$\theta = 8.2 \times 10^{-3} \text{ rad}$$

Because 1 rad = 57.3°,

$$\theta = 0.47°$$

Note: A general rule is that a shaft should not twist more than 1° in 20 diameters of length. In this example, 20 diameters = 20 × 5 cm = 100 cm = 1 m, which is the shaft length. The twist angle is well below the recommended limit of 1°.

Answer
The angle of twist is 0.47°.

EXAMPLE PROBLEM 12.4

A 4-in-diameter AISI stainless steel 301 full hard shaft is to transmit a torque of 13,000 lb-ft. Since this is an aircraft application, weight is important, so a core is to be drilled out of the shaft. Using a safety factor of 10 based on ultimate shear strength, what diameter core can be safely removed?

Solution

Using program 'Proptab', the ultimate tensile strength of AISI 301 full hard is 185 ksi (185,000 psi), so the ultimate shear strength (see Chapter 6) is 0.82 × 185 ksi = 151 ksi, and the allowable shear stress is 151,000 psi/10 = 15,100 psi.

$$s_s = T/Z_p$$

so $$Z_p = T/s_s = (13,000 \times 12 \text{ lb-in.})/15,100 \text{ psi} = 10.33 \text{ in.}^3$$

For a hollow circular shaft, $Z_p = \pi(D_o^4 - D_i^4)/16D_o$. Because $D_o = 4$ in.,

$$Z_p = \pi D_o^3/16 - \pi D_i^4/16D_o = \pi(4 \text{ in.})^3/16 - \pi D_i^4/16(4 \text{ in.})$$

$$Z_p = (12.57 - 0.0491D_i^4) \text{ in.}^3$$

So,

$$10.33 = 12.57 - 0.0491D_i^4$$

$$0.0491D_i^4 = 2.24$$

$$D_i^4 = 45.62$$

$$D_i = 2.60 \text{ in.}$$

Answer

A core diameter of 2.60 in. can be safely removed.

EXAMPLE PROBLEM 12.5

What diameter *solid* shaft, of the same material, would be required for the same conditions as those in Example Problem 12.4? Compare the weights of the two shafts.

Solution

For a solid shaft, $Z_p = \pi D^3/16 = 0.196D^3$. From Example Problem 12.4, $T/s_s = 10.33$ in.^{3}. Therefore, $0.196D^3 = 10.33$.

$$D^3 = 52.61$$

$$D = 3.75 \text{ in.}$$

The cross-sectional area of this shaft is $A_{SOLID} = (0.7854)(3.75 \text{ in.})^2$.

$$A_{SOLID} = 11.04 \text{ in.}^2$$

For the hollow shaft of Example Problem 12.3,

$$A_{HOLLOW} = (0.7854)[(4 \text{ in.})^2 - (2.60 \text{ in.})^2] \text{ in.}^2 = 7.26 \text{ in.}^2$$

So $A_{SOLID}/A_{HOLLOW} = 11.04 \text{ in.}^2/7.26 \text{ in.}^2 = 1.52$

Answer

The solid shaft is 52% heavier than the hollow shaft. Why?

As we discussed before, there is zero shear stress at the axis, and it increases linearly with radius, reaching a maximum at the surface of the shaft. So the inner core is contributing very little to torque resistance of the shaft. This is fortunate, not only from a weight-reduction point of view, but also because a central hole carrying lubricant to bearings would have little effect the shaft strength.

12.4 STRESS CONCENTRATIONS

Grooves, holes, and keyseats can reduce the shaft torque strength considerably. Experiments have been conducted to assess these effects, and the data have been expressed in terms of stress-concentration factors, k. These are similar to those we encountered in Section 6.4.1. To calculate the actual torsional shear stress, we first calculate it disregarding the discontinuity; then we multiply the answer by k.

12.4.1 Stepped Circular Shaft

Stress concentration factors for stepped circular shafts can be correlated by the equations

$$k = 0.90/(r/d)^n \qquad (12.16)$$
$$n = -8.438 \times 10^{-3}(D/d)^2 + 4.844 \times 10^{-2}(D/d) + 0.1336$$
$$2.5 \geq (D/d) \geq 1.25 \qquad (12.16(a))$$
$$n = 0.0024 + 0.1429(D/d) \qquad 1.25 \geq (D/d) \geq 1.11 \qquad (12.16(b))$$

The definitions of r, d, and D are the same as in Figure 6.6(a). These equations say that if D/d lies between 2.5 and 1.25, use equation (12.16(a)) to calculate n; then use equation (12.16). If D/d lies between 1.25 and 1.11, use equation (12.16(b)) for n; then use equation (10.16). The shear stress developed will be k times that calculated neglecting the discontinuity.

EXAMPLE PROBLEM 12.6

A 3-ft.-long hollow circular shaft is stepped down from a 2-in. diameter to a 1.7-in. diameter, with a fillet radius of 0.20 in. The shaft is made of AISI 1020 annealed steel and has an inside diameter of 1.0 in. The torsional shear stress is to be limited to one-tenth of the ultimate shear stress.

    a. What maximum torque can be applied?
    b. What will be the twist angle?
    c. Is the twist acceptable?

Solution
    a. Calculation of maximum torque.
        1. From program 'Proptab', the ultimate tensile strength of AISI 1020 annealed steel is 57 ksi; therefore, its ultimate shear strength is 0.82×57 ksi $=$ 46 ksi.

$$s_{max} = s_s/10 = 4.6 \text{ ksi} = 4600 \text{ psi}$$

2. Calculate the stress-concentration factor:

$$r/d = 0.20 \text{ in.}/1.7 \text{ in.} = 0.1176$$

$$D/d = 2.0 \text{ in.}/1.7 \text{ in.} = 1.176$$

Because this value of D/d lies between 1.25 and 1.11, use equation (12.16(b)) for n; then use equation (12.16).

$$n = 0.0024 + 0.1429(D/d)$$

$$= 0.0024 + 0.1429(1.176) = 0.170$$

$$k = 0.90/(r/d)^n$$

$$= 0.90/(0.1176)^{0.170}$$

$$= 1.295$$

3. Calculate the polar section modulus, Z_p:

$$Z_p = \pi(D_o^4 - D_i^4)/16D_o \qquad \text{from equation (12.14)}$$

Note: We base our calculations on the *smaller* outside diameter

$$D_o = 1.7 \text{ in.}, \qquad D_i = 1.0 \text{ in.}$$

$$Z_p = \frac{\pi[(1.7 \text{ in.})^4 - (1.0 \text{ in.})^4]}{(16)(1.7 \text{ in.})} = 0.849 \text{ in.}^3$$

4. If there were no step, the maximum allowable torque would be

$$T = s_{max}Z_p = (4600 \text{ psi})(0.849 \text{ in.}^3) = 3905 \text{ lb-in.}$$

but the step increases the stress by a factor of $k = 1.295$, so to stay within the required stress limit we must reduce the torque by the same amount.

$$T_{max} = 3905 \text{ lb-in.}/1.295 = 3015 \text{ lb-in.}$$

b. Find the twist angle.

1. From equation (12.15),

$$\theta = (TL)/(JG) \text{ radians}$$

2. $J = \pi (D_o^4 - D_i^4)/32$ from equation (12.13)
$= \pi[(1.7 \text{ in.})^4 - (1.0 \text{ in.})^4]/32 = 0.722 \text{ in.}^4$

3. $L = 36 \text{ in.}, \quad G = 11.5 \times 10^6 \text{ psi for alloy steel}$
$T = 3015 \text{ lb-in.}$

so

$$\theta = \frac{(3015 \text{ lb-in.})(36 \text{ in.})}{(0.722 \text{ in.}^4)(11.5 \times 10^6 \text{ psi})} \text{ rad}$$

$$= 0.013 \text{ rad} = (0.013 \text{ rad})(57.3°/\text{rad}) = 0.74°$$

c. Twenty times the diameter is $20 \times 1.7 \text{ in.} = 34 \text{ in.}$ We have a twist of 0.74° in 36 in., or $(0.74°) (34 \text{ in.}/36 \text{ in.}) = 0.70°$ in 20 diameters, which is well within the recommended limit of 1°.

Answers

a. The maximum torque that may be applied is 3015 lb-in.
b. The twist angle is 0.74°.
c. The twist is acceptable.

12.4.2   Grooved Circular Shaft

Stress-concentration factors for grooved circular shafts can be correlated by the equations

$$k = 0.331 + 0.6/(r/d)^n \qquad (12.17)$$

$$n = 0.3113(D/d) - 0.097 \qquad 1.2 \geq D/d \geq 1.05 \qquad (12.17(a))$$

$$n = 0.0163(D/d) + 0.2595 \qquad 2.0 \geq D/d \geq 1.2 \qquad (12.17(b))$$

The definitions of r, d, and D are the same as in Figure 6.6(b).

   These equations say that if D/d lies between 1.2 and 1.05, use equation (12.17(a)) to calculate n; then use equation (12.17). If D/d lies between 2.0 and 1.2, use equation (12.17(b)) for n; then use equation (12.17). The shear stress developed will be k *times that calculated neglecting the discontinuity.*

EXAMPLE PROBLEM 12.7

A 25-mm-diameter 1100-H12 aluminum alloy shaft has a retaining ring groove that is 23.1 mm in diameter with a 1.25-mm radius. Using a safety factor of 6 based on ultimate shear strength, what torque can be transmitted safely?

Solution

   1. From program 'Proptab', the ultimate shear strength of 1100-H12 is 69 MPa. The allowable shear stress, neglecting the groove, is 69 MPa/6 = 11.5 MPa. The actual usable shear stress is s_{max} = 11.5 MPa/k.

   2. Z_p, based on the *groove diameter*, is

$$Z_p = \pi d^3/16 \qquad \text{from equation (12.12)}$$

where d = 23.1 mm.

$$Z_p = \pi(23.1 \text{ mm})^3/16 = 2420 \text{ mm}^3$$

   3. Calculate the stress concentration factor.

$$r = 1.25 \text{ mm}, \qquad d = 23.1 \text{ mm}, \qquad D = 25 \text{ mm}$$

$$D/d = (25 \text{ mm})/(23.1 \text{ mm}) = 1.083$$

$$r/d = (1.25 \text{ mm})/(23.1 \text{ mm}) = 0.054$$

Using equation (12.17(a)),

$$n = 0.3113(D/d) - 0.097 = (0.3113)(1.083) - 0.097 = 0.240$$

Using equation (12.17),

$$k = 0.331 + 0.6/(r/d)^n = 0.331 + 0.6/(0.054)^{0.240}$$

$$= 0.331 + 0.6/0.4963 = 1.54$$

   4. Calculate the maximum usable stress:

$$s_{max} = (11.5 \text{ MPa})/k = (11.5 \text{ MPa})/1.54 = 7.47 \text{ MPa}$$

5. Calculate the maximum torque:

$$s_{max} = T_{max}/Z_p$$

So $T_{max} = s_{max}Z_p = (7.47 \text{ MPa})(2420 \text{ mm}^3) = 18077 \text{ N·mm} = 18.08 \text{ N·m}$

Answer
The maximum torque is 18.0 N·m.

12.4.3 Keyseats

Figure 12.4(a) and (b) shows two types of keyseats, namely, a sled runner and a profile type. The sled runner is cut with a circular milling cutter, which creates a shape like a runner on a sled. The profile type is cut with an end mill, leaving (almost) sharp corners. Which type do you think would create the larger stress concentration? Right, the one leaving the sharp corners.

For the sled runner type, $k = 1.6$.
For the profile type, $k = 2.0$.

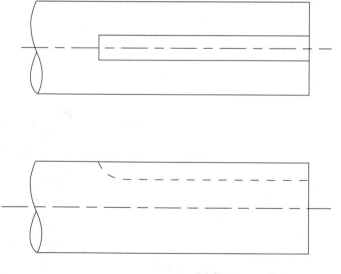

(a) Sled-runner keyseat

Figure 12.4   Keyseats.

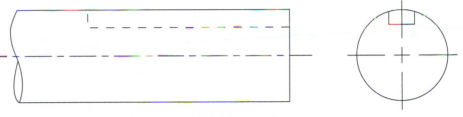

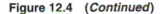

(b) Profile keyseat

Figure 12.4 (*Continued*)

EXAMPLE PROBLEM 12.8

A 6061-T6 aluminum alloy shaft transmits a torque of 1.0 kN·m under shock-loading conditions. The shaft has a profile keyseat. What diameter solid shaft is required?

Solution

From program 'Proptab', the ultimate shear strength of 6061-T6 is 207 MPa. Under shock-loading conditions it is recommended that, for ductile materials, the allowable stress should be limited to one-twelfth of the ultimate strength (see Section 7.3). The profile keyseat will impose a further limitation of a factor of 2. The actual allowable stress, then, is ultimate/24 = 207 MPa/24 = 8.6 MPa.

$$s_s = T/Z_p$$

so

$$Z_p = T/s_s$$

$$T = 10 \text{ kN·m}, \qquad s_s = 8.6 \text{ MPa}$$

$$Z_p = 1.0 \text{ kN·m}/8.6 \text{ MPa} = 0.116 \text{ (kN·m/MPa)}$$

What are these units?

$$1 \text{ kN·m} = (1000 \text{ N})(1000 \text{ mm}) = 10^6 \text{ N·mm}$$

$$1 \text{ MPa} = 1 \text{ N/mm}^2$$

$$1 \text{ (kN·m/MPa)} = 10^6 \text{ mm}^3$$

Therefore, $Z_p = 0.116 \times 10^6$ mm³.

$$Z_p = \pi D^3/16 \quad \text{from equation (12.12)}$$

or $\quad D^3 = 16Z_p/\pi = (16)(0.116 \times 10^6 \text{ mm}^3)/\pi = 1.86 \times 10^6 \text{ mm}^3$

Taking the cube root,

$$D = 123.0 \text{ mm}$$

Answer
The required shaft diameter is 123.0 mm.

12.5   POWER

*Power is the rate of doing work, and **work** is force times distance in the direction of the line of action of the force.*

For example, if you lift a 50-lb weight 6 ft, you have done (50 lb)(6 ft) = 300 ft-lb of work. It does not matter whether you accomplish this lift in 1 s, 30 s, or any amount of time, you have still done 300 ft-lb of work. Where the time comes in is in the amount of *power* you use.

Suppose you lift the 50-lb weight 6 ft in 2 s. Your rate of doing work (power) is 300 ft-lb in 2 s, or 150 ft-lb/s. To accomplish the task in 1 s would require twice the power.

One horsepower is defined as 550 ft-lb/s, or 33,000 ft-lb/min.

Lifting 50 lb 6 ft in 2 s would require (150 ft lb/s)/(550 ft lb/s) = 0.273 hp.
In the SI system of units, work is measured in *joules*.

One joule is the work done when a force of 1 N acts through a distance of 1 m.

$$1 \text{ J} = 1 \text{ N·m}$$

Power is work divided by time, i.e., J/s.

$$1 \text{ J/s} = 1 \text{ W}$$

Power is measured in watts (W) or kilowatts (kW).
The relationship between horsepower and watts is

$$1 \text{ hp} = 746 \text{ W} \quad \text{or} \quad 1 \text{ kW} = 1.340 \text{ hp}$$

EXAMPLE PROBLEM 12.9

A 1000-kg mass is to be raised 5 m in 3 s. How much horsepower is required?

Solution
First, we must convert the mass to weight.

$$1000 \text{ kg} = (1000 \text{ kg})(9.81 \text{ m/s}^2) = 9810 \text{ N}$$

$$\text{work} = \text{force} \times \text{distance} = (9810 \text{ N})(5 \text{ m}) = 49{,}050 \text{ N·m}$$

$$\text{power} = \text{work/time} = (49{,}050 \text{ N·m})/(3\text{s}) = 16{,}350 \text{ N·m/s}$$
$$= 16{,}350 \text{ W} = 16.35 \text{ kW}$$
$$\text{horsepower} = (16.35 \text{ kW})(1.340 \text{ hp/kW}) = 21.9 \text{ hp}$$

Answer

The horsepower required is 21.9 hp.

12.5.1 Rotary Motion

In Figure 12.5 we show a weight, W, attached to a cord that is wrapped around a pulley wheel of radius r.

In one revolution of the wheel, the weight moves a distance equal to the circumference of the wheel, i.e., $2\pi r$. Since the force, F, is equal to the weight, we can express the work done in one revolution as

$$\text{work} = \text{force} \times \text{distance} = (F)(2\pi r) = 2\pi r F$$

But rF is the torque, T, on the wheel, so

$$\text{work} = 2\pi T \qquad \text{in one revolution}$$

If T is in lb-ft, the work per revolution is $2\pi T$ ft-lb.

Note: It is common practice to express torque and moments in lb-ft, whereas work is usually expressed in ft-lb.

If the wheel is rotating at N revolutions per minute (rpm), then it completes one revolution in $(1/N)$ min. So, the work in one revolution is $2\pi T$ ft-lb, and it takes $(1/N)$ min. Therefore, the power is

$$\text{power} = (2\pi T \text{ ft-lb})/(1/N) \text{ min} = 2\pi T N \text{ ft-lb/min}$$

In terms of horsepower (1 hp = 33,000 ft-lb/min):

$$\text{hp} = 2\pi T N/33{,}000 = TN/5252 \qquad (T \text{ lb-ft}, N \text{ rpm}) \qquad (12.18(\text{a}))$$

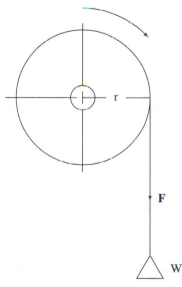

Figure 12.5 Wheel rotated by falling weight.

We often work in lb-in. for torque, in which case:
$$hp = 2\pi(T/12)N/33,000 = TN/63,025 \qquad (T \text{ lb-in.}, N \text{ rpm}) \quad (12.18(b))$$

In the SI system of units, the power unit is the watt, W, and 1 W is equal to 1 N·m/s. So, if the torque is in N·m and the rotational speed is in rpm,
$$\text{power} = 2\pi TN/60 = TN/9.55 \text{ W} \qquad (T \text{ N·m}, N \text{ rpm}) \qquad (12.19)$$

The relationships between horsepower and watts are
$$1 \text{ hp} = 746 \text{ W} \qquad 1 \text{ kW} = 1.340 \text{ hp} \qquad (12.20)$$

EXAMPLE PROBLEM 12.10

For a high-school project, a student makes a small electric generator. It is driven by a falling weight attached to a string wrapped around a pulley wheel. The wheel diameter is 5 in. and the weight is 10 lb. Just before hitting the floor, the weight is dropping at a rate of 9 ft/s. The overall efficiency of the system is 50%. What is its maximum output in watts?

Solution

$$\text{torque} = \text{force} \times \text{radius} = (10 \text{ lb})(2.5 \text{ in.}) = 25 \text{ lb-in.}$$

We can calculate the maximum rotational speed from the maximum speed of the falling weight. When the weight is falling at 9 ft/s, the rim of the wheel is moving at 9 ft/s. The wheel circumference is $\pi D = \pi(5 \text{ in.}) = 15.71 \text{ in.} = 1.309 \text{ ft}$. If the weight is traveling at 9 ft/s and it takes 1.309 ft for the wheel to complete one revolution, then the wheel is rotating at (9 ft/s)/(1.309 ft) = 6.88 rev/s, or 6.88 × 60 = 412.5 rpm.

$$hp = TN/63,025 \qquad (T \text{ lb-in}, N \text{ rpm}) \qquad \text{from equation} \qquad (12.18(b))$$
$$= (25 \text{ lb-in.})(412.5 \text{ rpm})/63,025 = 0.164 \text{ hp}$$

Because the system is only 50% efficient, the output horsepower is 0.164/2 = 0.08 hp.

$$1 \text{ hp} = 746 \text{ W}$$
$$\text{power} = (0.08 \text{ hp})(746 \text{ W/hp}) = 59 \text{ W}$$

Answer
The maximum output is 59 W.

12.6 POWER TRANSMISSION

The amount of power that can be transmitted safely by a shaft depends on many factors. Basically, it depends on the torque and the rpm, but the torque is limited by the maximum allowable stress or by the maximum allowable twist angle. If the limit is the twist angle, then the material ultimate shear strength does not enter into the calculation—only its modulus of rigidity. Concentration factors play an important role, as we show in the Example 12.11.

EXAMPLE PROBLEM 12.11

A 1-m-long, 50-mm-diameter AISI 501 OQT 1000 shaft has a sled-runner keyseat and a retaining-ring groove. The groove diameter is 47 mm and has 2.4-mm radii. The shaft is to be driven at 1800 rpm, and the angle of twist is not to exceed 1° over the length of the shaft. What maximum power may be transmitted?

The modulus of elasticity in shear (modulus of rigidity) is $G = 80 \times 10^3$ MPa.

Solution

1. Calculate the stress limit imposed by the angle of twist.

$$1° = 0.0175 \text{ rad}$$

From equation (12.6)

$$\theta = (s_s L)/(GR) \text{ radians}$$

Rearranging,

$$s_s = GR\theta/L$$

$$G = 80 \times 10^3 \text{ MPa}, \quad R = 25 \text{ mm}, \quad \theta = 0.0175 \text{ rad}, \quad L = 1000 \text{ mm}$$

So $s_s = (80 \times 10^3$ MPa$)(25$ mm$)(0.0175$ rad$)/1000$ mm

$$= 35 \text{ MPa}$$

2. Calculate the stress-concentration factor. There are two factors, one due to the keyseat, $k = 1.6$, and one due to the groove, which we now calculate:

$$r = 2.4 \text{ mm}, \quad D = 50 \text{ mm}, \quad d = 47 \text{ mm}$$

$$r/d = 2.4 \text{ mm}/47 \text{ mm} = 0.051, \quad D/d = 50 \text{ mm}/47 \text{ mm} = 1.064$$

Using equation (12.17 (a)),

$$n = 0.3113(D/d) - 0.097 \quad 1.2 \geq D/d \geq 1.05$$

$$n = (0.3113)(1.064) - 0.097 = 0.234$$

Then from equation (12.17),

$$k = 0.331 + 0.6/(r/d)^n = 0.331 + (0.6)/(.051)^{0.234}$$

$$= 0.331 + 0.6/0.498 = 1.54$$

The combined effect of the keyseat and the groove is to increase the stress by $(1.6)(1.54) = 2.46$. We cannot allow the stress to exceed 35 MPa, so before cutting the keyseat and groove, the stress would be limited to 35 MPa/2.46 = 14.2 MPa. This, then, is the stress we have to use in calculating the maximum torque.

3. Calculate the polar section modulus, Z_p. For a solid circular shaft,

$$Z_p = \pi D^3/16 \qquad \text{from equation (12.12)}$$

$$D = 50 \text{ mm}$$

$$Z_p = \pi(50 \text{ mm})^3/16 = 2.454 \times 10^4 \text{ mm}^3$$

4. Calculate the maximum torque.

$$T = s_s Z_p$$

$$s_s = 14.2 \text{ MPa}, \qquad Z_p = 2.454 \times 10^4 \text{ mm}^3$$

$$s_s = (14.2 \text{ MPa})(2.454 \times 10^4 \text{ mm}^3) = 3.485 \times 10^5 \text{ N·mm}$$

or $\qquad T = 348.5 \text{ N·m}$

5. Calculate the maximum power.

$$\text{power} = TN/9.55 \text{ W} \qquad (T \text{ N·m}, N \text{ rpm}) \qquad \text{from equation} \quad (12.19)$$

$$T = 348.5 \text{ N·m}, \qquad N = 1800 \text{ rpm}$$

$$\text{power} = (348.5 \text{ N·m})(1800 \text{ rpm})/9.55 = 65{,}690 \text{ W} = 65.7 \text{ kW}$$

Notice that the actual type of steel was not needed for this example, because the twist angle was the limiting factor.

Answer

The maximum power that may be transmitted is 65.7 kW.

12.7   SHAFT COUPLINGS

One of the most widely used ways in which power is transmitted between coaxial shafts is shown in Figure 12.6. The driving shaft transmits power to flange A through key B. Flange A then transmits its power to flange C through the bolts and then to the driven shaft through key D. The forces on the bolts (Figure 12.7) act at 90° to the radius.

In analyzing a system like this, we have to check the adequacy of

The keys in shear (Figure 12.8(a)),
The keys in bearing (Figure 12.8(b)),
Hub-web shear (Figure 12.9),
Bolt shear,
Bolt bearing.

Let's look at each of these.

Keys in Shear

Referring to Figure 12.8(a), there is a tendency to shear the key through a horizontal plane whose area is $(l_k w_k)$, where l_k is the length of one key and w_k is the key width. The force on the key is $F_k = T/r_k$, where T is the torque, and r_k is the distance from the shaft center to the key. This force varies over the height of the key, but the average value is $F_k = T/r_s$ where r_s is the shaft radius.

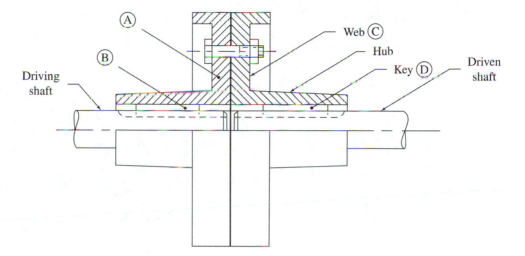

Figure 12.6   Power transmission between coaxial shafts.

So the shear stress on the key is

$$s_{sk} = \text{force/area}$$

where force $= (T/r_s)$ and area $= (l_k w_k)$. Therefore, $s_{sk} = (T/r_s)/(l_k w_k) = T/(r_s l_k w_k)$. In terms of shaft diameter, d_s—i.e., replacing r_s with $d_s/2$—

$$s_{sk} = 2T/(d_s l_k w_k) \qquad (12.21)$$

Keys in Bearing

The bearing area is $(l_k t_k/2)$ (see Figure 12.8(b)). The force on the key is T/r_s, as before, so the bearing stress is

$$s_{bk} = \text{force/area} = (T/r_s)/(l_k t_k/2) = 2T/(r_s l_k t_k)$$

In terms of shaft diameter, d_s,

$$s_{bk} = 4T/(d_s l_k t_k) \qquad (12.22)$$

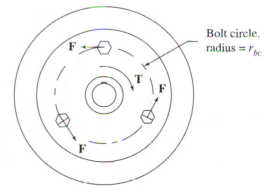

Figure 12.7   Bolt forces resist external torque.

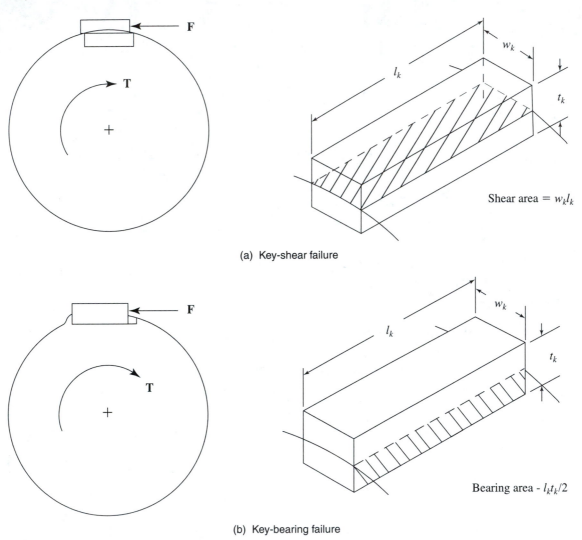

Shear area $= w_k l_k$

(a) Key-shear failure

Bearing area - $l_k t_k/2$

(b) Key-bearing failure

Figure 12.8   Types of key failure.

Hub-Web Shear

The shear area where the web attaches to the hub of the flange is the cylindrical surface ($\pi d_h t_w$), where d_h is the hub diameter and t_w is the web thickness (see Figure 12.9). The force acting on this shear area is the torque divided by the hub radius, or torque divided by one-half the hub diameter.

$$F_h = [T/(d_h/2)] = 2T/d_h$$

The shear stress on the hub-web section is

$$s_{sh} = (2T/d_h)/(\pi d_h t_w) = 2T/(\pi d_h^2 t_w) \tag{12.23}$$

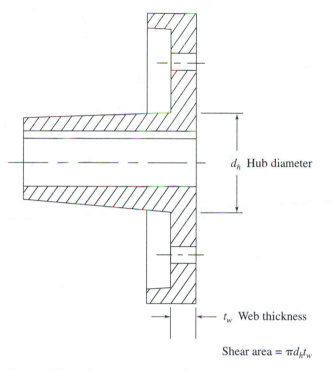

d_h Hub diameter

t_w Web thickness

Shear area $= \pi d_h t_w$

Figure 12.9   Hub-web shear failure.

Bolt Shear

The total force at radius r_{bc}, the bolt circle radius shown in Figure 12.7, is $F_b = T/r_{bc}$. This force is shared equally by the N_b bolts in that bolt circle. So the force acting on each bolt is $T/(N_b r_{bc})$. If the bolt diameter is D_b, then the cross-sectional area of one bolt is $(\pi D_b^2/4)$, and the shear stress is

$$s_{sb} = [T/(N_b r_{bc})]/(\pi D_b^2/4)$$

So,
$$s_{sb} = 4T/(N_b r_{bc} \pi D_b^2)$$

In terms of the bolt-circle diameter, d_{bc}, (i.e., $r_{bc} = d_{bc}/2$),

$$s_{sb} = 8T/(N_b d_{bc} \pi D_b^2) \tag{12.24}$$

Bolt Bearing

For one bolt on the bolt circle, the bearing area is $D_b t_w$, the bolt diameter times the web thickness, and the force acting on the bolt is $T/(N_b r_{bc})$. So the bearing stress is

$$s_{bb} = [T/(N_b r_{bc})]/(D_b t_w) = T/(N_b r_{bc} D_b t_w)$$

In terms of the bolt-circle diameter, d_{bc},

$$s_{bb} = 2T/(N_b d_{bc} D_b t_w) \tag{12.25}$$

Collecting these equations together, we have the following.

Key shear:

$$s_{sk} = 2T/(d_s l_k w_k) \tag{12.21}$$

Key bearing:

$$s_{bk} = 4T/(d_s l_k t_k) \tag{12.22}$$

Hub-web shear:

$$s_{sh} = 2T/(\pi d_h^2 t_w) \tag{12.23}$$

Bolt shear:

$$s_{sb} = 8T/(N_b d_{bc} \pi D_b^2) \tag{12.24}$$

Bolt bearing:

$$s_{bb} = 2T/(N_b d_{bc} D_b t_w) \tag{12.25}$$

EXAMPLE PROBLEM 12.12

Two 60-mm-diameter shafts are joined by a flange coupling that has 6 AISI 1020 hot-rolled, 6-mm bolts on a BC diameter (bolt-circle diameter) of 140 mm. The flange, which is made of AISI 1141 annealed steel, has a hub diameter of 100 mm and a web thickness of 25 mm. The 75-mm-long, 12-mm-square keys are made of AISI 1080 OQT 1300 steel. Using a safety factor of 4, based on ultimate shear strength, for all materials, what maximum horsepower can be safely transmitted at 500 rpm?

Solution

1. Run program 'Proptab' to get the ultimate tensile strengths (from which we will calculate the ultimate shear strengths) and the yield strengths for the bearing calculations.

The ultimate tensile strengths are

AISI 1020 hot-rolled steel	448 MPa	(bolts)
AISI 1141 annealed steel	600 MPa	(flange)
AISI 1080 OQT 1300 steel	807 MPa	(keys)

Multiplying each of these by 0.82 gives the following ultimate shear strengths:

AISI 1020 hot-rolled steel	367 MPa	(bolts)
AISI 1141 annealed steel	492 MPa	(flange)
AISI 1080 OQT 1300 steel	662 MPa	(keys)

Dividing each of these by the safety factor gives the allowable shear stresses:

AISI 1020 hot-rolled steel	91 MPa	(bolts)
AISI 1141 annealed steel	123 MPa	(flange)
AISI 1080 OQT 1300 steel	166 MPa	(keys)

The yield strengths are

AISI 1020 hot-rolled steel	331 MPa	(bolts)
AISI 1141 annealed steel	352 MPa	(flange)
AISI 1080 OQT 1300 steel	483 MPa	(keys)

Multiplying the ultimate strengths by 1.20 (see Section 9.3.1) for bearing strength and then dividing by 4 for the safety factor gives the allowable bearing stresses:

AISI 1020 hot-rolled steel	134 MPa	(bolts)
AISI 1141 annealed steel	180 MPa	(flange)
AISI 1080 OQT 1300 steel	242 MPa	(keys)

2. To calculate the maximum horsepower, we must first calculate the maximum torque that the coupling can safely handle. To do this, we have to determine the maximum torque that each component can withstand and then take the *smallest* of those. The known values are

$$d_s = 60 \text{ mm} \qquad d_{bc} = 140 \text{ mm} \qquad d_h = 100 \text{ mm} \qquad N_b = 6$$

$$t_w = 25 \text{ mm} \qquad l_k = 75 \text{ mm} \qquad w_k = 12 \text{ mm} \qquad t_k = 12 \text{ mm}$$

$$s_{sk} = 166 \text{ MPa} \qquad s_{sh} = 123 \text{ MPa} \qquad s_{sb} = 91 \text{ MPa}$$

$$s_{bk} = 242 \text{ MPa} \qquad s_{bh} = 180 \text{ MPa} \qquad s_{bb} = 134 \text{ MPa}$$

a. *Key shear.* Using equation (12.21)

$$s_{sk} = \frac{2T}{d_s l_k w_k}$$

or

$$T = \frac{s_{sk} d_s l_k w_k}{2}$$

$$= \frac{(166 \text{ MPa})(60 \text{ mm})(75 \text{ mm})(12 \text{ mm})}{2} = 4.48 \times 10^6 \text{ N·mm}$$

Since 1 N·mm = 10^{-3} N·m, the maximum torque that the key can handle in shear is 4480 N·m.

$$T = 4480 \text{ N·m} \qquad \text{key shear}$$

b. *Key bearing.* Using equation (12.22),

$$s_{bk} = \frac{4T}{d_s l_k t_k}$$

or

$$T = \frac{s_{bk} d_s l_k t_k}{4}$$

$$= \frac{(242 \text{ MPa})(60 \text{ mm})(75 \text{ mm})(12 \text{ mm})}{4} = 3.27 \times 10^6 \text{ N·mm} = 3270 \text{ N·m}$$

The maximum torque that the key can handle in bearing is 3270 N·m.

$$T = 3270 \text{ N·m} \qquad \text{key bearing}$$

c. *Hub-web shear.* Using equation (12.23),

$$s_{sh} = \frac{2T}{\pi d_h^2 t_w}$$

or $\quad T = \frac{s_{sh} \pi d_h^2 t_w}{2}$

$$= \frac{(123 \text{ MPa})(3.14)(100 \text{ mm})^2(25 \text{ mm})}{2} = 48.3 \times 10^6 \text{ N·mm} = 48,300 \text{ N·m}$$

The maximum torque that the hub-web shear can handle is 48,300 N·m.

$$T = 48,300 \text{ N·m} \qquad \text{hub-web shear}$$

d. *Bolt shear.* From equation (12.24),

$$s_{sb} = \frac{8T}{N_b d_{bc} \pi D_b^2}$$

or $\quad T = \frac{s_{sb} N_b d_{bc} \pi D_b^2}{8}$

$$= \frac{(91 \text{ MPa})(6)(140 \text{ mm})(3.14)(6 \text{ mm})^2}{8} = 1.08 \times 10^6 \text{ N·mm} = 1080 \text{ N·m}$$

The maximum torque the bolt shear can handle is 1080 N·m.

$$T = 1080 \text{ N·m} \qquad \text{bolt shear}$$

e. *Bolt bearing.* From equation (12.25),

$$s_{bb} = \frac{2T}{(N_b d_{bc} D_b t_w)}$$

or

$$T = \frac{s_{bb} N_b d_{bc} D_b t_w}{2}$$

Note: We need to check only the bolt bearing, because the allowable bearing stress for the web (180 MPa) is greater than that for the bolts (134 MPa).

$$T = \frac{s_{bb} N_b d_{bc} D_b t_w}{2}$$

$$= \frac{(134 \text{ MPa})(6)(140 \text{ mm})(6 \text{ mm})(25 \text{ mm})}{2} = 8.44 \times 10^6 \text{ N·mm} = 8440 \text{ N·m}$$

The maximum torque that can be handled by bolt bearing is 8440 N·m.

$$T = 8440 \text{ N·m} \qquad \text{bolt bearing}$$

Collecting the answers:

a. $T_{max} = 4480$ N·m

b. $T_{max} = 3270$ N·m

c. $T_{max} = 48,300$ N·m

d. $T_{max} = 1080$ N·m

e. $T_{max} = 8440$ N·m

The smallest of these, which is the limiting torque, is **1080 N·m** due to bolt shear.

3. Calculate the maximum power. Using equation (12.19),

$$\text{power} = TN/9.55 \text{ W} \quad (T \text{ N·m}, N \text{ rpm})$$
$$T = 1080 \text{ N·m}, \quad N = 500 \text{ rpm}$$

so,

$$\text{power} = (1080 \text{ N·m})(500 \text{ rpm})/9.55 = 56{,}545 \text{ W} = 56.5 \text{ kW}$$

4. Calculate the maximum horsepower.

1 kW = 1.340 hp,   so   56.5 kW = (56.5 kW)(1.340 hp/kW) = 75.8 hp

Answer

The power is 75.8 hp, limited by bolt shear.

Note: The bolts are limiting the maximum power. If we were to drill out the flange bolt holes and use 8-mm-diameter bolts of the same material, the maximum safe horsepower would increase to 102 hp. (Satisfy yourself that this is correct.)

12.8   COMPUTER ANALYSES

We present two programs in this section: one to calculate torque, power, and angle of twist (program 'Torque') and one to determine the strength of couplings (program 'Coupling').

12.8.1   Program 'Torque'

The required input values are

1. US or SI for units
2. Whether the shaft is solid (S) or hollow (H)
3. Shaft outside diameter
4. Shaft inside diameter, if it is hollow
5. Shaft length
6. Whether the shaft is continuous (C), stepped (S), or grooved (G)
7. Shaft minor diameter, if stepped, or groove diameter, if grooved.
8. Fillet radius, if stepped or grooved
9. Key type: N for none, S for sled, P for profile
10. Allowable torsional shear stress
11. Modulus of rigidity
12. Rotational speed, rpm

The output gives

1. Maximum allowable torque
2. Maximum allowable power
3. Twist angle

EXAMPLE PROBLEM 12.13

Determine the maximum allowable torque and horsepower of a hollow shaft having outside and inside diameters of 4.5 in. and 3.0 in. and a profile keyseat. The material's allowable stress is 8800 psi, and the shaft speed is 1800 rpm.

What would be the maximum allowable torque and horsepower if the keyseat were changed to a sled type?

Solution

1. Open program 'Torque'.
2. US or SI for units: Enter US
3. Is the shaft solid (S) or hollow (H)? Enter H
4. Shaft outside diameter: Enter 4.5
5. Shaft inside diameter, if it is hollow: Enter 3.
6. Shaft length: not needed for this problem.
7. Is the shaft continuous (C), stepped (S), or grooved (G)? Enter C
8. Shaft minor diameter, if stepped, or groove diameter, if grooved: not needed
9. Fillet radius, if stepped or grooved: not needed.
10. Key type; N for none, S for sled, P for profile: Enter P
11. Allowable torsional shear stress: Enter 8.8
12. Modulus of rigidity: not needed.
13. Rotational speed, rpm: Enter 1800.

The output gives

1. Maximum allowable torque: 63,175 lb-in.
2. Maximum allowable power: 1804 hp

Change the key type to S.
The output changes to

1. Maximum allowable torque: 78,969 lb-in.
2. Maximum allowable power: 2255 hp

Answers

With a profile keyseat:

maximum allowable torque = 63,200 lb-in.

maximum allowable power = 1800 hp.

With a sled-type keyseat:

maximum allowable torque = 79,000 lb-in.

maximum allowable power = 2250 hp

Do this calculation by hand and see if you get the same answers.

EXAMPLE PROBLEM 12.14

Determine the maximum allowable torque and power and the twist angle at that condition for a hollow shaft having outside and inside diameters of 50 mm and 10 mm. There is a 40-mm-diameter groove with 3-mm radii and a sled-type keyseat. The material's allowable torsional shear stress is 82.5 MPa, and its modulus of rigidity is 80×10^3 MPa. The shaft is 1.5 m long, and its speed is 500 rpm.

Solution
1. US or SI for units: Enter SI
2. Is the shaft solid (S) or hollow (H)? Enter H
3. Shaft outside diameter: Enter 50.
4. Shaft inside diameter, if it is hollow: Enter 10.
5. Shaft length: Enter 1500.
6. Is the shaft continuous (C), stepped (S), or grooved (G)? Enter G
7. Shaft minor diameter, if stepped, or groove diameter, if grooved: Enter 40.
8. Fillet radius, if stepped or grooved: Enter 3.
9. Key type, N for none, S for sled, P for profile: Enter S
10. Allowable torsional shear stress: Enter 82.5
11. Modulus of rigidity: Enter 80000.
12. Rotational speed, rpm: Enter 500.

Answer
The output gives

$$\text{maximum allowable torque} = 401 \text{ N·m}$$

$$\text{maximum allowable power} = 21 \text{ kW}$$

$$\text{twist angle} = 1.72°$$

12.8.3   Program 'Coupling'

This program calculates the maximum allowable torque and power for a flanged coupling. It determines the torque limits due to key shear, key bearing, flange hub-web shear, bolt shear, and bolt-web bearing. It then selects the minimum of these, which is the maximum limit for the coupling. We show the input/output sheet in Figure 12.10.

Starting at the top left, you can choose either US or SI units. Then enter the shaft diameter and rpm, the key length, width, thickness, and allowable shear and bearing stresses. Enter the hub diameter and allowable shear stress, the web thickness and allowable bearing stress, and the number of bolts, bolt diameter, bolt-circle diameter, and allowable shear and bearing stresses.

The output lists the maximum allowable toque, power, and the weakest (limiting) part and failure mode (limiting mode) of that part. Changing one (or more) of the parameters—for example, the number of bolts—gives the new output values immediately.

PROGRAM 'Coupling'

INPUT

Are you using US or SI units?	
Enter US or SI	SI

SHAFT

Diameter (mm)	0
Speed (rpm)	0

KEY

Length (mm)	0
Width (mm)	0
Thickness (mm)	0
Allow shear stress (MPa)	0
Allow bearing stress (MPa)	0

FLANGE HUB

Diameter (mm)	0
Allow shear stress (MPa)	0

FLANGE WEB

Thickness (mm)	0
Allow bearing stress (MPa)	0

BOLTS

Number of bolts	0
Diameter (mm)	0
BC Diameter (mm)	0
Allow shear stress (MPa)	0
Allow bearing stress (MPa)	0

OUTPUT - Don't make entries in these boxes

Max. allow. torque (Nm)	0.0
Max. allow. power (kW)	0.0
Limiting Part	#DIV/0!
Limiting Mode	#DIV/0!

Figure 12.10 Input/output sheet for program 'Coupling'.

Much of the program is concerned with straightforward input/output control, but if you are fairly new to spreadsheet programming you should study the following description of the table look-up part. If you are familiar with spreadsheet programming, skip to Example Problem 12.15.

The maximum torque for each part, such as the key in shear and the bolts in bearing, are collected into a table in the range H5 . . . J10 below the heading **Max. allow. torque**. The minimum value (which is the limiting torque for the coupling) is found using the function =MIN (15 . . I10) in cell I11. Having found the minimum value, we now want to identify which of the possible six conditions is limiting. We are going to use the =VLOOKUP function, but it is not quite that easy.

Suppose we were to rearrange the table in the following form:

TORQUE   PART   LIMITING MODE

Then we include the functions

=VLOOKUP (I11,H5 . . J10, 1)

and

=VLOOKUP (I11,H5 . . J10, 2)

What we are attempting to do is to search the table H5 . . J10 for the minimum torque, which is in cell I11, and then move across the table to the second column (the first column is column 1) to find the limiting part. Then we do it again, moving to the third column to find the limiting mode. Unfortunately, this won't work.

=VLOOKUP works by starting at the top of the table; it moves down the first column (column 1, containing the torque values) until it finds the *first value that is larger than the index* (the required value). It then uses the value immediately above it. If the table contains the values in increasing order from top to bottom, then it will use the actual index value if it is in the table. For example, suppose we have the following table in cells A1 . . B5:

2	red
4	blue
10	green
7	yellow
0.5	orange

and we write the function =VLOOKUP (4,A1 . . B5,2); then the search will start at 2, move down to 4, and then across to blue. So it will return the word *blue*.

Suppose now we write the function =VLOOKUP (7,A1 . . B5,2). Starting at the top, it will move down from 2 to 4 to 10. Since 10 is the first value greater than 7 in the table, it will drop back up to 4, and again give the result *blue*.

Let us apply this now to our problem of locating the minimum value in the table. Unless the minimum value happens to be the first number in the column, the first number will be greater than the index. So the search will stop at the top entry and try to drop back to the one above, but there is none above. It will return an error sign. A simple way to overcome this—and the one we use in program 'Coupling'—is to create a table of inverse values. For example, 2 becomes $\frac{1}{2}$, 4 becomes $\frac{1}{4}$, etc. Then the minimum becomes the maximum. We also take the inverse of the index, so our search is now for the maximum value rather than the minimum.

In our previous example, with the five colors, the function =VLOOKUP (0.5,A1 .. B5,1) would return an error because it would stop searching at 2, which is the first value greater than 0.5. Making a table of inverse values, we have

0.5	red
0.25	blue
0.1	green
0.1429	yellow
2.0	orange

Changing the function to

$$=\text{VLOOKUP } (2.0, \text{A1 .. B5}, 2)$$

would return the word *orange*.

EXAMPLE PROBLEM 12.15

A flange coupling connecting two 4.5-in.-diameter shafts has a hub diameter of 7 in. and a web thickness of 1.25 in. The webs are connected by six $\frac{5}{8}$-in.-diameter bolts on a 10-in. bolt-circle diameter. The keys connecting the hubs to the shafts are 6 in. in length, 1 in. wide, and 1 in. thick. The shafts are to rotate at 325 rpm. Determine the maximum allowable torque and horsepower and the limiting part and limiting mode, using the following allowable stresses:

key shear	11.67 ksi	key bearing	15.83 ksi
hub-web shear	10.83 ksi	web bearing	28.33 ksi
bolt shear	11.67 ksi	bolt bearing	15.83 ksi

Solution
Figure 12.11 shows the input/output sheet for this problem.

Answer
The maximum allowable torque is 10,700 lb-in. The maximum allowable horsepower is 550 hp. The limiting part is the key in bearing.

Try changing some of the input values and see what happens. For example, what is the effect of doubling the key length?

How would you modify the design so that the coupling could transmit 900 hp?

PROGRAM 'Coupling'

INPUT

Are you using US or SI units?	
Enter US or SI	US

SHAFT	
Diameter (in)	4.5
Speed (rpm)	325

KEY	
Length (in)	6
Width (in)	1
Thickness (in)	1
Allow shear stress (ksi)	11.67
Allow bearing stress (ksi)	15.83

FLANGE	
HUB	
Diameter (in)	7
Allow shear stress (ksi)	10.83

FLANGE	
WEB	
Thickness (in)	1.25
Allow bearing stress (ksi)	28.33

BOLTS	
Number of bolts	6
Diameter (in)	0.625
BC Diameter (in)	10
Allow shear stress (ksi)	11.67
Allow bearing stress (ksi)	15.83

OUTPUT - Don't make entries in these boxes

Max. allow. torque (lbin)	106852.5
Max. allow. power (hp)	551.0
Limiting Part	Key
Limiting Mode	Bearing

Figure 12.11   Input/output sheet for Example Problem 12.15.

315

12.9   PROBLEMS FOR CHAPTER 12

Hand Calculations

1. How much torque can be safely transmitted by a 4.25-in.-diameter solid shaft made of 7075-T6 aluminum alloy? Use a factor of safety of 4 based on the ultimate shear strength.

2. A 20-mm-diameter, 30-cm-long AISI 1045 steel rod is subjected to a torque of 200 N·m. What will be the twist angle in degrees? (Use $G = 80 \times 10^3$ MPa.)

3. A torque of 1.0 kN·m is applied to a 7.5-cm-diameter solid shaft made of 17-4PH H900 stainless steel. The modulus of elasticity in shear for steel is $G = 75 \times 10^3$ MPa. If the shaft is 500 cm long, calculate the twist angle in degrees. Is this amount of twist acceptable?

4. A hollow, AISI stainless steel 301 full hard shaft has an outside diameter of 100 mm and an inside diameter of 66 mm. It is to transmit a torque of 18 kN·m. Calculate the safety factor.

5. A 2.5-ft-long hollow circular shaft is stepped down from a 2.5-in.-diameter to a 2.0-in.-diameter, with a fillet radius of 0.20 in. The shaft is made of AISI 1141 OQT 1300 steel ($G = 11.5 \times 10^6$ psi) and has an inside diameter of 1.25 in. Use a safety factor of 6 based on the ultimate shear stress.
   a. What maximum torque can be applied?
   b. What will be the twist angle?
   c. Is the twist acceptable?

6. A 6061-T4 aluminum alloy solid shaft transmits a torque of 450 N·m under varying load conditions. The shaft has a sled-type keyseat. What diameter shaft is required?

7. A 36-in.-long, 2.0-in.-diameter AISI 430 annealed shaft has a profile-type keyseat and a retaining-ring groove. The groove diameter is 1.85 in. and has 0.1-in. radii. The shaft is to be driven at 1500 rpm, and the angle of twist is not to exceed 1° over the length of the shaft. What maximum power can be transmitted? (The modulus of elasticity in shear (modulus of rigidity) is $G = 11.5 \times 10^6$ psi.)

Computer Calculations

8. Determine the maximum allowable torque and power and the twist angle at that condition for a hollow shaft having outside and inside diameters of 7.5 cm and 3.0 cm. There is a 6.0-cm-diameter groove with 4-mm radii and a profile-type keyseat. The material's allowable torsional shear stress is 90.0 MPa, and its modulus of rigidity is 80×10^3 MPa. The shaft is 1.0 m long, and its rotational speed is 1800 rpm.

9. Repeat Problem 8 using a solid shaft of the same diameter.

10. A gear transmitting 15 kW at 150 rpm is fastened to a 7.0-cm-diameter shaft by a 20-mm by 12-mm key. The shaft and the key are AISI 1040 cold-drawn steel. What key length is required if the allowable shear stress is 50 MPa and the allowable bearing stress is 105 MPa? Is it the shear stress or the bearing stress that dictates the key length?

11. The rotational speed in the previous problem is increased to 200 rpm. What is the minimum key length now?

12. If the key in Problem 11 were changed to a square 20-mm × 20-mm key of the same material, what minimum key length would be required, and how would the failure mode be affected?

13. Two 50-mm diameter shafts are joined by a flange coupling that has 8 AISI 1020 hot-rolled 6-mm bolts on a bolt-circle diameter of 120 mm. The flange, which is made of AISI 1080 annealed steel, has a hub diameter of 80 mm and a web thickness of 15 mm. The 55-mm-long, 10-mm-square keys are made of AISI 1141 OQT 1300 steel. Use a safety

factor of 6, based on ultimate shear strength, for all materials. What maximum horsepower can be safely transmitted at 1000 rpm?

14. The horsepower rating of the coupling described in Problem 13 is to be increased to 189 hp. This is to be achieved by changing the bolt diameter and/or the key width and thickness (still a square key). What changes would you recommend?

15. Repeat Problem 13 using six bolts on a 140-mm diameter bolt circle.

16. The 55-mm-long, 10-mm-square keys in Problem 13 are replaced by 45-mm-long, 14-mm by 9-mm flat keys made of AISI 4140 annealed steel. What maximum horsepower can be safely transmitted at 800 rpm?

17. A flange coupling connecting two 5-in.-diameter shafts has a hub diameter of 7.5 in. and a web thickness of 1.125 in. The webs are connected by eight $\frac{1}{2}$-in.-diameter bolts on a 10.25-in. bolt-circle diameter. The keys connecting the hubs to the shafts are 6 in. in length, 1.25 in. wide, and 1.25 in. thick. The shafts are to rotate at 280 rpm. Determine the maximum allowable torque and horsepower and the limiting part and limiting mode, using the following allowable stresses:

key shear   12.5 ksi   key bearing   16.5 ksi
hub-web shear   10.5 ksi   web bearing   25.0 ksi
bolt shear   10.5 ksi   bolt bearing   15.8 ksi.

18. The allowable torque in the previous problem is to be increased to 12,000 lb-ft by increasing the bolt diameter. What minimum *standard* bolt diameter is required? What is the failure mode now?

19. Using the new bolt size found in Problem 18, we now want to increase the horsepower rating to 800 hp by increasing the key length. What new key length is required?

20. Returning to the original design of Problem 17 and changing only the bolt-circle diameter, what minimum diameter would be required to increase the horsepower rating to 450 hp?

21. Using the modification of the previous problem, what rpm would produce 500 hp?

13

Shear and Bending in Beams

In this chapter you will learn about

1. the different types of beams,
2. development of the *flexure formula* that is used in the calculation of bending stresses,
3. shear forces and bending moments, and
4. how to plot shear force and bending moment diagrams.

13.1 BEAM TYPES

A beam is a structural member that resists loads applied normal to its axis by bending. We often think of a beam as being a horizontal member simply supported at its ends, such as a plank lying on two sawhorses. Although this is a good example of a simply supported beam, there are many other configurations to be considered. Various types of beams are used in machines and control units; some are designed to remain rigid, and others are designed to function as flexible members. First, we have to be able to distinguish between *statically determinate* and *statically indeterminate* types.

Statically determinate beams are those whose unknown reactions and moments can be determined by applying the conditions for static equilibrium (Chapters 2 and 3). When the number of unknown values exceeds the number of equations provided by the static equilibrium conditions, the beam is statically indeterminate.

Figure 13.1 shows some examples of statically determinate beams:

> A simply supported beam
> An overhanging beam
> A cantilever beam

Each of these types may carry concentrated loads, distributed loads, or a combination of both concentrated and distributed loads.

Some examples of statically indeterminate beams are shown in Figure 13.2:

> Both ends fixed
> More than two supports
> Propped cantilever

Again, each of these beams may carry concentrated, distributed, or a combination of loads. Analyses of statistically indeterminate beams are discussed in Chapter 17.

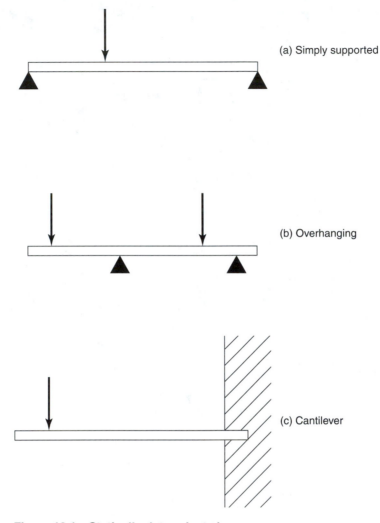

Figure 13.1   Statically determinate beams.

13.2   BENDING DEFORMATION

Figure 13.3 represents a simply supported beam carrying a single concentrated load. Suppose that the beam is a thin plank of wood and that the load is fairly heavy. There will be a perceptible bend in the beam, as shown. Although it would not be noticeable, a steel beam under the same conditions would also bend. The deflection may be only thousandths or millionths of an inch, but both beams are undergoing the same process, which is compression of the upper surface and tension in the lower. Because the effect changes from compression on top to tension on the bottom, there must be a line somewhere between the upper and lower surfaces where there is no load or stress

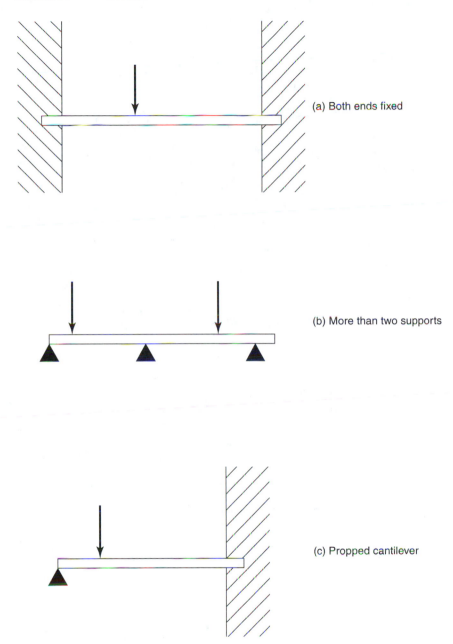

(a) Both ends fixed

(b) More than two supports

(c) Propped cantilever

Figure 13.2 Statically indeterminate beams.

on the beam. This location is the horizontal centroidal axis (Chapter 11) at each section of the beam; the plane created by the continuous connection of these axes along the beam is called the **neutral axis**.

There are basically two types of stresses developed within the beam to counteract the externally imposed loads. These are **shear stresses** and **bending stresses**. Figure

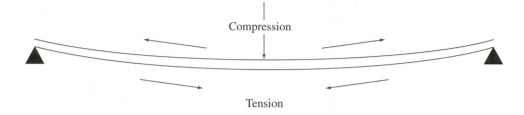

Figure 13.3   Beam bending under load.

13.4 shows a simply supported beam carrying a concentrated load. By the methods developed in Section 3.5, we calculate the reactions to be $R_1 = 80$ lb and $R_2 = 120$ lb. As explained in Section 2.4, any isolated part (free body) of a structure or member of a structure in equilibrium must itself be in equilibrium.

Taking the first (from the left end of the beam) 2 ft of the beam as a free body, as shown in Figure 13.5, we see that there must be an internal force, V, equal in magnitude and opposite in direction to the 80-lb reaction.

Immediately to the right of this free body there must be an equal and opposite reaction to balance the force V, and so on, along the beam. Each section then has forces that are tending to shear the beam apart, as shown in Figure 13.6. The sign convention adopted in this book and in most other texts, is that Figure 13.6 is a demonstration of *positive* shear—the section on the *right* tending to move *down* relative to the section on the left (*right down positive*).

Going back to Figure 13.5 we see that, even with the force V to balance the external reaction R_1, the free body is still not in equilibrium because there is an unbalanced moment about the right end of the free body of $-(80 \text{ lb})(2 \text{ ft}) = -160$ lb-ft. (Remember that clockwise moments are negative.) Therefore, there must be a counterclockwise moment, developed *inside* the beam, to balance the externally created moment. This is called the **bending moment**. See Figure 13.7. The sign convention for

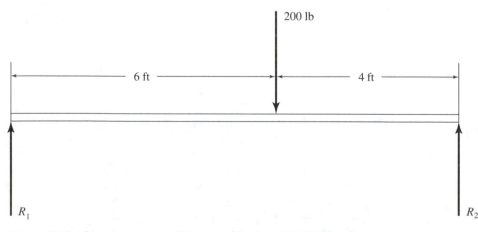

Figure 13.4   Simply supported beam with concentrated load.

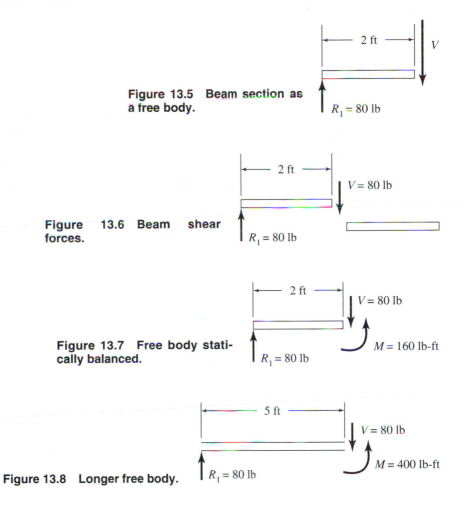

Figure 13.5 Beam section as a free body.

Figure 13.6 Beam shear forces.

Figure 13.7 Free body statically balanced.

Figure 13.8 Longer free body.

bending moments is as follows: The bending moment is considered to be *positive* if the sections of the beam above the neutral axis are in *compression* and *negative* if they are in *tension*. Figure 13.3 is an example of a positive bending moment.

Returning to Figure 13.4 and this time considering a 5-ft section (measured from the left end) to be a free body (Figure 13.8), the shear force required to balance the reaction R_1 is still 80 lb. The bending moment, however, must now balance a clockwise moment of $-(80 \text{ lb})(5 \text{ ft}) = -400$ lb-ft. Therefore, a positive (ccw) bending moment of 400 lb-ft will be developed inside the beam. If a 7-ft section is now taken for the free body (Figure 13.9), the shear force will have to balance the algebraic sum of the external loads, which is 80 lb − 200 lb = −120 lb. Taking moments about the right end of the free body, $-(80 \text{ lb})(7 \text{ ft}) - (200 \text{ lb})(1 \text{ ft}) = -360$ lb-ft. Therefore, the bending moment is 360 lb-ft counterclockwise.

We see from these simple calculations that, for a beam carrying concentrated loads, the shear force remains constant between loads, whereas the bending moment does not. We present detailed analyses of shear forces and bending moments later.

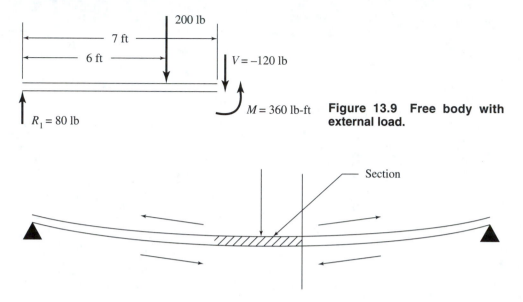

Figure 13.9 Free body with external load.

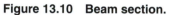

Figure 13.10 Beam section.

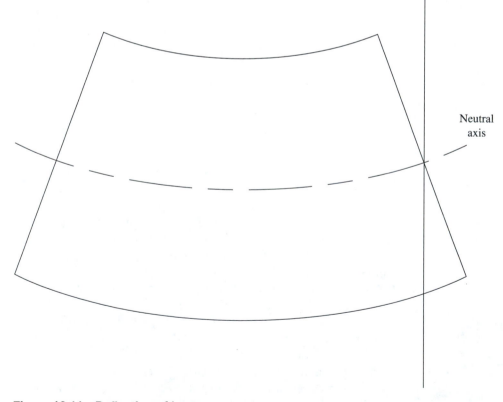

Figure 13.11 Deflection of beam segment.

To develop the classical beam theory, which will lead to the formulation of the flexure formula in Section 13.3, we assume that plane (flat) cross sections remain plane—but not parallel—during loading and that the cross sections rotate about their centroidal axes. Thus, a small section of the loaded beam shown in Figure 13.10 appears, greatly exaggerated, as in Figure 13.11. The ends remain straight but not parallel, and they are rotated about axes normal to the page.

13.3   THE FLEXURE FORMULA

The bending of the beam in Figure 13.10 causes any section above the neutral axis to contract and any section below to extend, as shown in Figure 13.11. The deflection of a thin section of area A at a distance y from the neutral axis is δ. (See Figure 13.12.) We made the basic assumption that the ends of the section remained straight and rotated about their centroidal (neutral) axes, which means that δ increases linearly with y, or, put another way, δ is directly proportional to y. Because the ratio of deflection to original length is defined as strain, we can see that the strain is also directly proportional to y. Assuming that the strain remains in the elastic range—i.e., stress is pro-

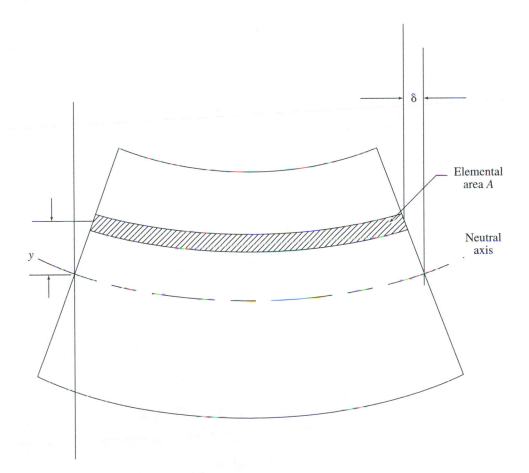

Figure 13.12   Segment of loaded beam.

portional to strain—then the stress must also be directly proportional to y. This may be expressed mathematically by the equation.

$$s = ky \qquad (13.1)$$

where k is a constant whose value is to be determined.

The stress, s, increases linearly from zero at the neutral axis (the axis through the centroids of the cross sections) to a maximum value, s_{max}, at distance c from the neutral axis, which is the distance to the outermost fibers. We can express this as

$$s = (y/c)s_{max} \qquad (13.1(a))$$

The force acting on area A is equal to the stress times area:

$$f = (ky)A = kAy \qquad (13.2)$$

The moment of this force about the centroidal axis is

$$m = fy = kAy^2 \qquad (13.3)$$

The sum of the moments of all the elements is, therefore,

$$M = \sum kAy^2 = k \sum Ay^2 \qquad (13.4)$$

$\sum Ay^2$ is the *moment of inertia* introduced in Chapter 11, so we can express equation (13.4) as

$$M = kI \qquad (13.5)$$

Rearranging,

$$k = M/I \qquad (13.6)$$

Replacing k in equation (13.1) yields the **flexure equation**:

$$s = My/I \qquad (13.7)$$

Equation (13.7) shows that

1. Bending stress increases linearly with bending moment,
2. Bending stress decreases with increasing moment of inertia,
3. Maximum bending stress occurs at the maximum value of y. This maximum value is at the outermost fibers of the beam, where $y = c$, the maximum distance from the neutral axis.

We will use the flexure equation later.

USING CALCULUS

We showed that the stress at any point can be expressed as

$$s = (y/c)s_{max} \qquad (13.1(a))$$

So the force on an element of area dA is

$$dF = sdA = (y/c)s_{max}dA$$

and the moment of that force about the neutral axis is

$$dM = ydF = (s_{max}/c)y^2dA$$

The total moment is, therefore,

$$M = (s_{max}/c) \int y^2 \, dA$$

But $\int y^2 \, dA = I$, the moment of inertia. Therefore,

$$M = (s_{max}/c)I$$

Rearranging,

$$s_{max} = Mc/I$$

and since $s = (y/c)s_{max}$, the stress at any point may be expressed as

$$s = My/I$$

13.4 BEAM SHEAR FORCE AND BENDING MOMENT DIAGRAMS

Shear stresses are calculated from shear forces, and bending stresses are calculated from bending moments.

We now study how to plot the distributions of shear forces and bending moments along the beam. These are known as *shear force diagrams* and *bending moment diagrams*.

13.5 SHEAR FORCES

We saw in Section 13.2 that there are basically two types of stresses that occur in loaded beams, shear and bending stresses. In Chapter 18, we will see that additional stresses may be developed under combined loading conditions, such as bending and compression, occurring at the same time. We will deal with shear forces (which create the shear stresses) first for two reasons. They are easier to calculate than bending moments, and one way to calculate bending moments is by using shear forces. Shear stresses are determined from shear forces and cross-sectional areas, whereas bending stresses are calculated from bending moments using the flexure equation developed in the previous section.

13.6 SHEAR FORCE DIAGRAMS

The shear force diagram is a graph of shear force plotted against beam location. First we draw the beam length to some convenient scale as the *x*-axis, and then we place the shear forces on the *y*-axis, also to some convenient scale. The shape of the shear force diagram depends on the type of loading and the type of beam supports. Examples covering each of the most common types are given next.

13.6.1 Simply Supported Beam with Concentrated Loads

Simply supported means that the supports cannot resist rotational movements of the beam. Roller, pinned, and hinged supports fall into this category. Roller supports permit horizontal and rotational movement but not vertical movement. Pinned and hinged supports permit no horizontal or vertical movement but do permit rotation.

Figure 13.13 (a) represents a 10-ft beam, simply supported at its ends, carrying a 200-lb load 4 ft from the left end and a 150-lb load 8 ft from the left end.

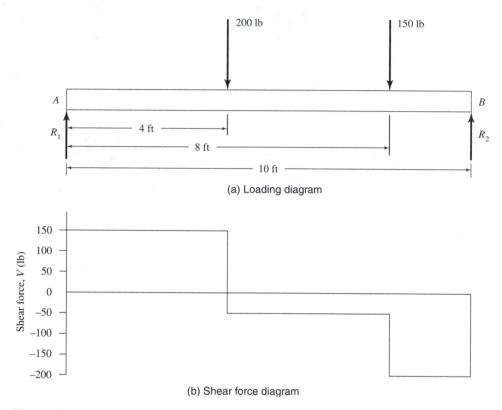

(a) Loading diagram

(b) Shear force diagram

Figure 13.13

We first calculate the reactions R_1 and R_2. Taking moments about the left end,

$$\sum M_A = 0 = -(200 \text{ lb})(4 \text{ ft}) - (150 \text{ lb})(8 \text{ ft}) + R_2(10 \text{ ft})$$

Therefore, $R_2 = 200$ lb. Similarly, we calculate R_1 by taking moments about B.

$$\sum M_B = 0 = (150 \text{ lb})(2 \text{ ft}) + (200 \text{ lb})(6 \text{ ft}) - R_1(10 \text{ ft})$$

from which $R_1 = 150$ lb.

Because the algebraic sum of the vertical forces must be zero;

$$\sum F_y = 0 = R_1 + R_2 + (-200 \text{ lb}) + (-150 \text{ lb})$$

That is, 150 lb + 200 lb + (−200 lb) + (−150 lb) = 0, which is a check that the calculations are correct.

*The **shear force** at any station along the beam is simply the algebraic sum of all the external forces to the left of that station. Up forces are positive, and down forces are negative.*

Draw a horizontal line 5 in. long to represent the beam. This is a scale of $\frac{1}{2}$ in. = 1 ft. Draw the $+y$-axis vertically up from the left end of the beam and the $-y$-axis vertically down. See Figure 13.13 (b). At any station between a fraction of an inch from the left end to a fraction of an inch short of 4 ft, the only external force acting on the beam is $R_1 = 150$ lb in an upward direction. Therefore, the shear force over this section is equal to $+150$ lb.

At a station 4 ft plus a fraction of an inch, the algebraic sum of the external forces to the left is 150 lb + (−200 lb), resulting in a negative shear force of −50 lb. This remains constant up to just short of the 8-ft location, after which the 150-lb down load changes the shear force to

$$V = 150 \text{ lb} + (-200 \text{ lb}) + (-150 \text{ lb}) = -200 \text{ lb}$$

which remains constant to the end of the beam, where the upward reaction R_2, which has a value of $+200$ lb, takes the shear force back to zero (Figure 13.13 (b)). Notice that the shear force diagram *must* go to zero at the right end because there are no shear forces beyond the end of the beam. This is not necessarily so in the case of a cantilever beam, where shear forces can exist in the support structure (wall).

EXAMPLE PROBLEM 13.1

A horizontal beam 12 ft long is simply supported at its ends and carries loads of 350 lb at 2 ft, 200 lb at 7 ft, and 150 lb at 10 ft, all measured from the left end of the beam. Calculate the shear forces at 1 ft, 6 ft, and 11 ft.

Solution

1. Make a sketch of the loaded beam. See Figure 13.14(a).
2. Calculate reaction R_2 by taking moments about A:

$$\sum M_A = 0 = -(350 \text{ lb})(2 \text{ ft}) - (200 \text{ lb})(7 \text{ ft}) - (150 \text{ lb})(10 \text{ ft}) + R_2(12 \text{ ft})$$

$$0 = -700 \text{ lb-ft} - 1400 \text{ lb-ft} - 1500 \text{ lb-ft} + 12R_2$$

Therefore, $R_2 = 300$ lb.

3. Calculate reaction R_1 by taking moments about B:

$$\sum M_B = 0 = (150 \text{ lb})(2 \text{ ft}) + (200 \text{ lb})(5 \text{ ft}) + (350 \text{ lb})(10 \text{ ft}) - R_1(12 \text{ ft})$$

$$0 = 300 \text{ lb-ft} + 1000 \text{ lb-ft} + 3500 \text{ lb-ft} - 12R_1$$

Therefore, $R_1 = 400$ lb.

4. Check that the algebraic sum of the vertical forces is zero:

$$\sum F_y = 0 = R_1 + L_1 + L_2 + L_3 + R_2$$

where R_1 and R_2 are the reactions and the loads are represented by the letter L.

$$\sum F_y = 400 \text{ lb} + (-350 \text{ lb}) + (-200 \text{ lb}) + (-150 \text{ lb}) + 300 \text{ lb}$$

$$= 0 \quad \text{Check}$$

5. Calculate the shear forces.

At $x = 1$ ft: The only force to the left of station $x = 1$ ft is the reaction R_1. So the shear force at that station is

$$V = R_1$$

$$V = 400 \text{ lb}$$

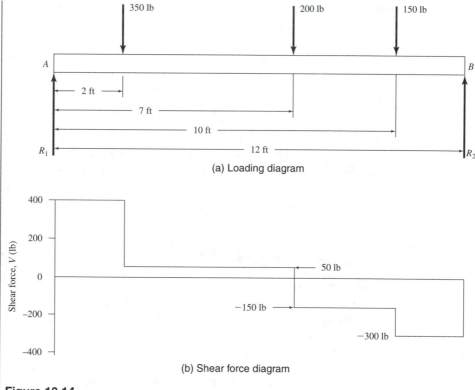

(a) Loading diagram

(b) Shear force diagram

Figure 13.14

At $x = 6$ ft: There are two forces acting to the left of this station. They are the reaction R_1 ($= 400$ lb) and the load L_1 ($= -350$ lb). So the total shear at station $x = 6$ ft is

$$V = 400 \text{ lb} + (-350 \text{ lb}) = 50 \text{ lb}$$

At $x = 11$ ft: The forces to the left of station $x = 11$ ft are R_1 ($= 400$ lb), L_1 ($= -350$ lb), L_2 ($= -200$ lb), and L_3 ($= -150$ lb). So

$$
\begin{aligned}
V &= R_1 \quad + \quad L_1 \quad + \quad L_2 \quad + \quad L_3 \\
&= 400 \text{ lb} + (-350 \text{ lb}) + (-200 \text{ lb}) + (-150 \text{ lb}) \\
&= -300 \text{ lb}
\end{aligned}
$$

Figure 13.14(b) is a plot of the complete shear force diagram.

Explanation
In the range $x_{L1} > x > x_{R1}$—i.e., the span between the left end of the beam and the first load L_1—the only contribution to the shear force is the upward (positive) *reaction* $R_1 = 400$ lb.

In the range $x_{L2} > x > x_{L1}$—i.e., the span between the first and second loads—the only *load* contributing to the shear force is L_1, which is a downward and, therefore, negative load equal to -350 lb. The *total shear force* acting on the beam between the loads L_1 and L_2 is the algebraic sum of all the vertical forces to the left, i.e., $V = 400$ lb $- 350$ lb $= 50$ lb.

In the range $x_{L3} > x > x_{L2}$, there are two loads (-350 lb and -200 lb) and the $+400$-lb reaction, all to the left of station x. So in this range $V = 400$ lb $+ (-350$ lb$) + (-200$ lb$) = -150$ lb.

Similarly, the shear force in the range $x_{L4} > x > x_{L3}$ is $V = 400$ lb $+ (-350$ lb$) + (-200$ lb$) + (-150$ lb$) = -300$ lb.

Finally, the reaction $R_2 = 300$ lb returns the shear force to zero.

Answer

At 1 ft, $V = 400$ lb; at 6 ft, $V = 50$ lb; at 11 ft, $V = -300$ lb.

EXAMPLE PROBLEM 13.2

A 10-m horizontal beam carries a mass of 100 kg at 3 m and 50 kg at 8 m, both measured from the left end of the beam. There is an upward force of 600 N in the middle of the span, and the beam is simply supported at its ends. Calculate the shear force at $x = 7$ m.

Solution

1. Convert the masses to forces and sketch the loaded beam.

$$(100 \text{ kg})(9.81 \text{ m/s}^2) = 981 \text{ N}, \qquad (50 \text{ kg})(9.81 \text{ m/s}^2) = 490 \text{ N}$$

See Figure 13.15(a).

2. Calculate reaction R_2 by taking moments about A:

$$\sum M_A = 0 = -(981 \text{ N})(3 \text{ m}) + (600 \text{ N})(5 \text{ m}) - 490 \text{ N})(8 \text{ m}) + 10R_2$$

$$0 = -2943 \text{ N·m} + 3000 \text{ N·m} - 3920 \text{ N·m} + 10R_2$$

Therefore, $R_2 = 386$ N (rounded off from 386.3 N).

3. Calculate reaction R_1 by taking moments about B:

$$\sum M_B = 0 = (490 \text{ N})(2 \text{ m}) - (600 \text{ N})(5 \text{ m}) + (981)(7 \text{ m}) - R_1(10 \text{ m})$$

$$0 = 980 \text{ N·m} - 3000 \text{ N·m} + 6867 \text{ N·m} - 10R_1$$

$$R_1 = 485 \text{ N} \qquad \text{(rounded off from 484.7 N)}$$

4. Check that the algebraic sum of the vertical forces is zero:

$$\sum F_y = 0 = R_1 + L_1 + L_2 + L_3 + R_2$$

$$= 485 \text{ N} + (-981 \text{ N}) + 600 \text{ N} + (-490 \text{ N}) + 386 \text{ N}$$

$$= 0 \qquad \text{Check}$$

5. Calculate the shear force at $x = 7$ m. Summing the forces to the left of station $x = 7$ m gives

$$V = R_1 + L_1 + L_2$$

$$= 485 \text{ N} + (-981 \text{ N}) + 600 \text{ N}$$

$$= 104 \text{ N}$$

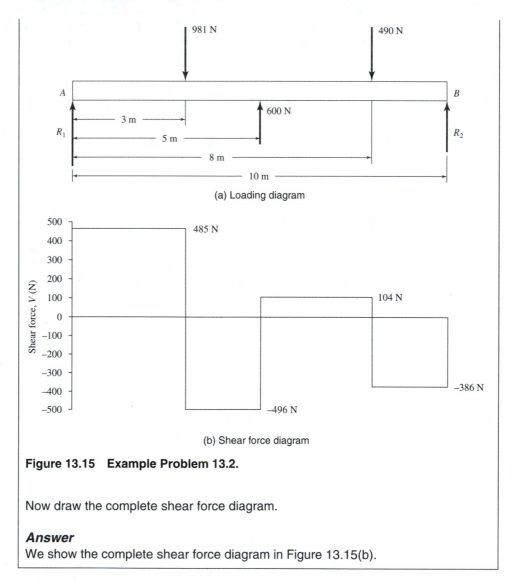

(a) Loading diagram

(b) Shear force diagram

Figure 13.15   Example Problem 13.2.

Now draw the complete shear force diagram.

Answer
We show the complete shear force diagram in Figure 13.15(b).

13.6.2   Simply Supported Beam with Distributed Loads

We take a 10-ft-long, simply supported beam as an example, but now the load is a distributed one of 50 lb/ft extending from a station 2 ft from the left end to 7 ft from the left end. See Figure 13.16(a). Because the span of the distributed load is 5 ft and it weighs 50 lb/ft, its total weight is 250 lb. As discussed in Chapter 3, for the purpose of calculating reactions this weight may be considered to be a concentrated load acting at the center of gravity of the distribution. The equivalent load, *for calculating reactions only*, is shown in Figure 13.16(b). It is very important to realize that this equivalent loading is used to calculate reactions *only*. Shear forces and bending moments are calculated from the *actual distributed loading*.

Calculating the reaction R_2 by taking moments about the left end of the beam and noting that the center of gravity of the distributed load is at station 2 ft + $(\frac{5}{2})$ ft = 4.5 ft,

$$\sum M_A = 0 = -(250 \text{ lb})(4.5 \text{ ft}) + R_2(10 \text{ ft})$$

from which $R_2 = 112.5$ lb.

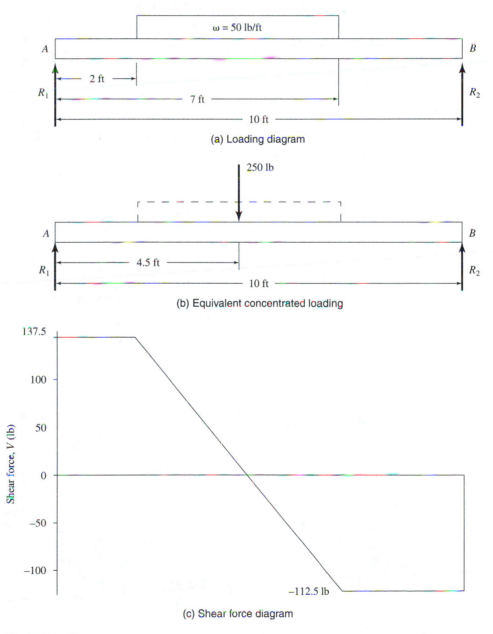

(a) Loading diagram

(b) Equivalent concentrated loading

(c) Shear force diagram

Figure 13.16

Again referring to Figure 13.16 (b), we calculate reaction R_1 by taking moments about B:

$$\sum M_B = 0 = (250 \text{ lb})(5.5 \text{ ft}) - R_1(10 \text{ ft})$$

Therefore, $R_1 = 137.5$ lb.

Starting just to the right of the left end of the beam, the only force acting is the reaction $R = 137.5$ lb. Thus, the shear force starts at this value and remains constant to the 2-ft station, where the distributed load begins. See Figure 13.16(c). At the 3-ft station, *part* of the distributed load (1 ft of it) is to the left and must be included in the shear force calculation. Because the distributed load is 50 lb/ft, this 1-ft section must weigh 50 lb. It is a down load and is, therefore, negative. Adding this *algebraically* to R_1 gives the shear force at the 3-ft station:

$$V_{3\text{ ft}} = 137.5 \text{ lb} + (-50 \text{ lb}) = +87.5 \text{ lb}$$

At the 4-ft station (4 ft from the left end of the beam) the distributed load contribution is $(2 \text{ ft})(-50 \text{ lb/ft}) = -100$ lb, because 2 ft of the distributed load are to the right of the 4-ft station. The shear force at this location is, therefore,

$$V_{4\text{ ft}} = 137.5 \text{ lb} + (-100 \text{ lb}) = +37.5 \text{ lb}$$

We see from this that the shear force is changing by -50 lb for every foot of span where the distributed load is acting, which means that the shear force is falling linearly over the span of the load. There is no need, therefore, to calculate station by station. It is sufficient to compute the shear force at the end of the distributed load and draw a straight line down to that point. In the present example the shear force changes by -50 lb/ft for a distance of 5 ft from the beginning to the end of the distributed load. The shear force at the 7-ft station (the end of the distributed load) is

$$V_{7\text{ ft}} = 137.5 \text{ lb} + (-50 \text{ lb/ft})(5 \text{ ft}) = -112.5 \text{ lb}$$

The reaction R_2, which is $+112.5$ lb, returns the shear force diagram to zero, as shown in Figure 13.16(c).

EXAMPLE PROBLEM 13.3

At which station is the shear force zero for the beam shown in Figure 13.16 (a)?

Solution
We calculated the reactions for this example previously and found that $R_1 = 137.5$ lb. So we have $R_1 = 137.5$ lb, $\omega_1 = -50$ lb/ft, and X_{D1s} (the x-location of the start of the distributed load) = 2 ft. Because the shear force is *dropping* at the rate of 50 lb/ft, it will take (137.5 lb)/(50 lb/ft) to drop from 137.5 lb to zero.

So $\qquad\qquad x = 2 \text{ ft} + (137.5 \text{ lb})/(50 \text{ lb/ft})$

$\qquad\qquad\qquad = 2 \text{ ft} + (137.5/50) \text{ ft}$

$\qquad\qquad\qquad = 2 \text{ ft} + 2.75 \text{ ft}$

Therefore, $x = 4.75$ ft.

We will see later that maximum and minimum bending moments usually occur at points of zero shear. (An exception to this is when the loading also includes a couple.)

EXAMPLE PROBLEM 13.4

Figure 13.17 (a) shows a horizontal beam carrying two distributed loads. Calculate the shear force at $x = 9$ m.

Solution

1. We must first change the 35 kg *mass* to a weight and then express it as a distributed load. Because this is the second (from the left) distributed load, we identify it with the subscript 2. It is a down load, and so it will be negative.

$$W_2 = (-35 \text{ kg})(9.81 \text{ m/s}^2) = -343.2 \text{ N}$$

which, as a distributed load, is

$$\omega_2 = (-343.2 \text{ N})/(3 \text{ m}) = -114.4 \text{ N/m}$$

2. Now we represent the distributed loads as equivalent concentrated loads for the purpose of calculating the reactions. The first load is $(-100 \text{ N/m})(6 \text{ m} - 1 \text{ m}) = -500$ N. This equivalent load is located at $1 \text{ m} + (\frac{5}{2}) \text{ m} = 3.5$ m from the left end of the beam.

The second equivalent load is -343.2 N, and it is located at $7 \text{ m} + (\frac{3}{2}) \text{ m} = 8.5$ m from the left end of the beam. The equivalent loading is shown in Figure 13.17 (b).

3. Calculate the reactions:

$$\sum M_A = 0 = (-500 \text{ N})(3.5 \text{ m}) + (-343.2 \text{ N})(8.5 \text{ m}) + R_2(10 \text{ m})$$
$$0 = -1750 \text{ N·m} - 2917 \text{ N·m} + 10R_2$$
$$R_2 = 466.7 \text{ N}$$
$$\sum M_B = 0 = (-343.2 \text{ N})(-1.5 \text{ m}) + (-500 \text{ N})(-6.5 \text{ m}) + R_1(-10 \text{ m})$$
$$0 = 514.8 \text{ N·m} + 3250 \text{ N·m} - 10R_1 = 0$$
$$R_1 = 376.5 \text{ N}$$

Figure 13.17(c) shows the reactions and the distributed loadings.

4. Check that the algebraic sum of the vertical forces is zero:

$$376.5 \text{ N} + (-500 \text{ N}) + (-343.2 \text{ N}) + 466.7 \text{ N} = 0 \quad \text{Check}$$

5. Calculate the shear force at $x = 9$ m. At the left end of the beam, the shear force is equal to the reaction R_1 ($= 376.5$ N) and remains constant for 1 m. It then drops at the rate of 100 N/m for 5 m. (See Figure 13.17(c) and (d).) So at the 6-m station, $V = 376.5 \text{ N} - (100 \text{ N/m})(5 \text{ m}) = -123.5$ N.

The shear force then remains constant for 1 m, after which it drops at the rate of 114.4 N/m. Therefore, from $x = 7$ m to $x = 9$ m there is a further drop of $(114.4 \text{ N/m})(9 \text{ m} - 7 \text{ m}) = 228.8$ N. So the total shear at $x = 9$ m is

$$V = (-123.5 \text{ N}) - (228.8 \text{ N}) = 352.3 \text{ N}$$

6. The complete shear force diagram is shown in Figure 13.17(d).

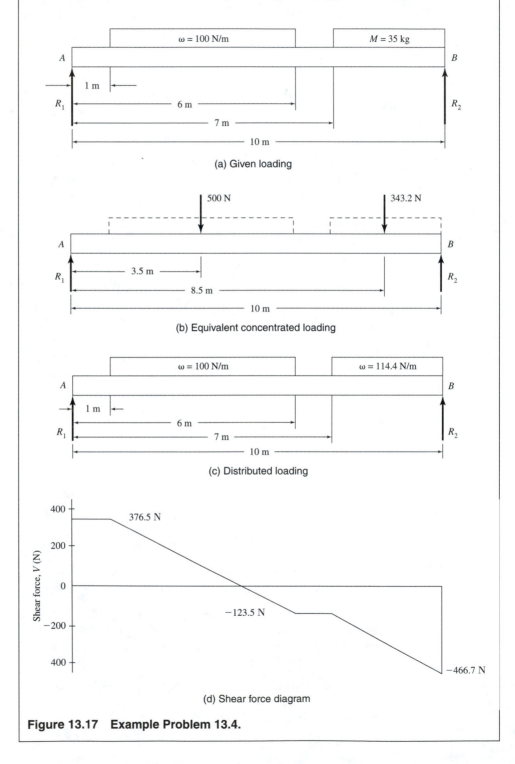

(a) Given loading

(b) Equivalent concentrated loading

(c) Distributed loading

(d) Shear force diagram

Figure 13.17   Example Problem 13.4.

13.6.3 Simply Supported Beam with Concentrated and Distributed Loads

Figure 13.18(a) shows a beam carrying a distributed load and two concentrated loads. It is actually a combination of the loads shown in Figures 13.13(a) and 13.16(a). First, to calculate the reactions, we have to temporarily replace the distributed load by its

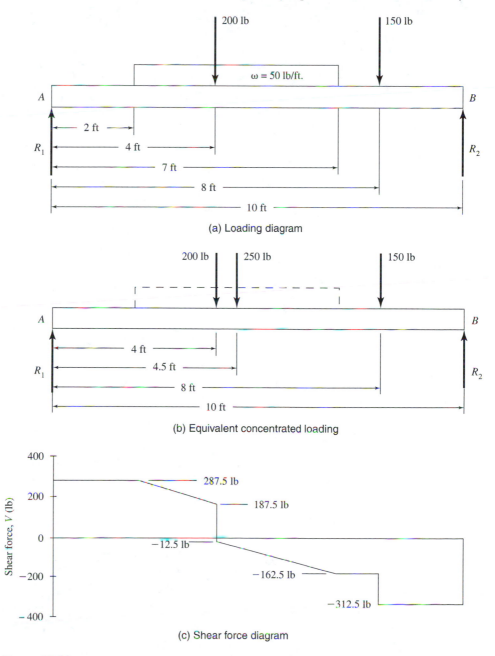

(a) Loading diagram

(b) Equivalent concentrated loading

(c) Shear force diagram

Figure 13.18

equivalent concentrated load of $(-50 \text{ lb/ft})(5 \text{ ft}) = -250 \text{ lb}$ at 4.5 ft from the left end of the beam. Then taking moments about A,

$$\sum M_A = 0 = -(200 \text{ lb})(4 \text{ ft}) - (250 \text{ lb})(4.5 \text{ ft}) - (150 \text{ lb})(8 \text{ ft}) + R_2(10 \text{ ft})$$

Solving, $R_2 = 312.5 \text{ lb}$.

Now taking moments about B,

$$\sum M_B = 0 = (150 \text{ lb})(2 \text{ ft}) + (250 \text{ lb})(5.5 \text{ ft}) + (200 \text{ lb})(6 \text{ ft}) - R_1(10 \text{ ft})$$

from which $R_1 = 287.5 \text{ lb}$.

Starting at the left end of the beam, the shear force is 287.5 lb up to the start of the distributed load at $x = 2$ ft. It then falls at the rate of 50 lb/ft up to the location of the first concentrated load, which is 200 lb at $x = 4$ ft. We see then, at $x = 4$ ft (which is 2 ft beyond the start of the distributed load), the shear force is 287.5 lb $-$ (50 lb/ft $\times$ 2 ft) = 187.5 lb. The 200-lb load then drops it a further 200 lb to -12.5 lb.

Notice that we are working with Figure 13.18(a), not (b). Figure 13.18(b) was used only to determine the reactions.

Between $x = 4$ ft and 7 ft the shear force continues to drop at the rate of 50 lb/ft— i.e., by (50 lb/ft) $\times$ (7 ft $-$ 4 ft) = 150 lb. So at $x = 7$ ft, $V = -12.5$ lb $+ (-150$ lb) $= -162.5$ lb.

There is no load between $x = 7$ ft and $x = 8$ ft, so the shear force remains constant over that 1-ft span. There is then a further drop of 150 lb due to the final concentrated load, bringing the shear force down to -162.5 lb $+ (-150$ lb) $= -312.5$ lb.

The reaction at B returns the shear force to zero.

Answer

Figure 13.18 (c) shows the complete shear force diagram. Of course, if there is more than one distributed load, we must include the contributions of them all.

EXAMPLE PROBLEM 13.5

The beam shown in Figure 13.19(a) has two concentrated loads and two distributed loads that overlap for part of the span. Calculate the shear force at $x = 6$ m and at $x = 9$ m.

Solution

1. The first thing we have to do is separate out the overlapping parts of the distributed loads. See Figure 13.19(b). Then draw the equivalent loading diagram (Figure 13.19(c)), from which we can calculate the reactions.

2. Calculate the reactions. Taking moments about A,

$$\sum M_A = 0 = -(225 \text{ kN})(2 \text{ m}) - (100 \text{ kN})(4 \text{ m}) - (300 \text{ kN})(6 \text{ m})$$
$$- (100 \text{ kN})(8 \text{ m}) - (300 \text{ kN})(8.5 \text{ m}) + R_2(12 \text{ m})$$

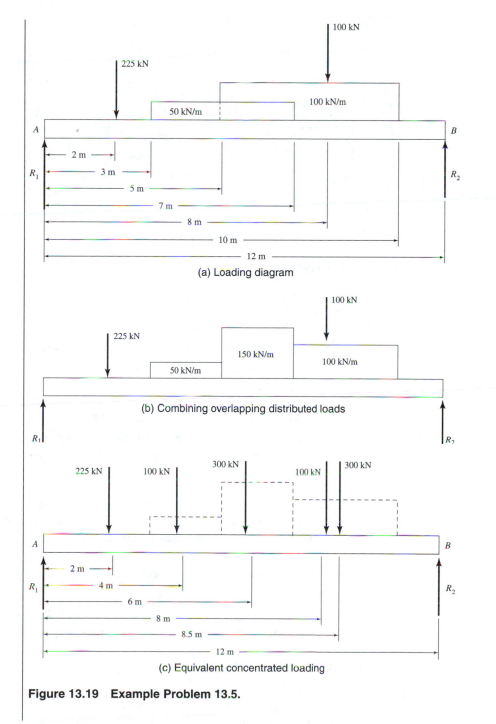

(a) Loading diagram

(b) Combining overlapping distributed loads

(c) Equivalent concentrated loading

Figure 13.19   Example Problem 13.5.

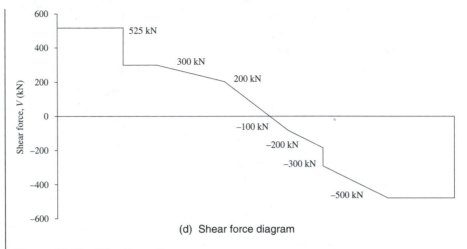

(d) Shear force diagram

Figure 13.19   (*Continued*)

Solving, R_2 = 500 kN.

Now taking moments about *B*,

$$\sum M_B = 0 = (300 \text{ kN})(3.5 \text{ m}) + (100 \text{ kN})(4 \text{ m}) + (300 \text{ kN})(6 \text{ m})$$
$$+ (100 \text{ kN})(8 \text{ m}) + (225 \text{ kN})(10 \text{ m}) - R_1(12 \text{ m})$$

from which, R_1 = 525 kN.

3. Check that the algebraic sum of the vertical forces is zero:

$$525 \text{ kN} + (-225 \text{ kN}) + (-100 \text{ kN}) + (-300 \text{ kN})$$
$$+ (-100 \text{ kN}) + (-300 \text{ kN}) + 500 \text{ kN} = 0 \qquad \text{Check}$$

4. Calculate the shear force.

At *x* = 6 m: See Figure 13.19 (b). The contributions to the shear force at this station are the reaction R_1 (= 525 kN), the load L_1 (= −225 kN), 2 m of the first distributed load of ω_1 = −50 kN/m, and 1 m of the overlapping loads with ω_{1+2} = −150 kN/m.

$$V = 525 \text{ kN} + (-225 \text{ kN}) + (-50 \text{ kN/m})(2 \text{ m}) + (-150 \text{ kN/m})(1 \text{ m})$$
$$= 50 \text{ kN}$$

At *x* = 9 m: There are two concentrated loads to the left of *x* = 9 m. Also there are, effectively, two complete distributed loads and part of a distributed load. See Figure 13.19(b).

The contributions to the shear force are

The reaction R_1 = 525 kN,

The first concentrated load, L_1 = −225 kN,

The first distributed load, ω_1 = −50 kN/m, for a span of 2 m, which gives −100 kN,

The overlapping loads ω_{1+2} = −150 kN/m for a span of 2 m, which gives −300 kN,

> The second distributed load, $\omega_2 = -100$ kN/m, for a span of 2 m, which gives
> $\qquad = -200$ kN, and
>
> The second concentrated load, $L_2 = -100$ kN.
> Summing these gives
>
> $\qquad V = 525$ kN $+ (-225$ kN$) + (-100$ kN$) + (-300$ kN$) + (-200$ kN$) + (-100$ kN$)$
>
> Therefore, $V = -400$ kN.
>
> **Answer**
> At 6 m, $V = 50$ kN; at 9 m; $V = -400$ kN.
>   For practice the student should calculate and plot the complete shear force diagram (see Figure 13.19(d)).

13.6.4   Overhanging Beams

We calculate shear forces for overhanging beams the same way as for simply supported beams. In determining the reactions, however, it is better to take moments about the points of support than about the ends of the beam. Doing this will eliminate one of the unknown reactions from the equations.

The loading shown in Figure 13.20(a) is the same as that for Figure 13.18(a), but we have moved the left support inboard by 3 ft. This has resulted in part of the load **overhanging** beyond the support. If we take moments about the left *support* (see Figure 13.20(b)):

$$\sum M_A = 0 = -(200\text{ lb})(1\text{ ft}) - (250\text{ lb})(1.5\text{ ft}) - (150\text{ lb})(5\text{ ft}) + R_2(7\text{ ft})$$

Solving for R_2 gives $R_2 = 189.3$ lb. Taking moments about B gives

$$\sum M_B = 0 = (150\text{ lb})(2\text{ ft}) + (250\text{ lb})(5.5\text{ ft}) + (200\text{ lb})(6\text{ ft}) - R_1(7\text{ ft})$$

from which $R_1 = 410.7$ lb.

Now we check that the algebraic sum of the forces is zero.

$$410.7\text{ lb} + (-200\text{ lb}) + (-250\text{ lb}) + (-150\text{ lb}) + 189.3\text{ lb} = 0 \qquad \text{Check}$$

Returning to Figure 13.20(a), we see that one foot of the distributed load is *overhanging* to the left of the support. The weight of this portion is $(50\text{ lb/ft})(1\text{ ft}) = 50$ lb. This is a down load, so the shear force diagram starts at zero at $x = 0$ and remains at zero until the presence of the distributed load is felt; then it drops linearly to -50 lb at $x = 3$ ft. At $x = 3$ ft, however, the upward reaction R_1 ($= 410.7$ lb) is also contributing to the shear, making $V = -50$ lb $+ 410.7$ lb $= 360.7$ lb (Figure 13.20(c)).

The distance from R_1 to the 200-lb concentrated load is 1 ft. That part of the distributed load between R_1 and the 200-lb load decreases the shear linearly by 50 lb, making it 360.7 lb $+ (-50$ lb$) = 310.7$ lb. Then, the 200-lb load drops it further to 310.7 lb $+ (-200$ lb$) = 110.7$ lb. The remaining portion of the distributed load is 3

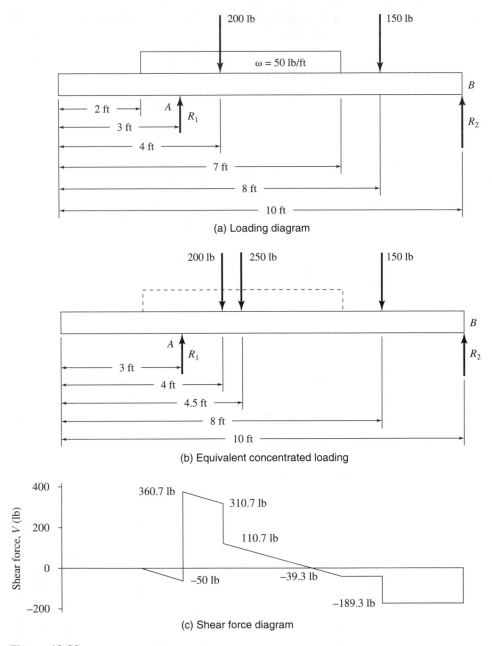

(a) Loading diagram

(b) Equivalent concentrated loading

(c) Shear force diagram

Figure 13.20

ft long, which at -50 lb/ft, drops the shear to $110.7 + (-50 \text{ lb/ft})(3 \text{ ft}) = -39.3$ lb at $x = 7$ ft. This stays constant until $x = 8$ ft, where the second concentrated load changes the shear to $-39.3 \text{ lb} + (-150 \text{ lb}) = -189.3$ lb. The reaction R_2 ($= 189.3$ lb) returns the shear force to zero at the right end of the beam.

EXAMPLE PROBLEM 13.6

Refer to Figure 13.20(a):

   a. Where is the shear force zero between $x = 4$ ft and $x = 8$ ft?
   b. What is the value of the shear force at $x = 2.5$ ft?
   c. What is the value of the shear force at $x = 4$ ft?

Solution

   a. Between $x = 4$ ft and $x = 8$ ft: The reaction R_1 ($= 410.7$ lb) is to the left of the region 4 ft to 8 ft and so makes a contribution to the shear force. There is one concentrated load to the left of the region, L_1 ($= -200$ lb), and part of the distributed load is $\omega = -50$ lb/ft.

   Let us define the station where the shear force is zero as $x_{SF=0}$. Then the length of the distributed load making a contribution to the shear force is ($x_{SF=0} - 2$) ft. (See Figure 13.20(a).) Therefore, the contribution to the shear force by the distributed load is $(-50$ lb/ft) ($x_{SF=0} - 2$) ft.
   Equating the sum of these shear force contributions to zero gives

$$V = 410.7 \text{ lb} + (-200 \text{ lb}) + (-50 \text{ lb/ft})(x_{SF=0} - 2) \text{ ft}$$

$$= 210.7 \text{ lb} - 50(x_{SF=0}) \text{ lb} + 100 \text{ lb} = 0$$

Therefore, $50 (x_{SF=0}) = 310.7$. So $x = 6.21$ ft.

   b. At $x = 2.5$ ft: There is no reaction to the left of this station. Also there is no concentrated load, but there is part of a distributed load.

$$V = (-50 \text{ lb/ft})(2.5 \text{ ft} - 2 \text{ ft}) = -25 \text{ lb}$$

   c. At $x = 4$ ft: Because there is a concentrated load at this station, there is a discontinuity in the shear. We will denote a station just to the left of $x = 4$ ft by $x = 4^-$ ft and one just to the right by $x = 4^+$ ft.

   1. At $x = 4^-$ ft, the reaction R_1 and part of the distributed load are to the left.
$$V = 410.7 \text{ lb} + (-50 \text{ lb/ft})(4 \text{ ft} - 2 \text{ ft}) = 310.7 \text{ lb}$$

   2. At $x = 4^+$ ft, we have to include the concentrated load $L_1 = -200$ lb, which drops the shear to 310.7 lb $+ (-200$ lb) $= 110.7$ lb. So at $x = 4$ ft, the shear force changes from 310.7 lb to 110.7 lb.

Answer

   a. The shear force zero, between $x = 4$ ft and $x = 8$ ft, is at $x = 6.21$ ft.
   b. The value of the shear force at $x = 2.5$ ft is $V = -25$ lb.
   c. The value of the shear force at $x = 4$ ft changes from 310.7 lb to 110.7 lb.

EXAMPLE PROBLEM 13.7

This problem is important because we now introduce a case where there are loads to the right of the second reaction.

Figure 13.21(a) shows a beam with loads overhanging at both ends. Calculate the shear force at $x = 9.5$ m.

Solution

1. Draw the equivalent concentrated load diagram (Figure 13.21(b)).
2. Calculate the reactions. Take moments about A, which is the station of the first reaction.

$$\sum M_A = 0 = (100 \text{ N})(2 \text{ m}) - (800 \text{ N})(1 \text{ m}) - (900 \text{ N})(4 \text{ m}) + R_2(5 \text{ m})$$
$$- (450 \text{ N})(6.5 \text{ m})$$
$$0 = 200 \text{ N} \cdot \text{m} - 800 \text{ N} \cdot \text{m} - 3600 \text{ N} \cdot \text{m} + 5R_2 - 2925 \text{ N} \cdot \text{m}$$

Therefore, $R_2 = 1425$ N.

Now, taking moments about B, which is the second reaction station,

$$\sum M_B = 0 = (100 \text{ N})(7 \text{ m}) - R_1(5 \text{ m}) + (800 \text{ N})(4 \text{ m}) + (900 \text{ N})(1 \text{ m})$$
$$- (450 \text{ N})(1.5 \text{ m})$$
$$0 = 700 \text{ N} \cdot \text{m} - 5R_1 + 3200 \text{ N} \cdot \text{m} + 900 \text{ N} \cdot \text{m} - 675 \text{ N} \cdot \text{m}$$

So $R_1 = 825$ N.

2. Check that the sum of the vertical forces is zero.

$$\sum F_y = (-100 \text{ N}) + 825 \text{ N} + (-800 \text{ N}) + (-900 \text{ N})$$
$$+ 1425 \text{ N} + (-450 \text{ N}) = 0 \qquad \text{Check}$$

3. Calculate the shear force at $x = 9.5$ m. Referring to Figure 13.21(a), the forces making contributions to the shear force are

Reaction $R_1 = 825$ N
Reaction $R_2 = 1425$ N
Concentrated load $L_1 = -100$ N
Concentrated load $L_2 = -900$ N
Distributed load $\omega_1 = -200$ N/m, creating a shear force of $(-200 \text{ N/m})(4 \text{ m})$
$= -800$ N
1.5 m of the second distributed load $\omega_2 = -150$ N/m, creating a shear force of $(-150 \text{ N/m})(1.5 \text{ m}) = -225$ N

Summing the contributions gives

$$V = 825 \text{ N} + 1425 \text{ N} + (-100 \text{ N}) + (-900 \text{ N}) + (-800 \text{ N}) + (-225 \text{ N})$$

So $V = 225$ N. We show the complete shear force diagram in Figure 13.21(c).

Note: A competent designer is always on the lookout for shortcuts. For example, with a little thought, the problem just completed could be solved in one line. We know that the shear must go to zero at the end of the last distributed load, i.e., at $x = 11$ m (see Figure 13.21(a)), which is 1.5 m beyond the station at which we are cal-

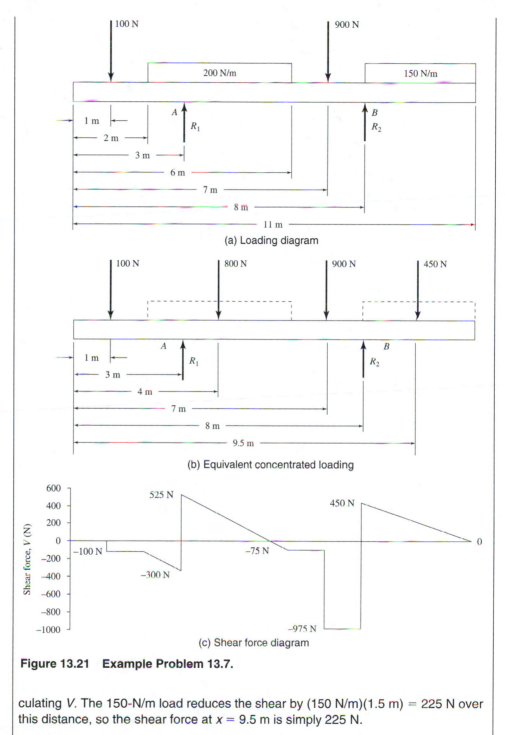

(a) Loading diagram

(b) Equivalent concentrated loading

(c) Shear force diagram

Figure 13.21 Example Problem 13.7.

culating V. The 150-N/m load reduces the shear by (150 N/m)(1.5 m) = 225 N over this distance, so the shear force at $x = 9.5$ m is simply 225 N.

Answer

The shear force at $x = 9.5$ m is $V = 225$ N.

13.6.5   Cantilever Beams

In the case of cantilever beams (such as diving springboards), the only reaction is at the wall. See Figure 13.22(a). Because the sum of the forces in the y-direction must be zero for equilibrium, the reaction is simply equal to the sum of the loads on the beam. Notice also that there must be a moment at the wall to counteract the moments created by the loads.

Caution: The wall can be drawn to the left or to the right of the figure. If you draw it to the left, realize that our sign conventions must change. The shear force at any station will be the algebraic sum of the forces to the *right* of the station. Later in this chapter we will be using computer programs to calculate and plot the shear force and bending moment diagrams. These programs have been written for the *wall on the right*. To avoid confusion and possible errors, we suggest *always drawing the wall to the right of the figure*.

Cantilever beams are somewhat easier because there is only one reaction, which exists at the wall. If all the loads are vertical, the reaction is simply their algebraic sum. If there are horizontal components, the reaction is the *vector sum* of the vertical and the horizontal components.

Let us draw the shear force diagram for the loading shown in Figure 13.22(a). Starting at the left end of the beam, the 100-kN concentrated load creates a shear of -100

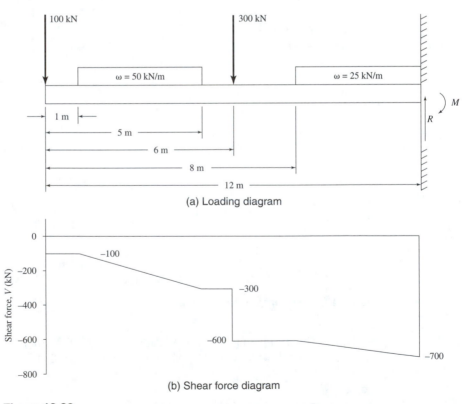

(a) Loading diagram

(b) Shear force diagram

Figure 13.22

kN that remains constant up to the start of the distributed load at $x = 1$ m. (See Figure 13.22(b).) The first distributed load drops the shear at the rate of 50 kN/m for a distance of 4 m, which is a drop of 200 kN to -300 kN at $x = 5$ m. The shear force stays constant until we reach the second concentrated load at $x = 6$ m. This load drops the shear by 300 kN to -600 kN, which remains constant up to the start of the second distributed load at $x = 8$ m. This 25-kN/m load drops the shear by 100 kN over the last 4 m, resulting in a shear force at the wall of -700 kN.

As a check, the wall reaction should return the shear to zero. We can calculate the value of this reaction by using the fact that the sum of the forces in the y-direction is zero:

$$(-100 \text{ kN}) + (-50 \text{ kN/m})(4 \text{ m}) + (-300 \text{ kN}) + (-25 \text{ kN/m})(4 \text{ m}) + R = 0$$
$$-100 \text{ kN} - 200 \text{ kN} - 300 \text{ kN} - 100 \text{ kN} + R = 0$$

Therefore, $R = 700$ kN, which is exactly the force required to return the shear to zero.

EXAMPLE PROBLEM 13.8

Draw the shear force diagram and calculate the wall reaction for the loading shown in Figure 13.23(a).

Solution
1. Redraw the figure with the wall to the right (Figure 13.23(b)).
2. Resolve the 1000-N force into its vertical and horizontal components. (See Figure 13.23(c).)

$$\text{vertical component} = F_y = (1000 \text{ N})(\sin 60°) = 866 \text{ N}$$

which is in the downward direction, so it will be negative: -866 N.

$$\text{horizontal component} = F_x = (1000 \text{ N})(\cos 60°) = 500 \text{ N}$$

This component acts in the positive x-direction, so it is positive.

3. Draw the shear force diagram. Notice that the horizontal component of the 1000-N force *does not contribute to the shear*.

Starting at the left (free) end of the beam, the 800-N load creates a -800-N shear force, which remains constant for a distance of 2 m, until the vertical component of the 1000-N load drops the shear force to $(-800 \text{ N} - 866 \text{ N}) = -1666$ N. Because there are no more external forces, the shear remains at this value to the wall. (See Figure 13.23(d).)

4. Calculate the wall reaction. The wall reaction must balance out the vertical and horizontal components imposed on the beam. The total vertical component on the beam is -1666 N, and the total horizontal component is 500 N. Therefore, there must be vertical and horizontal reaction components of 1666 N and -500 N, respectively. These (see Figure 13.23(e)) resolve into a resultant force of

$$R = \sqrt{(1666 \text{ N})^2 + (-500 \text{ N})^2} = 1739 \text{ N}$$

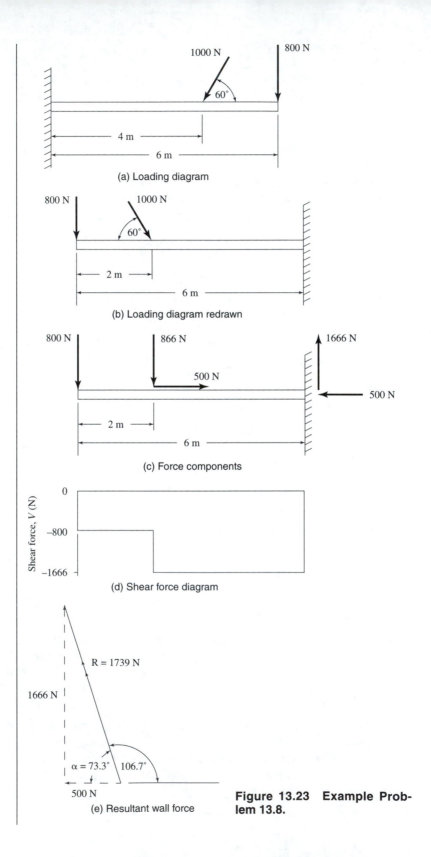

(a) Loading diagram

(b) Loading diagram redrawn

(c) Force components

(d) Shear force diagram

(e) Resultant wall force

Figure 13.23   Example Problem 13.8.

The resultant angle is

$$\alpha = \arctan[1666 \text{ N}]/[500 \text{ N}] = 73.3° \quad \text{or} \quad (180 - 73.3)° = 106.7°$$

from the conventional East $= 0°$ direction.

Answer
The wall reaction is 1739 N at an angle of 106.7°.

13.7   BENDING MOMENT DIAGRAMS

The bending moment at any beam station is the internal moment that is developed to *balance the sum of the external moments to the left of that station.* The bending moment diagram is a plot of the bending moment along the beam.

We begin again with a simply supported beam carrying only concentrated loads and build from there.

13.7.1   Simply Supported Beam with Concentrated Loads

Figure 13.24(a) and (b) repeats Figure 13.13(a) and (b), showing the loading and shear force diagrams for a beam carrying two concentrated loads. We first consider a station at $x = 1$ ft (i.e., 1 ft from the left end of the beam). The only force to the left of the station is the reaction $R_1 = 150$ lb. Taking the moment of this force about the station $x = 1$ ft, we have a positive force, 150 lb, but a negative distance, -1 ft, because we are measuring from right to left. This creates a negative (clockwise) moment of -150 lb-ft. There must be a moment inside the beam at this station whose magnitude is the same but in the opposite direction. This moment, M, is the bending moment at that station; i.e., at $x = 1$ ft the bending moment is M = 150 lb-ft.

If we do the same calculation for $x = 2$ ft, the moment to the left of the station is $-(150 \text{ lb})(2 \text{ ft}) = -300$ lb-ft. So at $x = 2$ ft, M = 300 lb-ft. Similarly, at $x = 3$ ft, M = 450 lb-ft, and at $x = 4$ ft, M = 600 lb-ft. Notice that the 200-lb concentrated load has not yet been included, because its moment about the 4-ft station is $-(200 \text{ lb})(0 \text{ ft}) = 0$. However, at $x = 5$ ft, the sum of the moments to the left of the station is $-(150 \text{ lb})(5 \text{ ft}) + (200 \text{ lb})(5 \text{ ft} - 4 \text{ ft}) = -550$ lb-ft, so M at this station is 550 lb-ft. Continuing in this way we see that M = 500 lb-ft at $x = 6$ ft, 450 lb-ft at $x = 7$ ft, and 400 lb-ft at $x = 8$ ft.

At $x = 9$ ft, the second concentrated load is included. The sum of the moments to the left of station $x = 9$ ft is

$$-(150 \text{ lb})(9 \text{ ft}) + (200 \text{ lb})(9 \text{ ft} - 4 \text{ ft}) + (150 \text{ lb})(9 \text{ ft} - 8 \text{ ft}) = -200 \text{ lb-ft}$$

So the bending moment is M = 200 lb-ft.

At the right end of the beam, $x = 10$ ft, the sum of the moments to the left is

$$-(150 \text{ lb})(10 \text{ ft}) + (200 \text{ lb})(10 \text{ ft} - 4 \text{ ft}) + (150 \text{ lb})(10 \text{ ft} - 8 \text{ ft}) = 0$$

We see that the bending moment is zero at both ends of the beam. It is usual to draw the bending moment diagram beneath—and lined up with—the shear force diagram, as we have done in Figure 13.24(c).

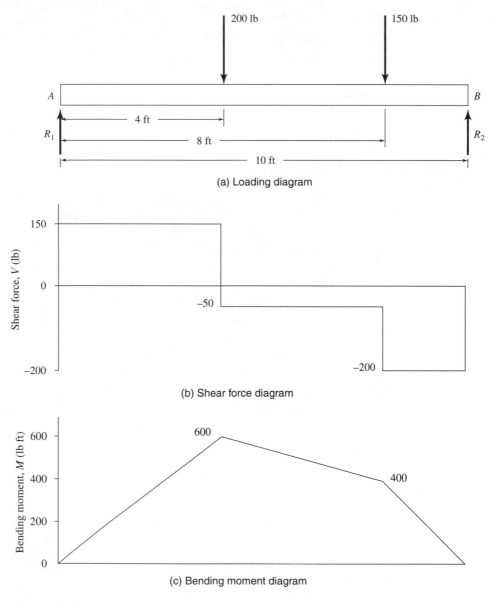

Figure 13.24

There are many things we should observe about the relationship between the shear force and the bending moment diagrams.

1. Where the shear force diagram is a horizontal line *above* the zero axis, the bending moment is a straight line with a *positive slope*.
2. Where the shear force diagram is a horizontal line *below* the zero axis, the bending moment is a straight line with a *negative slope*.

3. The maximum bending moment is at a station where the shear passes through zero. (Note that the shear may pass through zero more than once, but the maximum bending moment will be at one of those stations.)
4. The bending moment at any station is equal to the area under the shear diagram up to that station, with areas above the axis being positive and those below it being negative.

As an example of this last statement, let us consider any station, say, $x = 4$ ft. The area under the shear diagram is the area of a rectangle of height 150 lb and length 4 ft. So the area is $(150 \text{ lb})(4 \text{ ft}) = 600$ lb-ft, which agrees with the bending moment at that station. At $x = 8$ ft, we still have the 600-lb-ft area, but we also have an area of $(-50 \text{ lb})(4 \text{ ft}) = -200$ lb-ft. This indicates that the bending moment at $x = 8$ ft should be 600 lb-ft $-$ 200 lb-ft $= 400$ lb-ft, which it is.

EXAMPLE PROBLEM 13.9

Figure 13.25(a) and (b) repeats Figures 13.14(a) and (b).

a. Plot the complete bending moment diagram using the method of areas.
b. Calculate the bending moment at $x = 11.5$ ft.

Solution

a. The first part of the shear force diagram is a horizontal line above the zero axis. This part of the moment diagram will, therefore, be a straight line with a positive slope. It will start at zero and increase to $(400 \text{ lb})(2 \text{ ft})$, which is the area of the rectangle on the shear diagram from $x = 0$ to $x = 2$ ft. So $M = 800$ lb-ft at $x = 2$ ft.

The next section of the shear diagram is again a horizontal line above the axis and so will result in another straight line with positive slope on the moment diagram. The total area under the shear diagram from $x = 0$ to $x = 7$ ft consists of the previous 800 lb-ft plus an area of $(50 \text{ lb})(7 \text{ ft} - 2 \text{ ft})$, or 800 lb-ft $+$ 250 lb ft $= 1050$ lb-ft.

The shear diagram passes through zero at $x = 7$ ft, so we know we have reached the peak of the moment diagram. The shear diagram in the region $x = 7$ ft to $x = 10$ ft shows a horizontal line below the zero axis, indicating a linearly falling bending moment in this region. The shear diagram area here is $(-150 \text{ lb})(10 \text{ ft} - 7 \text{ ft}) = -450$ lb-ft. Adding this algebraically to the previously calculated area of 1050 lb-ft gives the bending moment at $x = 10$ ft as 1050 lb-ft $+ (-450 \text{ lb-ft}) = 600$ lb-ft.

The last section is another negative-area region, this time equal to $(-300 \text{ lb})(12 \text{ ft} - 10 \text{ ft}) = -600$ lb-ft. Adding this to the 600-lb-ft moment results in a zero bending moment at the right end of the beam, as it should be.

We show the complete bending moment diagram in Figure 13.25(c).

b. Bending moment at $x = 11.5$ ft: Station $x = 11.5$ ft is 0.5 ft from the end of the beam, where the bending moment must be zero. The shear force diagram in this region is a horizontal line below the zero axis, so the corresponding bending moment diagram in this region is a straight line with negative slope. The area under the shear

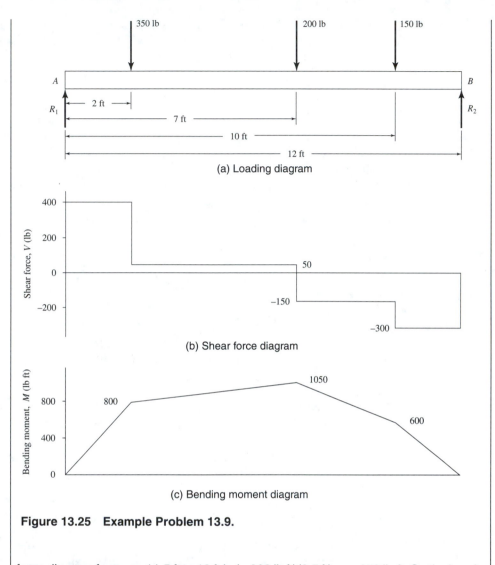

Figure 13.25   Example Problem 13.9.

force diagram from $x = 11.5$ ft to 12 ft is $(-300$ lb ft$)(0.5$ ft$) = -150$ lb-ft. So the bending moment falls from its value at $x = 11.5$ ft by 150 lb-ft to reach zero. Therefore, the bending moment at $x = 11.5$ ft is 150 lb-ft.

EXAMPLE PROBLEM 13.10

Calculate:
   a. The complete bending moment diagram, and
   b. The bending moment at $x = 9$ m for the loading shown in Figure 13.26(a), which is a repeat of Figure 13.15(a).

Solution

a. Calculate the complete bending moment diagram. We will use the area method to calculate the complete moment diagram. The first part of the shear diagram (Figure 13.26(b)) is a horizontal line above the zero axis, and the area under the diagram from $x = 0$ to $x = 3$ is $(484.7 \text{ N})(3 \text{ m}) = 1454 \text{ N} \cdot \text{m}$ (rounded off from $1454.1 \text{ N} \cdot \text{m}$) The bending moment starts at zero at the left end of the beam and increases linearly to $1454 \text{ N} \cdot \text{m}$ at $x = 3 \text{ m}$.

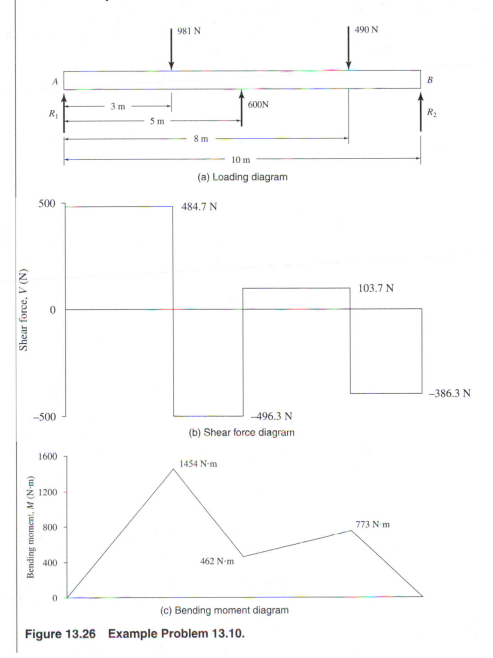

Figure 13.26   Example Problem 13.10.

Over the span 3 m to 5 m the shear indicates that the bending moment will be falling linearly because the shear in this region is a horizontal line below the zero axis. It will change by an amount equal to the area $(-496.3 \text{ N})(5 \text{ m} - 3 \text{ m}) = -992.6$ $\text{N} \cdot \text{m}$, resulting in a bending moment at $x = 5$ m of 1454.1 $\text{N} \cdot \text{m} + (-992.6 \text{ N} \cdot \text{m}) = $ 462 $\text{N} \cdot \text{m}$ (rounded off from 461.5 $\text{N} \cdot \text{m}$).

From 5 m to 8 m, the bending moment increases linearly by $(103.7 \text{ N})(3 \text{ m}) = $ 311.1 $\text{N} \cdot \text{m}$, resulting in a bending moment of $(461.5 \text{ N} \cdot \text{m} + 311.1 \text{ N} \cdot \text{m}) = 773 \text{ N} \cdot \text{m}$ (rounded off from 772.6 $\text{N} \cdot \text{m}$).

Finally, from 8 m to 10 m, the negative area, $(-386.3 \text{ N})(2 \text{ m}) = -772.6 \text{ N} \cdot \text{m}$, returns the bending moment to zero at the end of the beam. We show the complete bending moment diagram in Figure 13.26(c).

b. Calculate the bending moment at $x = 9$ m. The total moment of the external forces about the station $x = 9$ m consists of moments created by

reaction R_1: $M_{R1} = -(484.7 \text{ N})(9 \text{ m}) = -4362.3 \text{ N} \cdot \text{m}$

first load L_1: $M_{L1} = (981 \text{ N})(6 \text{ m}) = 5886 \text{ N} \cdot \text{m}$

second load L_2: $M_{L2} = -(600 \text{ N})(4 \text{ m}) = -2400 \text{ N} \cdot \text{m}$

third load L_3: $M_{L3} = (490 \text{ N})(1 \text{ m}) = 490 \text{ N} \cdot \text{m}$

The total moment of the external forces about station $x = 9$ m is the algebraic sum of these:

$$M = -386.3 \text{ N} \cdot \text{m}$$

The internal moment needed to balance this moment—i.e., the bending moment— is 386.3 $\text{N} \cdot \text{m}$.

13.7.2   Simply Supported Beam with Distributed Loads

One of the differences with distributed loads is that they cause *curved* bending moment diagrams. To see why this is so, let us study the loading diagram in Figure 13.27(a). This is a repeat of Figure 13.16(a). We calculated the reactions in Section 13.6.2: R_1 = 137.5 lb and R_2 = 112.5 lb. What we are going to do now is calculate the bending moment at each 1-ft station from the left end of the beam. Remember what the bending moment is. *It is the moment that* balances *the moment (at the station being considered) created by all the external forces to the left of that station.* Refer now to the series of diagrams in Figure 13.27(b).

At $x = 0$, the left end of the beam, there are no forces to the left, so M = 0.

The moment of R_1 about the station $x = 1$ ft is $-(137.5 \text{ lb})(1 \text{ ft}) = -137.5$ lb-ft. This is *not* the bending moment. The bending moment is the moment that balances this, i.e., $M_R = +137.5$ lb-ft. We used the subscript R to indicate that this is the bending moment created by the reaction.

The moment of R_1 about station $x = 2$ ft is equal to $-(137.5 \text{ lb})(2 \text{ ft}) = -275$ lb-ft. So the bending moment at $x = 2$ ft is $M_R = +275$ lb-ft.

At $x = 3$ ft the first 1 ft of the distributed load is to the left of the station, as is the reaction R_1. What is the contribution of this portion of the distributed load to the moment about the station $x = 3$ ft? Its weight is $(50 \text{ lb/ft})(1 \text{ ft}) = 50$ lb. This weight is acting effectively at its center of gravity, which is at its center, i.e., midway between the start ($x = 2$ ft) and the end ($x = 3$ ft) of the portion. So $x = (2 \text{ ft} + 3 \text{ ft})/2 = 2.5$ ft. The moment arm (the distance from the station to the force, normal to the line of action of the force) is 0.5 ft. It is negative, as it should be, because we are measuring from right to left, i.e., in a negative x-direction. So the moment of this portion of the load about $x = 3$ ft is $(-50 \text{ lb})(-0.5 \text{ ft}) = +25$ lb-ft. The bending moment caused by this portion of the load is the negative of this, because it is the balancing moment. Therefore, $M_D = -25$ lb-ft, where M_D is the bending moment due to the distributed load(s). We must not forget the moment due to the reaction. It is creating a moment of $-(137.5 \text{ lb})(3 \text{ ft}) = -412.5$ lb-ft about station $x = 3$ ft. Its contribution to the bending moment is, therefore, $+412.5$ lb-ft. The total bending moment at $x = 3$ ft is $M = M_R + M_D = 412.5 \text{ lb-ft} - 25 \text{ lb-ft} = 387.5$ lb-ft.

We present the remaining calculations in tabular form. The student should follow these calculations carefully in conjunction with the diagrams in Figure 13.27(b).

Station x ft	M_R lb-ft	M_D lb-ft	M lb-ft
0	0	0	0
1	(137.5×1)	0	137.5
2	(137.5×2)	0	275.0
3	(137.5×3)	$-(50 \times 0.5)$	387.5
4	(137.5×4)	$-(100 \times 1.0)$	450.0
5	(137.5×5)	$-(150 \times 1.5)$	462.5
6	(137.5×6)	$-(200 \times 2.0)$	425.0
7	(137.5×7)	$-(250 \times 2.5)$	337.5
8	(137.5×8)	$-(250 \times 3.5)$	225.0
9	(137.5×9)	$-(250 \times 4.5)$	112.5
10	(137.5×10)	$-(250 \times 5.5)$	0

Notice that as we proceed along the distributed load, both the effective force and the moment arm increase in magnitude, but from station $x = 7$ ft (the end of the distributed load) to the end of the beam the load remains constant. This makes the bending moment diagram curved in the region of the distributed load and straight in regions where there are no distributed loads. We show the complete bending moment diagram in Figure 13.27(c).

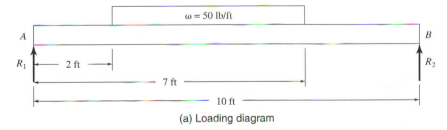

(a) Loading diagram

Figure 13.27

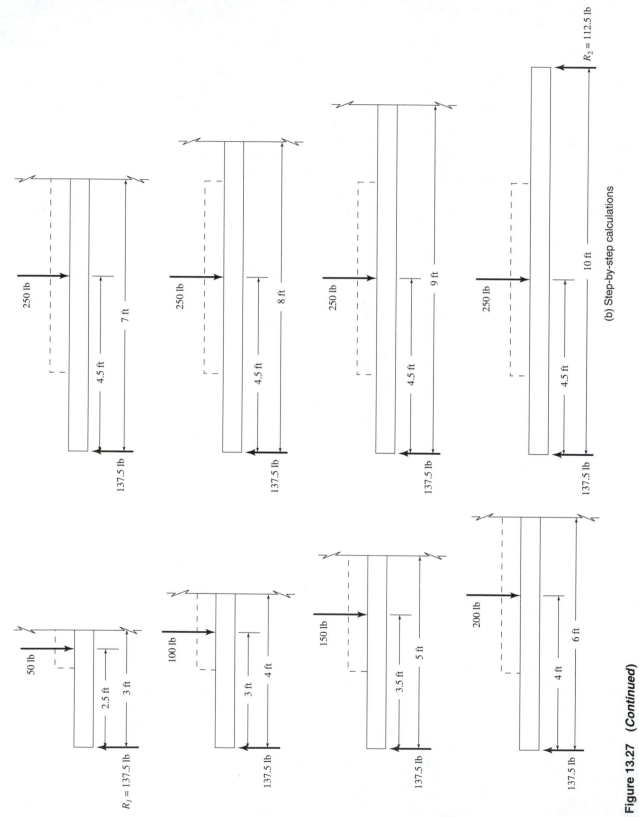

Figure 13.27 *(Continued)*

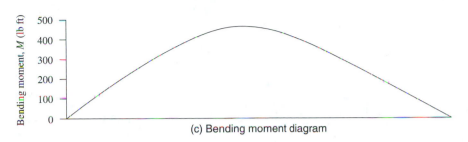

(c) Bending moment diagram

Figure 13.27   (*Continued*)

EXAMPLE PROBLEM 13.11

In this problem we calculate the bending moment using the step-by-step method and using areas. The loading diagram (Figure 13.28(a)) is a repeat of Figure 13.17(c) for Example Problem 13.4. We found for that example that $R_1 = 376.5$ N and $R_2 = 466.7$ N.

a. Determine the complete bending moment diagram using step-by-step calculations.

b. Determine the complete bending moment diagram by using areas from the shear diagram (Figure 13.28(b), which is a repeat of Figure 13.17(d)).

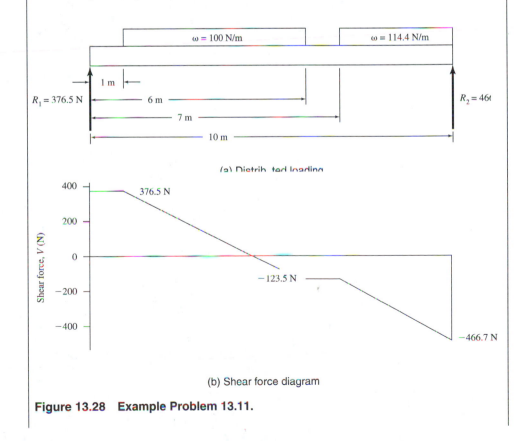

(b) Shear force diagram

Figure 13.28   Example Problem 13.11.

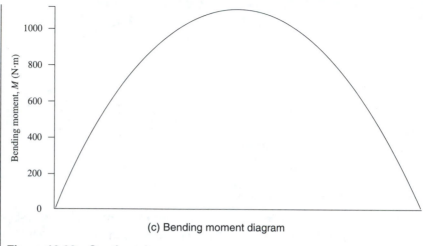

(c) Bending moment diagram

Figure 13.28   Continued

Solution

a. We present the step-by-step calculations in the following table. M_R is the contribution from the reaction R_1, M_{D1} is the contribution from the first distributed load, and M_{D2} is that from the second. The last column, M, is the total bending moment and is equal to the algebraic sum of $M_R + M_{D1} + M_{D2}$.

Station x (m)	M_R (N·m)	M_{D1} (N·m)	M_{D2} (N·m)	M (N·m)
0	0	0	0	0
1	(376.5×1) $= 376.5$	0	0	376.5
2	(376.5×2) $= 753$	$-(100 \times 0.5)$ $= -50$	0	703
3	(376.5×3) $= 1129.5$	$-(200 \times 1)$ $= -200$	0	929.5
4	(376.5×4) $= 1506$	$-(300 \times 1.5)$ $= -450$	0	1056
5	(376.5×5) $= 1882.5$	$-(400 \times 2)$ $= -800$	0	1082.5
6	(376.5×6) $= 2259$	$-(500 \times 2.5)$ $= -1250$	0	1009
7	(376.5×7) $= 2635.5$	$-(500 \times 3.5)$ $= -1750$	0	885.5
8	(376.5×8) $= 3012$	$-(500 \times 4.5)$ $= -2250$	$-(114.4 \times 0.5)$ $= -57.2$	704.8
9	(376.5×9) $= 3388.5$	$-(500 \times 5.5)$ $= -2750$	$-(228.8 \times 1)$ $= -228.8$	409.7
10	(376.5×10) $= 3765$	$-(500 \times 6.5)$ $= -3250$	$-(343.2 \times 1.5)$ $= -514.8$	0

Explanation:

Consider specific stations:

At $x = 3$ **m**, the moment created by R_1 is $-(376.5 \text{ N})(3 \text{ m})$. Because the bending moment is the negative of this, $M_R = (376.5 \text{ N})(3 \text{ m}) = +1129.5 \text{ N} \cdot \text{m}$. This station is 2 m into the first distributed load, and the moment arm of this force from station $x = 3$ m is $(3 \text{ m} - 2 \text{ m}) = 1$ m. The moment, being force $\times$ distance, is $(200 \text{ N})(1 \text{ m}) = +200 \text{ N} \cdot \text{m}$, but the bending moment is the negative of this: $M_{D1} = -200 \text{ N} \cdot \text{m}$. There is no contribution from the second distributed load, so $M_{D2} = 0$. The resulting bending moment is $M = M_R + M_{D1} = 1129.5 \text{ N} \cdot \text{m} + (-200 \text{ N} \cdot \text{m}) = 929.5 \text{ N} \cdot \text{m}$.

At $x = 7$ **m**, the moment created by R_1 is $-(376.5 \text{ N})(7 \text{ m}) = -2635.5 \text{ N} \cdot \text{m}$, so the contribution to the bending moment is $+2635.5 \text{ N} \cdot \text{m}$. At this station *all* the first distributed load is contributing to the bending moment. Its moment of $(500 \text{ N})(3.5 \text{ m}) = +1750 \text{ N} \cdot \text{m}$ results in a bending moment contribution of $-1750 \text{ N} \cdot \text{m}$. All the second distributed load is to the left of station $x = 7$ m, so it makes no contribution to the bending moment. The total bending moment is $M = M_R + M_{D1} = 2635.5 \text{ N} \cdot \text{m}. + (-1750 \text{ N} \cdot \text{m}) = 885.5 \text{ N} \cdot \text{m}$.

At $x = 9$ **m**, the R_1 contribution to the bending moment is $(376.5 \text{ N})(9 \text{ m}) = +3388.5 \text{ N} \cdot \text{m}$. All the first distributed load is to the left of the station, so the moment of this force about station $x = 9$ m is $(500 \text{ N})(5.5 \text{ m}) = +2750 \text{ N} \cdot \text{m}$, and the contribution to the bending moment is $-2750 \text{ N} \cdot \text{m}$. The station $x = 9$ m is 2 m into the second distributed load, resulting in an effective load of $(114.4 \text{ N/m})(2 \text{ m}) = 228.8$ N, acting at $x = 7 \text{ m} + (9 \text{ m} - 7 \text{ m})/2 = 8$ m. So the moment arm is 1 m, and its moment is $(228.8 \text{ N})(1 \text{ m}) = 228.8 \text{ N} \cdot \text{m}$. So the contribution to the bending moment is $-228.8 \text{ N} \cdot \text{m}$. Finally, $M = M_R + M_{D1} + M_{D2} = 3388.5 \text{ N} \cdot \text{m} + (-2750 \text{ N} \cdot \text{m}) + (-228.8 \text{ N} \cdot \text{m}) = 409.7 \text{ N} \cdot \text{m}$.

b. To find the bending moment from areas, we consider the following.

$x = 0$ **to 1 m**: Looking at Figure 13.28(b), we see that the shear force is horizontal and above the zero axis in this range. The bending moment will, therefore, be a straight line with positive slope in this region. We calculate the area under the shear diagram in this region: area $= (376.5 \text{ N})(1 \text{ m}) = 376.5 \text{ N} \cdot \text{m}$. This is the bending moment at $x = 1$ m, and it agrees with our previous calculation.

At $x = 2$ **m**: First we need to know what the shear force is at this station. We know that it is falling from 376.5 N at the rate of 100 N/m for 1 m. So $V = (376.5 \text{ N} - 100 \text{ N}) = 276.5$ N. The area of the trapezoid between $x = 1$ m and $x = 2$ m is $(1 \text{ m})(376.5 \text{ N} + 276.5 \text{ N})/2 = 326.5 \text{ N} \cdot \text{m}$ (see Appendix A). Adding this to the area between $x = 0$ and $x = 1$ m, i.e., $376.5 \text{ N} \cdot \text{m}$, gives $703 \text{ N} \cdot \text{m}$. This is the bending moment at $x = 2$ m, and it agrees with our previous calculation. Notice that the bending moment diagram begins to curve after $x = 1$ m because the presence of a distributed load is being felt.

We can continue in the same way up to $x = 4$ m. Somewhere between 4 m and 5 m the shear force crosses the zero axis. This is an important station, because the maximum bending moment—and, therefore, the maximum bending stress—occurs here. The question is, Where is this station?

Maximum Bending Moment

First, where does the shear force cross the zero axis? We know that $V = 376.5$ N at $x = 1$ m and falls linearly at a rate of 100 N/m after that. It will drop to zero in 376.5 N/(100 N/m), i.e., in 3.765 m. So $V = 0$ at $x = (1 \text{ m} + 3.765 \text{ m}) = 4.765$ m. The area of the triangle between $x = 1$ m and $x = 4.765$ m is 376.5 N $\times$ (4.76 m $- 1$ m)/2 $= 708.7$ N $\cdot$ m (see Appendix A). Adding this to the area between $x = 0$ and $x = 1$ m (376.5 N $\cdot$ m) gives 1085.2 N $\cdot$ m as the maximum bending moment.

At x = 6 m we have to add the triangular area between $x = 4.765$ m and 6 m (but it is a negative area, so we actually subtract it). This area is $(-123.5 \text{ N})(6 \text{ m} - 4.765$ m)/2 $= -76.3$ N $\cdot$ m. So the bending moment at $x = 6$ m is 1085.2 N $\cdot$ m $+ (-76.3$ N $\cdot$ m) $= 1009$ N $\cdot$ m, which agrees with our previous calculation.

The bending moment diagram can be completed in this way by taking incremental areas.

13.7.3   Simply Supported Beam with Concentrated and Distributed Loads

EXAMPLE PROBLEM 13.12

The beam shown in Figure 13.29(a) has two concentrated loads and two distributed loads that overlap for part of the span. It is a repeat of Figure 13.19(a). Calculate the bending moment (a) at $x = 6$ m and (b) at $x = 9$ m.

Solution

From Example Problem 13.5, $R_1 = 525$ kN, and $R_2 = 500$ kN.

a. Bending moment at $x = 6$ m: Referring to Figure 13.29(a), we see that station $x = 6$ m lies in the region where the two distributed loads overlap. We can calculate the bending moment in either of two ways. In the first way we regard the station as being between the start and the end of the first distributed load *and also* between the start and the end of the second distributed load. In this case there is no complete distributed load to the left of the station. The second way is to regard the region of overlapping loads as a separate load, as in Figure 13.29(b). Here the loads are combined to form a distributed load that is 2 m long, having a weight of 150 kN/m. In this case there *is* a complete distributed load to the left of the station.

First, the contribution from the reaction is

$$M_R = (525 \text{ kN})(6 \text{ m} - 0) = 3150 \text{ kN} \cdot \text{m}$$

There is one concentrated load to the left of the station. Therefore,

$$M_L = -(225 \text{ kN})(6 \text{ m} - 2 \text{ m}) = -900 \text{ kN} \cdot \text{m}$$

With regard to the first distributed load, it is $\omega = 50$ kN/m extending over a (6-m $-$ 3-m) = 3-m region, so its load is 150 kN. The center of gravity of this portion is at x

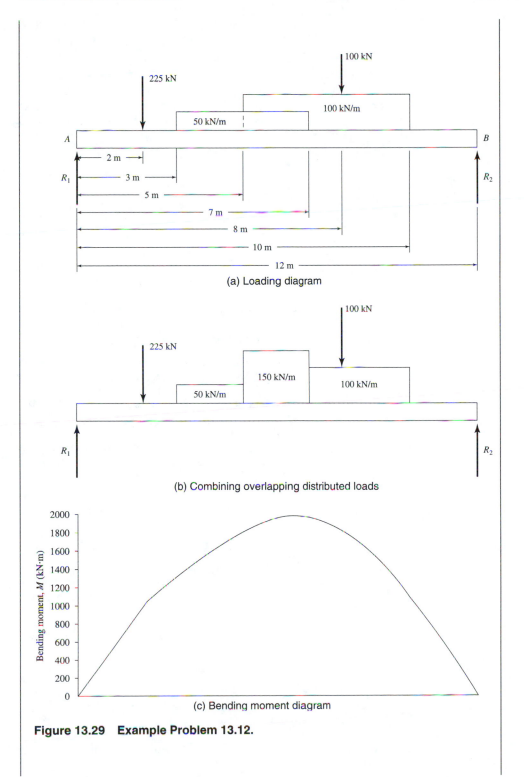

(a) Loading diagram

(b) Combining overlapping distributed loads

(c) Bending moment diagram

Figure 13.29   Example Problem 13.12.

= 3 m + 3 m/2 = 4.5 m. The distance from the 6-m station to the center of gravity of this part load is 1.5 m. Therefore, the moment is (150 kN)(1.5 m) = 225 kN · m, and the bending moment contribution is −225 kN · m. Similarly, with regard to the second distributed load, there is no complete distributed load to the left of the station. For this part load, ω_2 = 100 kN/m, for a span of 1 m, making a load of 100 kN. Its center of gravity is at x = 5.5 m, resulting in a moment about the 6-m station of (100 kN)(0.5 m) = 50 kN · m. Its contribution to the bending moment is, therefore, −50 kN · m.

The bending moment contribution of the distributed loads is

$$M_D = (-225 \text{ kN} \cdot \text{m}) + (-50 \text{ kN} \cdot \text{m}) = -275 \text{ kN} \cdot \text{m}$$

The total bending moment at x = 6 m is

$$M = M_R + M_L + M_D = 3150 \text{ kN} \cdot \text{m} + (-900 \text{ kN} \cdot \text{m}) + (-275 \text{ kN} \cdot \text{m})$$

So M = **1975 kN · m**.

b. Bending moment at x = 9 m: The moment created by the reaction is −(525 kN)(9 m) = −4725 kN · m. So the reaction contribution is M_R = 4725 kN · m.

There are two concentrated loads to the left of the station: L_1 = 225 kN at x = 2 m and L_2 = 100 kN at x = 8 m.

The moments they create about the 9-m station are, respectively,

$$(225 \text{ kN})(9 \text{ m} - 2 \text{ m}) = 1575 \text{ kN} \cdot \text{m}, \quad \text{and}$$

$$(100 \text{ kN})(9 \text{ m} - 8 \text{ m}) = 100 \text{ kN} \cdot \text{m}$$

Their contributions to the bending moment are the negatives of these, so:

$$M_L = -1575\text{k N} \cdot \text{m} + (-100 \text{ kN} \cdot \text{m}) = -1675 \text{ kN} \cdot \text{m}$$

With regard to the distributed load contribution (see Figure 13.29(a)), there is one complete distributed load to the left of station x = 9 m and one part load: ω_1 = 50 kN/m, with a span of (7 m − 3 m) = 4 m and center of gravity at x = (3 m + 4 m/2) = 5 m, and ω_2 = 100 kN/m, with a span of (9 m − 5 m) = 4 m and center of gravity at x = (5 m + 4 m/2) = 7 m. The moments of these loads about the 9-m station are, respectively,

$$(50 \text{ kN/m})(4 \text{ m})(9 \text{ m} - 5 \text{ m}) = 800 \text{ kN} \cdot \text{m}, \quad \text{and}$$

$$(100 \text{ kN/m})(4 \text{ m})(9 \text{ m} - 7 \text{ m}) = 800 \text{ kN} \cdot \text{m}$$

Remember that the bending moment contributions are the negatives of these:

$$M_D = -800 \text{ kN} \cdot \text{m} + (-800 \text{ kN} \cdot \text{m}) = -1600 \text{ kN} \cdot \text{m}$$

$$M = M_R + M_L + M_D = 4725 \text{ kN} \cdot \text{m} + (-1675 \text{ kN} \cdot \text{m}) + (-1600 \text{ kN} \cdot \text{m})$$

So M = 1450 kN · M.

Answers
a. The bending moment at x = 6 m is M = 1975 kN · m.
b. The bending moment at x = 9 m is M = 1450 kN · m.

We include the complete bending moment diagram for this case in Figure 13.29(c).

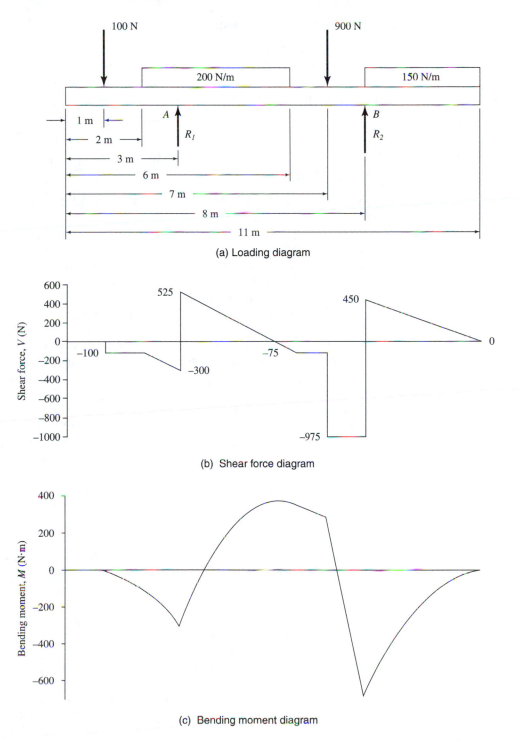

(a) Loading diagram

(b) Shear force diagram

(c) Bending moment diagram

Figure 13.30

13.7.4   Overhanging Beams

We calculate bending moments for overhanging beams the same way as for simply supported beams. The only difference is that the location of the left-end reaction may not be zero, or the right reaction may not coincide with the right end of the beam. The loading shown in Figure 13.30(a) is a repeat of Figure 13.21(a). The supports are not at the ends of the beam, making it an *overhanging beam*.

EXAMPLE PROBLEM 13.13

Calculate the complete bending moment diagram using the previously constructed shear force diagram (Figure 13.30(b)).

Solution

Because the shear force is zero from $x = 0$ to $x = 1$ m, the bending moment is also zero in this region. From $x = 1$ m to 2 m, the shear is constant at -100 N, so the bending moment will change linearly from 0 to $(-100 \text{ N})(1 \text{ m}) = -100 \text{ N} \cdot \text{m}$.

In the region $x = 2$ m to 3 m, the shear force changes linearly from -100 N to -300 N. This creates a curved bending moment line whose magnitude at $x = 3$ m is equal to the total area from $x = 0$ to $x = 3$ m. This area consists of the rectangle between $x = 1$ m and 2 m and the trapezoid from $x = 2$ m to 3 m. The rectangular area is $-100 \text{ N} \cdot \text{m}$, and the trapezoidal area is $(1 \text{ m})(-100 \text{ N} + (-300 \text{ N}))/2 = -200$ $\text{N} \cdot \text{m}$, making the bending moment at $x = 3$ m equal to $-300 \text{ N} \cdot \text{m}$.

The next area, from $x = 3$ m to the station where the shear crosses the zero axis, is a triangle. The question is, Where does the shear cross the zero axis? We see (Figure 13.30(b)) that there is a shear force of 525 N at $x = 3$ m and that it is falling at a rate of 200 N/m. The distance it will take to fall to zero is 525 N/(200 N/m) = 2.625 m. We see then that the shear force is zero at $x = 3$ m + 2.625 m = 5.625 m.

The area of the triangle (half the base times height) is $(525 \text{ N})(2.625 \text{ m})/2 = 689.06 \text{ N} \cdot \text{m}$. Adding this to the previous area gives $-300 \text{ N} \cdot \text{m} + 689.06 \text{ N} \cdot \text{m} = 389.06 \text{ N} \cdot \text{m}$. The bending moment diagram is curved in this region because the shear is a straight sloping line.

It continues as a curved line in the small region from $x = 5.625$ m to $x = 6$ m, which is the end of the first distributed load. The area of this small region is $(-75 \text{ N})(6 \text{ m} - 5.625 \text{ m})/2 = -14.06 \text{ N} \cdot \text{m}$. Adding this to the previous area gives 389.06 $\text{N} \cdot \text{m} + (-14.06 \text{ N} \cdot \text{m}) = 375 \text{ N} \cdot \text{m}$.

The bending moment in the next region, from $x = 6$ m to 7 m, is linear because the shear is a horizontal line. The area is $-75 \text{ N} \cdot \text{m}$, so the bending moment at $x = 7$ m is $375 \text{ N} \cdot \text{m} + (-75 \text{ N} \cdot \text{m}) = 300 \text{ N} \cdot \text{m}$.

The region from $x = 7$ m to 8 m also has a constant shear and so creates a linear bending moment. The area is $(-975 \text{ N})(1 \text{ m}) = -975 \text{ N} \cdot \text{m}$. The bending moment at $x = 8$ m is $300 \text{ N} \cdot \text{m} + (-975 \text{ N} \cdot \text{m}) = -675 \text{ N} \cdot \text{m}$.

Finally, from $x = 8$ m to the end of the beam at $x = 11$ m, the bending moment curves back up to zero. The area is $(450 \text{ N})(11 \text{ m} - 8 \text{ m})/2 = 675 \text{ N} \cdot \text{m}$. Adding this to the value at $x = 8$ m gives $M = 0$ at the end of the beam. We show the complete bending moment diagram in Figure 13.30(c).

13.7.5 Cantilever Beams

As we suggested in Section 13.6.5, *you should draw the wall on the right* to keep the sign conventions consistent.

EXAMPLE PROBLEM 13.14

Figure 13.31(a) and (b) repeats Figure 13.22(a) and (b), showing a cantilever beam with concentrated and distributed loads and the shear force diagram.
  a. Calculate the complete bending moment diagram using areas.
  b. Calculate the bending moment at $x = 7$ m and at $x = 10$ m.

Solution

a. The first area under the shear force diagram in Figure 13.31(b) is a rectangle equal to $(-100$ kN$)(1$ m$) = -100$ kN $\cdot$ m. So the bending moment diagram slopes down linearly from zero at $x = 0$ to -100 kN $\cdot$ m at $x = 1$ m. The area from $x = 1$ m to $x = 5$ m is a trapezoid. Its area is $(-100$ kN $+ (-300$ kN$))$ $(5$ m $- 1$ m$)/2 = -800$ kN $\cdot$ m. Due to the linearly downward sloping shear in this region, the bending moment is a downward curve. Adding this to the previous -100 kN $\cdot$ m gives the bending moment at $x = 5$ m as -900 kN $\cdot$ m. The next region, $x = 5$ m to $x = 6$ m, is a rectangular area equal to $(-300$ kN$)(6$ m $- 5$ m$) = -300$ kN $\cdot$ m, making the bending moment at $x = 6$ m equal to -300 kN $\cdot$ m $+ (-900$ kN $\cdot$ m$) = -1200$ kN $\cdot$ m. The shape of the bending moment diagram is linear, sloping down.

The shear force from $x = 6$ m to $x = 8$ m is also constant and equal to -600 kN. Its area is $(-600$ kN$)(8$ m -6 m$) = -1200$ kN $\cdot$ m, making the bending moment at $x = 8$ m equal to -1200 kN $\cdot$ m $+ (-1200$ kN $\cdot$ m$) = -2400$ kN $\cdot$ m. The bending moment will slope down linearly from -1200 kN $\cdot$ m at $x = 6$ m to -2400 kN $\cdot$ m at $x = 8$ m. The last region, from $x = 8$ m to $x = 12$ m, is a trapezoidal area equal to $(-600$ kN $+ (-700$ kN$))(12$ m $- 8$ m$)/2 = -2600$ kN $\cdot$ m. Adding this to the -2400 kN $\cdot$ m at $x = 8$ m gives $M = -5000$ kN $\cdot$ m.

We show the complete bending moment diagram in Figure 13.31(c).

b. We first find the bending moment at $x = 7$ m. Referring to Figure 13.31(a), there are two concentrated loads and one complete distributed load to the left of the station. The concentrated loads are $L_1 = 100$ kN at $x = 0$, creating a moment of $(100$ kN$)(7$ m$) = 700$ kN $\cdot$ m about station $x = 7$ m (and, therefore, having a contribution of -700 kN $\cdot$ m to the bending moment), and $L_2 = 300$ kN at $x = 6$ m, creating a moment of $(300$ kN$)(7$ m $- 6$ m$) = 300$ kN $\cdot$ m about the 7-m station, (and, therefore, having a bending moment contribution of -300 kN $\cdot$ m).

$$M_L = (-700 \text{ kN} \cdot \text{m}) + (-300 \text{ kN} \cdot \text{m}) = -1000 \text{ kN} \cdot \text{m}$$

The distributed load is $\omega_1 = 50$ kN/m, for a span of $(5$ m $- 1$ m$) = 4$ m and center of gravity at $(1$ m $+ 4$ m$/2) = 3$ m.

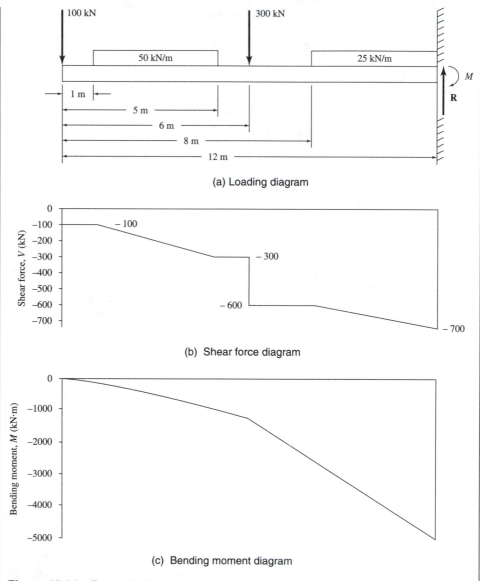

(a) Loading diagram

(b) Shear force diagram

(c) Bending moment diagram

Figure 13.31   Example Problem 13.14.

The moment of this load about the 7-m station is $(50 \text{ kN/m})(4 \text{ m})(7 \text{ m} - 3 \text{ m}) = 800 \text{ kN} \cdot \text{m}$. Its contribution to the bending moment is $M_D = -800 \text{ kN} \cdot \text{m}$.

Adding the contributions gives

$$M = M_L + M_D = (-1000 \text{ kN} \cdot \text{m}) + (-800 \text{ kN} \cdot \text{m}) = -1800 \text{ kN} \cdot \text{m}$$

Next, we find the moment at $x = 10$ m. There are two concentrated loads and one complete distributed load to the left of station $x = 10$ m, and the station lies within the second distributed load (Figure 13.31(a)).

For the concentrated loads, remember that the bending moment contributions are the negatives of the external moments:

$$M_L = -[(100 \text{ kN})(10 \text{ m}) + (300 \text{ kN})(4 \text{ m})]$$

$$= -[(1000 \text{ kN} \cdot \text{m}) + (1200 \text{ kN} \cdot \text{m})] = -2200 \text{ kN} \cdot \text{m}$$

For distributed loads, $\omega_1 = 50$ kN/m, for a span of (5 m $-$ 1 m) = 4 m and center of gravity at (1 m + 4 m/2) = 3 m. Also, $\omega_2 = 25$ kN/m, for a span of (10 m $-$ 8 m) = 2 m and center of gravity at (8 m + 2 m/2) = 9 m.

$$M_D = -[(50 \text{ kN/m})(4 \text{ m})(7 \text{ m}) + (25 \text{ kN/m})(2 \text{ m})(1 \text{ m})]$$

$$= -(1400 \text{ kN} \cdot \text{m} + 50 \text{ kN} \cdot \text{m}) = -1450 \text{ kN} \cdot \text{m}$$

Adding the contributions,

$$M = M_L + M_D = (-2200 \text{ kN} \cdot \text{m}) + (-1450 \text{ kN} \cdot \text{m}) = -3650 \text{ kN} \cdot \text{m}$$

Answer
a. The bending moment at $x = 7$ m is $M = -1800$ kN $\cdot$ m.
b. The bending moment at $x = 10$ m is $M = -3650$ kN $\cdot$ m.

EXAMPLE PROBLEM 13.15

Calculate the bending moment at the wall for the loading shown in Figure 13.32(a), which is a repeat of Figure 13.23(a).

Solution
First redraw the figure with the wall to the right, as in Figure 13.32(b). Now resolve the 1000-N force into its vertical and horizontal components, as we did in Example Problem 13.8. See Figure 13.32(c). The vertical component is 866 N.

Note: The horizontal component *does not have a contribution to the bending moment*. It merely pushes on the beam.

Taking moments about the wall,

$$(800 \text{ N})(6 \text{ m}) + (866 \text{ N})(4 \text{ m}) = 4800 \text{ N} \cdot \text{m} + 3464 \text{ N} \cdot \text{m} = 8264 \text{ N} \cdot \text{m}$$

The bending moment, being the negative of this, is -8264 N $\cdot$ m.

This is a clockwise moment within the wall, balancing the external moments imposed on the beam.

Answer
The bending moment at the wall is -8264 N $\cdot$ m.

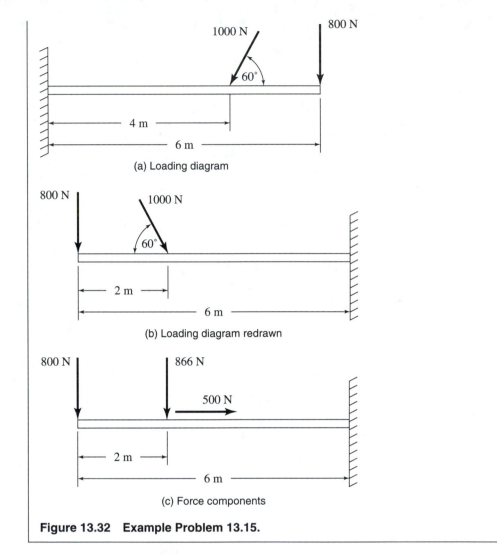

Figure 13.32   Example Problem 13.15.

13.8   GENERALIZED EQUATIONS

In order to write generalized computer programs for the calculation of shear forces and bending moments, we must first express the procedures outlined in the preceding sections in mathematical terms. We start with the development of shear force equations for simply supported and overhanging beams and then do those for cantilever beams. We then develop the equations for bending moments.

13.8.1   Shear Forces: Simply Supported and Overhanging Beams

With Concentrated Loads

Assuming that there are concentrated loads L_1, L_2, L_3, ... at distances x_{L1}, x_{L2}, x_{L3}, ... from the left end of the beam and reactions R_1 and R_2 at stations x_{R1} and x_{R2} (see Figure 13.33), then at any station x from the left end of the beam, the total shear force is

$$V = V_R + V_L \qquad (13.8)$$

where V_R is the shear force due to the reactions and V_L is due to the concentrated loads.

First let us look at V_R. If the station we are considering (station x) is to the left of the first reaction located at x_{R1}, there will be no contribution to the shear from the reactions. We can write this in equation form:

$$V_R = 0 \qquad x_{R1} > x \qquad (13.8(a))$$

If station x is between the reactions, then the first reaction, R_1, will contribute to the shear, but R_2 will not. We can write this as

$$V_R = R_1 \qquad x_{R2} > x > x_{R1} \qquad (13.8(b))$$

If x is to the right of the second reaction, then both reactions contribute to the shear:

$$V_R = R_1 + R_2 \qquad x > x_{R2} \qquad (13.8(c))$$

Now let us look at V_L. If x is to the left of the first concentrated load, there will be no contribution to the shear from the loads. Therefore, we can write

$$V_L = 0 \qquad x_{L1} > x \qquad (13.9(a))$$

If x is between L_1 and L_2, then L_1 will contribute, but L_2 will not. Writing this in symbols,

$$V_L = L_1 \qquad x_{L2} > x > x_{L1}$$

If x is between L_2 and L_3, then L_1 and L_2 will contribute to the shear, but L_3 will not. In symbols,

$$V_L = L_1 + L_2 \qquad x_{L3} > x > x_{L2}$$

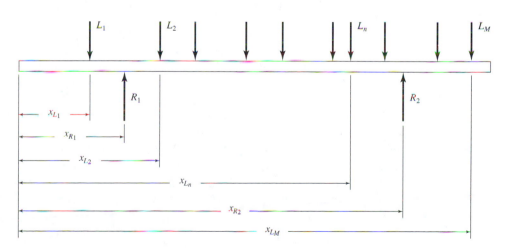

Figure 13.33   Concentrated loads: General case.

We can see from this that if x lies between the Nth and the $(N + 1)$st concentrated loads, then all the loads up to the Nth will contribute to the shear, but the $(N + 1)$st will not. So if we simply extend the previous equation we can derive a *general* equation:

$$V_L = L_1 + L_2 + L_3 + \cdots + L_n + \cdots + \cdots + L_N \qquad x_{L(N+1)} > x > x_{LN}$$

Let us check this out. Let $N = 2$. Then $x_{L(N+1)}$ is x_{L3} and x_{LN} is x_{L2}, so the equation becomes

$$V_L = L_1 + L_2 \qquad x_{L3} > x > x_{L2}$$

which is the same equation we had before.

We can write the general equation in compact form (see Appendix A) as

$$V_L = \sum_{n=1}^{n=N} Ln \qquad x_{L(N+1)} > x > x_{LN} \qquad \text{(13.9(b))}$$

Collecting these equations together, the total shear at any station x is given by

$$V = V_R + V_L \qquad\qquad\qquad\qquad \text{(13.8)}$$
$$V_R = 0 \qquad\qquad x_{R1} > x \qquad\qquad \text{(13.8(a))}$$
$$V_R = R_1 \qquad\qquad x_{R2} > x > x_{R1} \qquad \text{(13.8(b))}$$
$$V_R = R_1 + R_2 \qquad\qquad x > x_{R2} \qquad\qquad \text{(13.8(c))}$$
$$V_L = 0 \qquad\qquad x_{L1} > x \qquad\qquad \text{(13.9(a))}$$
$$V_L = \sum_{n=1}^{n=N} Ln \qquad x_{L(N+1)} > x > x_{LN} \qquad \text{(13.9(b))}$$

With Distributed Loads

Figure 13.34 shows the general case of a simply supported, or an overhanging beam carrying a number of distributed loads. As before, all distances are measured from the left end of the beam. The total shear force at any station will be the sum of the contributions by the reactions and the distributed loads to the left of that station.

$$V = V_R + V_D \qquad\qquad\qquad\qquad \text{(13.10)}$$

where V_R, the shear due to the reactions, is again given by equation (13.8(a)), (13.8(b)), or (13.8(c)), depending on whether x is to the left of R_1, between R_1 and R_2, or to the right of R_2. V_D is that part of the shear due to the distributed load(s) to the left of the station being considered.

Because more than one distributed load may be involved, we must be careful to distinguish between them. We always start from the left end of the beam, so *first* refers to the one nearest the left end. If the first distributed load extends from station x_{D1s} feet to x_{D1e} feet and it weighs ω_1 pounds per foot, then the amount of the load to the left of any station x—and, therefore, the contribution of the load to the shear force—depends on whether the station x is to the left of x_{D1s}, between x_{D1s} and x_{D1e}, or to the right of x_{D1e}. The subscripts mean the following:

D for distributed load
1 for the first distributed load

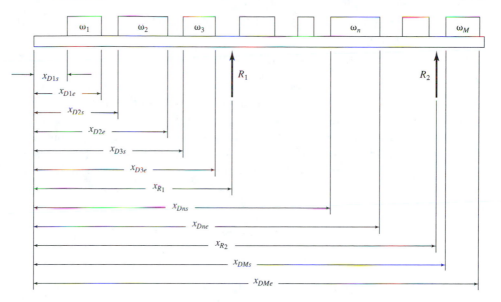

Figure 13.34   Distributed loads: General case.

2 for the second distributed load, etc.

n for the nth distributed load

s for the start of the distributed load

e for the end of the distributed load

M for the last complete distributed load to the left of station x

For example, x_{D3e} is the location of the end of the third distributed load measured from the left end of the beam. If x is to the left of x_{D1s}, *none* of the distributed load will have a contribution to the shear force. We write this in symbols as

$$V_D = 0 \qquad x_{D1s} > x \qquad (13.10(a))$$

What is the contribution if x lies somewhere between the start and the end of the *third* distributed load? All the first load will be included in the shear. Its weight is its weight per foot multiplied by the number of feet it extends. This amounts to

$$\omega_1(x_{D1e} - x_{D1s}) \quad \text{pounds}$$

Similarly, all of the second load will be included, which amounts to $\omega_2 (x_{D2e} - x_{D2s})$ pounds. Because station x lies between the start and the end of the third distributed load, only *part* of the load will have a contribution to the shear force, i.e., the part to the left of station x. The span of this effective part is $(x - x_{D3s})$ feet, and because it weighs ω_3 pounds per foot, its weight is $\omega_3 (x - x_{D3s})$ pounds. This is the contribution to the shear force at station x by the third distributed load. Note that if it is a down force, as it usually is, then it is negative. We put it in the equation as positive, however, because if it is a down load its value will be entered as a negative number. In this way we can account for both upward- and downward-acting forces. Collecting these terms together we see that the distributed load contribution to the shear is, in this case,

$$V_D = \omega_1(x_{D1e} - x_{D1s}) + \omega_2(x_{D2e} - x_{D2s}) + \omega_3(x - x_{D3s})$$

This is for the case where $x_{D3e} > x > x_{D3s}$, where station x is to the right of the start of the third distributed load but to the left of the end of that load; i.e., x lies between the start and the end of the third distributed load. To extend this to the case where x lies between the start and the end of the $(M + 1)$st distributed load, we have to include all the loads up to the Mth and that part of the $(M + 1)$st load to the left of x. We can again use the summation sign (see Appendix A) to write this in shorthand notation:

$$V_D = \sum_{n=1}^{M} \omega_n(x_{Dne} - x_{Dns}) + \omega_{(M+1)}(x - x_{D(M+1)s})$$

$$x_{D(M+1)e} > x > x_{D(M+1)s} \tag{13.10(b)}$$

We must also include the case where station x lies between the end of one distributed load and the start of the next. The last complete distributed load to the left of the station is identified by subscript M, so in this case x lies between x_{DMe} and $x_{D(M+1)s}$. There will no part-load contribution, so

$$V_D = \sum_{n=1}^{M} \omega_n(x_{Dne} - x_{Dns}) \qquad x_{D(M+1)s} > x > x_{DMe} \tag{13.10(c)}$$

In words, equations (13.10(a)), (13.10(b)) and (13.10(c)) say the following:

1. If the station being considered lies to the left of the start of the first distributed load, then the load makes no contribution to the shear. (*Only* the left-end reaction contributes.)
2. If the station lies *in the region* of the $(M + 1)$st distributed load, the load contribution to the shear is the sum of all the loads to the left of x plus the weight per foot of the $(M + 1)$st load multiplied by the number of feet of that load to the left of x.
3. If the station lies *between* two distributed loads, the contribution to the shear is the sum of all the loads to the left of x.

The total shear at any station, expressed by equation (13.10), is the sum of equations [(13.8(a)), (13.8(b)), or (13.8(c))] for V_R and [(13.10(a)), (13.10(b)), or (13.10(c))] for V_D, depending on the location of station x.

Collecting the governing equations together, we have

$$V = V_R + V_D \tag{13.10}$$
$$V_R = 0 \qquad x_{R1} > x \tag{13.8(a)}$$
$$V_R = R_1 \qquad x_{R2} > x > x_{R1} \tag{13.8(b)}$$
$$V_R = R_1 + R_2 \qquad x > x_{R2} \tag{13.8(c)}$$
$$V_D = 0 \qquad x_{D1s} > x \tag{13.10(a)}$$

$$V_D = \sum_{n=1}^{M} \omega_n(x_{Dne} - x_{Dns}) + \omega_{(M+1)}(x - x_{D(M+1)s})$$

$$x_{D(M+1)e} > x > x_{D(M+1)s} \tag{13.10(b)}$$

$$V_D = \sum_{n=1}^{M} \omega_n(x_{Dne} - x_{Dns}) \qquad x_{D(M+1)s} > x > x_{DMe} \tag{13.10(c)}$$

With Concentrated and Distributed Loads

We derived equations (13.8(a)), (13.8(b)) and (13.8(c)) for the shear force due to the reactions; equations (13.9(a)) and (13.9(b)) for the shear due to concentrated loads; and equations (13.10(a)), (13.10(b)) and (13.10(c)) for shear due to distributed loads. The total shear force is the sum of all the reactions, concentrated loads, and distributed loads (either full or partial) to the left of the station being considered.

$$V = V_R + V_L + V_D \tag{13.11}$$

Collecting the governing equations, we have

$$V_R = 0 \qquad\qquad x_{R1} > x \tag{13.8(a)}$$
$$V_R = R_1 \qquad\qquad x_{R2} > x > x_{R1} \tag{13.8(b)}$$
$$V_R = R_1 + R_2 \qquad\qquad x > x_{R2} \tag{13.8(c)}$$
$$V_L = 0 \qquad\qquad x_{L1} > x \tag{13.9(a)}$$

$$V_L = \sum_{n=1}^{n=N} Ln \qquad x_{L(N+1)} > x > x_{LN} \tag{13.9(b)}$$

$$V_D = 0 \qquad\qquad x_{D1s} > x \tag{13.10(a)}$$

$$V_D = \sum_{n=1}^{M} \omega_n(x_{Dne} - x_{Dns}) + \omega_{(M+1)}(x - x_{D(M+1)s})$$

$$x_{D(M+1)e} > x > x_{D(M+1)s} \tag{13.10(b)}$$

$$V_D = \sum_{n=1}^{M} \omega_n(x_{Dne} - x_{Dns}) \qquad x_{D(M+1)s} > x > x_{DMe} \tag{13.10(c)}$$

13.8.2 Shear Forces: Cantilever Beams

The same equations apply to cantilever beams as to simply supported and overhanging beams, because the same approach is used—calculate the sum of the forces to the left of the station—for all cases. *The only difference with cantilever beams is that V_R is always zero.*

13.8.3 Bending Moments: Simply Supported and Overhanging Beams

With Concentrated Loads

Refer again to Figure 13.33.

We will use the same symbols as for the shear force equations. Consider first M_R, which is the bending moment created by the reaction(s), and remember that it is equal in magnitude but opposite in direction to the moment caused by the reaction(s) to the left of the station.

First, if R_1 is to the right of the station being considered, it will not create any moment about the station. As before, we refer to the station being considered as x and the location of the reaction R_1 as x_{R1}. In symbols, then,

$$M_R = 0 \qquad x_{R1} > x \qquad \qquad (13.11(a))$$

If x lies between x_{R1} and x_{R2}, the moment of R_1 about x is equal to the force R_1 multiplied by the distance from x (traveling from right to left) to x_{R1}. This distance is ($x_{R1} - x$), so the moment of R_1 about x is $R_1(x_{R1} - x)$. Because the bending moment is the negative of this, we have $M_R = -R_1(x_{R1} - x)$, or

$$M_R = R_1(x - x_{R1}) \qquad x_{R2} > x > x_{R1} \qquad \qquad (13.11(b))$$

If the station x is to the right of both reactions, then the moment created by R_1 is $R_1(x_{R1} - x)$, and that created by R_2 is $R_2(x_{R2} - x)$, making the total moment about x equal to $R_1(x_{R1} - x) + R_2(x_{R2} - x)$. The bending moment is the negative of this, so

$$M_R = R_1(x - x_{R1}) + R_2(x - x_{R2}) \qquad x > x_{R2} \qquad \qquad (13.11(c))$$

Turning our attention now to the concentrated loads, the first is L_1 located at x_{L1}. If station x is to the left of x_{L1}, i.e., if $x_{L1} > x$, L_1 will not make a contribution to the bending moment. In symbols,

$$M_L = 0 \qquad x_{L1} > x \qquad \qquad (13.12(a))$$

If x lies between the Nth and the $(N + 1)$st concentrated loads, i.e., if $x_{L(N+1)} > x > x_{LN}$, then $L_1, L_2, L_3, \ldots, L_N$ contribute to the bending moment, but $L_{(N+1)}$, $L_{(N+2)}, \ldots$ do not. The distance from station x to the first concentrated load (measuring from right to left) is ($x_{L1} - x$), so the moment of L_1 about x is $L_1(x_{L1} - x)$, and the bending moment caused by this is $L_1 (x - x_{L1})$. In exactly the same way, the bending moment created by L_2 at x is $L_2(x - x_{L2})$. We see then that the nth concentrated load creates a bending moment at x of $L_n(x - x_{Ln})$. The sum of these is the bending moment at x created by all the concentrated loads.

$$M_L = \sum_{n=1}^{n=N} L_n(x - x_{Ln}) \qquad x_{L(N+1)} > x > x_{LN} \qquad \qquad (13.12(b))$$

Let us see how these equations apply to the example in Section 13.7.1 (Figure 13.24(a)) at station $x = 9$ ft.

The station is between the two reactions, so we use equation (13.11(b)) for M_R. $R_1 = 150$ lb, $x_{R1} = 0$, and $x = 9$ ft.

$$M_R = R_1(x - x_{R1}) = (150 \text{ lb})(9 \text{ ft} - 0) = 1350 \text{ lb-ft}$$

There are two concentrated loads to the left of the station, so $N = 2$. Equation (13.12(b)) reduces to

$$M_L = \sum_{n=1}^{n=N} L_n(x - x_{Ln}) = L_1(x - x_{L1}) + L_2(x - x_{L2})$$

$$L_1 = -200 \text{ lb}, \qquad L_2 = -150 \text{ lb}, \qquad x_{L1} = 4 \text{ ft}, \qquad x_{L2} = 8 \text{ ft}$$

Therefore,

$$M_L = (-200 \text{ lb})(9 \text{ ft} - 4 \text{ ft}) + (-150 \text{ lb})(9 \text{ ft} - 8 \text{ ft}) = -1150 \text{ lb-ft}$$

The total bending moment at $x = 9$ ft is the sum of M_R and M_L:

$$M = 1350 \text{ lb-ft} + (-1150 \text{ lb-ft}) = 200 \text{ lb-ft}$$

With Distributed Loads

We are now going to derive the general equations for the bending moment caused by distributed loads, but first let us consider the three distributed loads shown in Figure 13.35. We want to calculate the bending moment at station x, which is partway along the third load. We use the same symbols for the stations along the beam:

x_{D1s} is the start of the first distributed load.

x_{D1e} is the end of the first distributed load.

x_{D2s} is the start of the second distributed load.

x_{D2e} is the end of the second distributed load, and so on.

x_1 is the distance from station x to the effective line of action of the first distributed load, i.e., to the midpoint of that load.

x_2 is the distance from station x to the effective line of action of the second distributed load, i.e., to the midpoint of that load.

x_3 is the distance from station x to the effective line of action of the third distributed load. Because only a portion of the load is to the left of the station, x_3 is measured to the midpoint of that portion.

Now let us calculate the moment of the first load about station x. ω_1 is the weight per unit length (lb/ft or N/m) of the first distributed load. This creates a force of $(\omega_1 (x_{D1e} - x_{D1s}))$. The effective location of this force is at the midpoint of the load, i.e., at $x_{D1s} + (x_{D1e} - x_{D1s})/2$, which we can write as $(x_{D1e} + x_{D1s})/2$. The distance x_1,

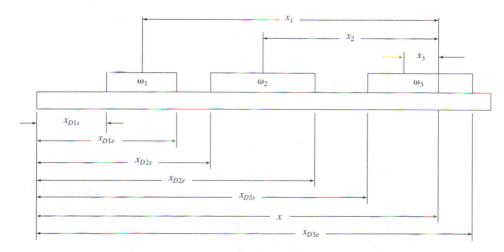

Figure 13.35   General case of distributed loads (three loads).

which is the moment arm, is (see Figure 13.35) the distance measured from station x to this midpoint. So $x_1 = (x_{D1e} + x_{D1s})/2 - x$. The moment of the first distributed load about station x is, therefore,

$$\omega_1(x_{D1e} - x_{D1s})[(x_{D1e} + x_{D1s})/2 - x]$$

The bending moment is the negative of this:

$$M_{D1} = \omega_1(x_{D1e} - x_{D1s})[x - (x_{D1e} + x_{D1s})/2]$$

In exactly the same way, the contribution to the bending moment at x by the second distributed load is

$$M_{D2} = \omega_2(x_{D2e} - x_{D2s})[x - (x_{D2e} + x_{D2s})/2]$$

In general, if the last complete distributed load to the left of station x is the Mth, its contribution is

$$M_{DM} = \omega_M(x_{DMe} - x_{DMs})[x - (x_{DMe} + x_{DMs})/2]$$

The sum of the contributions of all the complete distributed loads to the left of station x is, therefore,

$$M_D = \sum_{n=1}^{M} \omega_n(x_{Dne} - x_{Dns})[x - (x_{Dne} + x_{Dns})/2]$$

We now have to deal with the partial load. This is the third load in Figure 13.35. The portion of the load to the left of station x extends from $x = x_{D3s}$ to x, so its length is $(x - x_{D3s})$, and its weight is $\omega_3 (x - x_{D3s})$. The location of its effective force is midway between x_{D3s} and x, i.e., at $x = x_{D3s} + (x - x_{D3s})/2$. We can write this as $(x + x_{D3s})/2$. The distance x_3, which is the moment arm from x to this effective force, is $[(x + x_{D3s})/2 - x]$, which we can write as $(x_{D3s} - x)/2$. Multiplying this by the force gives

$$\omega_3(x - x_{D3s})(x_{D3s} - x)/2$$

The bending moment is the negative of this:

$$M_{D3} = \omega_3(x - x_{D3s})(x - x_{D3s})/2 = \omega_3(x - x_{D3s})^2/2$$

In general terms, this is the $(M + 1)$st load, so its contribution is $\omega_{(M+1)} (x - x_{D(M+1)s})^2/2$. Finally, the contribution of *all* the distributed loads to the bending moment at station x is

$$M_D = \sum_{n=1}^{M} \omega_n(x_{Dne} - x_{Dns})[x - (x_{Dne} + x_{Dns})/2] + \omega_{(M+1)}(x - x_{D(M+1)s})^2/2$$

We derived this equation for the general case, where station x lies somewhere between the start and the end of the $(M + 1)$st distributed load, i.e., for $x_{D(M+1)e} > x > x_{D(M+1)s}$. Of course, if there is no distributed load to the left of station x, then $M_D = 0$. In symbols, this is for $x_{D1s} > x$.

If station x lies *between* two distributed loads, there is no part-load contribution and

$$M_D = \sum_{n=1}^{M} \omega_n(x_{Dne} - x_{Dns})[x - (x_{Dne} + x_{Dns})/2] \qquad x_{D(M+1)s} > x > x_{DMe}$$

which says that if station x lies between the end of distributed load M (which is the last *complete distributed load* to the left of station x) and the start of the next distributed load, then there is no part-load contribution from the $(M + 1)$st load.

The total bending moment at station x is the sum of the bending moments due to the reactions and the distributed loads.

$$M = M_R + M_D \tag{13.13}$$

where

$$M_R = 0 \qquad\qquad x_{R1} > x \tag{13.11(a)}$$
$$M_R = R_1(x - x_{R1}) \qquad\qquad x_{R2} > x > x_{R1} \tag{13.11(b)}$$
$$M_R = R_1(x - x_{R1}) + R_2(x - x_{R2}) \quad x > x_{R2} \tag{13.11(c)}$$

and

$$M_D = 0 \qquad\qquad x_{D1s} > x \tag{13.13(a)}$$

$$M_D = \sum_{n=1}^{M} \omega_n(x_{Dne} - x_{Dns})\left(x - \frac{x_{Dne} + x_{Dns}}{2}\right) + \frac{\omega_{(M+1)}(x - x_{D(M+1)s})^2}{2}$$

$$x_{D(M+1)e} > x > x_{D(M+1)s} \tag{13.13(b)}$$

$$M_D = \sum_{n=1}^{M} \omega_n(x_{Dne} - x_{Dns})\left(x - \frac{x_{Dne} + x_{Dns}}{2}\right)$$

$$x_{D(M+1)s} > x > x_{DMe} \tag{13.13(c)}$$

With Concentrated and Distributed Loads

We derived equations (13.11(a)), (13.11(b)) and (13.11(c)) for the bending moment due to the reactions; equations (13.12(a)) and (13.12(b)) for the bending moment due to concentrated loads; and equations (13.13(a)), (13.13(b)) and (13.13(c)) for the bending moment due to distributed loads. The total bending moment is the sum of all of the contributions from the reactions, concentrated loads, and distributed loads (either full or partial) to the left of the station being considered. Collecting these equations together, we have

$$M_R = 0 \qquad\qquad x_{R1} > x \tag{13.11(a)}$$
$$M_R = R_1(x - x_{R1}) \qquad\qquad x_{R2} > x > x_{R1} \tag{13.11(b)}$$
$$M_R = R_1(x - x_{R1}) + R_2(x - x_{R2}) \quad x > x_{R2} \tag{13.11(c)}$$
$$M_L = 0 \qquad\qquad x_{L1} > x \tag{13.12(a)}$$

$$M_L = \sum_{n=1}^{n=N} Ln(x - x_{Ln}) \qquad x_{L(N+1)} > x > x_{LN} \tag{13.12(b)}$$

$$M_D = 0 \qquad\qquad x_{D1s} > x \tag{13.13(a)}$$

$$M_D = \sum_{n=1}^{M} \omega_n(x_{Dne} - x_{Dns})\left(x - \frac{x_{Dne} + x_{Dns}}{2}\right) + \frac{\omega_{(M+1)}(x - x_{D(M+1)s})^2}{2}$$

$$x_{D(M+1)\,e} > x > x_{D(M+1)s} \tag{13.13(b)}$$

$$M_D = \sum_{n=1}^{M} \omega_n (x_{Dne} - x_{Dns}) \cdot \left(x - \frac{x_{Dne} + x_{Dns}}{2} \right)$$

$$x_{D(M+1)s} > x > x_{DMe} \qquad (13.13(c))$$

13.8.4   Bending Moments: Cantilever Beams

The same equations apply, except that V_R is always zero.

EXAMPLE PROBLEM 13.16

This example shows how the foregoing equations might be used in a computer program. Referring to Figure 13.36(a), calculate the shear force and the bending moment at station $x = 6$ ft.

Solution

1. Determine the reactions: Reaction equations were developed in Chapter 3. They are

$$R_1 = \sum_{n=1}^{N} \frac{L_n(x_{Ln} - x_{R2})}{x_{R2} - x_{R1}} + \sum_{n=1}^{M} \omega_n (x_{Dne} - x_{Dns}) \frac{(x_{Dne} + x_{Dns})/2 - x_{R2}}{x_{R2} - x_{R1}} \quad (3.4(a))$$

and

$$R_2 = -\sum_{n=1}^{N} \frac{L_n(x_{Ln} - x_{R1})}{x_{R2} - x_{R1}} - \sum_{n=1}^{M} \omega_n (x_{Dne} - x_{Dns}) \frac{(x_{Dne} + x_{Dns})/2 - x_{R1}}{x_{R2} - x_{R1}} \quad (3.4.(b))$$

In the present example,

$$L_1 = -200 \text{ lb}, \qquad x_{L1} = 4 \text{ ft}, \qquad L_2 = -150 \text{ lb}, \qquad x_{L2} = 8 \text{ ft}$$
$$x_{R1} = 0, \qquad x_{R2} = 10 \text{ ft}, \qquad \omega_1 = -50 \text{ lb/ft}, \qquad x_{D1e} = 7 \text{ ft}, \qquad x_{D1s} = 2 \text{ ft}$$

Therefore, reaction R_1 is

$$R_1 = (-200 \text{ lb})(4 \text{ ft} - 10 \text{ ft})/(10 \text{ ft} - 0) + (-150 \text{ lb})(8 \text{ ft} - 10 \text{ ft})/10 \text{ ft}$$
$$+ (-50 \text{ lb/ft})(7 \text{ ft} - 2 \text{ ft})[(7 \text{ ft} + 2 \text{ ft})/2 - 10 \text{ ft}]/10 \text{ ft}$$
$$R_1 = 120 \text{ lb} + 30 \text{ lb} + 137.5 \text{ lb}$$
$$R_1 = 287.5 \text{ lb}$$

Reaction R_2 is

$$R_2 = - (-200 \text{ lb})(4 \text{ ft} - 0)/10 \text{ ft} - (-150 \text{ lb})(8 \text{ ft} - 0)/10 \text{ ft}$$
$$- (-50 \text{ lb/ft})(7 \text{ ft} - 2 \text{ ft})[(7 \text{ ft} + 2 \text{ ft})/2 - 0]/10 \text{ ft}$$
$$R_2 = 80 \text{ lb} + 120 \text{ lb} + 112.5$$
$$R_2 = 312.5 \text{ lb}$$

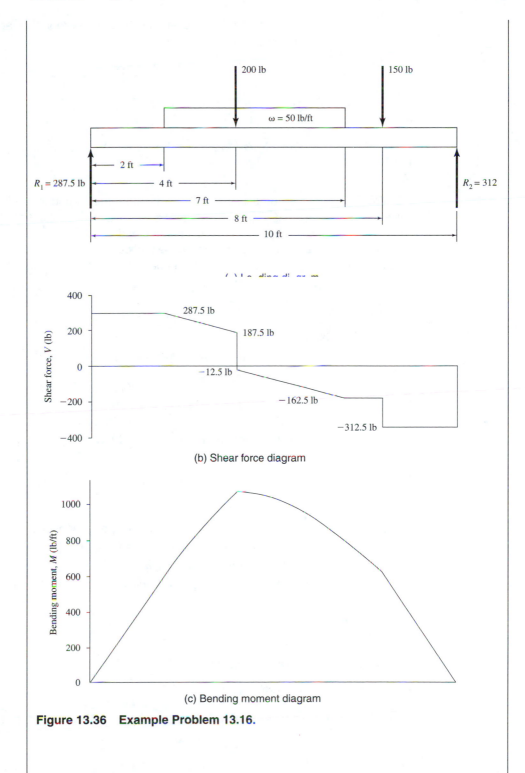

(a) Loading diagram

(b) Shear force diagram

(c) Bending moment diagram

Figure 13.36 Example Problem 13.16.

2. Calculate the shear force: The governing shear force equations are

$$V = V_R + V_L + V_D \tag{13.11}$$

$$V_R = R_1 \qquad x_{R2} > x > x_{R1} \tag{13.8(b)}$$

$$V_L = \sum_{n=1}^{n=N} Ln \qquad x_{L(N+1)} > x > x_{LN} \tag{13.9(b)}$$

$$V_D = \sum_{n=1}^{M} \omega_n(x_{Dne} - x_{Dns}) + \omega_{(M+1)}(x - x_{D(M+1)s})$$

$$x_{D(M+1)e} > x > x_{D(M+1)s} \tag{13.10(b)}$$

We selected these particular equations because the station $x = 6$ ft (1) lies between the two reactions R_1 and R_2, (2) lies between the two concentrated loads, and (3) lies within the boundaries of the distributed load.

Make sure that you understand why these particular equations were selected. Because there is only one distributed load, equation (13.10(b)) reduces to

$$V_D = \omega_{(M+1)}(x - x_{D(M+1)s})$$

which becomes

$$V_D = \omega_1(x - x_{D1s})$$

Inserting the known values,

$$L_1 = -200 \text{ lb} \qquad R_1 = 287.5 \text{ lb}$$
$$\omega_1 = -50 \text{ lb/ft} \qquad x_{D1s} = 2 \text{ ft}$$
$$V_R = R_1 = 287.5 \text{ lb}$$
$$V_L = L_1 = -200 \text{ lb}$$
$$V_D = (-50 \text{ lb/ft})(6 \text{ ft} - 2 \text{ ft}) = -200 \text{ lb}$$

Therefore,

$$V = V_R + V_L + V_D = 287.5 \text{ lb} - 200 \text{ lb} - 200 \text{ lb}$$
$$= -112.5 \text{ lb}$$

3. Calculate the bending moment: The governing equations are (satisfy yourself that these are the correct equations to use for this example):

$$M = M_R + M_L + M_D$$

$$M_R = R_1(x - x_{R1}) \qquad x_{R2} > x > x_{R1} \tag{13.11(b)}$$

$$M_L = \sum_{n=1}^{n=N} L_n(x - x_{Ln}) \qquad x_{L(N+1)} > x > x_{LN} \tag{13.12(b)}$$

$$M_D = \sum_{n=1}^{M} \omega_n(x_{Dne} - x_{Dns})\left(x - \frac{x_{Dne} + x_{Dns}}{2}\right) + \frac{\omega_{(M+1)}(x - x_{D(M+1)s})^2}{2}$$

$$x_{D(M+1)e} > x > x_{D(M+1)s} \tag{13.13(b)}$$

Because there is only one distributed load, equation (13.13 (b)) reduces to

$$M_D = \frac{\omega_{(M+1)}(x - x_{D(M+1)s})^2}{2}$$

which becomes

$$M_D = \frac{\omega_1(x - x_{D1s})^2}{2}$$

Inserting the known values,

$$L_1 = -200 \text{ lb}, \qquad x_{L1} = 4 \text{ ft}, \qquad\qquad R_1 = 287.5 \text{ lb}, \qquad X_{R1} = 0$$
$$X = 6 \text{ ft}, \qquad\qquad \omega_1 = -50 \text{ lb/ft}, \qquad x_{D1s} = 2 \text{ ft}$$
$$M_R = R_1(x - x_{R1}) = (287.5 \text{ lb})(6 \text{ ft} - 0) = 1725 \text{ lb-ft}$$
$$M_L = L_1(x - x_{L1}) = (-200 \text{ lb})(6 \text{ ft} - 4 \text{ ft}) = -400 \text{ lb-ft}$$
$$M_D = \omega_1(x - x_{D1s})^2/2 = (-50 \text{ lb/ft})(6 \text{ ft} - 2 \text{ ft})^2/2 = -400 \text{ lb-ft}$$

Therefore,

$$M = 1725 \text{ lb-ft} - 400 \text{ lb-ft} - 400 \text{ lb-ft} = 925 \text{ lb-ft}$$

Answer

The shear force at $x = 6$ ft is -112.5 lb. The bending moment at $x = 6$ ft is 925 lb-ft. We show the complete shear force and bending moment diagrams in Figure 13.36(b) and (c).

13.9   MOVING LOADS

In this section we consider the problem of determining the maximum shear force and the maximum bending moment, as well as the stations at which they occur, for loads that move along simple beams.

13.9.1   Vehicles with Two Axles

The load on axle 1 is W_1 and that on axle 2 is W_2, and the distance between the axles is d. See Figure 13.37. Axle 1 is distance x from A, A being the left end of the beam. The beam length is L, and the right end of the beam is identified as B.

The questions to be addressed are these:

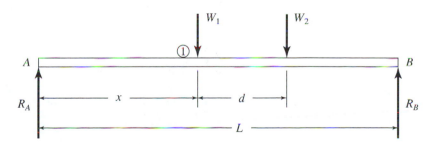

Figure 13.37   Two-axle vehicle.

1. How does the shear force vary with x, and what is the magnitude of the maximum shear force? That is, where does axle 1 have to be to create the maximum possible shear force in the beam?
2. How does the bending moment vary with x, and what is the magnitude of the maximum bending moment? That is, where does axle 1 have to be to create the maximum possible bending moment in the beam?

Maximum Shear Force

For simple beams (simply supported beams), the maximum shear force, V_{max}, occurs at and is equal to the maximum support reaction. We determine the reaction R_A at A by taking moments about B. (See Figure 13.37.)

$$\sum M_B = W_2(L - x - d) + W_1(L - x) - R_A L = 0$$

from which

$$R_A = (W_1 + W_2)(L - x)/L - W_2 d/L \tag{13.14}$$

What value of x will make this a maximum? The smaller x is, the larger the first (positive) term will be. Therefore, the maximum value of R_A occurs when $x = 0$—i.e., when axle 1 is over the left-end support. Inserting $x = 0$ into equation (13.14) gives the maximum value of R_A.

$$R_A)_{max} = (W_1 + W_2) - W_2 d/L \tag{13.15}$$

We do not know that this is the maximum shear force because $R_B)_{max}$ may be greater. We determine the reaction R_B at B by taking moments about A. (See Figure 13.37.)

$$\sum M_A = R_B L - W_2(x + d) - W_1 x = 0$$

from which

$$R_B = (W_1 + W_2)(x)/L + W_2 d/L \tag{13.16}$$

What value of x will make this a maximum? The largest possible value of x will maximize this expression. For the complete vehicle to remain on the beam, axle 2 cannot move beyond the right end of the beam. So the maximum possible value of x is $(L - d)$. Inserting this into equation (13.16) gives

$$R_B)_{max} = (W_1 + W_2)(L - d)/L + W_2 d/L$$
$$= (W_1 + W_2) - W_1 d/L \tag{13.17}$$

The maximum shear force will be the larger of $R_A)_{max}$, equation (13.15), and $R_B)_{max}$, equation (13.17):

$$R_A)_{max} = (W_1 + W_2) - W_2 d/L \tag{13.15}$$
$$R_B)_{max} = (W_1 + W_2) - W_1 d/L \tag{13.17}$$

Comparing these two equations, we see that if

$$W_1 > W_2, \quad \text{then} \quad R_A)_{max} > R_B)_{max}, \quad \text{so} \quad V_{max} = (W_1 + W_2) - W_2 d/L$$

and if

$$W_2 > W_1, \quad \text{then} \quad R_B)_{max} > R_A)_{max}, \quad \text{so} \quad V_{max} = (W_1 + W_2) - W_1 d/L$$

To summarize,

$$V_{max} = (W_1 + W_2) - W_2d/L \quad \text{at} \quad x = 0 \qquad W_1 > W_2 \quad (13.18)$$
$$V_{max} = (W_1 + W_2) - W_1d/L \quad \text{at} \quad x = (L - d) \qquad W_2 > W_1 \quad (13.19)$$

Maximum Bending Moment

First we determine the bending moment at x (the location of axle 1) and find what value of x will maximize it. Then we determine the bending moment at $(x + d)$ (the location of axle 2) and find what value of x will maximize it. The two expression are then compared to decide upon the condition that determines which is the maximum bending moment. Referring again to Figure 13.37, the bending moment at x is

$$M_x = R_A x$$

where from equation (13.14):

$$R_A = (W_1 + W_2)(L - x)/L - W_2d/L$$

so

$$M_x = (W_1 + W_2)(Lx - x^2)/L - W_2xd/L$$

or

$$M_x = (W_1 + W_2)x - (W_1 + W_2)x^2/L - W_2xd/L \qquad (13.20)$$

We can show that this a maximum (see Using Calculus at the end of this section) when

$$x = L/2 - W_2d/[2(W_1 + W_2)] \qquad (13.21)$$

Figure 13.38 shows the resultant $(W_1 + W_2)$ of the two axle loads, acting at distance ξ from axle 1. Taking moments about axle 1,

$$(W_1 + W_2)\xi = W_2d$$

Therefore, $\qquad\qquad\qquad \xi = W_2d/(W_1 + W_2) \qquad\qquad\qquad (13.22)$

Going back to equation (13.21) we see that x may be expressed as

$$x = L/2 - \xi/2 \qquad (13.23)$$

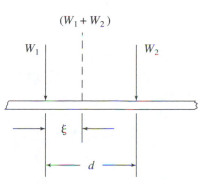

Figure 13.38   Resultant load.

In words, *for the maximum bending moment to be at the axle 1 location, the vehicle must be placed such that the center of the beam is midway between the resultant load and axle 1.*

We can determine the magnitude of the maximum bending moment by substituting for x in equation (13.20). After rearranging,

$$M_x)_{max} = [W_1 + W_2(1 - d/L)](L - \xi)/2 - (W_1 + W_2)(L - \xi)^2/4L$$

$$\text{(13.24)}$$

The bending moment at the axle 2 location, i.e., at $(x + d)$, is (see Figure 13.37):

$$M_{(x+d)} = R_A(x + d) - W_1 d$$

where, from equation (13.14),

$$R_A = (W_1 + W_2)(L - x)/L - W_2 d/L$$

so

$$M_{(x+d)} = (W_1 + W_2)(L - x)(x + d)/L - W_2 d(x + d)/L - W_1 d \quad \text{(13.25)}$$

We can show that this a maximum (see Using Calculus at the end of this section) when

$$x = L/2 - W_2 d/[2(W_1 + W_2)] - d/2 \qquad \text{(13.26)}$$

The second term is equal to $\xi/2$. So

$$x = L/2 - \xi/2 - d/2 \qquad \text{(13.27)}$$

The location of axle 2 is $(x + d) = L/2 - \xi/2 - d/2 + d = L/2 + (d - \xi)/2$. In words, *for the maximum bending moment to be at the axle 2 location, the vehicle must be placed such that the center of the beam is midway between the resultant load and axle 2.*

Note: The same conditions can be shown to hold for any number of axles. We can determine the magnitude of the maximum bending moment by substituting for x in equation (13.25). The derivation involves quite a lengthy series of algebraic manipulations, so we present only the result. If you enjoy algebra, work through it yourself. The end result is:

$$M_{(x+d)})_{max} = M_x)_{max} - (W_1 - W_2)(d/2)(1 - d/2L) \qquad \text{(13.28)}$$

d is certainly smaller than $2L$, so the term $(1 - d/2L)$ is always positive. If $W_1 > W_2$, the term $(W_1 - W_2)(d/2)(1 - d/2L)$ will be positive, making $- (W_1 - W_2)(d/2)(1 - d/2L)$ negative. In this case $M_x)_{max}$ will be greater than $M_{(x+d)})_{max}$. If $W_1 < W_2$, the term $(W_1 - W_2)(d/2)(1 - d/2L)$ will be negative, making $- (W_1 - W_2)(d/2)(1 - d/2L)$ positive. In this case $M_{(x+d)})_{max}$ will be greater than $M_x)_{max}$. To summarize: If $W_1 > W_2$,

$$M_{max} = [W_1 + W_2(1 - d/L)](L - \xi)/2$$
$$- (W_1 + W_2)(L - \xi)^2/4L \quad \text{at} \quad L/2 - \xi/2 \quad \text{(13.29)}$$

If $W_1 < W_2$

$$M_{max} = [W_1 + W_2(1 - d/L)](L - \xi)/2 - (W_1 + W_2)(L - \xi)^2/4L$$
$$- (W_1 - W_2)(d/2)(1 - d/2L) \quad \text{at} \quad L/2 + (d - \xi)/2 \quad \text{(13.30)}$$

Axle 1 will be at $x = L/2 - \xi/2 - d/2$.

EXAMPLE PROBLEM 13.17

A two-axle vehicle drives from left to right over a simply supported 10-m-long bridge. The load on the front axle is 20 kN, and on the rear axle it is 40 kN. The distance between the axles is 3 m.

   a. Determine the maximum shear force and where the vehicle will be when it occurs.

   b. Find the maximum bending moment and where the vehicle will be when it occurs.

Solution

Let the rear axle be axle 1 and the front axle be axle 2 (see Figure 13.39).

$$W_1 = 40 \text{ kN}, \quad W_2 = 20 \text{ kN}, \quad L = 10 \text{ m}, \quad d = 3 \text{ m}$$

   a. Because $W_1 > W_2$, we use equation (13.18) to determine the maximum shear force:

$$\begin{aligned}
V_{max} &= (W_1 + W_2) - W_2 d/L \quad \text{at} \quad x = 0 \\
&= (40 \text{ kN} + 20 \text{ kN}) - (20 \text{ kN})(3 \text{ m}/10 \text{ m}) \\
&= 54 \text{ kN} \\
x &= 0
\end{aligned}$$

   b. Because $W_1 > W_2$, we use equation (13.29) to determine the maximum bending moment:

$$M_{max} = [W_1 + W_2(1 - d/L)](L - \xi)/2 - (W_1 + W_2)(L - \xi)^2/4L \quad \text{at} \quad x = L/2 - \xi/2$$

where $\xi = W_2 d/(W_1 + W_2)$ from equation (13.22). Therefore

$$\begin{aligned}
\xi &= (20 \text{ kN})(3 \text{ m})/(40 \text{ kN} + 20 \text{ kN}) \\
&= 1 \text{ m} \\
M_{max} &= [40 \text{ kN} + (20 \text{ kN})(1 - 3 \text{ m}/10 \text{ m})](10 \text{ m} - 1 \text{ m})/2 \\
&\quad - (40 \text{ kN} + 20 \text{ kN})(10 \text{ m} - 1 \text{ m})^2/(40 \text{ m}) \\
&= [40 \text{ kN} + (20 \text{ kN})(0.7)](9 \text{ m})/2 - (60 \text{ kN})(9 \text{ m})^2/(40 \text{ m}) \\
&= (54 \text{ kN})(9 \text{ m})/2 - (60 \text{ kN})(81 \text{ m}^2)/(40 \text{ m}) \\
&= 243 \text{ kN·m} - 121.5 \text{ kN·m} \\
&= 121.5 \text{ kN·m} \\
x &= L/2 - \xi/2 = 10 \text{ m}/2 - 1 \text{ m}/2 = 4.5 \text{ m}
\end{aligned}$$

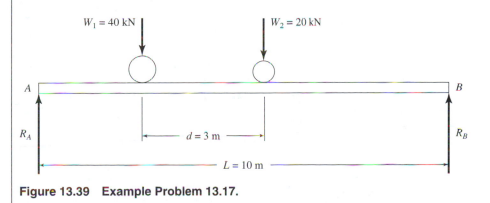

Figure 13.39   Example Problem 13.17.

Answer
a. The maximum shear force is 54 kN, and it occurs when the rear axle is directly over the left support.
b. The maximum bending moment is 121.5 kN · m, and it occurs when the rear axle is 4.5 m from the left support.

13.9.2   Vehicles with Three Axles

Analysis of the three-axle case is similar to that of the two-axle case.

Maximum Shear Force

We determine the reaction R_A at A by taking moments about B. (See Figure 13.40.)

$$\sum M_B = W_3(L - x - d_1 - d_2) + W_2(L - x - d_1) + W_1(L - x) - R_AL = 0$$

from which

$$R_A = W_3(L - x - d_1 - d_2)/L + W_2(L - x - d_1)/L + W_1(L - x)/L \quad (13.31)$$

What value of x will make this a maximum?

The smaller x is, the larger each term will be. Therefore, the maximum value of R_A occurs when $x = 0$. That is when axle 1 is over the left-end support.
Inserting $x = 0$ into equation (13.31) gives the maximum value of R_A.

$$R_A)_{max} = W_3(L - d_1 - d_2)/L + W_2(L - d_1)/L + W_1$$
$$R_A)_{max} = (W_1 + W_2 + W_3) - W_2d_1/L - W_3(d_1 + d_2)/L \quad (13.32)$$

The total load is $(W_1 + W_2 + W_3)$, so we can determine the reaction at B by summing the vertical forces:

$$R_B = (W_1 + W_2 + W_3) - R_A$$

Substituting for R_A from equation (13.31):

$$R_B = (W_1 + W_2 + W_3) - W_3(L - x - d_1 - d_2)/L$$
$$- W_2(L - x - d_1)/L - W_1(L - x)/L \quad (13.33)$$

Because of the double negative values of x, R_B will be maximum when x takes its maximum value. For the vehicle to remain on the beam, the maximum value of x is

$$x = (L - d_1 - d_2) \quad (13.34)$$

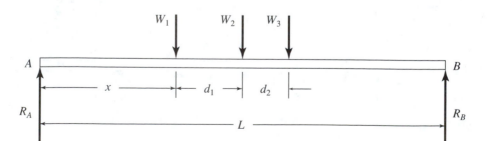

Figure 13.40   Three-axle vehicle.

(See Figure 13.40.) Substituting this value into equation (13.33) gives

$$R_B)_{max} = (W_1 + W_2 + W_3) - W_2 d_2/L - W_1(d_1 + d_2)/L \qquad (13.35)$$

Comparing this with $R_A)_{max}$,

$$R_A)_{max} = (W_1 + W_2 + W_3) - W_2 d_1/L - W_3(d_1 + d_2)/L \qquad (13.32)$$

If we rewrite equation (13.32) as

$$R_A)_{max} = (W_1 + W_2 + W_3) - [W_2 d_1/L + W_3(d_1 + d_2)/L] \qquad (13.36)$$

and equation (13.35) as

$$R_B)_{max} = (W_1 + W_2 + W_3) - [W_2 d_2/L + W_1(d_1 + d_2)/L] \qquad (13.37)$$

we see that if

$$[W_2 d_2/L + W_1(d_1 + d_2)/L] > [W_2 d_1/L + W_3(d_1 + d_2)/L]$$

then $R_A)_{max} > R_B)_{max}$. Multiplying through by L, the condition to be satisfied for $R_A)_{max} > R_B)_{max}$ is

$$W_2 d_2 + W_1(d_1 + d_2) > W_2 d_1 + W_3(d_1 + d_2)$$

or

$$(W_1 - W_3)(d_1 + d_2) > W_2(d_1 - d_2)$$

or

$$(W_1 - W_3)/W_2 > (d_1 - d_2)/(d_1 + d_2)$$

V_{max} will be the larger of $R_A)_{max}$ and $R_B)_{max}$. In summary,

If $(W_1 - W_3)/W_2 > (d_1 - d_2)/(d_1 + d_2)$,

$$V_{max} = (W_1 + W_2 + W_3) - [W_2 d_1/L + W_3(d_1 + d_2)/L] \quad \text{at} \quad x = 0 \quad (13.36)$$

If $(W_1 - W_3)/W_2 < (d_1 - d_2)/(d_1 + d_2)$,

$$V_{max} = (W_1 + W_2 + W_3) - [W_2 d_2/L + W_1(d_1 + d_2)/L]$$
$$\text{at} \quad x = (L - d_1 - d_2) \qquad (13.37)$$

Maximum Bending Moment

The location of the maximum bending moment must be at the location of one of the axles. Although we can write equations for the bending moments at each of these locations, it is difficult to determine analytically which is the largest. We will derive the bending moment equations and determine which is the largest in each specific problem by entering known values and comparing the results. This type of problem is easily solved by a spreadsheet program using the =MAX( ) function.

The maximum bending moment for a simply supported beam carrying any number of concentrated loads must occur at one of the loads. You can satisfy yourself that this is true by sketching the shear force diagram and then the bending moment diagram for a simply supported beam carrying, say, three loads.

We derive the bending moments at each of the three axle locations and show that *for each case the maximum bending moment will occur when the vehicle is placed such that the center of the beam is midway between that axle and the resultant load.*

Let ξ be the distance between the resultant load and axle 1. See Figure 13.41. Taking moments about point 1,

$$\xi(W_1 + W_2 + W_3) = W_2 d_1 + W_3(d_1 + d_2)$$

So
$$\xi = \frac{W_2 d_1 + W_3(d_1 + d_2)}{W_1 + W_2 + W_3} \tag{13.38}$$

Reaction R_A: We could use equation (13.31) for R_A but introducing ξ and the resultant load leads to a more compact solution for the bending moment. Figure 13.42 shows the three-axle loads replaced by the resultant load $(W_1 + W_2 + W_3)$ at distance $(x + \xi)$ from the left end of the beam. Taking moments about B gives

$$R_A L = (W_1 + W_2 + W_3)(L - x - \xi)$$

So
$$R_A = (W_1 + W_2 + W_3)(L - x - \xi)/L \tag{13.39}$$

We now consider each of the three cases.

Case 1: Bending moment at the axle 1 location.   Referring again to Figure 13.42,

$$M_1 = R_{A1} x$$

Substituting for R_A from equation (13.39)

$$M_1 = (W_1 + W_2 + W_3)(L - x - \xi)x/L \tag{13.40}$$

We can show (see Using Calculus at the end of this section) that M_1 will be a maximum when

$$x = (L/2 - \xi/2) \tag{13.41}$$

This says that axle 1 is at location $(L/2 - \xi/2)$. See Figure 13.43. That is, *the maximum bending moment at the axle 1 location will occur when the vehicle is placed such that the center of the beam is midway between axle 1 and the resultant load.*

Substituting for x from equation (13.41) into equation (13.40) gives

$$
\begin{aligned}
M_1)_{\max} &= (W_1 + W_2 + W_3)[L - (L/2 - \xi/2) - \xi](L/2 - \xi/2)/L \\
&= (W_1 + W_2 + W_3)(L/2 - \xi/2)^2/L \\
&= (W_1 + W_2 + W_3)(L - \xi)^2/4L \tag{13.42}
\end{aligned}
$$

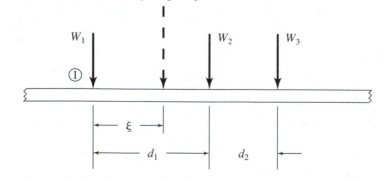

Figure 13.41   Resultant load.

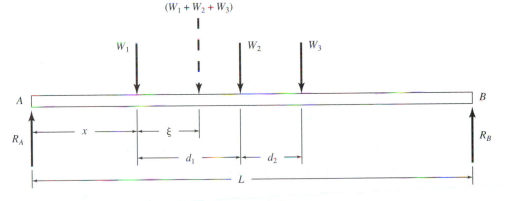

Figure 13.42   Figure for determining reaction R_A.

Case 2: Bending moment at the axle 2 location: Referring again to Figure 13.42,

$$M_2 = R_{A2}(x + d_1) - W_1 d_1$$

Substituting for R_A from equation (13.39) gives

$$M_2 = (W_1 + W_2 + W_3)(L - x - \xi)(x + d_1)/L - W_1 d_1 \qquad (13.43)$$

We can show (see Using Calculus at the end of this section) that M_2 will be a maximum when

$$x = L/2 - (d_1 + \xi)/2 \qquad (13.44)$$

The location of axle 2 is $(x + d_1)$, i.e., at $L/2 - d_1/2 - \xi/2 + d_1$, so axle 2 is at location $[L/2 + (d_1 - \xi)/2]$. See Figure 13.44.

That is, *the maximum bending moment at the axle 2 location will occur when the vehicle is placed such that the center of the beam is midway between axle 2 and the resultant load.*

Substituting for x from equation (13.44) into equation (13.43) gives an expression for the maximum bending moment at the axle 2 location:

$$M_2)_{max} = (W_1 + W_2 + W_3)\{L - [L/2 - (d_1 + \xi)/2] - \xi\}$$
$$\times [L/2 + (d_1 - \xi)/2]/L - W_1 d_1$$
$$= (W_1 + W_2 + W_3)(L + d_1 - \xi)^2/4L - W_1 d_1 \qquad (13.45)$$

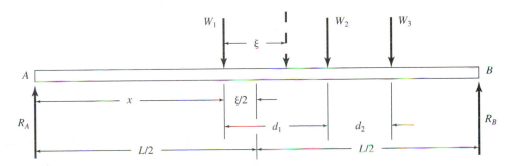

Figure 13.43   Case 1.

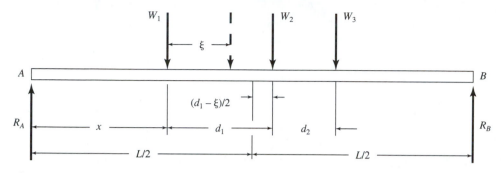

Figure 13.44   Case 2.

Case 3: Bending moment at the axle 3 location: Referring again to Figure 13.42:

$$M_3 = R_{A3}(x + d_1 + d_2) - W_1(d_1 + d_2) - W_2 d_2$$

Substituting for R_A from equation (13.39):

$$M_3 = (W_1 + W_2 + W_3)(L - x - \xi)(x + d_1 + d_2)/L$$
$$- W_1(d_1 + d_2) - W_2 d_2 \qquad (13.46)$$

We can show (see Using Calculus) that M_3 will be a maximum when

$$x = L/2 - (d_1 + d_2 + \xi)/2 \qquad (13.47)$$

The location of axle 3 is $(x + d_1 + d_2)$, i.e., at $L/2 + d_1/2 + d_2/2 - \xi/2$, so axle 3 is at location $[L/2 + (d_1 + d_2 - \xi)/2]$. See Figure 13.45.

That is, *the maximum bending moment at the axle 3 location will occur when the vehicle is placed such that the center of the beam is midway between axle 3 and the resultant load.*

Substituting for x from equation (13.47) into equation (13.46) gives an expression for the maximum bending moment at the axle 3 location:

$$M_3)_{\max} = (W_1 + W_2 + W_3)\{L - [L/2 - (d_1 + d_2 + \xi)/2] - \xi\}$$
$$\times [L/2 + (d_1 + d_2 - \xi)/2]/L - W_1(d_1 + d_2) - W_2 d_2$$

which reduces to

$$M_3)_{\max} = (W_1 + W_2 + W_3)[L/2 + (d_1 + d_2 - \xi)/2]^2/L - W_1(d_1 + d_2) - W_2 d_2$$
$$M_3)_{\max} = (W_1 + W_2 + W_3)(L + d_1 + d_2 - \xi)^2/4L - W_1(d_1 + d_2) - W_2 d_2$$
$$(13.48)$$

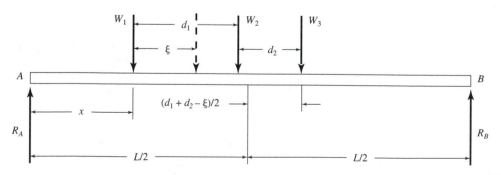

Figure 13.45   Case 3.

In summary, the maximum bending moment is the largest of

1. $M)_{max} = (W_1 + W_2 + W_3)(L - \xi)^2/4L$ at $(L/2 - \xi/2)$ (13.49)

2. $M)_{max} = (W_1 + W_2 + W_3)(L + d_1 - \xi)^2/4L - W_1 d_1$
   at $[L/2 + (d_1 - \xi)/2]$

   Axle 1 will be at

$$x = L/2 - (d_1 + \xi)/2 \qquad (13.50)$$

3. $M)_{max} = (W_1 + W_2 + W_3)(L + d_1 + d_2 - \xi)^2/4L - W_1(d_1 + d_2)$
   $- W_2 d_2$ at $[L/2 + (d_1 + d_2 - \xi)/2]$

   Axle 1 will be at

$$x = L/2 - (d_1 + d_2 + \xi)/2 \qquad (13.51)$$

where $\qquad \xi = [W_2 d_1 + W_3(d_1 + d_2)]/(W_1 + W_2 + W_3)$ (13.52)

EXAMPLE PROBLEM 13.18

A three-axle vehicle drives from left to right over a simply supported 50-ft-long bridge. The load on the front axle is 2000 lb, the load on the second axle is 4000 lb, and the load on the rear axle is 6000 lb. The distance between the front and second axles is 5 ft, and between the second and rear axles it is 8 ft.

a. Determine the maximum shear force and where the vehicle will be when it occurs.

b. Find the maximum bending moment and where the vehicle will be when it occurs.

Solution

Let the rear axle be axle 1, the second axle be 2, and the front axle be axle 3. (See Figure 13.46.)

$$W_1 = 6000 \text{ lb}, \qquad W_2 = 4000 \text{ lb}, \qquad W_3 = 2000 \text{ lb}$$
$$d_1 = 8 \text{ ft}, \qquad d_2 = 5 \text{ ft}, \qquad L = 50 \text{ ft}$$

a.

$$(W_1 - W_3)/W_2 = (6000 \text{ lb} - 2000 \text{ lb})/4000 \text{ lb} = 1$$
$$(d_1 - d_2)/(d_1 + d_2) = (8 \text{ ft} - 5 \text{ ft})/(8 \text{ ft} + 5 \text{ ft}) = 0.23$$

Therefore, $(W_1 - W_3)/W_2 > (d_1 - d_2)/(d_1 + d_2)$. From equation (13.36),

$$V_{max} = (W_1 + W_2 + W_3) - [W_2 d_1/L + W_3(d_1 + d_2)/L] \quad \text{at} \quad x = 0$$
$$= (6000 \text{ lb} + 4000 \text{ lb} + 2000 \text{ lb}) - [(4000 \text{ lb})(8 \text{ ft})/50 \text{ ft}$$
$$+ (2000 \text{ lb})(8 \text{ ft} + 5 \text{ ft})/50 \text{ ft}]$$
$$= (12{,}000 \text{ lb}) - (640 \text{ lb}) - (520 \text{ lb})$$
$$= 10{,}840 \text{ lb} \quad \text{at} \quad x = 0$$

b. First we need the value of ξ from equation (13.52):

$$\xi = [W_2 d_1 + W_3(d_1 + d_2)]/(W_1 + W_2 + W_3)$$
$$= (4000 \text{ lb} \times 8 \text{ ft} + 2000 \text{ lb} \times 13 \text{ ft})/12,000 \text{ lb}$$
$$= 4.83 \text{ ft}$$

Case 1.   *Maximum bending moment at axle* 1:

$$M)_{max} = (W_1 + W_2 + W_3)(L - \xi)^2/4L \quad \text{at} \quad (L/2 - \xi/2) \qquad (13.49)$$
$$= (12,000 \text{ lb})(50 \text{ ft} - 4.83 \text{ ft})^2/200 \text{ ft} \quad \text{at} \quad (50 \text{ ft}/2 - 4.83 \text{ ft}/2)$$

Therefore, $M)_{max} = 122,400$ lb-ft   at   22.6 ft

Case 2.   *Maximum bending moment at axle* 2:

$$M)_{max} = (W_1 + W_2 + W_3)(L + d_1 - \xi)^2/4L - W_1 d_1 \quad \text{at} \quad L/2 + (d_1 - \xi)/2$$

Axle 1 will be at

$$x = L/2 - (d_1 + \xi)/2 \qquad (13.50)$$
$$M)_{max} = (12,000 \text{ lb})(50 \text{ ft} + 8 \text{ ft} - 4.83 \text{ ft})^2/200 \text{ ft} - 6000 \text{ lb}$$
$$\times 8 \text{ ft} \quad \text{at} \quad 50 \text{ ft}/2 + (8 \text{ ft} - 4.83 \text{ ft})/2$$

Axle 1 will be at

$$x = 50 \text{ ft}/2 - (8 \text{ ft} + 4.83 \text{ ft})/2$$

So $M)_{max} = 121,600$ lb-ft   at   26.6 ft, $x = 18.6$ ft

Case 3.   *Maximum bending moment at axle* 3:

$$M)_{max} = (W_1 + W_2 + W_3)(L + d_1 + d_2 - \xi)^2/4L - W_1(d_1 + d_2)$$
$$- W_2 d_2 \quad \text{at} \quad [L/2 + (d_1 + d_2 - \xi)/2]$$

Axle 1 will be at

$$x = L/2 - (d_1 + d_2 + \xi)/2 \qquad (13.51)$$
$$M)_{max} = (12,000 \text{ lb})(50 \text{ ft} + 8 \text{ ft} + 5 \text{ ft} - 4.83 \text{ ft})^2/200 \text{ ft} - (6000 \text{ lb})(8 \text{ ft} + 5 \text{ ft})$$
$$- (2000 \text{ lb})(5 \text{ ft}) \quad \text{at} \quad [50 \text{ ft}/2 + (8 \text{ ft} + 5 \text{ ft} - 4.83 \text{ ft})/2]$$
$$x = 50 \text{ ft}/2 - (8 \text{ ft} + 5 \text{ ft} + 4.83 \text{ ft})/2$$

So $M)_{max} = 115,000$ lb-ft   at   29.1 ft, $x = 16.1$ ft

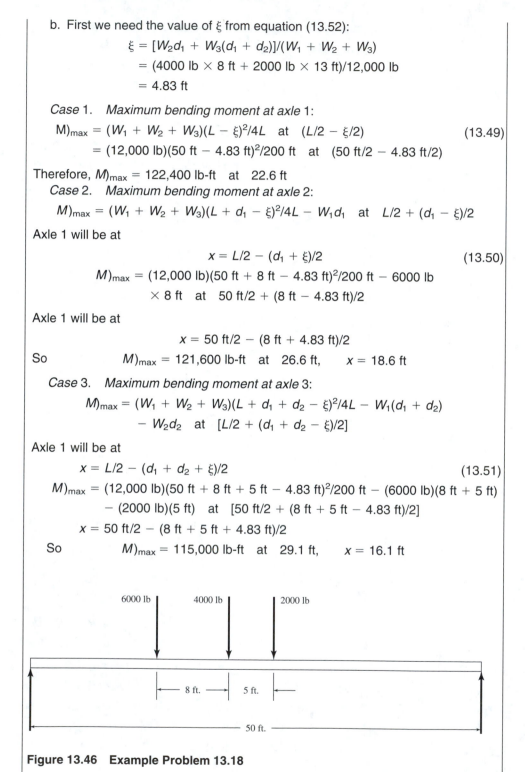

Figure 13.46   Example Problem 13.18

The maximum of these is case 1:

$$M)_{max} = 122{,}400 \text{ lb-ft} \quad \text{at} \quad 22.6 \text{ ft}$$

Answer

The maximum shear force is $V_{max} = 10{,}840$ lb. This occurs when the rear axle is directly over the left beam support. The maximum bending moment is $M)_{max} = 122{,}400$ lb-ft. This occurs at the rear-axle location when the rear axle is 22.6 ft from the left end of the beam.

USING CALCULUS

Two-axle Vehicle

Equation (13.20) gave the following expression for the bending moment:

$$M_x = (W_1 + W_2)x - (W_1 + W_2)x^2/L - W_2xd/L \qquad (13.20)$$

Differentiating with respect to x gives

$$dM_x/dx = (W_1 + W_2) - 2(W_1 + W_2)x/L - W_2d/L$$

For maximum or minimum, $dM_x/dx = 0$, so

$$(W_1 + W_2) - 2(W_1 + W_2)x/L - W_2d/L = 0$$

$$2(W_1 + W_2)x/L = (W_1 + W_2) - W_2d/L$$

$$x = L[(W_1 + W_2) - W_2d/L]/2\,(W_1 + W_2)$$

or

$$x = L/2 - W_2d/[2(W_1 + W_2)]$$

Three-axle Vehicle

Case 1.   The bending moment is given by equation (13.40):

$$M_1 = (W_1 + W_2 + W_3)(L - x - \xi)x/L \qquad (13.40)$$

Differentiating with respect to x gives

$$dM_1/dx = (W_1 + W_2 + W_3)(L - 2x - \xi)/L$$

Equating this to zero, we have

$$(L - 2x - \xi) = 0$$

$$x = (L - \xi)/2$$

Case 2.   The bending moment is given by equation (13.43):

$$M_2 = (W_1 + W_2 + W_3)(L - x - \xi)(x + d_1)/L - W_1d_1 \qquad (13.43)$$

Differentiating with respect to x gives

$$dM_2/dx = (W_1 + W_2 + W_3)(L - x - \xi)/L - (W_1 + W_2 + W_3)(x + d_1)/L$$

Equating to zero,

$$(L - x - \xi) = (x + d_1)$$
$$2x = (L - d_1 - \xi)$$
$$x = L/2 - (d_1 + \xi)/2$$

Case 3.   The bending moment is given by equation (13.46):

$$M_3 = (W_1 + W_2 + W_3)(L - x - \xi)(x + d_1 + d_2)/L - W_1(d_1 + d_2) - W_2 d_2$$

$$(13.46)$$

Differentiating with respect to *x*, we have

$$dM_3/dx = (W_1 + W_2 + W_3)(L - x - \xi)/L - (W_1 + W_2 + W_3)(x + d_1 + d_2)/L$$

Equating this to zero,

$$(L - x - \xi) = (x + d_1 + d_2)$$
$$2x = L - d_1 - d_2 - \xi$$
$$x = L/2 - (d_1 + d_2 + \xi)/2$$

13.10   COMPUTER ANALYSES

13.10.1   Programs 'Shr&mmt' and 'Shrmmtcl'

Program 'Shr&mmt' is a program for calculating the shear force and bending moment diagrams for simply supported and for overhanging beams. Any number, up to six each, of concentrated loads and distributed loads may be entered.

Program 'shrmmtcl' is a similar program for cantilever beams.

EXAMPLE PROBLEM 13.19

This problem shows the ease with which you can calculate the maximum shear force, the shear force diagram, the maximum bending moment, and the bending moment diagram for any loading. The loading, shown in Figure 13.47(a), is for an overhanging beam carrying four concentrated loads and three distributed loads.

Solution
Open program 'Shr&mmt.' You will see the input/output sheet (Figure 13.47(b)).

Enter 2 in cell A10 for x_1.
Enter 10 in cell B10 for x_2.
Enter 13 in cell D9 for beam length.

Enter the concentrated loads, −250, −115, −465, and −300, in cells A17, 18, 19, and 20, respectively.

Note: They are *downward-acting* loads and therefore are *negative*.

Enter the concentrated load locations, 4, 7.5, 8, and 13, in cells B17, 18, 19, and 20, respectively.
Enter the distributed loads, −150, −275, and −130, in cells C17, 18, and 19 (*negative values*).
Enter the start of each load, 0, 5, and 8.5, in cells D17, 18, and 19.
Enter the end of each load, 3, 7, and 12, in cells E17, 18, and 19.
Enter N in cell L3.
Enter N/m in cell L4.
Enter m in cell L5.

Your screen should now look like Figure 13.47(c).

Answer

The support reactions are 966 N and 1619 N. The maximum shear force is 1043 N. The maximum bending moment is 1202 N · m. The shear force and bending moment diagrams in Figure 13.47(d) also appear on the screen.

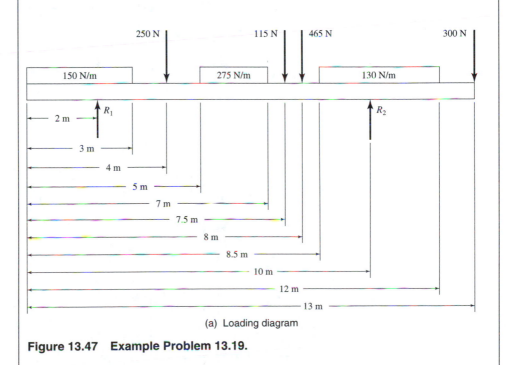

(a) Loading diagram

Figure 13.47   Example Problem 13.19.

SHEAR FORCE AND BENDING MOMENT DIAGRAM
Program 'Shr&mmt'

NOTE: DO NOT ERASE OR ENTER ANY VALUES IN ANY SPACE SURROUNDED BY DOUBLE LINES.

UNITS

CONCENTR'D LOAD	N
DISTRIBUTED LOAD	N/m
DISTANCE	m

SUPPORTS

X1	X2
m	m
0	0

BEAM LENGTH

m
0

SUPPORT REACTIONS

R1	R2
N	N
#DIV/0!	#DIV/0!

MAXIMUM SHEAR FORCE

N
#DIV/0!

MAXIMUM BENDING MOMENT

Nm
#DIV/0!

CONCENTRATED LOADS / **DISTRIBUTED LOADS**

LOAD	X	LOAD	START	END
N	m	N/m	m	m
0	0	0	0	0
0	0	0	0	0
0	0	0	0	0
0	0	0	0	0
0	0	0	0	0
0	0	0	0	0
0				

DISTANCE FROM LEFT END OF BEAM	SHEAR FORCE	BENDING MOMENT
m	N	Nm
0	#DIV/0!	#DIV/0!
0	#DIV/0!	#DIV/0!
0	#DIV/0!	#DIV/0!
0	#DIV/0!	#DIV/0!
0	#DIV/0!	#DIV/0!
0	#DIV/0!	#DIV/0!
0	#DIV/0!	#DIV/0!
0	#DIV/0!	#DIV/0!
0	#DIV/0!	#DIV/0!
0	#DIV/0!	#DIV/0!
0	0	#DIV/0!

(b) Input/output sheet for program 'Shr&mmt'

Figure 13.47 (*Continued*)

SHEAR FORCE AND BENDING MOMENT DIAGRAM
Program 'Shr&mmt'

NOTE: DO NOT ERASE OR ENTER ANY VALUES IN ANY SPACE SURROUNDED BY DOUBLE LINES.

UNITS

CONCENTR'D LOAD	N
DISTRIBUTED LOAD	N/m
DISTANCE	m

MAXIMUM BENDING MOMENT

Nm	1202.00

BEAM LENGTH

m	13

SUPPORTS

X1	X2
m	m
2	10

SUPPORT REACTIONS

R1	R2
N	N
966	1619

MAXIMUM SHEAR FORCE

N	1043

CONCENTRATED LOADS

LOAD	X
N	m
-250	4
-115	7.5
-465	8
-300	13
0	0
0	0
-1130	

DISTRIBUTED LOADS

LOAD	START	END
N/m	m	m
-150	0	3
-275	5	7
-130	8.5	12
0	0	0
0	0	0
0	0	0

DISTANCE FROM LEFT END OF BEAM	SHEAR FORCE	BENDING MOMENT
m	N	Nm
0	0	0
1.3	-195	-127
2.6	576	73
3.9	516	756
5.2	211	1121
6.5	-146	1163
7.8	-399	794
9.1	-942	-260
10.4	508	-946
11.7	339	-396
13	0	4.54747E-13

(c) Input/output sheet

Figure 13.47 *(Continued)*

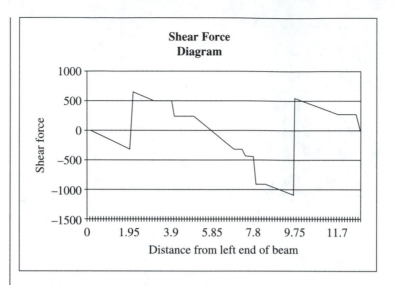

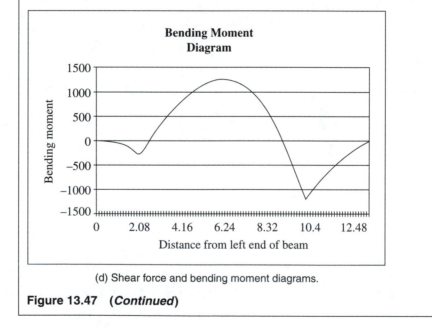

(d) Shear force and bending moment diagrams.

Figure 13.47   (*Continued*)

EXAMPLE PROBLEM 13.20

This problem uses program 'Shrmmtcl.' Plot the shear force and bending moment diagrams for the cantilever beam loading shown in Figure 13.48(a).

Solution

Notice that the wall is on the left in Figure 13.48(a). The first thing we have to do is to view it from the other side, so that the wall is on the right, as in Figure 13.48(b). Be careful to check that you have redimensioned it correctly. One of the concentrated loads, the 1000-lb load, is an up load. This *does not* make it a propped cantilever, which is a statically indetermine configuration we study in Chapter 17. If it were a support (reaction) whose magnitude we did not know, then it would be a propped cantilever.

Open program 'Shrmmtcl'. You will see the input/output sheet (Figure 13.48(c)).

Enter 10 in A9 for beam length.
Enter lb in cell L3.
Enter lb/ft in cell L4.
Enter ft in cell L5.
Enter the concentrated loads, −500, 1000 (this is a positive value), and
−750, in cells A17, 18, and 19, respectively.
Enter the concentrated load locations, 0, 3, and 7, in cells B17, 18, and 19.
Enter the distributed loads, −150, and −100, in cells C17 and 18.
Enter the start of each load, 1 and 6, in cells D17 and 18.
Enter the end of each load, 5 and 9, in cells E17 and 18.

Your screen should now look like Figure 13.48(d).

Answer

The maximum shear force is 1150 lb. The maximum bending moment is 5200 lb-ft. The shear force and bending moment diagrams (Figure 13.48(e)) also appear on the screen.

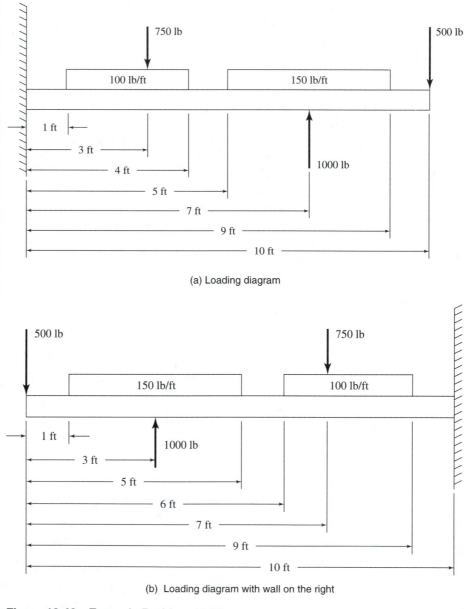

(a) Loading diagram

(b) Loading diagram with wall on the right

Figure 13.48   Example Problem 13.20.

SHEAR FORCE AND BENDING MOMENT DIAGRAMS
CANTILEVER BEAM
Program 'Shrmmtcl'

**NOTE: DO NOT ERASE OR ENTER ANY VALUES IN
ANY SPACE SURROUNDED BY DOUBLE LINES.**

UNITS		
CONCENTR'D LOAD	lb	
DISTRIBUTED LOAD	lb/ft	
DISTANCE	ft	

BEAM LENGTH
ft
0

WALL REACT
RW
lb
0

MAXIMUM S.F.
lb
0

MAXIMUM B.M.
lbft
0

CONCENTRATED LOADS		DISTRIBUTED LOADS		
LOAD	X	LOAD	START	END
lb	ft	lb/ft	ft	ft
0	0	0	0	0
0	0	0	0	0
0	0	0	0	0
0	0	0	0	0
0	0	0	0	0
0	0			
0				

DISTANCE FROM LEFT END OF BEAM	SHEAR FORCE	BENDING MOMENT
ft	lb	lbft
0	0	0
0	0	0
0	0	0
0	0	0
0	0	0
0	0	0
0	0	0
0	0	0
0	0	0
0	0	0
0	0	0

(c) Input/output sheet for program 'Shrmmtcl'

Figure 13.48 *(Continued)*

401

SHEAR FORCE AND BENDING MOMENT DIAGRAMS
CANTILEVER BEAM
Program 'Shrnmtcl'

NOTE: DO NOT ERASE OR ENTER ANY VALUES IN
ANY SPACE SURROUNDED BY DOUBLE LINES.

UNITS	
CONCENTR'D LOAD	lb
DISTRIBUTED LOAD	lb/ft
DISTANCE	ft

BEAM LENGTH
ft: 10

WALL REACT RW
lb: 1150

MAXIMUM S.F.
lb: 1150

MAXIMUM B.M.
lbft: 5200

CONCENTRATED LOADS	
LOAD (lb)	X (ft)
-500	0
1000	3
-750	7
0	0
0	0
0	0
-250	

DISTRIBUTED LOADS		
LOAD (lb/ft)	START (ft)	END (ft)
-150	1	5
-100	6	9
0	0	0
0	0	0
0	0	0
0	0	0

DISTANCE FROM LEFT END OF BEAM (ft)	SHEAR FORCE (lb)	BENDING MOMENT (lbft)
0	-500	0
1	-500	-500
2	-650	-1075
3	200	-1800
4	50	-1675
5	-100	-1700
6	-100	-1800
7	-200	-1950
8	-1050	-2950
9	-1150	-4050
10	-1150	-5200

(d) Input/output sheet

Figure 13.48 *(Continued)*

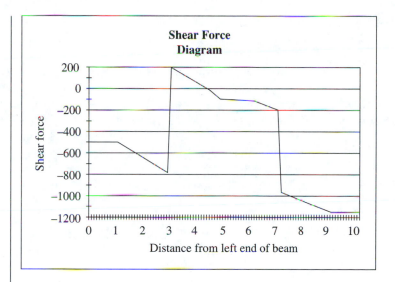

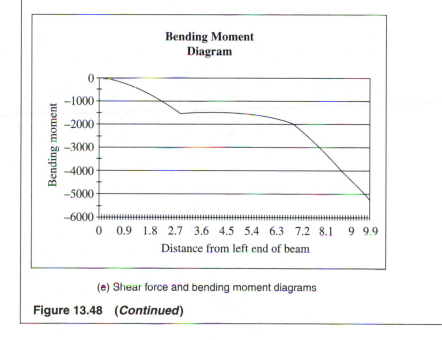

(e) Shear force and bending moment diagrams

Figure 13.48 (*Continued*)

USING CALCULUS

Shear Forces and Bending Moments

Figure 13.49(a) shows an arbitrary distributed load, $\omega(x)$, acting on a beam. The shaded area is an element of length dx of the distributed load. There can also be concentrated loads (for example, F_1 and F_2 in Figure 13.49(a)), but because these cause discontinuities in the shear and bending moments, we restrict the selection of the element dx to stations where there are no concentrated loads.

We show an expanded view of the element in Figure 13.49(b). There is a shear force, V, at the left end of the element and $V + (dV/dx)\, dx$ at the right end. Similarly, there is a moment M at the left end of the element and $M + (dx/dx)\, dM$ at the right end.

Note: If the distributed load is a download, i.e., a negative load, then dV/dx will be negative and dM/dx will be positive. For the general analysis we assume positive loads, and in actual examples we write down loads as negative values.

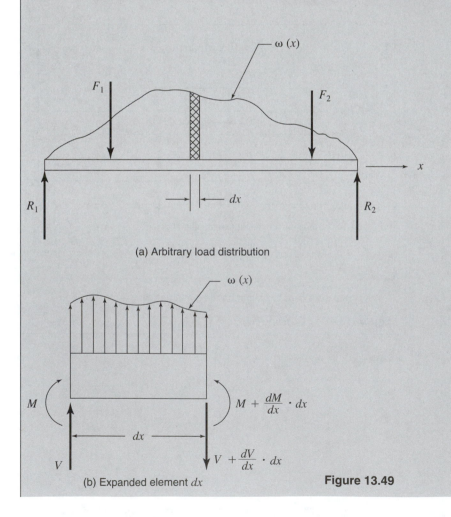

(a) Arbitrary load distribution

(b) Expanded element dx

Figure 13.49

Summing the vertical forces,

$$V - [V + (dV/dx)dx] + \omega dx = 0$$
$$-(dV/dx)dx + \omega dx = 0$$

Therefore,

$$dV/dx = \omega$$

Taking moments about the right end of the element and remembering the convention that cw is negative and ccw is positive,

$$-M - Vdx - \omega dx(dx/2) + [M + (dM/dx)dx] = 0$$

The third term, being the product of two differentials, goes to zero.

$$-M - Vdx + [M + (dM/dx)dx] = 0$$
$$-Vdx + (dM/dx)dx = 0$$

Therefore,

$$V = dM/dx$$

The conclusions are

$$dV/dx = \omega, \qquad V = dM/dx$$

Integrating,

$$V = \int \omega \, dx + C, \qquad M = \int V \, dx + D$$

where C and D are constants of integration.

EXAMPLE

A 9-m simply supported beam is shown in Figure 13.50(a). It carries one concentrated load of -300 kN at $x = 3$ m and a distributed load described by $\omega(x) = -5x$ kN/m. Plot the shear force and bending moment diagrams.

Solution

1. Determine the reactions R_1 and R_2. First, we need to find the location of the centroid of the distributed load to draw the equivalent loading diagram in Figure 13.50(b). Because the loading is triangular, we know that the centroid is one-third of the height up from the base, i.e., at $x = 6$ m, but for more complicated loadings we would have to use the equation developed in Chapter 11:

$$\bar{x} = (1/A) \int_{x_1}^{x_2} x(y_2 - y_1) \, dx$$

In the present example,

$$A = \int_0^9 5x \, dx = (\tfrac{5}{2})x^2 \Big|_0^9 = (\tfrac{5}{2}) \big| (9^2 - 0) \text{ kN}$$

$$= 202.5 \text{ kN}$$
$$y_2 = 5x, \qquad y_1 = 0$$

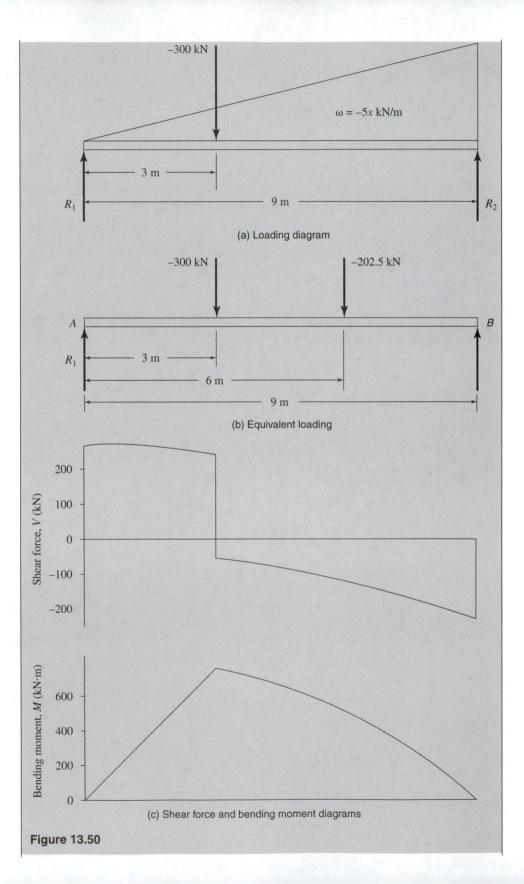

(a) Loading diagram

(b) Equivalent loading

(c) Shear force and bending moment diagrams

Figure 13.50

Therefore,

$$\bar{x} = \int_0^9 (1/202.5)5x^2 \, dx = (1/202.5)(\tfrac{5}{3}) \, x^3 \big|_0^9$$

$$= (1/202.5)(\tfrac{5}{3})(9^3 - 0) = (1/202.5)(\tfrac{5}{3})(729) \text{ m}$$

$$= 6 \text{ m}$$

Taking moments about A in Figure 13.50(b) and equating the sum to zero gives

$$-(300 \text{ kN})(3 \text{ m}) - (202.5 \text{ kN})(6 \text{ m}) + 9R_2 = 0$$

$$9R_2 = 900 \text{ kN} \cdot \text{m} + 1215 \text{ kN} \cdot \text{m} = 2115 \text{ kN} \cdot \text{m}$$

$$R_2 = 2115 \text{ kN} \cdot \text{m}/9 \text{ m} = 235 \text{ kN}$$

Taking moments about B in Figure 13.50(b) and equating the sum to zero gives

$$(300 \text{ kN})(6 \text{ m}) + (202.5 \text{ kN})(3 \text{ m}) - 9R_1 = 0$$

from which

$$R_1 = 267.5 \text{ kN}$$

Check: Summing the vertical forces,

$$267.5 \text{ kN} - 300 \text{ kN} - 202.5 \text{ kN} + 235 \text{ kN} = 0 \qquad \text{Check}$$

2. Determine the shear force equations. We have to determine two equations, one for each side of the discontinuity at $x = 3$ m.

In the region $x = 0$ to $x = 3$ m:

$$V = \int \omega \, dx + C_1 = \int -5x \, dx + C_1 = -(\tfrac{5}{2})x^2 + C_1$$

when $x = 0$, $V = R_1 = 267.5$ kN. Therefore,

$$V = -(\tfrac{5}{2})x^2 + 267.5 \text{ kN} \tag{1}$$

In the region $x = 3$ m to $x = 9$ m,

$$V = \int \omega \, dx + C_2 = \int -5x \, dx + C_2 = -(\tfrac{5}{2})x^2 + C_2$$

When $x = 9$ m, $V = -R_2 = -235$ kN. Therefore,

$$V = -(\tfrac{5}{2})(9)^2 + C_2 = -235 \text{ kN}$$

$$-202.5 \text{ kN} + C_2 = -235 \text{ kN}$$

$$C_2 = -32.5 \text{ kN}$$

Hence,

$$V = -(\tfrac{5}{2})x^2 - 32.5 \text{ kN} \tag{2}$$

3. Calculate shear forces. Using equation (1) for the region $x = 0$ to $x = 3$ m and equation (2) for the region $x = 3$ m to $x = 9$ m:

x m	V kN	x m	V kN
0	267.5	3^+	−55.0
1	265.0	4	−72.5
2	257.5	6	−122.5
3^-	245.0	8	−192.5
		9	−235.0

4. Determine the bending moment equations.

$$M = \int V \, dx + D$$

For the region $x = 0$ to $x = 3$ m (equation (1)):

$$V = -\left(\tfrac{5}{2}\right)x^2 + 267.5 \text{ kN}$$

Therefore,

$$M = -\left(\tfrac{5}{2}\right)\int x^2 \, dx + 267.5 \int dx + D_1$$
$$= -\left(\tfrac{5}{2}\right)\left(\tfrac{1}{3}\right)x^3 + 267.5x + D_1$$
$$= -\left(\tfrac{5}{6}\right)x^3 + 267.5x + D_1$$

When $x = 0$, $M = 0$; therefore, $D_1 = 0$. So

$$M = -\left(\tfrac{5}{6}\right)x^3 + 267.5x \qquad (3)$$

For the region $x = 3$ m to $x = 9$ m (equation (2)):

$$V = -\left(\tfrac{5}{2}\right)x^2 - 32.5 \text{ kN}$$

Therefore,

$$M = -\left(\tfrac{5}{2}\right)\int x^2 \, dx - 32.5 \int dx + D_2$$
$$= -\left(\tfrac{5}{6}\right)x^3 - 32.5x + D_1$$

When $x = 9$ m, $M = 0$, so

$$0 = -\left(\tfrac{5}{6}\right)(9)^3 - (32.5)(9) + D_1$$
$$0 = -607.5 - 292.5 + D_1$$
$$D_1 = 900 \text{ kN} \cdot \text{m}$$

So

$$M = -\left(\tfrac{5}{6}\right)x^3 - 32.5x + 900 \text{ kN} \cdot \text{m} \qquad (4)$$

5. Calculate bending moments. Using equation (3) for the region $x = 0$ to $x = 3$ m and equation (4) for the region $x = 3$ m to $x = 9$ m gives the following.

x m	M kN · m	x m	M kN · m
0	0	3^+	780.0
1	266.7	4	716.7
2	528.3	6	525.0
3^-	780.0	8	213.3
		9	0

Answer

Figure 13.50(c) shows the shear force and bending moment diagrams.

13.11   COMPUTER PROGRAM FOR MOVING LOADS

Program 'Moveload' is actually two programs. One is for two-axle vehicles and one is for three-axle vehicles. Both programs calculate the maximum shear force, the vehicle location for maximum shear force, the maximum bending moment, and the vehicle location for maximum bending moment. The programs are based on the equations developed in Section 13.9.

Two-Axle Program

The required input values are

US or SI for units
W_1 the load on the left or rear axle
W_2 the load on the right or front axle
d the distance between the axles
L the beam length.

Three-Axle Program

The required input values are

US or SI for units
W_1 the load on the left or rear axle
W_2 the load on the second axle
W_3 the load on the right or front axle
d_1 the distance between the rear and second axles
d_2 the distance between the second and front axles
L the beam length.

Figure 13.51 shows the input/output sheets for the two-axle problem of Example Problem 13.17 and for the three-axle problem of Example Problem 13.18.

Program 'Moveload'

Two Axle Vehicle				Three Axle Vehicle		
Input:				**Input:**		
Enter US or SI for units	SI			Enter US or SI for units	US	
Rear axle load, W1	40000	N		Rear axle load, W1	6000	lb
Front axle load, W2	20000	N		Second axle load, W2	4000	lb
Dist between axles, d	3	m		Front axle load, W3	2000	lb
Beam length, L	10	m		Dist between rear & second axles, d1	8	ft
				Dist. Between second and front axles, d2	5	ft
Output:				Beam length, L	50	ft
Max. shear, Vmax	54000	N				
Rear axle location, x	0	m		**Output:**		
				Max. shear, Vmax	10840	lb
Max B.M, Mmax	121500	N m		Rear axle location, x	0	ft
Beam location	4.5	m				
Rear axle location, x	4.5	m		Max B.M, Mmax	122401.7	lb ft
				Beam location	22.58333	ft
				Rear axle location, x	22.58333	ft

Figure 13.51   Input/output for Example Problems 13.17 and 13.18.

EXAMPLE PROBLEM 13.21

A two-axle steam roller is crossing a 30-ft-long simply supported bridge from left to right. The load on the rear axle is 18,000 lb, and that on the front axle is 12,000 lb. The axles are 9 ft apart.

　a. Determine the maximum shear force and the vehicle position for maximum shear force.
　b. Find the maximum bending moment and the vehicle position for maximum bending moment.

Solution
Call up program 'Moveload' and enter the following values in the two-axle program:

$$W_1 = 18{,}000, \qquad W_2 = 12{,}000, \qquad d = 9, \qquad L = 30$$

Answer
　a. The maximum shear force is 26,400 lb when the rear axle is over the left support.
　b. The maximum bending moment is 174,000 lb-ft at 13.2 ft, with the rear axle at 13.2 ft.

EXAMPLE PROBLEM 13.22

The total load on the three axles of a vehicle is 240 kN. The distance between the rear axle and the second axle is 5 m, and that between the front and second axles is 3 m. The vehicle is to cross a simply supported 20-m-long bridge. Show that the maximum shear force and the maximum bending moment will both be the lowest if the load is distributed equally over the three axles. What are these values of shear force and bending moment?

Solution
　Call up program 'Moveload' and enter the following values in the three-axle program:

$$d_1 = 5, \qquad d_2 = 3, \qquad L = 20$$

Keeping these entries, enter any combination of W_1, W_2, and W_3 that adds up to 240,000, for example, $W_1 = 120{,}000$, $W_2 = 80{,}000$, and $W_3 = 40{,}000$. Note V_{max} and M_{max}, and try another combination. Finally, enter $W_1 = W_2 = W_3 = 80{,}000$.

Answer
The least maximum shear force is 196 kN. The least maximum bending moment is 881 kN m at 10.3 m, with the rear axle at 5.3 m.

13.12   PROBLEMS FOR CHAPTER 13

Hand Calculations

1. Calculate and plot the shear force diagram and the bending moment diagram for the loading shown in Figure 13.52.

2. Calculate and plot the shear force diagram and the bending moment diagram for the loading shown in Figure 13.53.

3. Calculate and plot the shear force diagram and the bending moment diagram for the loading shown in Figure 13.54.

4. Calculate and plot the shear force diagram and the bending moment diagram for the loading shown in Figure 13.55.

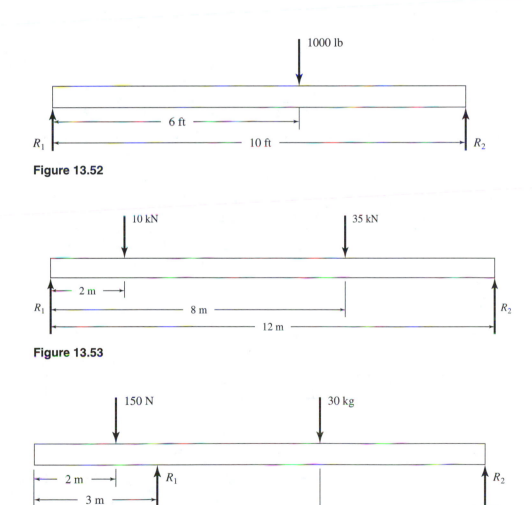

Figure 13.52

Figure 13.53

Figure 13.54

5. Calculate and plot the shear force diagram and the bending moment diagram for the loading shown in Figure 13.56.

6. Calculate and plot the shear force diagram and the bending moment diagram for the loading shown in Figure 13.57.

7. Calculate and plot the shear force diagram and the bending moment diagram for the loading shown in Figure 13.58.

8. Calculate and plot the shear force diagram and the bending moment diagram for the loading shown in Figure 13.59.

9. Calculate and plot the shear force diagram and the bending moment diagram for the loading shown in Figure 13.60.

10. Calculate and plot the shear force diagram and the bending moment diagram for the loading shown in Figure 13.61.

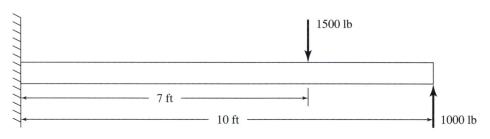

Figure 13.55

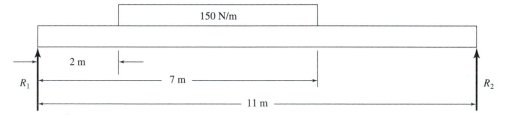

Figure 13.56

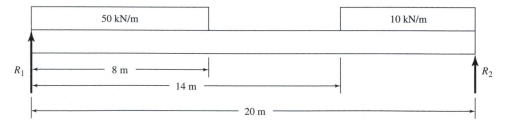

Figure 13.57

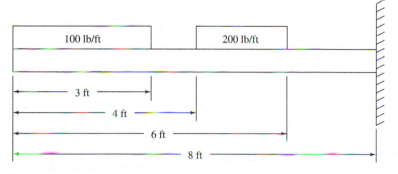

Figure 13.58

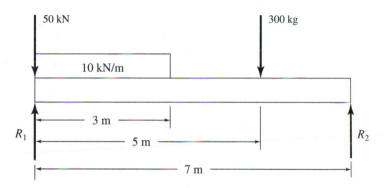

Figure 13.59

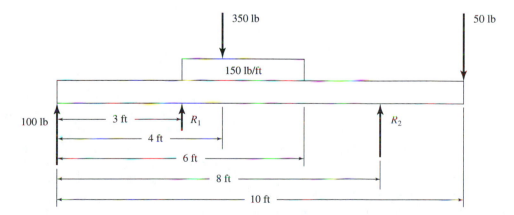

Figure 13.60

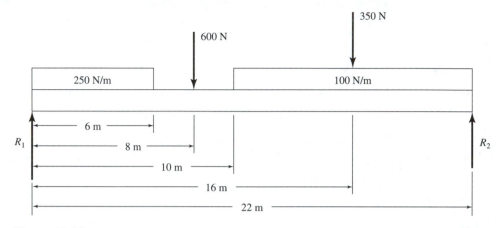

Figure 13.61

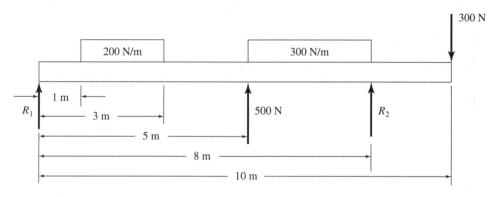

Figure 13.62

11. Calculate and plot the shear force diagram and the bending moment diagram for the loading shown in Figure 13.62.

12. Calculate and plot the shear force diagram and the bending moment diagram for the loading shown in Figure 13.63.

13. Calculate and plot the shear force diagram and the bending moment diagram for the loading shown in Figure 13.64.

14. Calculate and plot the shear force diagram and the bending moment diagram for the loading shown in Figure 13.65.

15. Calculate and plot the shear force diagram and the bending moment diagram for the loading shown in Figure 13.66.

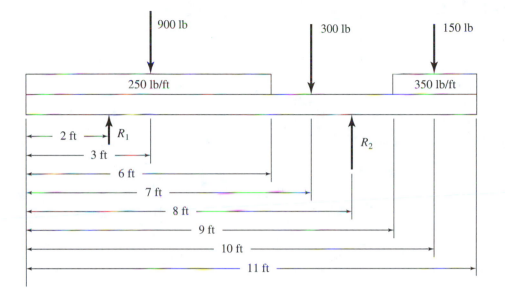

Figure 13.63

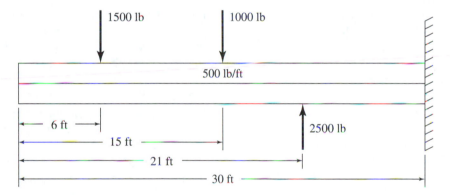

Figure 13.64

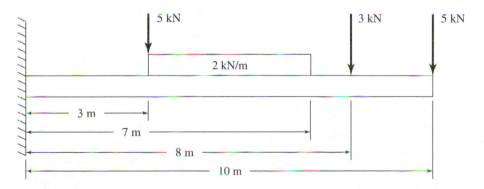

Figure 13.65

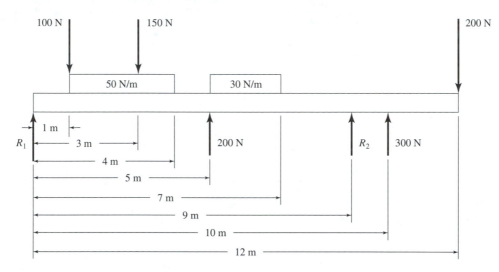

Figure 13.66

Computer Calculations

16. For the loading shown in Figure 13.61, calculate the shear forces and the bending moments at x-locations 5 m, 10 m, 15 m, and 20 m.

17. For the loading shown in Figure 13.62, calculate the shear forces and the bending moments at x-locations 2, 4, 6, 8, and 10, and determine the maximum bending moment.

18. For the loading shown in Figure 13.63, calculate the shear forces and the bending moments at x-locations 2, 4, 6, 8, and 10, and determine the maximum bending moment.

19. For the loading shown in Figure 13.65, calculate the shear forces and the bending moments at x-locations 2, 4, 6, 8, and 10, and determine the maximum bending moment.

20. For the loading shown in Figure 13.66, calculate the shear forces and the bending moments at x-locations 2, 4, 6, 8, and 10, and determine the maximum bending moment.

21. For the loading shown in Figure 13.64 , calculate the shear forces and the bending moments at x-locations 5 ft, 10 ft, 15 ft, 20 ft, and 25 ft, and find the magnitude and location of the maximum bending moment.

22. Plot the shear force diagram and the bending moment diagram for the loading shown in Figure 13.61.

23. Plot the shear force diagram and the bending moment diagram for the loading shown in Figure 13.62.

24. Plot the shear force diagram and the bending moment diagram for the loading shown in Figure 13.63.

25. Plot the shear force diagram and the bending moment diagram for the loading shown in Figure 13.64.

26. Plot the shear force diagram and the bending moment diagram for the loading shown in Figure 13.65.

27. Plot the shear force diagram and the bending moment diagram for the loading shown in Figure 13.66.

28. Write a computer program to calculate the shear force at any station of a simply supported beam carrying up to three concentrated and two distributed loads.

29. Write a computer program to calculate the bending moment at any station of a simply supported beam carrying up to three concentrated and two distributed loads.

30. Write a computer program to plot the shear force diagram and the bending moment diagram for a simply supported beam carrying up to three concentrated and two distributed loads.

31. A two-axle vehicle moves from left to right across a 75-ft-long simply supported bridge. The total vehicle weight is 12,000 lb, and the distance between the axles is 12 ft. Determine the maximum bending moments for the following three cases:
    a. One-fourth of the load on the front axle
    b. One-third of the load on the front axle
    c. One-half of the load on the front axle

32. A three-axle vehicle moves from left to right across a 25-m-long simply supported bridge. The axle loads are

    rear: $W_1 = 20$ kN,   second: $W_2 = 50$ kN,

    front: $W_3 = 30$ kN

    and the distances between the axles are

    rear and second: $d_1 = 5$ m,
    front and second: $d_2 = 4$ m

    a. Determine the maximum shear force and the vehicle position for maximum shear force.
    b. Find the maximum bending moment and the vehicle position for the maximum bending moment.

33. A farmer has to drive a load across an old wooden 30-ft-long bridge. The farmer has a two-axle truck, whose axles are 6 ft apart, and a three-axle truck, whose second axle is 7 ft from the rear axle and 6 ft from the front axle. When loaded, the two-axle vehicle would have a load of 7000 lb on the rear axle and 2000 lb on the front axle. The three-axle truck, being heavier, would be 12,000 lb when loaded, but the farmer would be able to distribute the load equally over all three axles. Being concerned about the strength of the bridge, which truck should the farmer use?

Using Calculus

34. A 10-ft simply supported beam carries one concentrated load of -1000 lb at $x = 6$ ft and a distributed load over the entire span described by $\omega(x) = -2x^2$ lb/ft. Plot the shear force and bending moment diagrams.

35. Move the concentrated load in Problem 34 to $x = 4$ ft, and repeat the problem.

14

Shear and Bending Stresses in Beams

In this chapter you will learn about *horizontal shear stresses* and *bending stresses* that develop in loaded beams.

14.1 HORIZONTAL SHEAR STRESSES

We discussed the existence of shear and bending stresses in beams in Chapter 13. Those shear stresses were vertical shear stresses, acting in a vertical plane. Horizontal shear stresses also exist.

The development of horizontal shear stresses can be readily understood with the help of Figure 14.1(a) and 14.1(b). We show three unconnected planks placed on top of each other and simply supported at the ends. With no load (Figure 14.1(a)), their ends are all level. If we force them to bend by placing a load on them, each plank will slide on its neighbor, and the ends will move from their original positions (Figure 14.1(b)). You can see the same effect by taking, say, 100 pages of this book and bending them. The pages slide over each other. If the planks in Figure 14.1 were bonded together, stresses would develop in the glue as the layers tried to slide over each other.

Extending this reasoning to the case of a solid plank or metal beam, you can see that horizontal stresses are developed between the fibers or molecules when bending occurs.

Figure 14.2 shows a simply supported beam carrying a load. We have zoomed in on a very small part of the beam with dimensions Δx, Δy, Δz. $\mathbf{F_H}$ and $\mathbf{F_V}$ are the horizontal and vertical forces created by the bending. This element is a free body and so must be in equilibrium. Therefore, the sum of the moments about any axis must be zero. Taking moments about AB:

$$\sum M)_{AB} = 0$$
$$F_H \cdot \Delta y - F_v \cdot \Delta x = 0$$
$$F_H \cdot \Delta y = F_V \cdot \Delta x$$
$$F_H/\Delta x = F_V/\Delta y$$

Divide both sides by Δz:

$$F_H/(\Delta x \Delta z) = F_V/(\Delta y \Delta z)$$

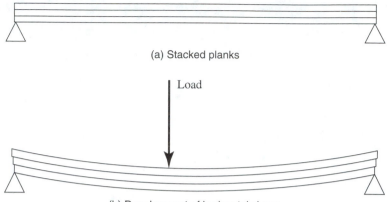

(a) Stacked planks

(b) Development of horizontal shear

Figure 14.1

But $(\Delta x \Delta z)$ is the area over which F_H acts, and $(\Delta y \Delta z)$ is the area over which F_V acts. Since force/area is stress, we have $S_H = S_V$, i.e., *the horizontal shear stress at a point in a beam is equal to the vertical shear stress at that point.*

We show later that the horizontal shear stress varies across the beam from zero at the upper and lower surfaces to a maximum value at the neutral axis.

Consider the simply supported beam shown in Figure 14.3(a). We are going to isolate a section between the two planes $ABCD$ and $EFGH$. We see from Figure 14.3(b) that the bending moment increases across the section. Our objective is to determine the horizontal shear stress on the infinitesimally thin element $JKLM$ shown in Figure

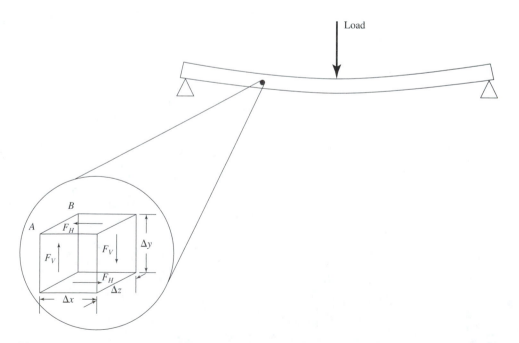

Figure 14.2   Horizontal and vertical stresses at a point are equal.

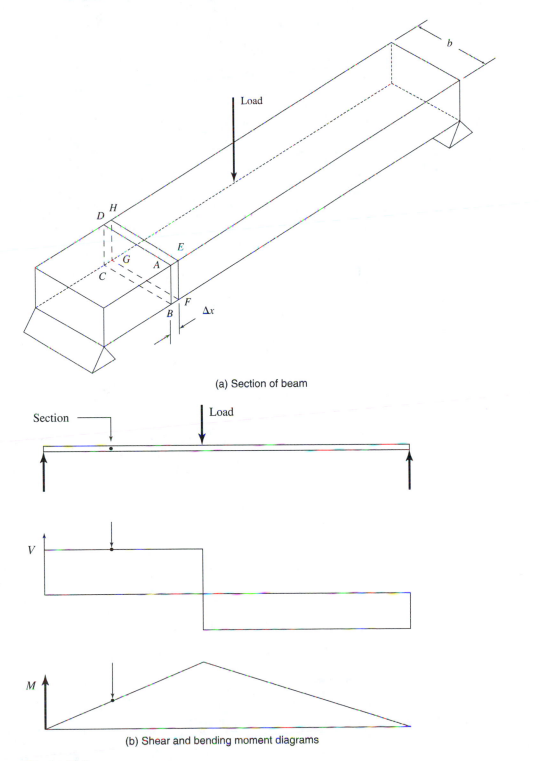

(a) Section of beam

(b) Shear and bending moment diagrams

Figure 14.3

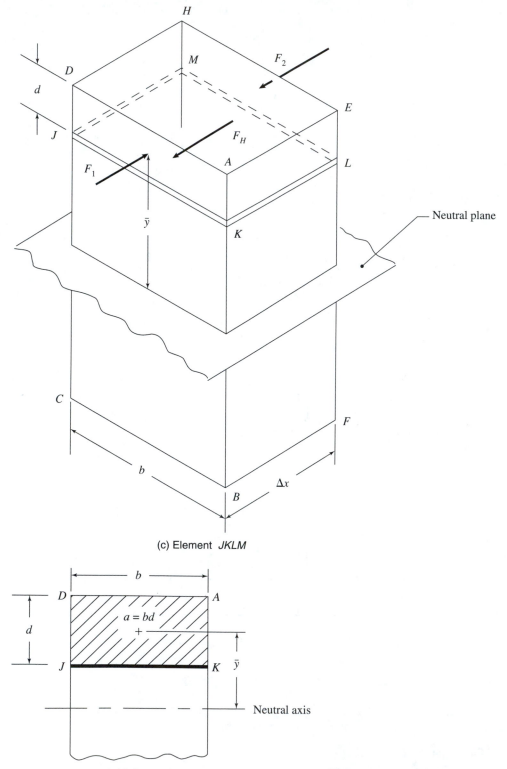

(c) Element *JKLM*

(d) Area *a*

Figure 14.3   Continued

14.3(c). As discussed before, the horizontal shear stress arises due to the bending, and because the bending moment is increasing across the section, the force $\mathbf{F}_2$ (Figure 14.3(c)) is greater than the force $\mathbf{F}_1$. For the block $ADHELKJM$ to be in equilibrium, there must be a force $\mathbf{F}_H$ to balance out the difference between $\mathbf{F}_1$ and $\mathbf{F}_2$. This is the horizontal shear force. So,

$$F_H = F_2 - F_1 \tag{14.1}$$

The force $\mathbf{F}_1$ acts at the centroid of the area $AKJD$, which is a distance $\bar{y}$ from the neutral axis. Similarly, $\mathbf{F}_2$ acts at the centroid of the area $ELMH$, which is the same distance, $\bar{y}$, from the neutral axis. We know that the stresses associated with these forces can be calculated by using the flexure equation (13.7), derived in Chapter 13. This equation says that the stress due to bending is equal to the bending moment times the distance from the neutral axis divided by the moment of inertia.

The stress associated with $\mathbf{F}_1$ is

$$s_1 = M_1\bar{y}/I$$

and the stress associated with $\mathbf{F}_2$ is

$$s_2 = M_2\bar{y}/I$$

Let us denote the area of $AKJD$ by the letter a. This is the shaded area in Figure 14.3(d). Area $ELMH$ also is a. So we can write

$$F_1 = as_1 \quad \text{and} \quad F_2 = as_2$$

So the horizontal shear force, F_H, which is $(F_2 - F_1)$, can be expressed as

$$\begin{aligned} F_H &= as_2 - as_1 \\ &= a\bar{y}M_2/I - a\bar{y}M_1/I \\ &= a\bar{y}(M_2 - M_1)/I \end{aligned} \tag{14.2}$$

Note: The product $(a\bar{y})$ is called the **static moment** and is often given the symbol Q.

If the section width is Δx, then the bending moment increases by an amount $V \cdot \Delta x$ across the section:

$$M_2 = M_1 + V \cdot \Delta x$$

or

$$(M_2 - M_1) = V \cdot \Delta x \tag{14.3}$$

Substituting this into equation (14.2) gives

$$F_H = a\bar{y}V \cdot \Delta x/I \tag{14.4}$$

Now the horizontal shear stress, s_H, is equal to the horizontal shear force, F_H, divided by the area $KJML$ (Figure 14.3(c)), which is $(b \cdot \Delta x)$. That is,

$$s_H = F_H/(b \cdot \Delta x) = (a\bar{y}V \cdot \Delta x/I)/(b \cdot \Delta x)$$

So, finally,

$$s_H = a\bar{y}V/(Ib) \tag{14.5}$$

where

a = the cross-sectional area between the element being considered and the nearest outermost fiber.

$\bar{y}$ = the distance from the neutral axis of the entire cross section to centroid of area a.

V = the vertical shear force at the section being considered
I = the moment of inertia of the entire cross section of the beam
b = the width of the fiber being considered

We see from equation (14.5) that when $\bar{y} = 0$—i.e., on the neutral axis—area a is maximum, making the horizontal shear stress maximum.* When $\bar{y} = c$, the distance to the outermost fibers, area a is zero, making the horizontal shear stress zero.

USING CALCULUS

Referring to Figure 14.3(c), Δx becomes dx. From the flexure formula, the bending stress at any point on the area $ADJK$ is $s_b = My/I$, where M is the bending moment at that section (section 1) and y is the distance from the neutral axis. So the force

$$F_1 = \int s_b \, da = (M/I) \int y \, da$$

But the definition of $\bar{y}$, the y-location of the centroid, is

$$\bar{y} = (1/a) \int y \, da$$

so

$$\int y \, da = a\bar{y}$$

Therefore,

$$F_1 = (a\bar{y}/I)M$$

The bending moment at section 2, which is at a distance dx from section 1, is $M + (dM/dx) \, dx$. Therefore,

$$F_2 = (a\bar{y}/I) \, [M + (dM/dx)dx]$$

$F_2 - F_1 = F_H$, the horizontal shear stress on the surface bdx, so

$$F_H = (a\bar{y}/I) \, [M + (dM/dx)dx] - (a\bar{y}/I)M$$

$$F_H = (a\bar{y}/I)(dM/dx)dx$$

The horizontal shear stress on the surface bdx is

$$s_H = F_H/(bdx) = (a\bar{y})(dM/dx)/(Ib)$$

But $dM/dx = V$, so

$$s_H = a\bar{y}V/(Ib)$$

EXAMPLE PROBLEM 14.1

A simply supported 10-m beam carries a load of 100 kN at a station 8 m from the left end (see Figure 14.4(a)). The beam is 50 mm wide by 200 mm deep. Calculate the horizontal shear stress at a point that is 4 m from the left end of the beam and 30 mm from the upper surface (see Figure 14.4(b)).

*This is true unless the thickness at some other axis is smaller than at the centroidal axis.

Solution

1. Calculate area a.

$$a = (50 \text{ mm})(30 \text{ mm}) = 1500 \text{ mm}^2$$

2. Calculate $\bar{y}$. The centroid of area a is 15 mm below the upper surface. The neutral axis is 100 mm below the upper surface. Therefore,

$$\bar{y} = 100 \text{ mm} - 15 \text{ mm} = 85 \text{ mm}$$

3. Calculate the vertical shear force, V. Because V is equal to the sum of the external forces to the left of the station being considered, $V = R_A$, the reaction at A. Taking moments about B,

$$\sum M)_B = 0 = (1 \times 10^5 \text{ N})(2 \text{ m}) - R_A(10 \text{ m})$$
$$R_A = (2 \times 10^5 \text{ N} \cdot \text{m})/(10 \text{ m}) = 2 \times 10^4 \text{ N}$$

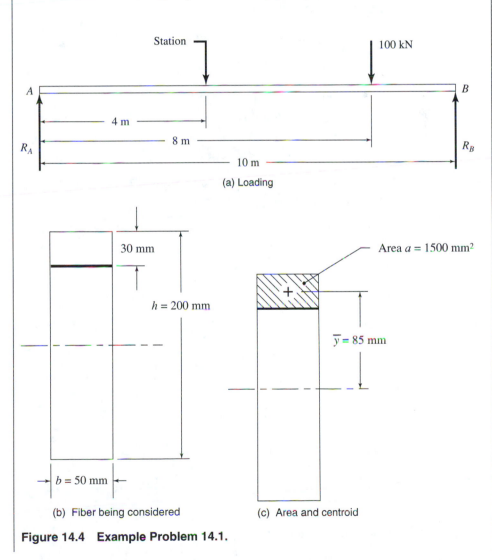

(a) Loading

(b) Fiber being considered

(c) Area and centroid

Figure 14.4 Example Problem 14.1.

Therefore, $V = 2 \times 10^4$ N.

4. Calculate the moment of inertia, I, of the entire section.

$$I = bh^3/12 = (50 \text{ mm})(200 \text{ mm})^3/12 = 33.33 \times 10^6 \text{ mm}^4$$

5. Calculate the horizontal shear stress.

$$s_H = a\bar{y}V/(Ib)$$
$$= (1500 \text{ mm}^2)(85 \text{ mm}) (2 \times 10^4 \text{ N})/[(33.33 \times 10^6 \text{ mm}^4)(50 \text{ mm})]$$
$$= 1.53 \text{ N/mm}^2 = 1.53 \text{ MPa}$$

Answer

The horizontal shear stress is 1.53 MPa.

EXAMPLE PROBLEM 14.2

Calculate the distribution of horizontal shear stress across the section of Example Problem 14.1.

Solution

We know the values of V, I, and b, from Example Problem 14.1.

$$V = 2 \times 10^4 \text{ N}, \qquad I = 33.33 \times 10^6 \text{ mm}^4, \qquad b = 50 \text{ mm}$$

So $s_H = a\bar{y}V/(Ib) = a\bar{y}(2 \times 10^4 \text{ N})/[(33.33 \times 10^6 \text{ mm}^4)(50 \text{ mm})]$

$$= a\bar{y}/(83,325) \text{ MPa}$$

We now take a series of elements starting 10 mm below the upper surface, then 20 mm, and so on. See Figure 14.5.

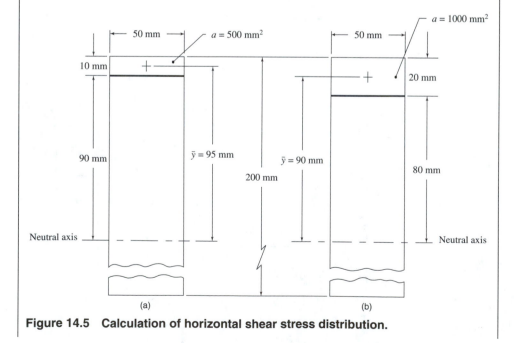

Figure 14.5   Calculation of horizontal shear stress distribution.

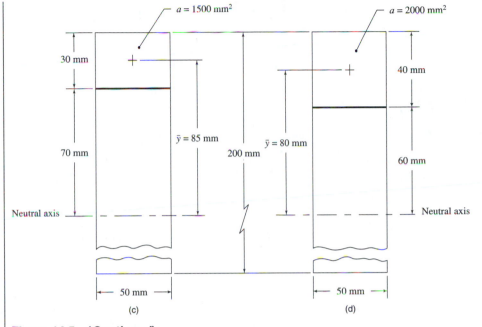

Figure 14.5 (*Continued*)

For the first case, the area is a = (10 mm)(50 mm) = 500 mm², and the distance from the neutral axis to the centroid of area a is $\bar{y}$ = 95 mm. Therefore, $(a\bar{y})$ = (500 mm²)(95 mm) = 47,500 mm³.

So $\quad s_H = a\bar{y}/(83,325)$ MPa = 47,500/83,325 MPa = 0.57 MPa

We show the first four cases in Figure 14.5. Tabulating the calculations gives the following.

Distance from Neutral Axis (mm)	Area a (mm²)	$\bar{y}$ (mm)	s_H (MPa)
100	0	100	0
90	500	95	0.57
80	1000	90	1.08
70	1500	85	1.53
60	2000	80	1.92
50	2500	75	2.25
40	3000	70	2.52
30	3500	65	2.73
20	4000	60	2.88
10	4500	55	2.97
0	5000	50	3.00

This would be repeated below the centerline.

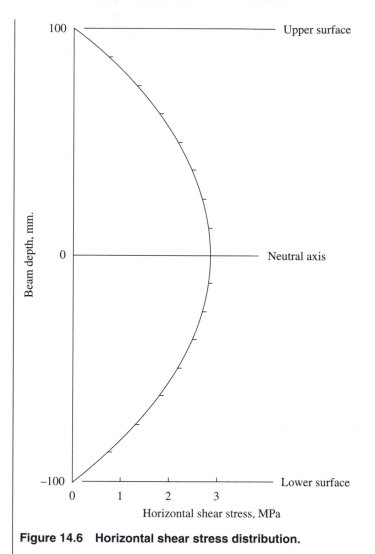

Figure 14.6   Horizontal shear stress distribution.

Answer

We show this distribution in Figure 14.6. This is a parabolic distribution, going from zero at the surface and increasing to a maximum at the neutral axis.

14.1.1   Rectangular Cross Sections

We can derive a simple equation for the special case of a beam having a rectangular cross section. Figure 14.7 shows the cross section of a rectangular beam having depth h and width b. At any distance y from the neutral axis, the depth of the element from y to the outermost fibers is $(h/2 - y)$. Therefore, the area a is

$$a = b(h/2 - y) = (bh/2 - by)$$

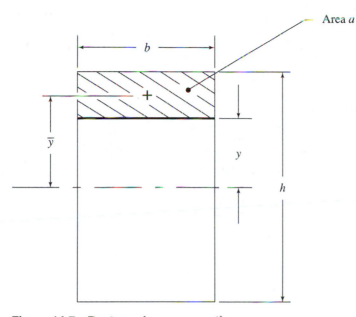

Figure 14.7   Rectangular cross section.

The centroid of area a is at the center of the area, so its distance from the neutral axis is $\bar{y} = y + (h/2 - y)/2$, which we can write as

$$\bar{y} = (y/2 + h/4)$$

The moment of inertia of the entire section is

$$I = bh^3/12$$

$$s_H = a\bar{y}V/(Ib) \qquad \text{from equation (14.5)}$$

Substituting the expressions for a, $\bar{y}$, and I into this equation gives

$$a = (bh/2 - by)$$
$$\bar{y} = (y/2 + h/4)$$
$$I = bh^3/12$$

Therefore,

$$\begin{aligned} s_H &= (bh/2 - by)(y/2 + h/4)V/(b^2h^3/12) \\ &= 6V(h/2 - y)(y + h/2)/(bh^3) \\ &= (6V/bh^3)(h^2/4 - y^2) \end{aligned} \qquad (14.6)$$

At the surface of the beam, $y = h/2$, so $y^2 = h^2/4$, making $s_H = 0$. At the neutral axis, $y = 0$, making $S_H = (6V/bh^3)(h^2/4) = 3V/(2bh)$. So

$$s_{Hmax} = 3V/(2bh)$$

But $bh = A$, the cross-sectional area of the beam, so

$$s_{Hmax} = \tfrac{3}{2}V/A \qquad (14.7)$$

EXAMPLE PROBLEM 14.3

Returning to Example Problem 14.2, calculate the horizontal shear stress at $y = 50$ mm.

Solution

$$V = 2 \times 10^4 \text{ N}, \qquad b = 50 \text{ mm}, \qquad h = 200 \text{ mm}$$
$$s_H = (6V/bh^3)(h^2/4 - y^2) \qquad \text{from equation (14.6)}$$
$$= [(6 \times 2 \times 10^4 \text{ N})/(50 \text{ mm})(200 \text{ mm})^3][(200 \text{ mm})^2/4 - (50 \text{ mm})^2]$$
$$= (3 \times 10^{-4} \text{ N/mm}^4)(7500 \text{ mm}^2)$$
$$= 2.25 \text{ N/mm}^2 = 2.25 \text{ MPa}$$

Answer

The horizontal shear stress is 2.25 MPa.

14.1.2   Circular Cross Sections

We will develop an equation for the *maximum* horizontal shear stress, which *occurs at the neutral axis*.

Figure 14.8 shows the cross section of a beam of diameter d. The following expressions apply (see Figure 11.4).

For the shaded area, $\bar{y} = 0.425r$. The actual expression for this, achieved by integration, is

$$\bar{y} = 2d/(3\pi)$$

The moment of inertia of the whole circle is

$$I = \pi d^4/64$$

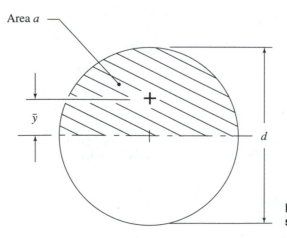

Figure 14.8 Circular cross section.

The area, a, is the area of a semicircle:

$$a = \pi d^2/8$$

and the width, b, is equal to d.

$$
\begin{aligned}
s_{Hmax} &= a\bar{y}V/(Ib) \qquad \text{from equation (14.5)} \\
&= (\pi d^2/8)(2d/3\pi)V/[(\pi d^4/64)d] \\
&= (64 \times 2V)/(8 \times 3 \times \pi d^2) \\
&= \tfrac{16}{3}V/(\pi d^2) \\
&= \tfrac{4}{3}V/(\pi d^2/4)
\end{aligned}
$$

But $(\pi d^2/4) = A$, the cross-sectional area of the beam, so

$$s_{Hmax} = \tfrac{4}{3}V/A \tag{14.8}$$

EXAMPLE PROBLEM 14.4

A 10-m-long Douglas fir rod has a diameter of 35 mm. It is loaded as shown in Figure 14.9. With regard to *horizontal shear only,* will it fail? The allowable horizontal shear stress for Douglas fir is 0.59 MPa.

Solution

1. Calculate the maximum vertical shear force. Because this is a cantilever beam, the maximum vertical shear force will occur at the wall, i.e., at $x = 10$ m.

$$
\begin{aligned}
V_{10} &= -200 \text{ N} - (50 \text{ N/m})(4 \text{ m}) - (30 \text{ N/m})(3 \text{ m}) \\
&= -490 \text{ N}
\end{aligned}
$$

2. Calculate the maximum horizontal shear force. For a round bar,

$$s_{Hmax} = \tfrac{4}{3}V/A$$

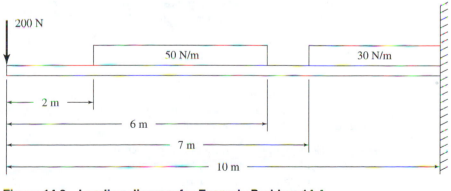

Figure 14.9   Loading diagram for Example Problem 14.4.

A, the cross-sectional area of the rod, is $\pi(35 \text{ mm})^2/4 = 962 \text{ mm}^2$. Therefore, s_{Hmax} $= \frac{4}{3}(490 \text{ N})/962 \text{ mm}^2 = 0.68 \text{ MPa}$.

This is greater than the allowable, so the rod will fail.

Answer
The rod will fail.

Note: It is to be understood that this was an exercise to determine the maximum *horizontal shear stress. The beam would actually fail due to the bending stress,* which is much greater. *Short, heavily loaded beams* are more likely to fail due to horizontal shear.

14.2   BENDING STRESSES

The *flexure equation*, equation (13.7), developed in Chapter 13, expresses the stress due to bending at a station where the bending moment is M and the distance from the neutral axis is y for a beam with moment of inertia I.

$$s_B = My/I$$

The maximum bending stress at any station occurs where $y = c$, the distance from the neutral axis to the outermost fibers, i.e., to the surface of the beam. It follows that the maximum bending stress in the beam is on the surface (the one farther away from the neutral axis if the beam is not symmetrical), at the station where the bending moment is a maximum. On the neutral axis, where $y = 0$, the bending stress is zero.

Unlike shear stress, which goes from a maximum at the neutral axis to zero at the surface, the bending stress goes from zero at the neutral axis to a maximum at the surface. Both stresses must be considered in beam analyses.

EXAMPLE PROBLEM 14.5

Calculate the maximum bending stress in the beam shown in Figure 14.10(a). The beam cross section is shown in Figure 14.10(b).

Solution

$$s_B = Mc/I$$

where M is the maximum bending moment, c is the larger distance from the neutral axis to the beam surface, and I is the moment of inertia of the beam cross section. There are four steps in this calculation:

1. Calculate the maximum bending moment.
2. Calculate the position of the neutral axis.
3. Calculate the moment of inertia.
4. Calculate the bending stress.

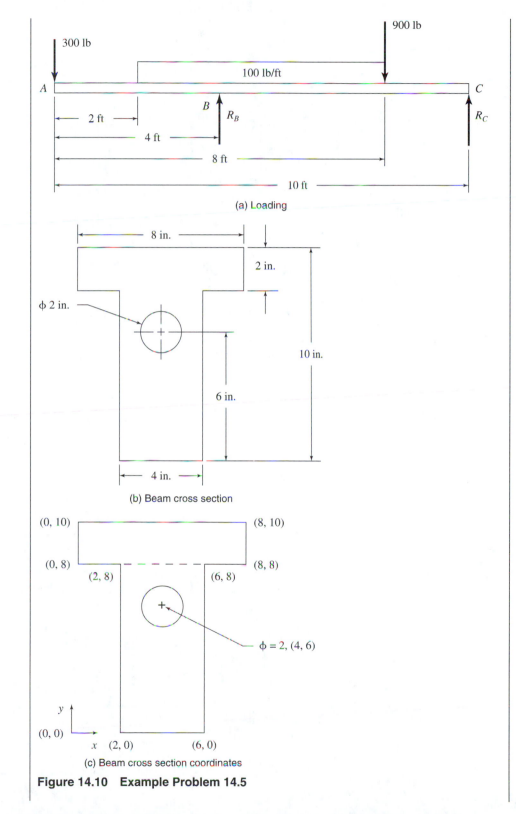

(a) Loading

(b) Beam cross section

(c) Beam cross section coordinates

Figure 14.10   Example Problem 14.5

CENTROID and MOMENTS OF INERTIA CALCULATION   PROGRAM 'Centroid'

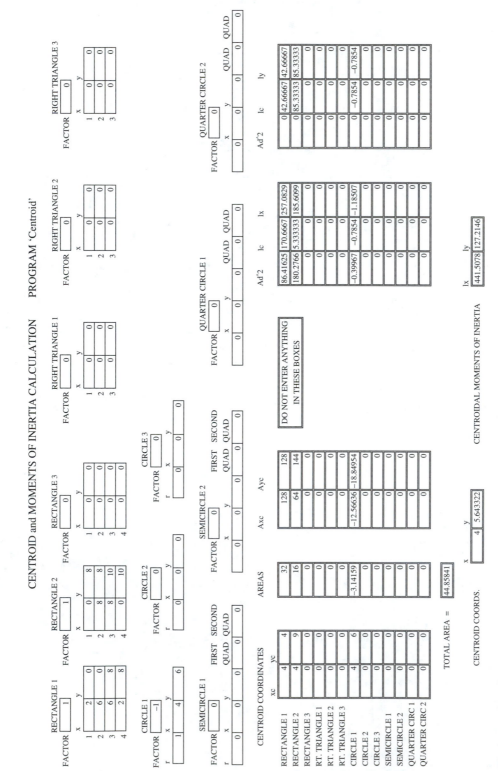

Figure 14.11 Centroid and moment of inertia for Example Problem 14.5.

1. Calculate the maximum bending moment. Open program 'Shr&mmt'. Refer to Example Problem 13.19 if necessary.

The maximum bending moment is given as -1400 lb-ft at $x = 4$ ft.

2. Calculate the position of the neutral axis.
3. Calculate the moment of inertia. We can do both of these by using program 'Centroid'. Open program 'Centroid'.

Set up x- and y-axes as shown in Figure 14.10(c), and divide the tee into two rectangles. Then mark the corners with their (x, y)-coordinates.

Enter the coordinates and the circle radius ($r = 1$) in the cells shown in Figure 14.11. The two rectangles are solid, so they both have FACTOR = 1. The circle is cut out, so its FACTOR is -1.

The results are $\bar{y} = 5.643$ in. and $I_x = 441.5$ in⁴.

4. Calculate the bending stress. First, we need the value of c.

The total height of the tee is 10 in. and $\bar{y} = 5.643$ in. So the distances of the neutral axis from the upper and lower surfaces are (10 in. $-$ 5.643 in.) = 4.357 in. and 5.643 in., respectively.

So
$$c = 5.643 \text{ in.}$$

$$s_B = Mc/I = (1400 \text{ lb-ft})(5.643 \text{ in.})/441.5 \text{ in.}^4$$

Be careful with the units. This result shows that you must carry the units throughout all your calculations.

We have to convert the bending moment from lb-ft to lb-in. To change from feet to inches we multiply by 12, so to change from lb-ft to lb-in., we also multiply by 12.

$$s_B = (1400 \text{ lb-ft})(12 \text{ lb-in./lb-ft})(5.643 \text{ in.})/441.5 \text{ in.}^4$$

$$= 215 \text{ lb/in.}^2$$

Answer
The stress due to bending is 215 lb/in^2.

14.3 STRESS CONCENTRATIONS

We may tend to think of a beam as being a wooden member supporting a floor or a ceiling or a steel I-beam in a building or a bridge. There are, however, many forms a beam can take. For example, the railroad car axle shown in Figure 14.12 and the internal combustion engine wrist pin shown in Figure 14.13 are also *beams*. The axle is subjected to two concentrated loads, and the wrist pin has a varying continuous load.

Any member that is supported at one, two, or more stations and subjected to transverse loads may be regarded as a beam. Sometimes, a round bar acting as a beam is held from moving sideways by retaining rings let into grooves in the rod.

Any groove will increase the rod's bending stress, and again as in Chapters 6 and 12, we apply stress-concentration factors to allow for the increased stress. Experiments conducted on grooved round bars in bending have shown that the stress increases with

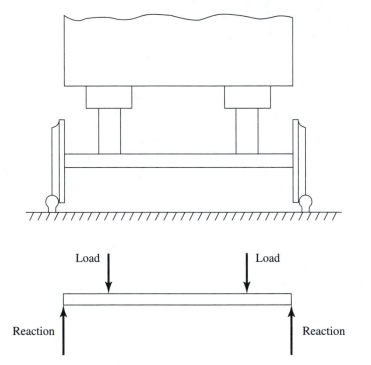

Figure 14.12   Railroad car axle.

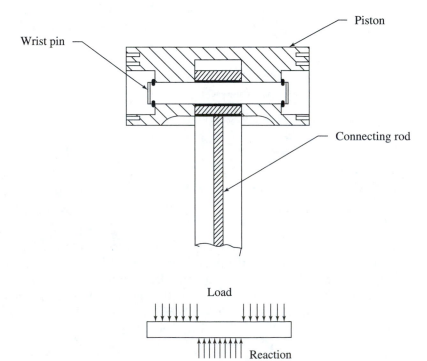

Figure 14.13   Engine wrist pin.

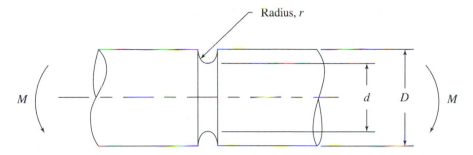

Figure 14.14   Grooved round bar in bending.

increasing D/d and decreases with increasing r/d. Here, D is the diameter of the bar, d is the diameter of the groove, and r is the groove radius. (See Figure 14.14.)

To calculate the maximum bending stress, we first calculate the bending stress based on the *groove* diameter, d, and then multiply by the stress concentration factor, k, which is given by

$$k = 1.055/(r/d)^n \tag{14.9(a)}$$

where

$$n = \begin{cases} (D/d)/3 - 0.125 & 1.2 \geq D/d \geq 1.05 & (14.9(b)) \\ 0.01625(D/d) + 0.2555 & 2.0 \geq D/d \geq 1.2 & (14.9(c)) \end{cases}$$

EXAMPLE PROBLEM 14.6

Calculate the bending stress at the retaining-ring locations for the support rod shown in Figure 14.15(a). The rod diameter is 20 mm, and external series 11-410 retaining rings are used to prevent sideways motion of the hanger. (For a 20-mm diameter, a series 11-410 retaining ring groove has a diameter of 18.1 mm.) To reduce the stress concentration, we are including a groove fillet radius of 1.3 mm.

Solution

The 50-N load creates a distributed load of 50 N/m on the beam (see Figure 14.15(b)).

1. Calculate the bending moment at $x = 2.5$ m. We leave it to the student to show that $M_{max} = 62.5$ N·m.

2. Calculate the section moment of inertia.

$$I = \pi d^4/64 = \pi(18.1 \text{ mm})^4/64 = 5268 \text{ mm}^4$$

3. Calculate the nominal stress *without the concentration factor*. (Notice that we use the *groove* diameter).

$$s_B = Mc/I = (62.5 \text{ N·m})(18.1 \text{ mm}/2)/5268 \text{ mm}^4$$

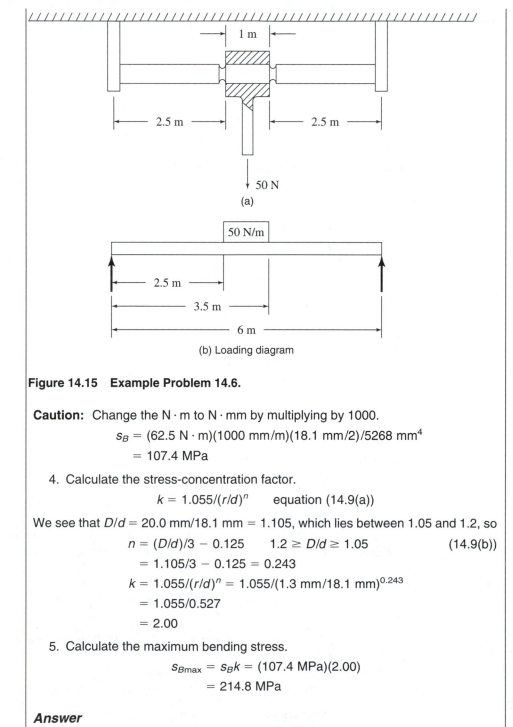

Figure 14.15 Example Problem 14.6.

Caution: Change the N · m to N · mm by multiplying by 1000.

$$s_B = (62.5 \text{ N} \cdot \text{m})(1000 \text{ mm/m})(18.1 \text{ mm/2})/5268 \text{ mm}^4$$
$$= 107.4 \text{ MPa}$$

4. Calculate the stress-concentration factor.

$$k = 1.055/(r/d)^n \qquad \text{equation (14.9(a))}$$

We see that $D/d = 20.0$ mm/18.1 mm $= 1.105$, which lies between 1.05 and 1.2, so

$$n = (D/d)/3 - 0.125 \qquad 1.2 \geq D/d \geq 1.05 \qquad (14.9(b))$$
$$= 1.105/3 - 0.125 = 0.243$$
$$k = 1.055/(r/d)^n = 1.055/(1.3 \text{ mm/18.1 mm})^{0.243}$$
$$= 1.055/0.527$$
$$= 2.00$$

5. Calculate the maximum bending stress.

$$s_{Bmax} = s_B k = (107.4 \text{ MPa})(2.00)$$
$$= 214.8 \text{ MPa}$$

Answer

The maximum bending stress is 215 MPa. For this case, the retaining ring groove *doubled* the bending stress.

14.4   COMPOSITE BEAMS

We know from the flexure equation, equation (13.7),

$$s_B = My/I$$

that bending stress is a maximum at the upper and/or lower surfaces of a beam. We can use this knowledge to construct more efficient, lighter beams of two or more materials. These are called *composite* beams. For example, Figure 14.16 shows a wooden beam with steel plates securely fastened to the upper and lower surfaces.

Referring again to the development of the flexure formula, Section 13.3, our basic assumption (borne out by measurements) was that the strain is directly proportional to the distance from the neutral axis. Based on this, we show the strain distribution across a composite beam in Figure 14.17(a). The same wooden beam without steel plates would deflect more under the same loading, so the strain distribution would be more severe.

Now, stress is equal to strain times the elastic modulus,

$$s = \varepsilon E$$

and E for steel is about 27 times that for wood. This results in a very abrupt change in stress distribution at the wood-steel interface. See Figure 14.17(b). What we have done is transfer part of the stress in the wood to the steel plates.

The simplest way to analyze a beam like this is to use what is called **effective areas**, or **transformed sections**. With this method, we imagine that we have replaced the beam with an equivalent beam made entirely of steel and then analyze that beam. This equivalent beam is to have the same strain distribution and the same stress distribution as the composite beam. The question is, What does that beam cross section look like?

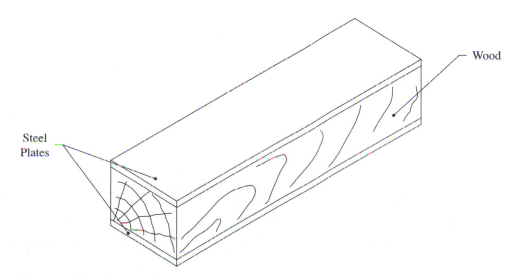

Figure 14.16   Composite beam.

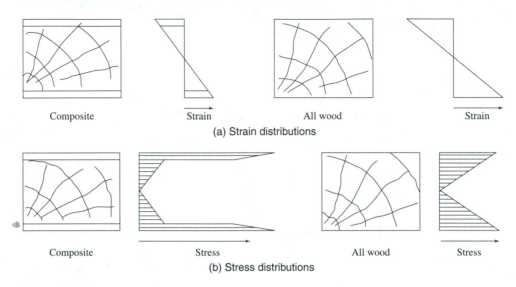

Composite   Strain   All wood   Strain

(a) Strain distributions

Composite   Stress   All wood   Stress

(b) Stress distributions

Figure 14.17

Using the subscript w for wood and the subscript st for steel, we have, at any point in a wooden beam,

$$s_w = \varepsilon_w E_w$$

or

$$\varepsilon_w = s_w/E_w$$

and at any point in a steel beam,

$$s_{st} = \varepsilon_{st} E_{st}$$

If we are to replace the wooden part by a steel part that has the same strain, i.e., $\varepsilon_{st} = \varepsilon_w$, then the stress in this equivalent steel beam must be related to the stress in the wood by

$$s_{st} = \varepsilon_{st} E_{st} = (s_w/E_w)E_{st} = s_w E_{st}/E_w$$

Therefore, the stress at any point in the equivalent steel beam must be the same as that in the wood, modified by the ratio of the elastic modulii.

Denoting force by F and cross-sectional area by A, we have

$$s_w = F_w/A_w \quad \text{and} \quad s_{st} = F_{st}/A_{st}$$

Therefore,

$$F_{st}/A_{st} = (F_w/A_w)(E_{st}/E_w)$$

$$A_{st}/A_w = (F_{st}/F_w)/(E_{st}/E_w)$$

But both the composite beam and its equivalent steel beam have the same force distribution, so $F_{st} = F_w$; therefore,

$$A_{st}/A_w = 1/(E_{st}/E_w)$$

(E_{st}/E_w) is called the **effectiveness factor** and is generally given the symbol n:

$$E_{st}/E_w = n$$

So $\qquad\qquad\qquad\qquad\qquad\qquad s_{st}/s_w = n$

or $\qquad\qquad\qquad\qquad\qquad\qquad s_w = s_{st}/n$ $\qquad\qquad\qquad$ (14.10(a))

and $\qquad\qquad\qquad\qquad\qquad\quad A_{st} = A_w/n$ $\qquad\qquad\qquad$ (14.10(b))

These equations tell us that

1. The equivalent all-steel beam would be $1/n$ as thick (same depth) as the wooden beam, and
2. The stress in the wood would be $1/n$ times the stress in the equivalent steel.

An example should clarify this.

EXAMPLE PROBLEM 14.7

Calculate the bending stress in the steel and in the wood just each side of the wood-steel interface for the composite beam shown in Figure 14.18(a). The wood is Sitka spruce, for which $E_w = 9.0 \times 10^3$ MPa. For steel, $E_{st} = 200 \times 10^3$ MPa. The bending moment at the station under analysis is 75 kN · m.

Solution
1. Calculate the equivalent steel beam thickness.
$$n = E_{st}/E_w$$
where $\qquad\qquad\qquad E_{st} = 200 \times 10^3$ MPa

$\qquad\qquad\qquad\qquad E_w = 9.0 \times 10^3$ MPa

So $\qquad\qquad\qquad n = (200 \times 10^3 \text{ MPa})/(9.0 \times 10^3 \text{ MPa}) = 22.22$

So the thickness of the vertical steel member (to replace the wood) is 200 mm/22.22 = 9 mm. See Figure 14.18(b).

2. To calculate the equivalent beam section moment of inertia, open the program 'Centroid'.

We show the input coordinates in Figure 14.18(c).

Enter the data as three rectangles, and read the moment of inertia about the horizontal centroidal axis as $I_x = 2.6$ E+08, i.e., $I_x = 2.6 \times 10^8$ mm^4. See Figure 14.19.

3. Calculate the bending stress in the steel.
$$M = 75 \text{ kN} \cdot \text{m} = 75{,}000 \text{ N} \cdot \text{m} = 75 \times 10^6 \text{ N} \cdot \text{mm}$$
$$y = (200 \text{ mm} - 15 \text{ mm}) = 185 \text{ mm} \qquad (\text{Figure } 14.18(\text{b}))$$
$$s_{Bst} = My/I = (75 \times 10^6 \text{ N mm})(185 \text{ mm})/(2.6 \times 10^8 \text{ mm}^4)$$
$$= 53.4 \text{ MPa}$$

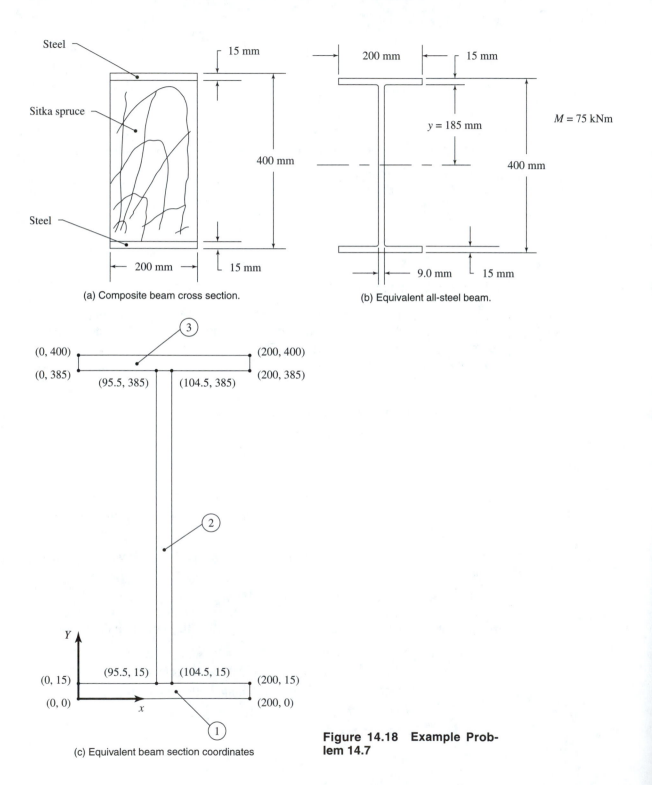

Steel

Sitka spruce

Steel

15 mm

400 mm

200 mm   15 mm

(a) Composite beam cross section.

200 mm   15 mm

$y = 185$ mm

400 mm

9.0 mm   15 mm

$M = 75$ kNm

(b) Equivalent all-steel beam.

③

(0, 400)   (200, 400)

(0, 385)   (200, 385)

(95.5, 385)   (104.5, 385)

②

Y

(95.5, 15)   (104.5, 15)

(0, 15)   (200, 15)

(0, 0)   (200, 0)

x

①

(c) Equivalent beam section coordinates

Figure 14.18   Example Problem 14.7

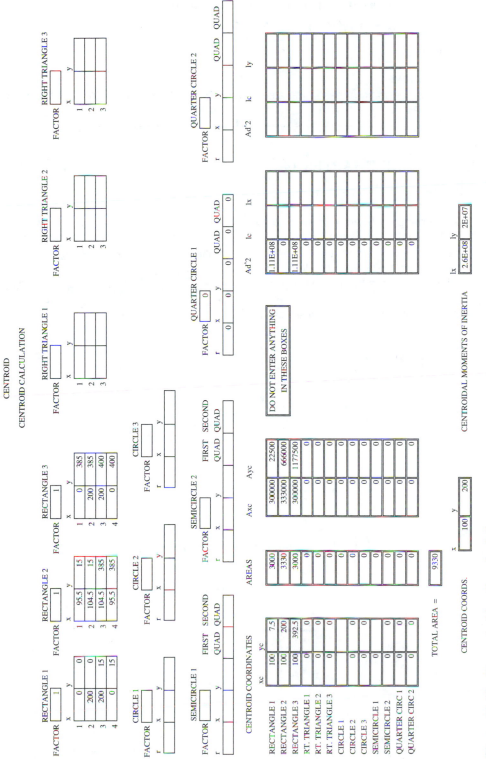

Figure 14.19 Moment of inertia calculation.

443

4. Calculate the bending stress in the wood.

$$s_{Bw} = (1/n)(My/I) = (1/22.22)(75 \times 10^6 \text{ N mm})(185 \text{ mm})/(2.6 \times 10^8 \text{ mm}^4)$$
$$= 2.4 \text{ MPa}$$

Answer

The bending stress in the steel is 53.4 MPa. The bending stress in the wood is 2.4 MPa.

14.5   PROBLEMS FOR CHAPTER 14

Hand Calculations

1. A 5-ft-long overhanging beam carries loads of 1000 lb at the left end of the beam and 3000 lb at a station 3.5 ft from the left end. The supports are at 1.5 ft from the left end and at the right end of the beam (see Figure 14.20). The beam cross section is also shown in Figure 14.20. Calculate the hori-zontal shear stress at the midpoint of the beam and 2 in from the upper surface.

2. Determine the maximum horizontal shear stress in the beam shown in Figure 14.20.

3. Calculate the maximum horizontal shear stress in the round beam shown in Figure 14.21.

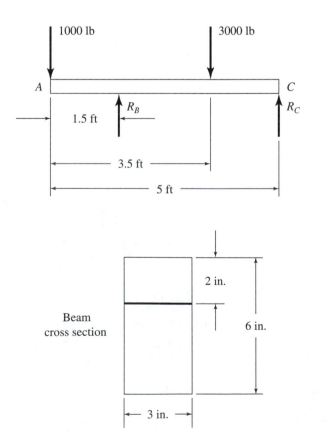

Figure 14.20

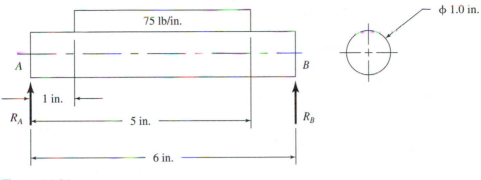

Figure 14.21

4. Calculate the bending stress in the beam shown in Figure 14.20 at the same location as in Problem 1.

5. Determine the maximum bending stress in the beam shown in Figure 14.20.

6. Determine the maximum bending stress in the round beam shown in Figure 14.21.

7. Calculate the maximum bending stress in the round beam shown in Figure 14.22.

8. What should be the maximum length of a 10-

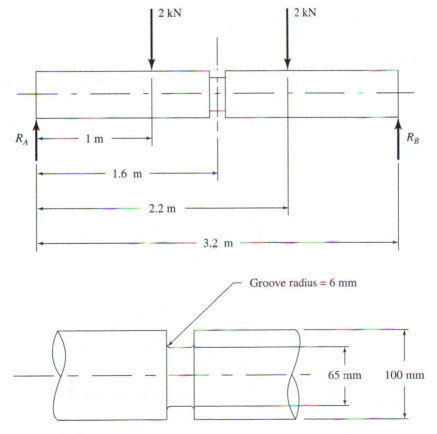

Figure 14.22

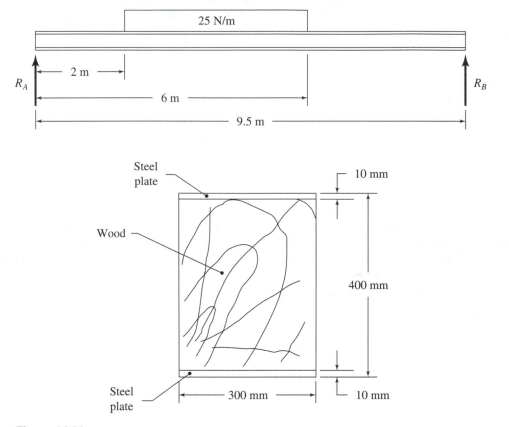

Figure 14.23

by 12-in. wooden beam carrying a uniformly distributed load of 1750 lb/ft over its entire length? The beam is simply supported at its ends. The allowable shear stress is 110 psi, and the allowable bending stress is 1400 psi. The standard dressed dimensions of a 10- by 12-in. wooden beam are 9.5 in. by 11.5 in.

Computer Calculations

11. Write a program to calculate the stress concentration factor for bending for a grooved round bar. The input variables are D, d, and r.
12. Repeat Problem 1 using any computer program(s) to assist in its solution.
13. Repeat Problem 2 using any computer program(s) to assist in its solution.

9. Repeat Problem 8 for a cantilevered beam.
10. Calculate the maximum bending stress in the steel and in the wood for the composite beam shown in Figure 14.23. The wood is ponderosa pine, for which $E_w = 7.6 \times 10^3$ MPa. For steel, $E_{st} = 200 \times 10^3$ MPa.

14. Repeat Problem 4 using any computer program(s) to assist in its solution.
15. Repeat Problem 5 using any computer program(s) to assist in its solution.
16. Compute the maximum bending stress in the aluminum alloy and in the epoxy for the composite beam shown in Figure 14.24. E for

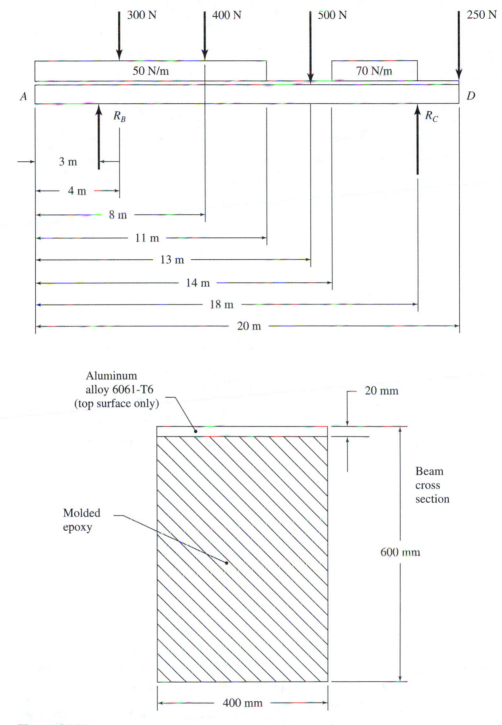

Figure 14.24

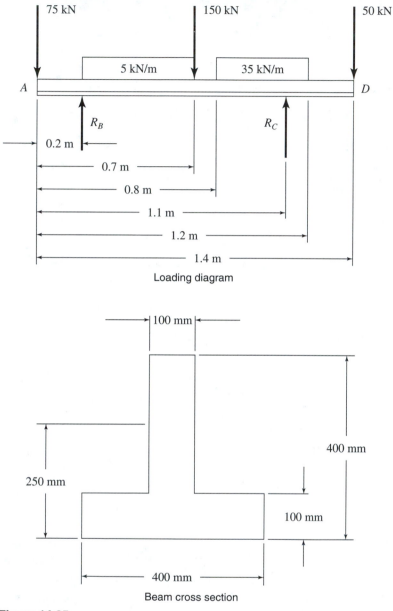

Loading diagram

Beam cross section

Figure 14.25

molded epoxy is 20.7 GPa (20.7 × 10³ MPa), and for 6061-T6 aluminum alloy, E = 69 GPa.

Note: The aluminum alloy is bonded to the *top surface* only.

Hint: The equivalent all-aluminum alloy beam cross section will be a tee.)

17. Determine the horizontal shear stress at the midstation of the beam shown in Figure 14.25 at a height of 250 mm from the lower surface.

18. Repeat the previous problem to determine the horizontal shear stress at the midstation at a height of 100 mm from the lower surface.

19. Write a computer program to calculate and to plot the distribution of horizontal shear stress across any section of a rectangular-cross-section beam that is simply supported and carries one concentrated load.

20. Use the program developed in Problem 19 to check the results of Example Problem 14.2.

15

Deflection of Beams

■ *Objectives*

In this chapter you will learn about beam deflections and how to calculate the deflections of beams with a variety of loadings.

Up to this point in our study of beams, we have learned how to calculate shear and bending stresses. These calculations, together with data giving material allowable stresses, enable us to design beams that will be fail-safe. There is, however, another aspect to beam design, and that is the amount the beam bends. For example, the beams supporting a bowling alley floor must be designed for the absolute minimum of deflection. Ceiling beams supporting plaster should be limited to a maximum deflection of $\frac{1}{360}$ of the span to avoid cracking of the plaster. For high-precision machine parts, the maximum deflection should be limited to $\frac{1}{100,000}$ of the span. You can see that for certain designs the stresses are well below the allowable, because deflection is the design criterion.

15.1 BEAM DEFLECTION THEORIES

One method available for deriving beam deflection formulas is known as the *double integration method* (see Using Calculus).

Another method, the *moment-area method*, uses trigonometry, and its development is presented here. We demonstrate how it can be used to derive some of the more common deflection formulas.

Remembering Chapter 13, we used the area under the shear force diagram to calculate bending moments. The moment-area method for deflections is similar, but it uses the area under the *bending moment diagram* to calculate deflections.

15.1.1 The Moment-Area Method

It is important to understand that actual beam deflections are generally very small compared to the length of the beam and that all the following diagrams are greatly exaggerated. With this in mind, Figure 15.1 shows a small part of a loaded beam. We are going to consider what has happened to the element $PQRS$, which was originally a rectangular block, now that the beam has bent. Point P has moved to P', due to the compression of the upper surface, and point Q has moved to Q' due to the stretching of the lower surface. The length between T and U, Δx, which is on the neutral axis, has not changed.

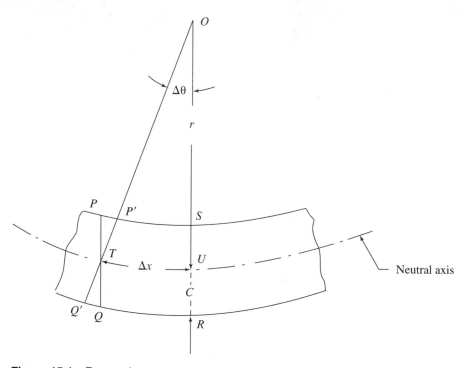

Figure 15.1   Beam element.

We now extend the lines RS and $Q'P'$ (the sides of the element) until they meet at point O. In reality, the angle at point O, $\Delta\theta$, is very small. If we call the distance from O to the neutral axis r, then the length Δx is

$$\Delta x = r \cdot \Delta\theta \qquad (15.1)$$

where $\Delta\theta$ is measured in radians.

The distance from O to the lower surface of the beam (see Figure 15.1) is $(r + c)$, so the length RQ' is

$$RQ' = (r + c) \cdot \Delta\theta$$

Before the beam was loaded, the length of RQ' was RQ, which is the same length as Δx. So the amount that RQ has stretched is

$$\begin{aligned} \delta &= RQ' - RQ \\ &= RQ' - \Delta x \\ &= (r + c) \cdot \Delta\theta - r \cdot \Delta\theta \\ &= c \cdot \Delta\theta \end{aligned}$$

The strain, which is the extension divided by the original length, is

$$\begin{aligned} \varepsilon &= \delta/l \\ &= \delta/\Delta x \\ &= (c \cdot \Delta\theta)/(r \cdot \Delta\theta) \end{aligned}$$

so

$$\varepsilon = c/r$$

Relating this now to the stress being created, we have

$$\text{stress} = \text{strain} \times \text{elastic modulus}$$

$$s = \varepsilon E$$

$$= cE/r$$

But we know from the flexure formula that

$$s = Mc/I$$

Equating these,

$$cE/r = Mc/I$$

$$E/r = M/I$$

so

$$r = EI/M \qquad (15.2)$$

The variable r is called the *radius of curvature* and is the radius of the circle that would match the shape of the curve at and in the immediate neighborhood of the point. In words, it tells us that the radius of curvature at any beam location can be expressed as the elastic modulus times the moment of inertia divided by the bending moment at that location.

Because deflections are usually very small, we would expect the radius of curvature to be large. Let us put this in perspective with a simple example.

EXAMPLE PROBLEM 15.1

Calculate the radius of curvature of the hemlock beam shown in Figure 15.2 at its midpoint. Its modulus of elasticity is 1300 ksi.

Solution

The expression for the radius of curvature is

$$r = EI/M$$

1. Calculate the moment of inertia about the neutral axis. For a rectangular cross section,

$$I = bd^3/12$$

$$b = 2 \text{ in.} \qquad d = 4 \text{ in.}$$

$$I = (2 \text{ in.})(4 \text{ in.})^3/12 = 10.67 \text{ in.}^4$$

2. Calculate the bending moment at $x = 7$ ft. The reactions R_1 and R_2 are equal because the load is at the midpoint of the beam. They share the load equally, so

$$R_1 = 300 \text{ lb}$$

The bending moment at the midpoint of the beam is

$$M = (7 \text{ ft})(300 \text{ lb}) = 2100 \text{ lb-ft}$$

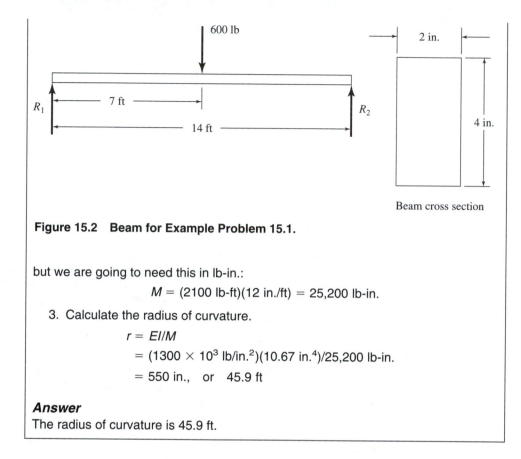

Figure 15.2   Beam for Example Problem 15.1.

but we are going to need this in lb-in.:

$$M = (2100 \text{ lb-ft})(12 \text{ in./ft}) = 25,200 \text{ lb-in.}$$

3. Calculate the radius of curvature.

$$r = EI/M$$
$$= (1300 \times 10^3 \text{ lb/in.}^2)(10.67 \text{ in.}^4)/25,200 \text{ lb-in.}$$
$$= 550 \text{ in., or } 45.9 \text{ ft}$$

Answer
The radius of curvature is 45.9 ft.

Figure 15.3(a) shows the deflection curve (the neutral axis curve) of a loaded beam, and again we have marked two infinitesimally close points, T and U. The subtended angle (that is, the angle made between the two sides of the element) is again $\Delta\theta$. If we draw two tangents to the deflection curve at T and at U, the angle between them will also be $\Delta\theta$. This is because a tangent at a point is always at right angles to the radius of curvature at that point, so as the angle at O goes from zero to $\Delta\theta$ as we move from U to T, then the tangent angle goes through the same angle, $\Delta\theta$. Projecting these tangents back to the left end of the beam, the vertical distance between them is Δy (see Figure 15.3(a)). If we move back along the beam, taking little elements and projecting their tangents back to A, then the sum of all the Δy's will be y_{max}, the maximum deflection (see Figure 15.3(c)).

The height is $\Delta y = (x \cdot \Delta\theta)$, where x is the distance from the start of the beam to the element being considered. So,

$$y_{max} = \sum(x \cdot \Delta\theta) \tag{15.3}$$

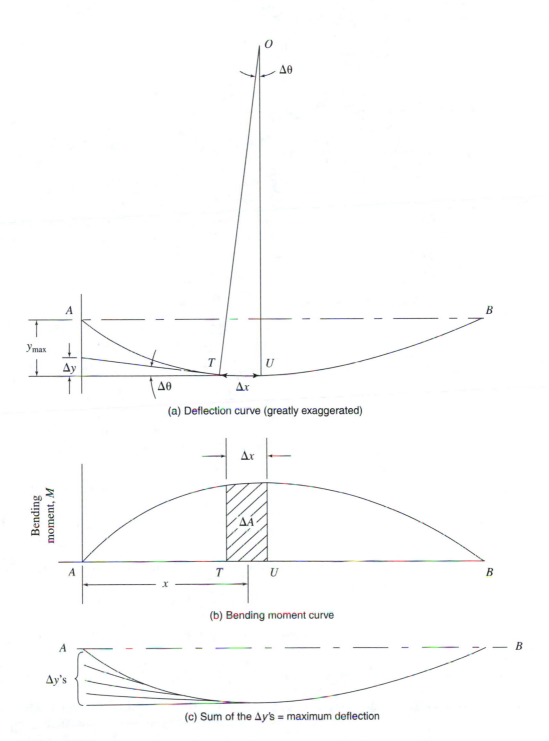

(a) Deflection curve (greatly exaggerated)

(b) Bending moment curve

(c) Sum of the Δy's = maximum deflection

Figure 15.3

Earlier we derived the equations

$$\Delta x = r \cdot \Delta \theta, \tag{15.1}$$

and

$$r = EI/M \tag{15.2}$$

From equation (15.1) we have

$$r = \Delta x / \Delta \theta$$

Substituting for r in equation (15.2):

$$\Delta x / \Delta \theta = EI/M$$

so

$$\Delta \theta = M \cdot \Delta x/EI \tag{15.4}$$

Looking now at Figure 15.3(b), we see that the product ($M \cdot \Delta x$) is the shaded area ΔA, because, with Δx being infinitesimally thin, the shaded area is a rectangle of height M and width Δx. We can, therefore, write

$$\Delta \theta = \Delta A/EI$$

which, on substituting into equation (15.3), gives

$$y_{max} = \sum(x \cdot \Delta)/EI$$

For a homogeneous beam of constant cross section, both E and I are constant over the whole length, so the preceding equation becomes

$$y_{max} = \sum(x \cdot \Delta A/EI)$$

We know from equation (11.1) that

$$\sum(x \cdot \Delta A) = \bar{x} \cdot \sum \Delta A = \bar{x}A$$

where $\bar{x}$ is the distance to the centroid of area A.

A is the total area under the bending moment curve from the start to the station where the deflection is required. Therefore, the formula for maximum deflection is

$$y_{max} = \bar{x}A/EI \tag{15.5}$$

where

$\quad A =$ the area under the bending moment diagram between the left reaction
$\qquad$ and the station of maximum deflection.
$\quad \bar{x} =$ the horizontal distance from the left reaction to the centroid of area A
$\quad E =$ the material modulus of elasticity
$\quad I =$ the section moment of inertia of the beam

15.2   DEFLECTIONS OF SIMPLY SUPPORTED BEAMS

We will use equation (15.5) to derive some of the basic deflection formulas for simply supported beams. The method is the same in each case—draw the bending moment diagram, determine the product ($A\bar{x}$), and use equation (15.5).

EXAMPLE PROBLEM 15.2

Derive a formula for the maximum deflection of a simply supported beam of length L that carries a single concentrated load, F, at its midstation (Figure 15.4(a)).

Solution

1. Determine the reactions. Because the loading is symmetrical, $R_1 = R_2 = F/2$.
2. Draw the shear force diagram. The shear force value $F/2$ is constant from the left support to the midstation, where it drops to $-F/2$. It remains constant at $-F/2$ until the right reaction brings it back to zero. (See Figure 15.4(b)).

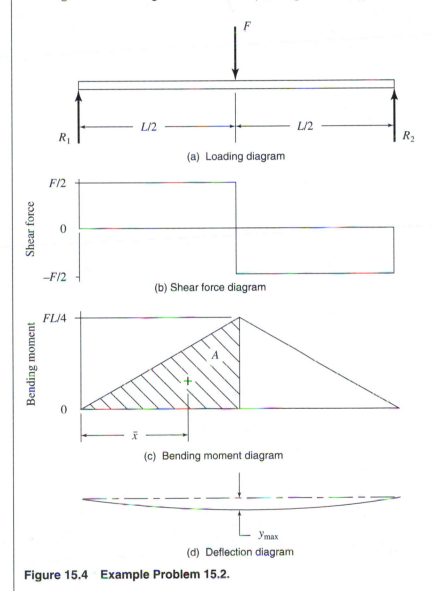

(a) Loading diagram

(b) Shear force diagram

(c) Bending moment diagram

(d) Deflection diagram

Figure 15.4 Example Problem 15.2.

3. Draw the bending moment diagram. Using the area method, the area under the first half of the shear force diagram is $(F/2)(L/2) = FL/4$. So the bending moment diagram (Figure 15.4(c)) starts at zero, increases linearly to $FL/4$ at the mid-station, and then falls linearly to zero at the right end of the beam.

4. Determine the area A and the horizontal position of its centroid. The shaded area, A (Figure 15.4(c)), is

$$A = \text{(base)(height)}/2 = (L/2)(FL/4)/2 = FL^2/16$$

The horizontal position of its centroid, i.e., $\bar{x}$ is

$$\bar{x} = \tfrac{2}{3}(L/2) = L/3$$

Note: The centroid of a triangle is one-third of the length from the base, or two-thirds the length from the apex.

5. Derive the deflection formula.

$$y_{max} = \bar{x}A/EI$$
$$= (L/3)(FL^2/16)/EI$$
$$y_{max} = FL^3/48EI \tag{15.6}$$

Answer
$y_{max} = FL^3/48EI$

EXAMPLE PROBLEM 15.3

Derive a formula for the maximum deflection of a simply supported beam of length L that carries two concentrated loads at stations $L/3$ and $2L/3$. See Figure 15.5(a).

Solution

1. Determine the reactions. Since the loading is symmetrical, $R_1 = R_2 = F$.

2. Draw the shear force diagram. The shear force value of F is constant from the left support to station $L/3$, where it drops to zero. It remains at zero until station $2L/3$, where it drops to $-F$. It remains constant at $-F$ until the right reaction brings it back to zero (see Figure 15.5(b)).

Note: There is no shear force—and, therefore, no shear stress—over the midspan. This condition is known as *pure bending*.

3. Draw the bending moment diagram. Using the area method, the area under the shear force diagram from station 0 to $L/3$ is $(F)(L/3) = FL/3$. So the bending moment diagram (Figure 15.5(c)) starts at zero, increases linearly to $FL/3$ at station $L/3$, remains constant to station $2L/3$, and then falls linearly to zero at the right end of the beam.

4. Determine the area A and the horizontal position of its centroid. Divide area A into two regions, 1 and 2, as in Figure 15.5(c).

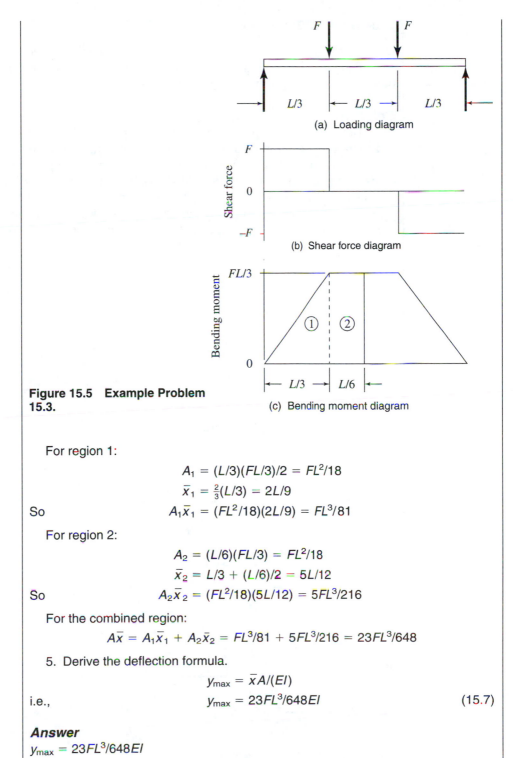

Figure 15.5   Example Problem 15.3.

(a) Loading diagram

(b) Shear force diagram

(c) Bending moment diagram

For region 1:
$$A_1 = (L/3)(FL/3)/2 = FL^2/18$$
$$\bar{x}_1 = \tfrac{2}{3}(L/3) = 2L/9$$
So
$$A_1\bar{x}_1 = (FL^2/18)(2L/9) = FL^3/81$$

For region 2:
$$A_2 = (L/6)(FL/3) = FL^2/18$$
$$\bar{x}_2 = L/3 + (L/6)/2 = 5L/12$$
So
$$A_2\bar{x}_2 = (FL^2/18)(5L/12) = 5FL^3/216$$

For the combined region:
$$A\bar{x} = A_1\bar{x}_1 + A_2\bar{x}_2 = FL^3/81 + 5FL^3/216 = 23FL^3/648$$

5. Derive the deflection formula.
$$y_{max} = \bar{x}A/(EI)$$
i.e.,
$$y_{max} = 23FL^3/648EI \qquad\qquad (15.7)$$

Answer
$$y_{max} = 23FL^3/648EI$$

EXAMPLE PROBLEM 15.4

Derive a formula for the maximum deflection of a simply supported beam of length *L*, carrying two concentrated loads, one at a distance *a* from the left end and the other at a distance *a* from the right end. See Figure 15.6(a).

Solution

1. Determine the reactions. Since the loading is symmetrical, $R_1 = R_2 = F$.

2. Draw the shear force diagram. The only difference between this loading (Figure 15.6(a)) and that of the previous problem (Figure 15.5(a)) is the spacing of the loads. The shear force, *F* (Figure 15.6(b)), will be constant for a distance *a*. It will be zero in the region between the two forces and then $-F$ to the end of the beam.

3. Draw the bending moment diagram. The bending moment will increase linearly from zero to the value (*aF*) at station *a*, remain constant to station ($L - a$), and then fall linearly to zero at the end of the beam.

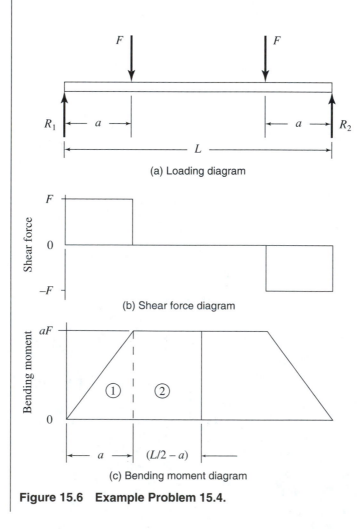

(a) Loading diagram

(b) Shear force diagram

(c) Bending moment diagram

Figure 15.6   Example Problem 15.4.

4. Determine the area A and the horizontal position of its centroid. Divide area A into two regions, 1 and 2, as in Figure 15.6(c).

For region 1:

$$A_1 = (a)(aF)/2 = a^2F/2$$
$$\bar{x}_1 = \tfrac{2}{3}(a) = 2a/3$$
$$A_1\bar{x}_1 = (a^2F/2)(2a/3) = a^3F/3$$

For region 2:

$$A_2 = (aF)(L/2 - a)$$
$$\bar{x}_2 = a + (L/2 - a)/2 = (L/2 + a)/2$$
$$A_2\bar{x}_2 = (aF)(L/2 - a)(L/2 + a)/2 = (aF)(L^2/4 - a^2)/2$$

For the combined region:

$$Ax = A_1\bar{x}_1 + A_2\bar{x}_2 = a^3F/3 + (aF)(L^2/4 - a^2)/2$$
$$= (aF)(a^2/3 + L^2/8 - a^2/2)$$
$$= (aF)(L^2/8 - a^2/6)$$
$$= (aF)(3L^2 - 4a^2)/24$$

5. Derive the deflection formula.

$$y_{max} = \bar{x}A/EI$$
$$= (aF)(3L^2 - 4a^2)/24EI \qquad (15.8)$$

Answer

$$y_{max} = (aF)(3L^2 - 4a^2)/24EI$$

EXAMPLE PROBLEM 15.5

Derive a formula for the maximum deflection of a simply supported beam of length L that carries a distributed load of ω per unit length. See Figure 15.7(a).

Solution

The bending moment at any station x (see Figure 15.7(a)) consists of the difference between the moment (R_1x) and the moment due to that part of the load to the left of station x. The latter is a load of (ωx) whose center of gravity is at station $x/2$. The moment due to this load is $-(\omega x)(x/2) = -\omega x^2/2$. The bending moment at x is, therefore,

$$M = R_1x - \omega x^2/2$$

Because the beam is loaded symmetrically, $R_1 = R_2 =$ half the total load. So $R_1 = \omega L/2$. Therefore,

$$M = \omega Lx/2 - \omega x^2/2 \qquad (15.9)$$

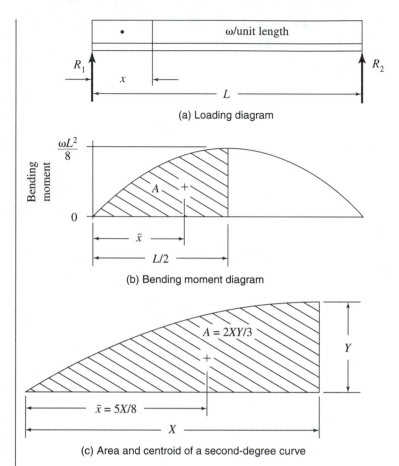

(a) Loading diagram

(b) Bending moment diagram

(c) Area and centroid of a second-degree curve

Figure 15.7   Example Problem 15.5.

The maximum bending moment occurs at $x = L/2$, so

$$M_{max} = \omega L(L/2)/2 - \omega(L/2)^2/2 = \omega L^2/8 \qquad \text{(Figure 15.7(b))}$$

We now need to determine the area under the bending moment diagram from 0 to $L/2$ and the position of the centroid of that area.

Equation (15.9) is a *second-degree* curve, which means that the highest power of x is 2. The area under a second-degree curve is $2XY/3$, and its centroid is at a horizontal distance $5X/8$, as shown in Figure 15.7(c). Referring to Figure 15.7(b),

$$X = L/2 \quad \text{and} \quad Y = \omega L^2/8$$

So

$$A = 2(L/2)(\omega L^2/8)/3 = \omega L^3/24$$

and

$$\bar{x} = 5(L/2)/8 = 5L/16$$

Therefore,

$$A\bar{x} = (\omega L^3/24)(5L/16) = 5\omega L^4/384$$

This gives

$$y_{max} = 5\omega L^4/(384EI) \tag{15.10}$$

We can express this in terms of the total load as

$$y_{max} = 5WL^3/(384EI) \tag{15.10(a)}$$

where W is the total load, i.e., $W = \omega L$.

Answer
$y_{max} = 5WL^3/(384EI)$

EXAMPLE PROBLEM 15.6

A 10-ft pine beam 2 in. $\times$ 4 in. is simply supported and carries a distributed load of 10 lb/ft. What will the maximum deflection be if

(a) It is placed flat,
(b) It is placed on its edge?

The modulus of elasticity for pine is 1.1×10^6 psi.

Note: A standard 2 in. $\times$ 4 in. beam has dressed dimensions $1\frac{1}{2}$ in. $\times$ $3\frac{1}{2}$ in.

Solution
(a) $I = bh^3/12 = (3.5 \text{ in.})(1.5 \text{ in.})^3/12 = 0.984 \text{ in.}^4$

From equation (15.10(a)),

$y_{max} = 5WL^3/(384EI)$

$W = -100 \text{ lb}, \quad L = 120 \text{ in.}, \quad E = 1.1 \times 10^6 \text{ lb/in.}^2, \quad I = 0.984 \text{ in.}^4$

$$y_{max} = \frac{5(-100 \text{ lb})(120 \text{ in.})^3}{(384)(1.1 \times 10^6 \text{ lb/in.}^2)(0.984 \text{ in.}^4)}$$

$$= -2.08 \text{ in.}$$

(b) $I = bh^3/12 = (1.5 \text{ in.})(3.5 \text{ in.})^3/12 = 5.36 \text{ in.}^4$

$y_{max} = 5WL^3/(384EI)$

$= 5(-100 \text{ lb})(120 \text{ in.})^3/[(384)(1.1 \times 10^6 \text{ lb/in.}^2)(5.36 \text{ in.}^4)]$

$= -0.38 \text{ in.}$

which is less than one-fifth the deflection of the flat beam.

Answer
(a) $y_{max} = -2.08 \text{ in.}$
(b) $y_{max} = -0.38 \text{ in.}$

15.3   DEFLECTIONS OF CANTILEVER BEAMS

We now derive some of the basic deflection formulas for cantilever beams.

EXAMPLE PROBLEM 15.7

Derive a formula for the maximum deflection of a cantilever beam of length L, carrying a single concentrated load, F, at its free end (Figure 15.8(a)).

Solution

The shear force diagram, Figure 15.8(b), is simply a rectangle whose area is $-FL$. Therefore, the bending moment diagram, Figure 15.8(c), is a triangle going from zero to $-FL$. The negative bending moment indicates that the top of the beam is in tension, not compression, as for a simply supported beam.

$$A = (L)(FL)/2 = FL^2/2$$
$$\bar{x} = \tfrac{2}{3}L = 2L/3$$
$$A\bar{x} = FL^3/3$$

So
$$y_{max} = FL^3/(3EI) \tag{15.11}$$

Answer

$y_{max} = FL^3/(3EI)$

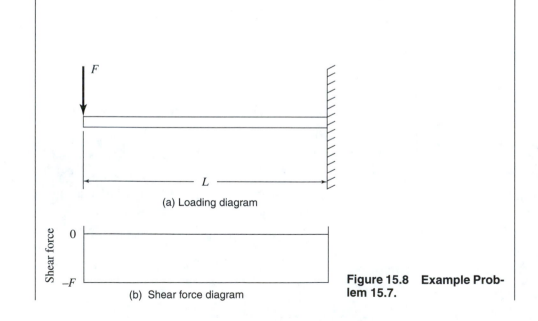

(a) Loading diagram

(b) Shear force diagram

Figure 15.8   Example Problem 15.7.

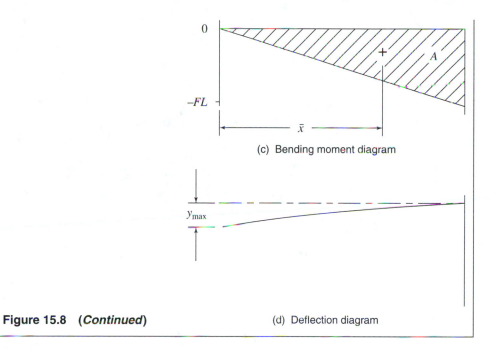

(c) Bending moment diagram

Figure 15.8 (*Continued*)   (d) Deflection diagram

EXAMPLE PROBLEM 15.8

Derive a formula for the maximum deflection of a cantilever beam of length L, carrying a single concentrated load, F, at a distance a from the free end (Figure 15.9(a)).

Solution

The area under the shear force diagram, Figure 15.9(b), is $-F(L - a)$. The corresponding bending moment diagram, Figure 15.9(c), is zero up to station $x = a$; then

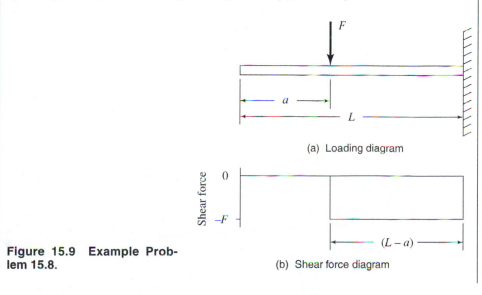

(a) Loading diagram

Figure 15.9   Example Problem 15.8.

(b) Shear force diagram

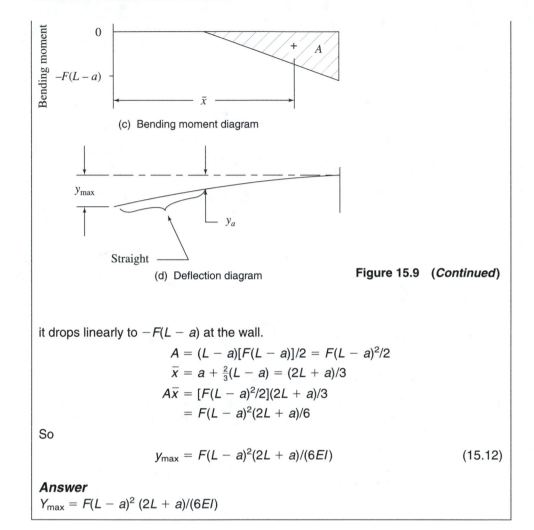

(c) Bending moment diagram

(d) Deflection diagram

Figure 15.9 (*Continued*)

it drops linearly to $-F(L - a)$ at the wall.

$$A = (L - a)[F(L - a)]/2 = F(L - a)^2/2$$
$$\bar{x} = a + \tfrac{2}{3}(L - a) = (2L + a)/3$$
$$A\bar{x} = [F(L - a)^2/2](2L + a)/3$$
$$= F(L - a)^2(2L + a)/6$$

So

$$y_{max} = F(L - a)^2(2L + a)/(6EI) \qquad (15.12)$$

Answer
$$Y_{max} = F(L - a)^2 (2L + a)/(6EI)$$

Note: The formula for the deflection at a, y_a, can be derived by modifying the formula in Example Problem 15.8. The only modification required is to measure $\bar{x}$ from station a instead of from the end of the beam.

$$\bar{x} = \tfrac{2}{3}(L - a)$$

So
$$A\bar{x} = [F(L - a)^2/2][\tfrac{2}{3}(L - a)] = F(L - a)^3/3$$
$$y_a = F(L - a)^3/(3EI) \qquad (15.13)$$

Notice that, since there are no shear or bending stresses from station a to the free end, there is nothing to bend the beam in this region. It is a straight line from y_a to y_{max}, Figure 15.12(d).

EXAMPLE PROBLEM 15.9

Derive a formula for the maximum deflection of a cantilever beam of length L, carrying a distributed load of ω per unit length over the entire span (Figure 15.10(a)).

Solution

The bending moment at station x (Figure 15.10(a)) is the load to the left of x multiplied by the distance of the center of gravity of that portion of the load to x. This is,

$$M = -(\omega x)(x/2) = -\omega x^2/2 \tag{15.14}$$

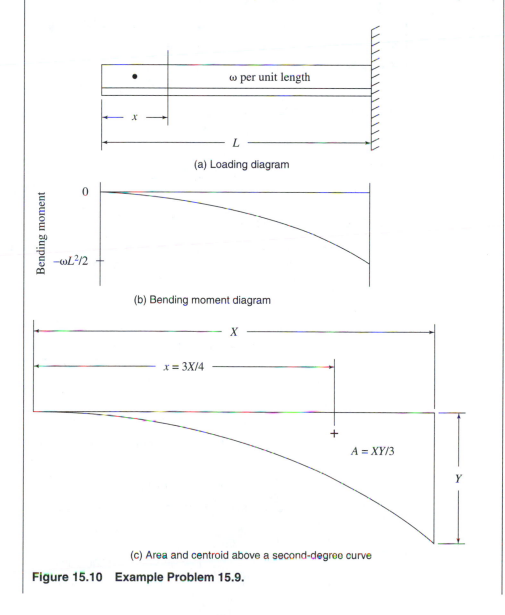

(a) Loading diagram

(b) Bending moment diagram

(c) Area and centroid above a second-degree curve

Figure 15.10   Example Problem 15.9.

The bending moment at the wall, i.e., at $x = L$, is

$$M_{max} = -\omega L^2/2$$

Equation (15.14) is a second-degree equation, but this time we need the area above the curve (see Figure 15.10(c)), which shows the area and the position of the centroid for the region above a second-degree curve.)

$$A = XY/3 \qquad \bar{x} = 3X/4$$

Here $X = L$ and $Y = \omega L^2/2$. Therefore, $A = (L)(\omega L^2/2)/3 = \omega L^3/6$

and

$$\bar{x} = 3L/4$$

$$A\bar{x} = (\omega L^3/6)(3L/4) = \omega L^4/8$$

So

$$y_{max} = \omega L^4/(8EI) \tag{15.15}$$

Expressing this in terms of the total load,

$$y_{max} = WL^3/(8EI) \tag{15.15(a)}$$

where $W = \omega L$, the total load.

Answer

$y_{max} = WL^3/(8EI)$

EXAMPLE PROBLEM 15.10

The 2-m cantilever beam shown in Figure 15.11(a) carries a mass of 100 kg at its midstation. It is a structural steel I-beam with a section moment of inertia of 212×10^6 mm^4. Determine the deflection 25 cm from the free end.

Solution
Because there is no load between the free end of the beam and the midpoint, this 1-m portion of the beam will be straight. The calculation procedure is to determine the deflection at the midpoint and at the free end and then linearly interpolate to find the deflection at the 25-cm station.

1. Calculate the load.

$$F = (-100 \text{ kg})(9.81 \text{ m}^2/\text{s}) = -981 \text{ N}$$

2. Calculate the deflection at the load. From equation (15.13),

$$y_a = F(L - a)^3/(3EI)$$

E for structural steel is 200 GPa (200×10^3 MPa).

$$L = 2 \text{ m} = 2000 \text{ mm}, \qquad a = 1 \text{ m} = 1000 \text{ mm}$$

$$y_{1\,m} = \frac{F(L - a)^3}{3EI}$$

$$= \frac{(-981 \text{ N})(2000 \text{ mm} - 1000 \text{ mm})^3}{(3)(200 \times 10^3 \text{ N/mm}^2)(212 \times 10^6 \text{ mm}^4)}$$

$$= -0.0077 \text{ mm}$$

3. Calculate the maximum deflection (i.e., at the free end). From equation (15.12):

$$y_{max} = \frac{F(L - a)^2(2L + a)}{(6EI)}$$

$$= \frac{(-981 \text{ N})(2000 \text{ mm} - 1000 \text{ mm})^2(2 \times 2000 \text{ mm} + 1000 \text{ mm})}{(6)(200 \times 10^3 \text{ N/mm}^2)(212 \times 10^6 \text{ mm}^4)}$$

$$= -0.0193 \text{ mm}$$

4. Calculate the deflection at $x = 25$ cm. The deflection diagram (Figure 15.11(b)) is a straight line from $x = 0$ (the free end) to $x = 1$ m. 25 cm (0.25 m) is one-fourth of the way from the free end to the load, so the deflection there is

$$y_{25 \text{ cm}} = y_{max} - (y_{max} - y_{1 \text{ m}})/4$$

$$y_{1 \text{ m}} = -0.0077 \text{ mm} \quad \text{and} \quad y_{max} = -0.0193 \text{ mm}$$

$$y_{25 \text{ cm}} = -0.0193 \text{ mm} - [-0.0193 \text{ mm} - (-0.0077 \text{ mm})]/4$$

$$= -0.0193 \text{ mm} - (-0.0029 \text{ mm})$$

$$= -0.0193 \text{ mm} + 0.0029 \text{ mm}$$

$$= 0.016 \text{ mm}$$

Answer

The deflection at $x = 25$ cm is 0.016 mm.

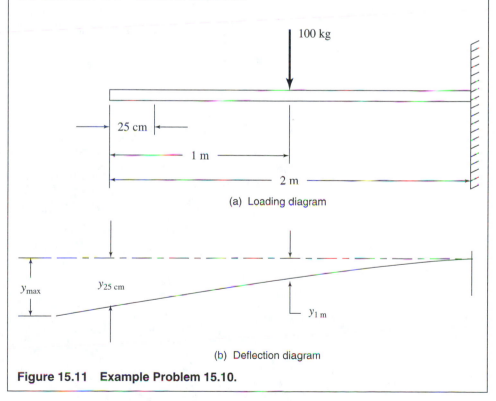

(a) Loading diagram

(b) Deflection diagram

Figure 15.11   Example Problem 15.10.

USING CALCULUS

It can be shown that the radius of curvature, r, at any point (x, y) of a curve is given by the equation

$$1/r = (d^2y/dx^2)/[1 + (dy/dx)^2]^{3/2} \qquad (1)$$

We make the assumption that the slopes are small, i.e., dy/dx is small. The term $(dy/dx)^2$ is, therefore, negligible and equation (1) becomes

$$1/r = (d^2y/dx^2) \qquad (2)$$

We have shown (equation (15.2)) that

$$r = EI/M$$

Therefore,

$$d^2y/dx^2 = M/EI \qquad (3)$$

Integrating once gives

$$dy/dx = \int M/(EI)\ dx + C \qquad (4)$$

This is the basic equation for the slope. C is a constant of integration. Integrating again gives the basic equation for the deflection:

$$y = \iint M/EI\ dx\ dx + \int C\ dx + D \qquad (5)$$

where D is the second integration constant.

EXAMPLE

Derive the slope and deflection equations for the loading shown in Figure 15.4(a).

Solution

Because the load, F, is at the midpoint of the beam, the reaction $R_1 = F/2$. Therefore, the shear force has a constant value of $V = F/2$ from $x = 0$ to $x = L/2$. In this region the bending moment, being the integral of the shear, is $M = (F/2)x$.

From equation (4) the slope at any station is

$$dy/dx = \int M/EI\ dx + C = \int (F/2/EI)x\ dx + C$$

If the beam is uniform, EI will be a constant value; therefore,

$$dy/dx = (F/2/EI) \int x\ dx + C$$
$$dy/dx = (F/2/EI)x^2/2 + C = Fx^2/4EI$$

The slope is zero at the midpoint of the beam, i.e., at $x = L/2$:

$$0 = F(L/2)^2/4EI + C$$
$$C = -FL^2/16EI$$

So

$$dy/dx = Fx^2/4EI - FL^2/16EI$$
$$dy/dx = (F/4EI)(x^2 - L^2/4)$$

The deflection at any point is given by equation (5):

$$y = \iint M/(EI) \, dx \, dx + \int C \, dx + D$$

$$= (F/2EI) \iint x \, dx \, dx - (FL^2/16EI) \int dx + D$$

$$= (F/2EI) \int (x^2/2) \, dx - (FL^2/16EI)x + D$$

$$= (F/12EI)x^3 - (FL^2/16EI)x + D$$

The deflection $= 0$ at $x = 0$; therefore, $D = 0$.

$$y = (F/12EI)x^3 - (FL^2/16EI)x$$

$$= (Fx/48EI)(4x^2 - 3L^2)$$

The maximum deflection occurs at $x = L/2$ (where $dy/dx = 0$):

$$y_{max} = (FL/96EI)(L^2 - 3L^2) = -FL^3/48EI$$

Answer

The slope equation is $dy/dx = (F/4EI)(x^2 - L^2/4)$. The deflection equation is $y = (Fx/48EI)(4x^2 - 3L^2)$. The maximum deflection is $y_{max} = -FL^3/48EI$ at $x = L/2$.

15.4 PRINCIPLE OF SUPERPOSITION

The cantilever beam shown in Figure 15.12(a) carries a concentrated load and a distributed load. The **principle of superposition** tells us that that the deflection at any station on the beam can be determined by calculating the deflection caused by the concentrated load alone and the deflection caused by the distributed load alone and adding the two together.

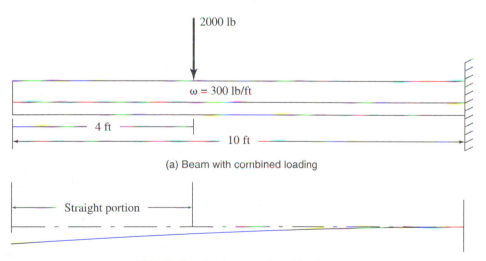

(a) Beam with combined loading

(b) Deflection due to concentrated load only

Figure 15.12   Example Problem 15.11.

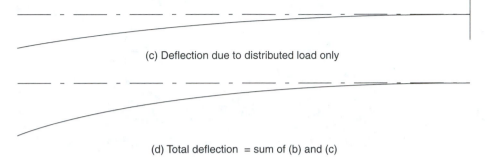

(c) Deflection due to distributed load only

(d) Total deflection = sum of (b) and (c)

Figure 15.12   (*Continued*)

EXAMPLE PROBLEM 15.11

Determine the deflection at the free end of the beam shown in Figure 15.12(a). The section moment of inertia is 152 in.^{4} and the modulus of elasticity is 30×10^6 psi.

Solution

1. Calculate the deflection due to the concentrated load alone. In Example Problem 15.8 we derived equation (15.12):

$$y_{max} = F(L - a)^2 (2L + a)/(6EI)$$

where

   F = concentrated load = -2000 lb
   L = beam length = 10 ft
   a = distance from the free end of the beam to the concentrated load = 4 ft
   y_{max} = $(-2000$ lb)(10 ft $-$ 4 ft)^2(2 $\times$ 10 ft + 4 ft)/6EI
      = $(-2000$ lb)(36 ft^2)(24 ft/6)/EI
      = $(-2.88 \times 10^5$ lb-ft^3)/EI

2. Calculate the deflection due to the distributed load alone. In Example Problem 15.9 we derived equation (15.15):

$$y_{max} = \omega L^4/8EI$$

where

   ω = distributed load = -300 lb/ft
   L = beam length = 10 ft
   y_{max} = $(-300$ lb/ft)(10 ft)^4/8EI
   y_{max} = $(-3.75 \times 10^5$ lb-ft^3)/EI

3. Use the principle of superposition to determine the total deflection. y_{max} for the combined loading is

   y_{max} = $(-2.88 \times 10^5$ lb-ft^3)/EI + $(-3.75 \times 10^5$ lb-ft^3)/EI
      = $(-6.63 \times 10^5$ lb-ft^3)/EI
      = -1.146×10^9 lb-in^3/EI

   E = 30×10^6 lb/in.^{2} I = 152 in.^{4}

Therefore,

$$(EI) = (30 \times 10^6 \text{ lb/in.}^2)(152 \text{ in.}^4) = 4.56 \times 10^9 \text{ lb-in.}^2$$

Finally,

$$y_{max} = (-1.146 \times 10^9 \text{ lb-in.}^3)/(4.56 \times 10^9 \text{ lb-in.}^2)$$
$$= -0.251 \text{ in.}$$

Answer

The deflection is -0.251 in.

15.5 DEFLECTION EQUATIONS

Figure 15.13 shows 16 beam loading configurations and their deflection equations. Configuration 1, for example, is a simply supported beam carrying a single concentrated load, and configuration 16 is a built-in beam carrying a distributed load.

Although these 16 configurations cover a wide variety of loading and support conditions, many more can be analyzed by using the principle of superposition. Equations for the maximum deflection and the location of the maximum deflection are given for many of the configurations.

15.5.1 Maximum Deflection

The maximum deflection of a *cantilever beam* carrying all downloads is at the free end, regardless of the locations of the loads. We may calculate the maximum deflection due

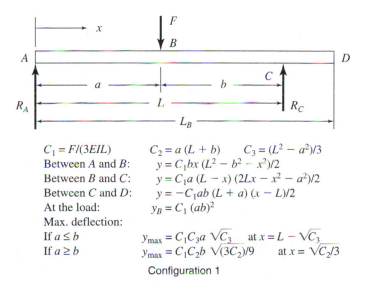

$C_1 = F/(3EIL)$ $C_2 = a (L + b)$ $C_3 = (L^2 - a^2)/3$

Between A and B: $y = C_1bx (L^2 - b^2 - x^2)/2$

Between B and C: $y = C_1a (L - x) (2Lx - x^2 - a^2)/2$

Between C and D: $y = -C_1ab (L + a) (x - L)/2$

At the load: $y_B = C_1 (ab)^2$

Max. deflection:

If $a \leq b$ $y_{max} = C_1C_3a \sqrt{C_3}$ at $x = L - \sqrt{C_3}$

If $a \geq b$ $y_{max} = C_1C_2b \sqrt{(3C_2)}/9$ at $x = \sqrt{C_2/3}$

Configuration 1

Figure 15.13

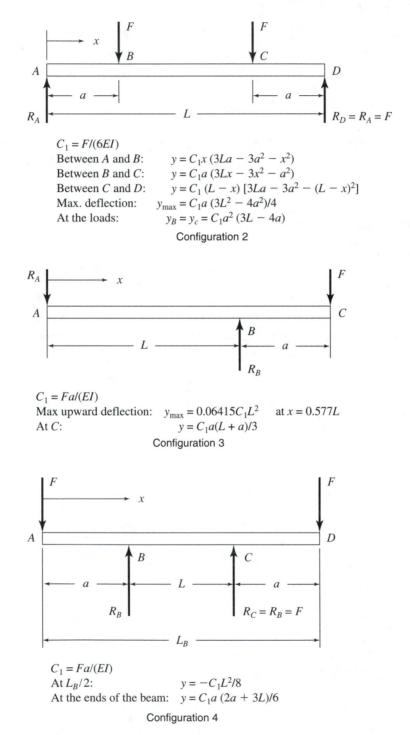

$C_1 = F/(6EI)$
Between A and B: $y = C_1x\,(3La - 3a^2 - x^2)$
Between B and C: $y = C_1a\,(3Lx - 3x^2 - a^2)$
Between C and D: $y = C_1\,(L - x)\,[3La - 3a^2 - (L - x)^2]$
Max. deflection: $y_{max} = C_1a\,(3L^2 - 4a^2)/4$
At the loads: $y_B = y_c = C_1a^2\,(3L - 4a)$

Configuration 2

$C_1 = Fa/(EI)$
Max upward deflection: $y_{max} = 0.06415C_1L^2$ at $x = 0.577L$
At C: $y = C_1a(L + a)/3$

Configuration 3

$C_1 = Fa/(EI)$
At $L_B/2$: $y = -C_1L^2/8$
At the ends of the beam: $y = C_1a\,(2a + 3L)/6$

Configuration 4

Figure 15.13 (*Continued*)

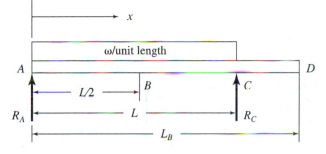

$C_1 = \omega/(24EI)$

Between A and B: $y = C_1 x(L^3 - 2Lx^2 + x^3)$

Between B and C: $y = C_1(L - x)[L^3 - 2L(L - x)^2 + (L - x)^3]$

Between C and D: $y = -C_1 L^3(x - L)$

At $L/2$: $y = 5C_1 L^4/16$

Configuration 5

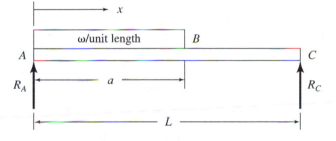

$C_1 = \omega/(24\,EI)$ $C_2 = a(2L - a)$

Between A and B: $y = C_1 x(C_2^2 - 2C_2 x^2 + Lx^3)/L$

Between B and C: $y = C_1 a(L - x)(4Lx - 2x^2 - a^2)/2$

Configuration 6

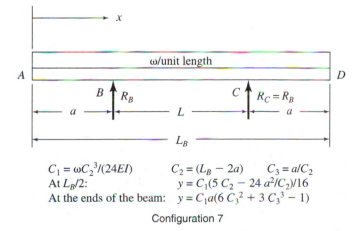

$C_1 = \omega C_2^3/(24EI)$ $C_2 = (L_B - 2a)$ $C_3 = a/C_2$

At $L_B/2$: $y = C_1(5\,C_2 - 24\,a^2/C_2)/16$

At the ends of the beam: $y = C_1 a(6\,C_3^2 + 3\,C_3^3 - 1)$

Configuration 7

Figure 15.13   *(Continued)*

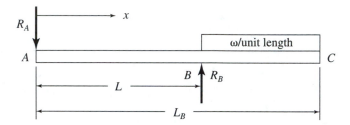

$C_1 = \omega(L_B - L)^2/EI$

Max. upward deflection: $y = -0.03208 C_1 L^2$ at $x = 0.577L$

At C: $y = C_1(L_B - L)(L + 3L_B)/24$

Configuration 8

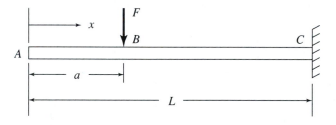

$C_1 = F/(6EI)$

Between A and B: $y = C_1(L - a)^2(2L - 3x + a)$

Between B and C: $y = C_1(L - x)^2(2L - 3a + x)$

Max deflection: $y_{max} = C_1(L - a)^2(2L + a)$ at $x = 0$

At the load: $y_B = 2C_1(L - a)^3$

Configuration 9

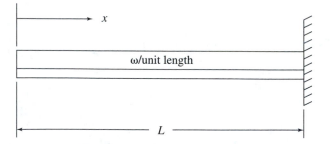

$C_1 = \omega/(8EI)$

$y = C_1(L - x)^2[2L^2 + (L + x)^2]/3$

Max. deflection: $y_{max} = C_1 L^4$ at $x = 0$

Configuration 10

Figure 15.13   (Continued)

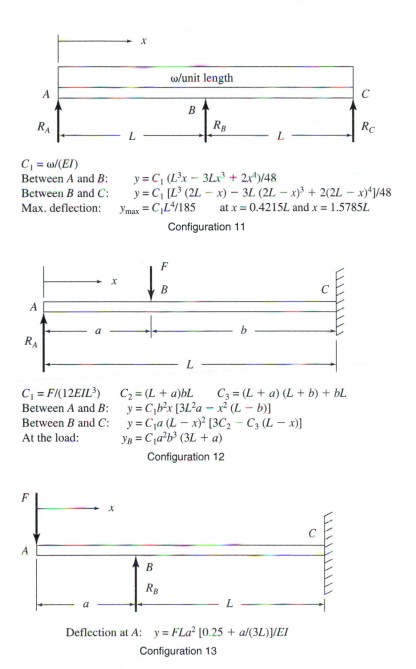

$C_1 = \omega/(EI)$
Between A and B: $y = C_1\,(L^3x - 3Lx^3 + 2x^4)/48$
Between B and C: $y = C_1\,[L^3\,(2L - x) - 3L\,(2L - x)^3 + 2(2L - x)^4]/48$
Max. deflection: $y_{max} = C_1L^4/185$ at $x = 0.4215L$ and $x = 1.5785L$

Configuration 11

$C_1 = F/(12EIL^3)$ $C_2 = (L + a)bL$ $C_3 = (L + a)\,(L + b) + bL$
Between A and B: $y = C_1b^2x\,[3L^2a - x^2\,(L - b)]$
Between B and C: $y = C_1a\,(L - x)^2\,[3C_2 - C_3\,(L - x)]$
At the load: $y_B = C_1a^2b^3\,(3L + a)$

Configuration 12

Deflection at A: $y = FLa^2\,[0.25 + a/(3L)]/EI$

Configuration 13

Figure 15.13   *(Continued)*

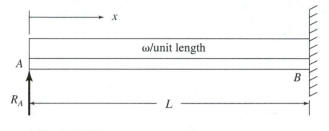

$C_1 = \omega L/(EI)$
$\quad y = C_1 x\,(L - x)^2\,(L + 2x)/(48L)$
Max. deflection: $\quad y_{max} = C_1 L^3/185 \qquad$ at $x = 0.421L$
At $L/2$: $\qquad\qquad\qquad y = C_1 L^3/192$

Configuration 14

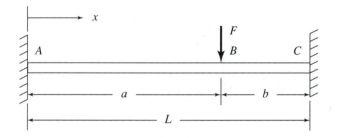

Condition: $a \geq b$
$C_1 = F/(3EI)$
Between A and B: $\qquad y = C_1\,(xb)^2\,[2a\,(L - x) + L\,(a - x)]/(2L^3)$
Between B and C: $\qquad y = C_1\,(L - x)^2 a^2\,[2bx + L\,(b - L + x)]/(2L^3)$
Max. deflection: $\qquad y_{max} = 2C_1 a^3 b^2/(3a + b)^2 \qquad$ at $x = 2aL/(3a + b)$
At the load: $\qquad\qquad y_B = C_1\,(ab/L)^3$

Configuration 15

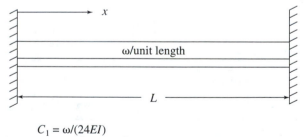

$C_1 = \omega/(24EI)$
$\quad y = C_1 x^2\,(L - x)^2$
Max. deflection: $\quad y_{max} = C_1 L^4/16 \qquad$ at $x = L/2$

Configuration 16

Figure 15.13   (*Continued*)

to each load and then use the principle of superposition to determine the total maximum deflection at the free end.

Now let us consider a *simply supported beam* carrying a number of downloads. Suppose we have a 100-in. beam carrying a 10,000-lb load at 70 in. from the left end, and $E = 30 \times 10^6$ psi and $I = 100$ in.[4] (Figure 15.14(a)). Call this Case 1.

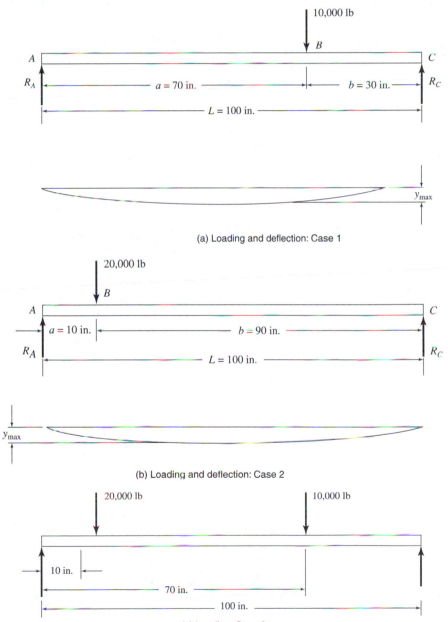

(a) Loading and deflection: Case 1

(b) Loading and deflection: Case 2

(c) Loading: Case 3

Figure 15.14

Case 1

We can calculate the maximum deflection and where it occurs by using the equations given in Figure 15.13, configuration 1. Because a (70 in.) is greater than b (30 in.),

$$y_{max} = C_1 C_2 b \sqrt{3C_2}/9 \qquad \text{at } x = \sqrt{C_2/3}$$

where
$$C_1 = F/(3EIL), \qquad C_2 = a(L + b)$$
$$F = -10{,}000 \text{ lb}, \qquad E = 30 \times 10^6 \text{ psi} \qquad I = 100 \text{ in.}^4$$
$$L = 100 \text{ in.}, \qquad b = 30 \text{ in.}$$

Therefore

$$C_1 = -10{,}000 \text{ lb}/(3 \times 30 \times 10^6 \text{ psi} \times 100 \text{ in.}^4 \times 100 \text{ in.})$$
$$= -1.111 \times 10^{-8} \text{ in.}^{-3}$$
$$C_2 = (70 \text{ in.})(100 \text{ in.} + 30 \text{ in.})$$
$$= 9100 \text{ in.}^2$$
$$y_{max} = C_1 C_2 \, b \sqrt{3C_2}/9$$
$$= (-1.111 \times 10^{-8} \text{ in.}^{-3})(9100 \text{ in.}^2)(30 \text{ in.}) \sqrt{3 \times 9100 \text{ in.}^2}/9$$
$$= -(3.033 \times 10^{-3})(165.23 \text{ in.})/9$$
$$= -0.056 \text{ in.}$$

The station at which the maximum deflection occurs is

$$x = \sqrt{C_2/3} = \sqrt{9100 \text{ in.}^2/3} = \sqrt{3033.3 \text{ in.}^2}$$
$$= 55.08 \text{ in.}$$

So for Case 1,

$$y_{max} = -0.056 \text{ in.} \qquad \text{at } x = 55.08 \text{ in.}$$

The location of the maximum deflection is to the right of the beam's midpoint (Figure 15.14(a)).

Case 2

Now suppose we have the same beam, but it is loaded as shown in Figure 15.14(b). This time a (10 in.) is less than b (90 in.); therefore, from the equations in Figure 15.13, configuration 1,

$$y_{max} = C_1 C_3 a \sqrt{C_3} \qquad \text{at } x = L - \sqrt{C_3}$$

where

$$C_1 = F/(3EIL), \qquad C_3 = (L^2 - a^2)/3$$
$$F = -20{,}000 \text{ lb}, \qquad E = 30 \times 10^6 \text{ psi}, \qquad I = 100 \text{ in.}^4$$
$$a = 10 \text{ in.}, \qquad L = 100 \text{ in.}$$
$$C_1 = -20{,}000 \text{ lb}/(3 \times 30 \times 10^6 \text{ psi} \times 100 \text{ in.}^4 \times 100 \text{ in.})$$
$$= -2.222 \times 10^{-8} \text{ in.}^{-3}$$
$$C_3 = [(100 \text{ in.})^2 - (10 \text{ in.})^2]/3$$
$$= 3300 \text{ in.}^2$$
$$y_{max} = C_1 C_3 a \sqrt{C_3}$$
$$= (-2.222 \times 10^{-8} \text{ in.}^{-3})(3300 \text{ in.}^2)(10 \text{ in.}) \sqrt{3300 \text{ in.}^2}$$
$$= 0.042 \text{ in.}$$

The station at which the maximum deflection occurs is

$$x = L - \sqrt{C_3} = 100 \text{ in.} - 3300 \text{ in.}^2$$
$$= 100 \text{ in.} - 57.45 \text{ in.}$$
$$= 42.55 \text{ in.}$$

So for Case 2,

$$y_{max} = -0.042 \text{ in.} \qquad \text{at } x = 42.55 \text{ in.}$$

The location of the maximum deflection is to the left of the beam's midpoint (Figure 15.14(b)).

Case 3

Figure 15.14(c) shows Case 3, which is a combination of Case 1 and Case 2. In this case, we probably cannot use the principle of superposition with certainty to say that the maximum deflection is

$$y_{max} = -0.056 \text{ in.} + (-0.042 \text{ in.}) = -0.098 \text{ in.}$$

because the maximum deflections for the two cases occur at different beam locations. And how are we going to determine where the maximum deflection occurs for the combined loading case?

What we have to do is calculate the deflection at, say, 100 locations along the beam for Case 1, do the same for Case 2, add the results at each location, and then determine the maximum deflection. This process is long and tedious by hand—but simple by computer. Program 'Deflect', introduced in the next section, does exactly that.

EXAMPLE PROBLEM 15.12

A 10-ft-long, 4-in.-diameter shaft carries a 150-lb pulley midway between its end supports. There is a belt pull of 300 lb on the pulley. How much will the shaft deflect at its midpoint? Regard the supports as being simple supports. *Ignore the shaft weight for this problem.*

Solution

1. The elastic modulus for steel is 30×10^6 psi.
2. Calculate the moment of inertia. For more complicated cross sections we could use program 'Centroid', but for a solid round shaft,

$$I = \pi d^4/64 = \pi(4 \text{ in.})^4/64 = 12.57 \text{ in.}^4$$

3. The concentrated load is $F = 150 \text{ lb} + 300 \text{ lb} = 450 \text{ lb}$, which—being a down-load—is entered in the equation as -450 lb.
4. Determine the configuration. It is configuration 1 with $L = L_B = 10 \text{ ft} = 120 \text{ in.}$ and $a = b = 60$ in.
5.

$$C_1 = F/(3EIL) = (-450 \text{ lb})/[3(30 \times 10^6 \text{ psi})(12.57 \text{ in.}^4)(120 \text{ in.})]$$
$$= -3.315 \times 10^{-9} \text{ in.}^{-3}$$

Using the equation $y_B = C_1 (ab)^2$ for the deflection at the load (see Figure 15.13, configuration 1) gives

$$y_B = (-3.315 \times 10^{-9} \text{ in.}^{-3})(60 \text{ in.} \times 60 \text{ in.})^2 = -0.043 \text{ in.}$$

Answer
The maximum deflection is 0.043 in. downward.

EXAMPLE PROBLEM 15.13

Realistically, the weight of the shaft cannot be ignored in the previous problem. Repeat the calculation, *including* the shaft weight.

Solution
1. The density of carbon and alloy steels is 0.283 lb/in.^{3}.
2. The shaft volume is $[\pi(4 \text{ in.})^2/4]$ (120 in.) = 1508 in^3.
3. Therefore, the shaft weight is (0.283 lb/in.^{3}) (1508 in.^{3}) = 427 lb. This must be regarded as a *distributed* load of -427 lb/120 in., or -3.558 lb/in.
4. Configuration 5 applies this time.

$$L = 120 \text{ in.,} \qquad E = 30 \times 10^6 \text{ psi,} \qquad I = 12.57 \text{ in.}^4$$

$$\omega = -3.558 \text{ lb/in.}$$

$$C_1 = \omega/(24EI) = (-3.558 \text{ lb/in.})/[24(30 \times 10^6 \text{ psi})(12.57 \text{ in.}^4)]$$

$$= -3.931 \times 10^{-10} \text{ in.}^{-3}$$

At $x = L/2$, $y = y_{max} = 5C_1L^4/16$:

$$y_{max} = 5(-3.931 \times 10^{-10} \text{ in.}^{-3})(120 \text{ in.})^4/16$$

$$= -0.025 \text{ in.}$$

5. From the theory of superposition, the total deflection is

$$y = -0.043 \text{ in.} + (-0.025 \text{ in.}) = -0.068 \text{ in.}$$

Answer
The maximum deflection is 0.068 in. downward.

EXAMPLE PROBLEM 15.14

More realistically still, the bearings act as built-in supports rather than as simple supports. What effect does this have on the maximum deflection?

Solution
We will employ configuration 15 and Configuration 16 and then add the resulting deflections.

Configuration 15:

$$F = -450 \text{ lb}, \quad L = 120 \text{ in.}, \quad a = b = 60 \text{ in.}$$

$$C_1 = F/(3EI) = (-450 \text{ lb})/[3(30 \times 10^6 \text{ psi})(12.57 \text{ in.}^4)]$$

$$= -3.978 \times 10^{-7} \text{ in.}^{-2}$$

$$y_{max} = y_B = C_1(ab/L)^3 = (-3.978 \times 10^{-7} \text{ in.}^{-2})(60 \text{ in.} \times 60 \text{ in.}/120 \text{ in.})^3$$

$$= -0.0107 \text{ in.}$$

Configuration 16:

$$\omega = -3.558 \text{ lb/in.}, \quad L = 120 \text{ in.}$$

$$C_1 = \omega/(24EI) = (-3.558 \text{ lb/in.})/[24(30 \times 10^6 \text{ psi})(12.57 \text{ in.}^4)]$$

$$= -3.9313 \times 10^{-10} \text{ in.}^{-3}$$

$$y_{max} = C_1 L^4/16 = (-3.9313 \times 10^{-10} \text{ in.}^{-3})(120 \text{ in.})^4/16$$

$$= -0.0051 \text{ in.}$$

Using superposition, the deflections are −0.011 in. and −0.005 in., respectively. Therefore, the total deflection is $y = -0.016$ in, which is considerably less than −0.068 in. obtained by assuming simple supports. This demonstrates the importance of defining the correct model as closely as possible in all engineering calculations.

Answer
The maximum deflection is 0.016 in. downward.

15.6 COMPUTER ANALYSES

15.6.1 Program 'Deflect'

'Deflect' calculates the maximum deflection and plots the deflection curve for a simply supported beam carrying any or all of the following loads:

1. A uniformly distributed load across the span
2. One, two, three, or four concentrated loads at any locations on the beam

The input/output sheet for this program is shown in Figure 15.15.
First you are asked to enter US or SI. This controls the units and indicates the units to use in entering the input values.
The required inputs are as follows:

1. Modulus of elasticity
2. Section moment of inertia
3. Beam length
4. Distributed load, if there is one (If not, enter 0.)
5. Up to four concentrated loads and their locations (distances from the left end of the beam)

Program 'Deflect'
Simply Supported

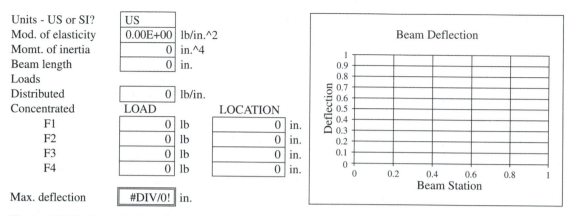

Units - US or SI? | US |
Mod. of elasticity | 0.00E+00 | lb/in.^2
Momt. of inertia | 0 | in.^4
Beam length | 0 | in.
Loads
Distributed | 0 | lb/in.
Concentrated   LOAD   LOCATION
F1 | 0 | lb | 0 | in.
F2 | 0 | lb | 0 | in.
F3 | 0 | lb | 0 | in.
F4 | 0 | lb | 0 | in.

Max. deflection | #DIV/0! | in.

Figure 15.15   Input/output for program 'Deflect'.

The program divides the beam length into 100 equally spaced locations and calculates the deflection at each location due to each of the loads. These are then summed at each location, and the @MIN function is used to determine the maximum deflection. The @MIN function, not the @MAX function, is used, because the deflections are negative.

The beam deflection curve is, of course, greatly exaggerated.

Moving to the right of the screen, the program lists the x-locations, the deflection due to each load, and the total deflections. This is where to look if you want to know exactly where the maximum deflection occurs.

EXAMPLE PROBLEM 15.15

In Section 15.5.1 we discussed the problem of trying to determine the maximum deflection and the location of the maximum deflection of a simply supported beam carrying multiple concentrated loads. We now revisit that problem using program 'Deflect'.

Refer to Figure 15.14(c) for the loading. The beam is 100 in. long, carrying a concentrated load of 20,000 lb at $x = 10$ in. and another concentrated load of 10,000 lb at $x = 70$ in. The modulus of elasticity of the beam material (steel) is 30×10^6 psi, and the moment of inertia is 100 in.^{4}.

Solution

Open program 'Deflect'. Enter the values shown in Figure 15.16. The maximum deflection is given as -0.096 in. In Section 15.5.1 we calculated a maximum deflection of -0.098 in. by adding the individual maximum deflections. Move to the right of the screen and see that the maximum deflection occurs at $x = 50.0$ in.

Program 'Deflect'
Simply Supported

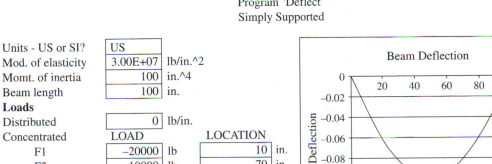

Units - US or SI?	US	
Mod. of elasticity	3.00E+07	lb/in.^2
Momt. of inertia	100	in.^4
Beam length	100	in.

Loads

| Distributed | 0 | lb/in. |

Concentrated	LOAD		LOCATION	
F1	−20000	lb	10	in.
F2	−10000	lb	70	in.
F3	0	lb	0	in.
F4	0	lb	0	in.

| Max. deflection | −0.096111 | in. |

Figure 15.16 Input/output for Example Problem 15.15.

Answer

The maximum deflection is −0.096 in. What do you think the maximum deflection would be if you doubled the moment of inertia? Enter the new value and see if you are correct.

EXAMPLE PROBLEM 15.16

Determine the maximum deflection and its location for the loading shown in Figure 15.17(a). The beam is made of ponderosa pine and is 10 cm wide by 20 cm deep. E for ponderosa pine is 7.6×10^3 MPa.

Solution

1. Calculate the moment of inertia. For a rectangle,

$$I = bd^3/12$$

$$b = 10 \text{ cm} = 100 \text{ mm}$$

$$d = 20 \text{ cm} = 200 \text{ mm}$$

$$I = (100 \text{ mm})(200 \text{ mm})^3/12$$

$$= 66.667 \times 10^6 \text{ mm}^4$$

2. Open program 'Deflect'. Enter the following:

    SI for units
    7600 for the modulus of elasticity
    66.667E06 for the moment of inertia
    8000 for the beam length
    −0.1 for the distributed load

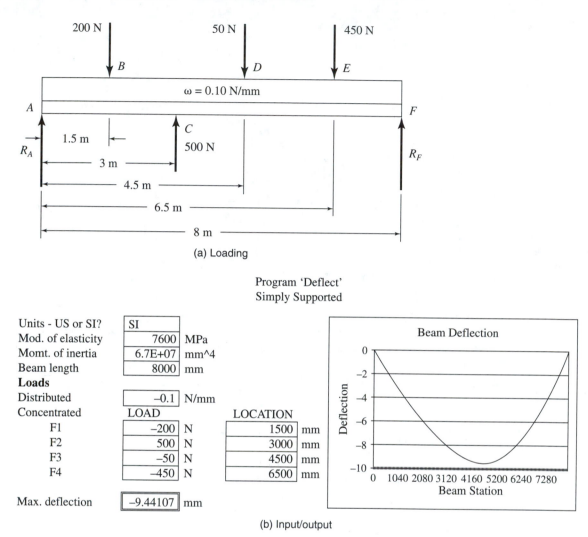

(a) Loading

Program 'Deflect'
Simply Supported

Units - US or SI?	SI	
Mod. of elasticity	7600	MPa
Momt. of inertia	6.7E+07	mm^4
Beam length	8000	mm
Loads		
Distributed	−0.1	N/mm

Concentrated	LOAD		LOCATION	
F1	−200	N	1500	mm
F2	500	N	3000	mm
F3	−50	N	4500	mm
F4	−450	N	6500	mm

Max. deflection   −9.44107   mm

(b) Input/output

Figure 15.17   Example Problem 15.16.

$$F1 = -200 \quad at \quad 1500$$
$$F2 = 500 \quad at \quad 3000$$
$$F3 = -50 \quad at \quad 4500$$
$$F4 = -450 \quad at \quad 6500$$

We show the input/output sheet in Figure 15.17(b).

$$y_{max} = -9.44 \text{ mm}$$

We find the location of the maximum deflection (to the right of the screen):

$$x = 4480 \text{ mm}$$

Answer

The maximum deflection is −9.44 mm at $x = 4480$ mm.

Program 'Deflection'

INPUT:
Units: Enter US or SI | US |

Configuration:
Enter number, 1 to 10 | 1 |

Enter: a | 0 | in.
         L | 0 | in.
         LB | 0 | in.
Enter: F | 0 | lb   (Downloads are negative)
Enter: E | 0.00E+00 | psi
Enter: I | 0.00E+00 | in.^4
Enter: x | 0 | in.

OUTPUT:
Deflection at x | #DIV/0! | in.
Deflection at the load | #DIV/0! | in.
Max. deflection | #DIV/0! | in.

Figure 15.18   Input/output sheet for program 'Deflection'.

15.6.2   Program 'Deflection'

'Deflection' is designed to handle all the Figure 15.13 configurations from 1 to 10. The input/output sheet is shown in Figure 15.18. US or SI units are selected; then the configuration number is entered. The input variables and the output are both keyed to the configuration number. For example, if you enter configuration 1 (see Figure 15.13, configuration 1), you will be directed to enter a, L, and L_B in addition to the load, F, the elastic modulus, E, the moment of inertia, I, and the x-location where the deflection is to be calculated. The output will consist of the deflections at x, at the load, and the maximum deflection. Now if you select configuration 10, say, the input will change to L, ω, E, I, and x, and the output will give the deflection at x and the maximum deflection.

EXAMPLE PROBLEM 15.17

A 3-m steel cantilever beam carries a full-span distributed load of 100 N/m and a concentrated load of 150 N at 1 m from the free end. $E = 207 \times 10^3$ MPa and $I = 3.6 \times 10^5$ mm⁴. Determine the deflection at the concentrated load and the maximum deflection.

Solution
Referring to Figure 15.13 we see that a cantilever beam carrying a concentrated load is configuration 9, and that with a full-span distributed load is configuration 10.

For configuration 9,

$$a = 1\text{ m} = 1000\text{ mm}$$
$$L = 3\text{ m} = 3000\text{ mm}$$
$$F = -150\text{ N}$$

Figure 15.19(a) shows the input/output sheet.

$$\text{deflection at the load} = -5.37\text{ mm}$$
$$\text{maximum deflection} = -9.39\text{ mm}$$

Program 'Deflection'

INPUT:

Units:	Enter US or SI	SI	
Configuration:			
Enter number, 1 to 10		9	
Enter:	a	1000	mm
	L	3000	mm
	************	0	mm
Enter:	F	−150	N (Downloads are negative)
Enter:	E	2.07E+05	MPa
Enter:	I	3.60E+05	mm^4
Enter:	x	0	mm

OUTPUT:

Deflection at x	−9.39345	mm
Deflection at the load	−5.36769	mm
Max. deflection	−9.39345	mm

(a) Input/output sheet

Program 'Deflection'

INPUT:

Units:	Enter US or SI	SI	
Configuration:			
Enter number, 1 to 10		10	
Enter:	L	3000	mm
	************	0	mm
	************	0	mm
Enter:	w	−0.1	N/mm (Downloads are negative)
Enter:	E	2.07E+05	MPa
Enter:	I	3.60E+05	mm^4
Enter:	x	1000	mm

OUTPUT:

Deflection at x	−7.60422	mm
Max. deflection	−13.587	mm
*******************	***********	mm

(b) Input/output sheet

Figure 15.19   Example Problem 15.17.

For configuration 10,

$$L = 3000 \text{ mm}$$
$$\omega = -100 \text{ N/m} = -0.10 \text{ N/mm}$$
$$x = 1000 \text{ mm}$$

Figure 15.19(b) shows the input/output sheet.

$$\text{deflection at the load} = -7.60 \text{ mm}$$
$$\text{maximum deflection} = -13.59 \text{ mm}$$

Add the two maximum deflections. At the load,

$$y = -5.37 \text{ mm} + (-7.60 \text{ mm}) = -12.97 \text{ mm}$$
$$y_{max} = -9.39 + (-13.59 \text{ mm}) = -22.98 \text{ mm}$$

Answer
Deflection at the concentrated load is -12.97 mm; maximum deflection is -22.98 mm.

15.7 PROBLEMS FOR CHAPTER 15

Hand Calculations

1. For the loading shown in Figure 15.20:
   a. Draw the shear force diagram.
   b. Draw the bending moment diagram.
   c. Show, by using the *moment-area method,* that the maximum deflection is $y_{max} = 19FL^3/(384. EI)$.

2. For the loading shown in Figure 15.20, show that the maximum deflection is $y_{max} = 19 FL^3/(384 EI)$, by using the *theory of superposition* applied to Figure 15.13, configurations 1 and 2.

3. Calculate the maximum deflection of the simply supported Sitka spruce round bar shown in Figure 15.21. It is 3 m long and 10 cm in diameter and carries a mass of 100 kg at its midpoint. Sitka spruce has an elastic modulus of 9000 MPa. Ignore the weight of the bar.

4. A 200-lb diver stands on the end of a springboard. See Figure 15.22. The board is 2.5 ft wide and 2 in thick. Its elastic modulus is $E = 1 \times 10^6$ psi. Calculate the deflection at the end of the board.

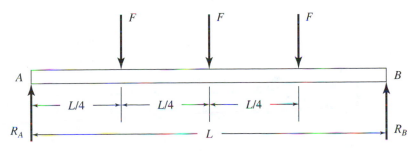

Figure 15.20

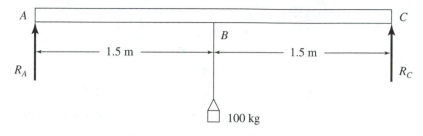

Figure 15.21

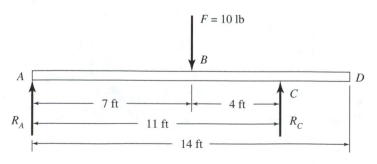

Figure 15.22

Computer Calculations

5. The carbon-steel beam shown in Figure 15.23 has a 1-in. × 1-in. cross section. Determine the deflections at 3.5 ft, 7.0 ft, 10.5 ft and 14.0 ft from the left end. What is the maximum deflection, and where does it occur?

6. Repeat Problem 5 with the load moved to a new location of 9 ft from the left end of the beam.

7. For the beam shown in Figure 15.24, $E = 9000$ MPa and $I = 4000$ mm^4. Determine the maximum downward deflection.

8. What is the maximum upward deflection for the beam in the Problem 7, and where does it occur?

9. The steel cantilever beam shown in Figure 15.25 has a modulus of elasticity of 30×10^6

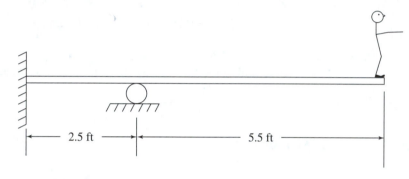

Figure 15.23

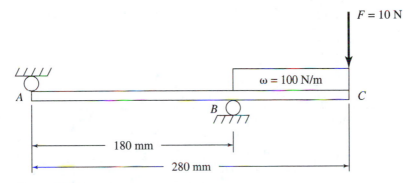

Figure 15.24

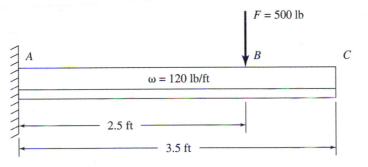

Figure 15.25

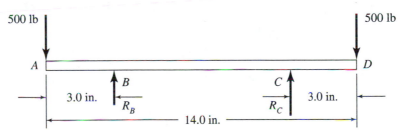

Figure 15.26

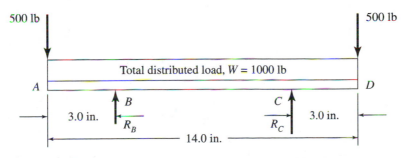

Figure 15.27

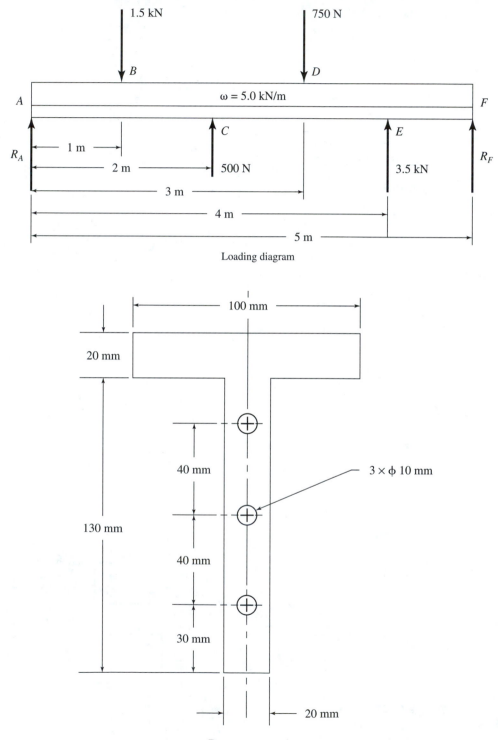

Loading diagram

Beam cross section

Figure 15.28

psi and a section moment of inertia of 32.5 in^4. Calculate the deflection at *C*.

10. If the beam in Problem 9 were made of aluminum alloy 6061, what would be the deflection at *C*?

11. Figure 15.26 shows a 6061-T4 aluminum alloy beam that is 2 in. $\times$ 2 in. in cross section. $E = 10 \times 10^6$ psi.
    a. What is the deflection at the center of the beam?
    b. What is the deflection at each end of the beam?

12. What loads would be needed to replace the 500-lb loads in Figure 15.26 to increase the midpoint deflection to 0.0030 in.?

13. With the new load found in Problem 12, what would the midpoint deflection be if the beam were made of steel?

14. The beam shown in Figure 15.27 is the same as that for Problem 11, but it carries an additional 1000 lb as an evenly distributed load.
    a. What is the deflection at the center of the beam?

b. What is the deflection at each end of the beam?

15. Referring to Problem 14, what would the deflection be at the midpoint and at the ends of the beam if the beam were made of cast magnesium ($E = 6.5 \times 10^6$ psi)?

16. The aluminum alloy 6061-T4, 5-m-long simply supported beam shown in Figure 15.28 carries a distributed load of 5.0 kN/m, two downloads of 1.5 kN and 750 N, and two uploads of 500 N and 3.5 kN. Its cross section is a tee with three lightening holes. Determine its maximum deflection and where it occurs. The elastic modulus for 6061-T4 is 69 GPa.

17. *Using calculus,* derive the equations for slope, deflection, and maximum deflection for a simply supported beam carrying a uniformly distributed load over its entire span.

18. *Using calculus,* derive the equations for slope, deflection, and maximum deflection for a cantilever beam carrying a uniformly distributed load over its entire span.

16

Beam Design

■ *Objectives*

Structural beams are available in many shapes, sizes, and materials. In this chapter you will learn how to select the best beam that will satisfy the allowable stresses and the acceptable deflection.

16.1   GENERAL CONSIDERATIONS

As we have seen from previous chapters, a beam must be designed to safely take the maximum bending moment and the maximum shear force and is often limited to a maximum deflection.

The fundamental parameters that affect these are as follows:

1. The type of supports
2. The type of loading
3. The magnitude of the loads
4. The beam length
5. The allowable stresses (type of material)
6. The beam cross-sectional shape and area
7. The permissible deflection

16.1.1   Allowable Stresses—Structural Steel

The AISC (American Institute of Steel Construction) code gives the allowable stresses in steel beams for static loads as

$$\text{allowable bending stress} = 0.66 \text{ times the yield strength}$$

and

$$\text{allowable shear stress} = 0.40 \text{ times the yield strength}$$

The yield stress values can be found in program 'Proptab'; however, for convenience, we list the allowable stresses for the most commonly used structural steels in the following table.

Structural Steel	Allowable Bending Stress		Allowable Shear Stress	
	psi	MPa	psi	MPa
A 36	24,000	165	14,500	100
A 242	33,000	228	20,000	138
A 441	30,400	210	18,400	127
A 529	27,700	191	16,800	116
A 572				
Grade 42	27,700	191	16,800	116
Grade 50	33,000	228	20,000	138
Grade 60	39,600	273	24,000	165
Grade 65	42,900	296	26,000	179
A 588	33,000	228	20,000	138

Modulus of elasticity is $E = 29 \times 10^6$ psi $= 200 \times 10^3$ MPa.

16.1.2   Allowable Stresses—Aluminum Alloys

The Aluminum Association recommends using the following for allowable bending stresses for beams with static loads:

allowable bending stress = the lower of (0.61 times the yield strength) *or* (0.51 times the ultimate strength)

The allowable shear stress should be approximately **0.25 times the yield strength**. For convenience, we tabulate these values for the most commonly used aluminum alloys.

Aluminum Alloy	Allowable Bending Stress		Allowable Shear Stress	
	psi	MPa	psi	MPa
1100-H12	8,120	56	3,750	26
1100-H18	12,180	84	5,500	38
2014-0	8,560	59	3,500	24
2014-T4	25,670	177	10,500	72
2014-T6	35,680	246	15,000	103
3003-0	3,630	25	1,500	10
3003-H12	9,720	67	4,500	31
3003-H18	14,790	102	6,750	47
5154-0	10,300	71	4,250	29
5154-H32	18,270	126	7,500	52
5154-H38	23,790	164	9,750	67
6061-0	4,930	34	2,000	14
6061-T4	12,910	89	5,250	36
6061-T6	22,920	158	10,000	69
7075-0	9,140	63	3,750	26
7075-T6	42,350	292	18,250	126
204.0-T4	17,690	122	7,250	50
356.0-T6	13,490	93	5,500	38

Modulus of elasticity is $E = 10 \times 10^6$ psi $= 69 \times 10^3$ MPa for 1100, 3003, 6061. $E = 10.2 \times 10^6$ psi $= 70 \times 10^3$ MPa for 5154. $E = 10.4 \times 10^6$ psi $= 72 \times 10^3$ MPa for 7075. $E = 10.6 \times 10^6$ psi $= 73 \times 10^3$ MPa for 2014.

16.1.3 Allowable Stresses—Timber

The allowable stresses for timber depend on the moisture content, cross-sectional dimensions, and the grade. The following table is a guide, giving the average values for medium grades.

	Allowable Stress					
	Bending		Shear		Elastic Modulus	
Species	psi	MPa	psi	MPa	10^6 psi	10^3 MPa
California redwood	1350	9.3	100	0.69	1.3	9.0
Douglas fir	1450	10.0	95	0.66	1.7	11.7
Hemlock	1250	8.6	75	0.52	1.4	9.7
Ponderosa pine	1000	6.9	65	0.45	1.1	7.6
Sitka spruce	1150	7.9	70	0.48	1.3	9.0
Southern pine	1000	6.9	70	0.48	1.3	9.0

16.1.4 Section Modulus

We derived the flexure equation for stress due to bending in equation (13.7):

$$s = My/I$$

In this equation, M is the bending moment at the section of interest, y is the distance from the neutral axis to the fiber of interest, and I is the moment of inertia of the beam's cross section. The maximum bending stress at any station occurs at the outermost fibers, that is, at the surface, where $y = c$. So the *maximum* bending stress is given by

$$s_{max} = Mc/I$$

Now both I and c are dependent *only* on the cross-sectional shape and size of the beam. Therefore, for a beam of a given cross-sectional shape and size, I/c is a constant value. For example, a rectangle that is b wide and d deep has moment of inertia

$$I = bd^3/12$$

The centroid (position of the neutral axis) is in the middle of the rectangle, so $c = d/2$; therefore,

$$I/c = (bd^3/12)/(d/2)$$
$$= bd^2/6$$

We call the ratio I/c the *section modulus* and give it the symbol S.

$$S = I/c \qquad (16.1)$$

So we can write the maximum bending stress as

$$s_{max} = M/S \qquad (16.2)$$

Writing this as $S = M/s_{max}$ gives us a particularly useful equation. If we know the maximum allowable bending stress and the maximum bending moment, we can calculate

S. Then we can determine the best cross-sectional shape for our beam. The next section tells how.

16.2   STRUCTURAL SHAPES

Steel manufacturers list a wide variety of standard beam shapes from which to choose. I-beams can be either wide flange or standard flange. Wide-flange beams (Figure 16.1(a)) are known as W-shapes. They are identified by W$xxx \times yyy$, where xxx is the approximate depth of the beam in inches or millimeters and yyy is the weight in lb/ft or mass in kg/m. For example, a W36 $\times$ 160 is a wide-flange beam approximately 36 in. deep and weighing 160 lb/ft. The equivalent metric beam is W920 $\times$ 238, which is approximately 920 mm deep and has a *mass* of 238 kg/m.

Standard-flange beams (Figure 16.1(b)) are called S-shapes. The designation is the same as for W-beams. An S24 $\times$ 100 is a standard-flange beam that is approximately 24 in. deep and weighs 100 lb/ft.

Standard channels (Figure 16.2) are referred to as C-shapes. Again, the designations are the same, giving the approximate depth and lb/ft or kg/m. C380 $\times$ 60 is a standard channel that is approximately 380 mm deep and has a mass of 60 kg/m.

Standard angles, called L-shapes, can have either equal-length legs (Figure 16.3(a)) or unequal legs (Figure 16.3(b)). A typical designation for an equal-leg channel might be L4 $\times$ 4 $\times$ $\frac{3}{4}$, which is a channel in which both legs are 4 in. long and the thickness is $\frac{3}{4}$ in. You get the weight per foot from a table.

Other standard structural steel shapes include square tubing, rectangular tubing, and pipe. Square and rectangular shapes are identified by depth $\times$ height $\times$ wall thickness.

Pipe sizes are *nominal* sizes. For example, what we call a 4-in. pipe has an *outside* diameter of 4.500 in. No, the inside diameter is *not* 4.00 in. either! The inside diameter depends on the *schedule*, which is related to the wall thickness. As an example of this,

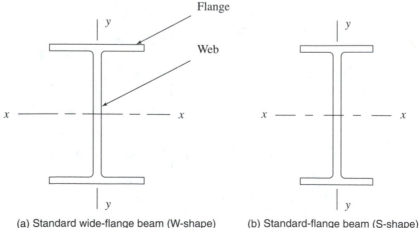

(a) Standard wide-flange beam (W-shape)   (b) Standard-flange beam (S-shape)

Figure 16.1   Structural shapes.

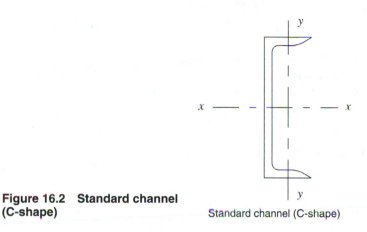

Figure 16.2 Standard channel (C-shape)

Standard channel (C-shape)

a 4-in. steel pipe has an outside diameter of 4.500 and can have any of the following inside diameters, depending on the schedule.

Schedule	Inside Diameter (in.)
5S	4.334
10S	4.260
40S	4.026
80S	3.826
120	3.626
160	3.438
XXS	3.152

There are more. All you need to know is that the outside diameter remains constant and the wall thickness increases as the schedule increases. Nominal sizes of standard pipes range from $\frac{1}{8}$ in. (O.D. = 0.405 in.) to 42 in. (O.D. = 42 in.) For pipes of 14 in. in diameter and greater, the nominal diameter *is* the outside diameter.

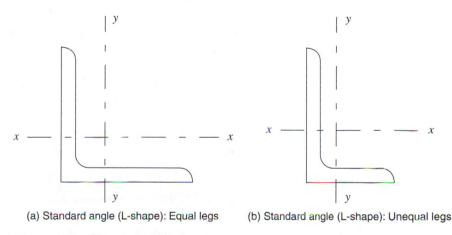

(a) Standard angle (L-shape): Equal legs   (b) Standard angle (L-shape): Unequal legs

Figure 16.3 Structural shapes.

There are also standard aluminum alloy *I*-beams and channels.

Each of the standard beams discussed has a specific section modulus, which is what we are looking for once we have determined the maximum bending moment and the allowable stress. Later in this chapter we will use program 'Shapes', which will do a lot of the work for us, but we need to do some hand calculations first to fully understand what is involved in selecting a beam shape.

16.3   BEAM SELECTION

First we lay out the procedure; then we do a detailed example. The procedure is as follows:

1. Draw the *loading, shear force,* and *bending moment diagrams.*
2. Determine the *maximum shear force,* V_{max}, and the *maximum bending moment,* M_{max}.
3. Determine the *allowable bending stress,* s_{max}, for the beam material.
4. Calculate the *minimum acceptable section modulus,* S_{min}, from the equation

$$S_{min} = M_{max}/s_{max}$$

A section modulus less than this would result in a stress that exceeds the allowable.

5. Find the *lightest* beam whose section modulus is *greater* than S_{min}. We would do this by looking in standard beam-shape tables. Later, program 'Shapes' will do it for us.
6. Check the web shear stress.

$$s_{web} = V_{max}/A_{web}$$

where

$$A_{web} = td$$
$$t = \text{beam web thickness}$$
$$d = \text{beam web depth}$$

If s_{web} exceeds the allowable shear stress, select the next beam in the series whose web area and section modulus are larger than those of the first choice.

7. Check the maximum deflection. If it is not specified, use the following guide:
   a. For general machine parts and beams carrying plaster, the maximum deflection should not exceed $\frac{1}{360}$ of the span.
   b. For moderate precision parts, the maximum deflection should not exceed $\frac{1}{2000}$ of the span.
   c. For high-precision parts, the maximum deflection should not exceed $\frac{1}{100,000}$ of the span. If the maximum deflection is exceeded, select the beam having the next higher *moment of inertia.*

In summary, what we are searching for is the *lightest* beam whose

1. Section modulus satisfies the bending requirement,
2. Web area satisfies the shear requirement, and
3. Moment of inertia satisfies the deflection requirement.

EXAMPLE PROBLEM 16.1

Select an A36 structural steel cantilever I-beam with a span of 12 ft carrying a uniformly distributed load of 700 lb/ft and a concentrated load of 16,000 lb at its midpoint. The deflection is not to exceed $\frac{1}{360}$ of the span. See Figure 16.4.

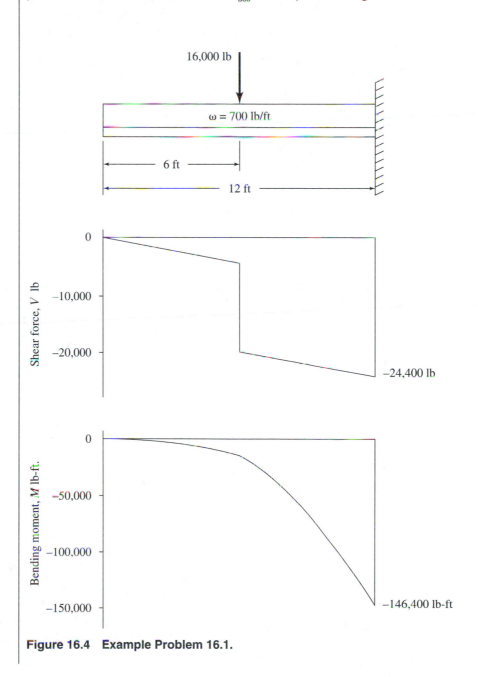

Figure 16.4   Example Problem 16.1.

Solution

1. Loading, shear force, and bending moment diagrams: We show the shear force and bending moment diagrams in Figure 16.4, but in the case of cantilever beams we can calculate the maximum values directly.

2. Maximum shear force and bending moment: Both occur at the wall. The maximum shear force is simply the total load to the left of the wall, which *is* the total load.

$$V_{max} = (700 \text{ lb/ft})(12 \text{ ft}) + 16{,}000 \text{ lb} = 24{,}400 \text{ lb}$$

The total distributed load is (700 lb/ft)(12 ft) = 8400 lb. Its center of gravity is 6 ft from the wall. The concentrated load of 16,000 lb is also 6 ft from the wall, so, in effect, there is a (8400-lb + 16,000-lb) = 24,400-lb load 6 ft from the wall. This creates a bending moment of (24,400 lb)(6 ft) = 146,400 lb-ft at the wall. That is,

$$M_{max} = 146{,}400 \text{ lb-ft} = 1.757 \times 10^6 \text{ lb-in.}$$

3. Allowable bending stress: For A36 structural steel, $s_{max} = 24{,}000$ psi in bending.
4. Minimum section modulus:

$$S_{min} = M_{max}/s_{max} = 1.757 \times 10^6 \text{ lb-in./24,000 psi} = 73.2 \text{ in.}^3$$

5. Find the *lightest* beam whose section modulus is *greater* than S_{min}.

The following table lists seven steel I-beam shapes that might be considered.

Designation	Section Modulus S in.^{3}	Moment of Inertia I in.^{4}	Web Thickness in.	Web Depth in.
W12 × 58	78.0	475	0.360	12.19
W14 × 38	54.6	385	0.310	14.10
W14 × 53	77.8	541	0.370	13.92
S15 × 50	64.8	486	0.550	15.00
W16 × 50	81.0	659	0.380	16.26
W18 × 46	78.8	712	0.360	18.06
S18 × 54.7	89.4	804	0.461	18.00

With regard to bending, the required section modulus of 73.2 in.^{3} is met by all but the W14 × 38 and the S15 × 50 beams. The beam whose section modulus is closest to (but greater than) the required section modulus is the W14 × 53 beam. The lightest, however, is the W18 × 46, being 46 lb/ft. This is 7 lb/ft lighter than the W14 × 53 beam and, therefore, cheaper too. So we will tentatively select the W18 × 46 shape and check if it satisfies the other requirements.

6. Web shear stress:

$$s_{web} = V_{max}/A_{web}$$

where $A_{web} = td$. For the W18 × 46 beam, $t = 0.36$ in. and $d = 18.06$ in. Therefore, $A_{web} = (0.36 \text{ in.})(18.06 \text{ in.}) = 6.50 \text{ in.}^2$

$$s_{web} = V_{max}/A_{web} = 24{,}400 \text{ lb/6.50 in.}^2 = 3754 \text{ psi}$$

The allowable shear stress for A36 steel is 14,500 psi, so the web is certainly sufficiently strong in shear.

7. Maximum deflection: The moment of inertia of the tentatively selected beam is 712 in.^{4}.

Deflection due to the concentrated load: Referring to Figure 15.13, configuration 9,

$$y_{max} = C_1(L - a)^2(2L + a)$$

where

$$C_1 = F/(6EI)$$
$$F = -16,000 \text{ lb}, \qquad E = 29 \times 10^6 \text{ psi}, \qquad I = 712 \text{ in.}^4$$

Therefore,

$$C_1 = -16,000 \text{ lb}/[6(29 \times 10^6 \text{ psi})(712 \text{ in.}^4)]$$
$$C_1 = -1.291 \times 10^{-7} \text{ in.}^{-2}$$
$$a = 6 \text{ ft} = 72 \text{ in.} \qquad L = 12 \text{ ft} = 144 \text{ in.}$$

Therefore,

$$\begin{aligned} y_{max} &= C_1(L - a)^2(2L + a) \\ &= (-1.291 \times 10^{-7} \text{ in.}^{-2})(144 \text{ in.} - 72 \text{ in.})^2(2 \times 144 \text{ in.} + 72 \text{ in.}) \\ &= (-1.291 \times 10^{-7} \text{ in.}^{-2})(5184 \text{ in.}^2)(360 \text{ in.}) \\ &= -0.241 \text{ in.} \end{aligned}$$

Deflection due to the distributed load: The distributed load, ω, is -700 lb/ft $= -58.33$ lb/in. Referring to Figure 13.15, configuration 10,

$$y_{max} = C_1 L^4$$

where

$$\begin{aligned} C_1 &= \omega/(8EI) \\ &= (-58.33 \text{ lb/in.})/[8(29 \times 10^6 \text{ lb/in.}^2)(712 \text{ in.}^4)] \\ &= -3.531 \times 10^{-10} \text{ in.}^{-3} \\ y_{max} = C_1 L^4 &= (-3.531 \times 10^{-10} \text{ in.}^{-3})(144 \text{ in.})^4 \\ &= -0.152 \text{ in.} \end{aligned}$$

The total maximum deflection is

$$y_{max} = (-0.241 \text{ in.}) + (-0.152 \text{ in.}) = -0.393 \text{ in.}$$

(The negative sign simply means that it is bending *down*.)

The maximum allowable deflection is $L/360 = 144$ in./360 $= 0.4$ in.

Summary

The selected W18 × 46 beam will satisfy all the requirements and is the lightest of those identified as possible candidates. Notice that the selected beam weighs 46 lb/ft. It is 12 ft long, so its total weight is (46 lb/ft)(12 ft) = 552 lb. The load carried by the beam is (8400 lb + 16,000 lb) = 24,400 lb. So the beam weight is 2.26% of the load and can be neglected. Generally, beam weights are neglected unless they exceed 10% of the load.

Answer

Select a W18 × 46 beam.

16.4   BASIC DEFLECTION

If the deflection in the preceding example had been too great, we would have selected another beam whose moment of inertia was higher and then recalculated for that beam. It is important to realize however that *beam deflections are inversely proportional to the moment of inertia for all loading configuration*. This means that if we double the moment of inertia, we halve the deflection.

Using this fact, we can define a *basic deflection*, which is the theoretical deflection that would result if the moment of inertia were **1.0 in.[4] for U.S. customary units or 1 × 10[6] mm[4] for SI**. In most cases the basic deflection would be ridiculously large, but it is just a number from which we can determine the actual required moment of inertia.

For example, suppose the maximum deflection is limited to $y_{max} = 5$ mm for a certain beam. Using the known loading, the elastic modulus, and a moment of inertia of $I = 1 \times 10^6$ mm^4, we calculate the basic deflection, y_{basic}. For the purpose of this illustration, suppose $y_{basic} = 55$ mm^4. Then the required moment of inertia is

$$I = (y_{basic}/y_{max})(1 \times 10^6 \text{ mm}^4)$$
$$= (55 \text{ mm}^4/5 \text{ mm}^4)(1 \times 10^6 \text{ mm}^4)$$
$$= 11 \times 10^6 \text{ mm}^4$$

EXAMPLE PROBLEM 16.2

A 6-ft-long A36 structural steel pipe is to be used as a cantilever beam to support a 1690-lb load at its free end. The maximum allowable deflection is $\frac{5}{8}$ in. What are the diameter and schedule of the lightest pipe that should be selected?

Solution
We have to determine the outside and the inside diameters and then look in a table of pipe sizes to find the schedule.

1. Determine the required section modulus:

$$S = M_{max}/S_{allow}$$

where M_{max}, the maximum bending moment, is (1690 lb) (6 ft):

$$M_{max} = 10{,}140 \text{ lb-ft} = 121{,}680 \text{ lb-in.}$$
$$S_{allow} = 24{,}000 \text{ lb/in.}^2 \text{ for A36 structural steel}$$

Therefore,

$$S = 121{,}680 \text{ lb-in./24,000 lb/in.}^2 = 5.07 \text{ in.}^3$$

2. Determine the minimum allowable moment of inertia. The maximum allowable deflection is

$$y_{max} = \tfrac{5}{8} \text{ in.} = 0.625 \text{ in.}$$

The *basic deflection* is (see Figure 15.13, configuration 9)

$$y_{basic} = C_1(L - a)^2(2L + a)$$

where $a = 0$ and $I = 1$ in.^{4} in the formula for C_1. Therefore, $y_{basic} = C_1L^2$ $(2L) = 2C_1L^3$.

$$C_1 = F/(6EI)$$

where $F = -1690$ lb, $E = 29 \times 10^6$ lb/in.^{2}, and $I = 1$ in.^{4} for basic deflection.

$$C_1 = -1690 \text{ lb}/(6 \times 29 \times 10^6 \text{ lb/in.}^2)(1 \text{ in.}^4)$$
$$= -9.72 \times 10^{-6} \text{ in.}^4$$
$$y_{basic} = 2C_1L^3$$

where $L = 6$ ft $= 72$ in.

$$y_{basic} = 2(-9.72 \times 10^{-6} \text{ in.}^4)(72 \text{ in.})^3$$
$$= 7.256 \text{ in.}$$
$$I = (y_{basic}/y_{max})(1 \text{ in.}^4)$$
$$= (7.256 \text{ in.}/0.625 \text{ in.})(1 \text{ in.}^4)$$
$$= 11.61 \text{ in.}^4$$

So far we have determined the required minimum section modulus ($S = 5.07$ in.^{3}) and the minimum moment of inertia ($I = 11.61$ in.^{4}).

3. Determine the *minimum possible* pipe diameter. The moment of inertia for a pipe with outside diameter D and inside diameter d is

$$I = \pi(D^4 - d^4)/64 \tag{16.3}$$

This has to be equal to or greater than 11.61 in.^{4} to satisfy the maximum deflection requirement.

$$\pi(D^4 - d^4)/64 \geq 11.61 \text{ in.}^4$$

Therefore, $(D^4 - d^4) \geq (11.61 \text{ in.}^4)(64)/\pi$, or

$$(D^4 - d^4) \geq 236.52 \text{ in.}^4$$

Suppose $d = 0$, i.e., the pipe is solid; then

$$D^4 \geq 236.52 \text{ in.}^4$$

or

$$D \geq 3.92 \text{ in.}$$

Section modulus for a pipe is

$$S = I/c$$
$$I = \pi(D^4 - d^4)/64$$

where $c = D/2$ (the distance from the center of the pipe to the outer surface), so

$$S = [\pi(D^4 - d^4)/64]/(D/2)$$
$$S = \pi(D^4 - d^4)/(32D) \tag{16.4}$$

We have determined that this must be equal to or greater than 5.07 in.^{3} to satisfy the maximum allowable stress requirement.

$$\pi(D^4 - d^4)/(32D) \geq 5.07 \text{ in.}^3$$

so

$$(D^4 - d^4)/D \geq (5.07 \text{ in.}^3)(32)/\pi$$

Therefore,

$$(D^4 - d^4)/D \geq 51.64 \text{ in.}^3$$

Again, the minimum possible value for D is when $d = 0$. If $d = 0$, then

$$D^3 \geq 51.64 \text{ in.}^3$$

so

$$D \geq 3.72 \text{ in.}$$

But to satisfy the maximum deflection requirement, $D \geq 3.92$ in. The nearest standard pipe size is 4.000 in. This is a *nominal* 3.5-in. pipe.

4. Consider the available standard pipes. We are looking for the lightest standard pipe whose outside diameter is 4.000 in. or greater, whose moment of inertia is 11.61 in.^{4} or greater, and whose section modulus is 5.07 in.^{3} or greater. The following table lists pipe data in the 4-in. to 8.625-in. range.

Nominal size (in.)	Actual size (in.)	Schedule	I (in.^{4})	S (in.^{3})	Weight per ft (lb)
3.5	4.000	XXS	9.848	4.924	22.85
4.0	4.500	XS	9.61	4.27	14.98
4.0	4.500	120	11.65	5.18	18.96
5.0	5.563	10S	8.43	3.03	7.77
5.0	5.563	40	15.17	5.45	14.62
6.0	6.625	10S	14.40	4.35	9.29
6.0	6.625	10S (thick wall)	22.66	6.84	15.02
8.0	8.625	5S	26.45	6.13	9.91

5. Select a pipe. Although the nominal 4.0-in. schedule 120 pipe is the closest in section modulus and moment-of-inertia requirements, the nominal 8.0-in. schedule 5S pipe also satisfies the requirements and is almost half the weight. So we will tentatively select the 8.0-in. 5S pipe, but we have to check its strength in shear.

6. Determine shear strength. The maximum shear force acting on the pipe is 1690 lb at the wall. The selected pipe has an outside diameter of 8.625 in. and an inside diameter of 8.407 in. (from standard pipe tables). So the cross-sectional area of the steel is

$$A = \pi(D^2 - d^2)/4$$
$$= \pi[(8.625 \text{ in.})^2 - (8.407 \text{ in.})^2]/4$$
$$= 2.916 \text{ in.}^2$$

The shear stress is force/area = 1690 lb/2.916 in.^{2} = 580 psi. This is well below the allowable 14,500 psi for A36 steel.

Answer
The lightest standard pipe that will satisfy all of the requirements is a nominal 8.0-in. schedule 5S pipe.

EXAMPLE PROBLEM 16.3

What is the maximum acceptable length of a simply supported 10- $\times$ 12-in. dressed-size oak beam designed to carry a uniformly distributed load of 1800 lb/ft across the entire span? The allowable bending stress is 1450 psi, the allowable shear stress is 110 psi, and the maximum deflection must not exceed $\frac{1}{360}$ of the span. Use an elastic modulus of 1.2×10^6 psi.

Solution

1. Calculate the moment of inertia. The standard dressed size of a 10-in. $\times$ 12-in. wooden beam is 9.5 in. $\times$ 11.5 in.

$$I = bd^3/12$$

We will place the beam with its longer face vertical for maximum moment of inertia.

$$b = 9.5 \text{ in.}, \qquad d = 11.5 \text{ in.}$$
$$I = (9.5 \text{ in.})(11.5 \text{ in.})^3/12$$
$$= 1204 \text{ in.}^4$$

2. Calculate the section modulus.

$$S = I/c = 1204 \text{ in.}^4/(11.5 \text{ in.}/2)$$
$$= 209 \text{ in.}^3$$

3. Determine the shear stress. Denote the beam length as L inches. Then the total load is $(L \text{ in.}) (1800 \text{ lb/ft})/(12 \text{ in./ft}) = (150L) \text{ lb}$, so the maximum shear force (see Figure 16.5) is $V_{max} = 75L$ lb. The beam's cross-sectional area is $(9.5 \text{ in.})(11.5 \text{ in.}) = 109.3 \text{ in}^2$. The maximum shear stress is, therefore,

$$s_{max} = (75L \text{ lb})/109.3 \text{ in.}^2 = (0.69L) \text{ psi}$$

Because the maximum allowable shear stress is 110 psi,

$$0.69L = 110$$

or

$$L = 110/0.69 \text{ in.} = 159.4 \text{ in.}$$

4. Find the bending stress. The maximum bending moment (see Figure 16.5) is

$$M_{max} = \text{triangular area of height } (75L) \text{ lb and base } L/2 \text{ in.}$$
$$= (75L)(L/2)/2$$
$$= (18.75L^2) \text{ lb-in.}$$

The maximum bending stress is, therefore:

$$s_{max} = M_{max}/S = (18.75L^2) \text{ lb-in./209 in.}^3$$
$$= (0.09L^2) \text{ psi}$$

Since the maximum allowable bending stress is 1450 psi,

$$0.09L^2 = 1450$$
$$L^2 = 16{,}110 \text{ in.}^2$$
$$L = 126.9 \text{ in.}$$

So far, we see that the maximum beam length is dictated by the maximum allowable bending stress.

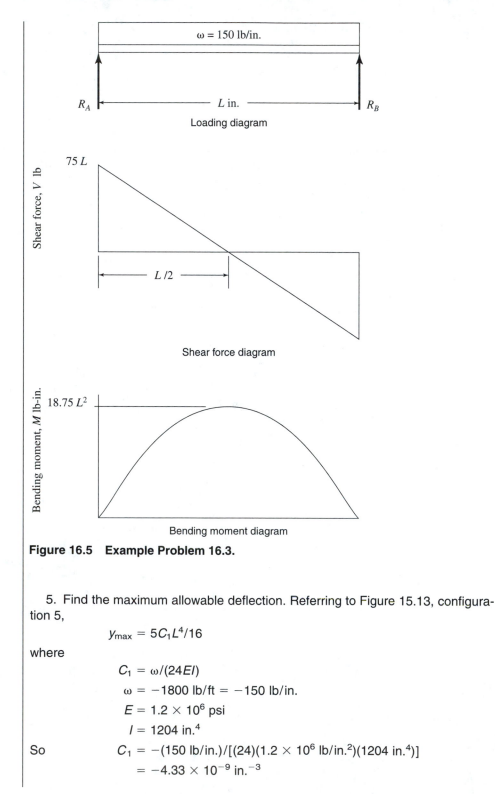

Figure 16.5   Example Problem 16.3.

5. Find the maximum allowable deflection. Referring to Figure 15.13, configuration 5,

$$y_{max} = 5C_1L^4/16$$

where

$$C_1 = \omega/(24EI)$$
$$\omega = -1800 \text{ lb/ft} = -150 \text{ lb/in.}$$
$$E = 1.2 \times 10^6 \text{ psi}$$
$$I = 1204 \text{ in.}^4$$

So

$$C_1 = -(150 \text{ lb/in.})/[(24)(1.2 \times 10^6 \text{ lb/in.}^2)(1204 \text{ in.}^4)]$$
$$= -4.33 \times 10^{-9} \text{ in.}^{-3}$$

Therefore,

$$y_{max} = -5(4.33 \times 10^{-9} \text{ in.}^{-3})L^4/16$$
$$= -(1.35 \times 10^{-9})L^4 \text{ in.}$$

The maximum allowable deflection is $-L/360$ in., so

$$-(1.35 \times 10^{-9})L^4 = -L/360$$
$$L^3 = 2.055 \times 10^6 \text{ in.}^2$$
$$L = 127.1 \text{ in.}$$

6. The maximum allowable beam lengths are

dictated by shear stress, $L_{max} = 159.4$ in.

dictated by bending stress, $L_{max} = 126.9$ in.

dictated by deflection, $L_{max} = 127.1$ in.

Answer

The maximum beam length is 126.9 in., limited by bending stress.

You should realize that there is no difference between the lengths required for shear stress and for deflection in the preceding example. The difference of 0.2 in. means nothing. After conducting a few experiments in the Strengths Lab you will be happy to come to within 10% of your predictions. This is especially true when wood is the material, because moisture content, grain configuration, cracks, etc., all affect the mechanical properties.

16.5   COMPUTER ANALYSES

16.5.1   Programs 'Shapes1' and 'Shapes2'

Both 'Shapes1' and 'Shapes2' are designed to select 10 candidate beams for a given set of input values. The selected beams are in two sets of five each, the first set satisfying the section modulus requirement, i.e., satisfying the bending stress, and the second set satisfying the moment of inertia requirement, i.e., satisfying the deflection. The operator then chooses the best of the 10 designs.

Program 'Shapes1' is for the determination of the optimum beam for steel W, S, and C types.

Program 'Shapes2' is for the determination of the optimum beam for steel L-types, aluminum alloy I- and C-types, and timber beams.

The input parameters are

1. The type of units, US or SI,
2. The maximum bending moment,
3. The maximum shear force,
4. The maximum allowable deflection,
5. The type of material, and
6. The basic deflection (see Section 16.4).

Figure 16.6(a) shows the input/output sheet for program 'Shapes1', and Figure 16.6(b), (c), and (d) shows the input/output sheets for program 'Shapes2'.

The maximum bending moment and shear force may be calculated by hand for simple loadings or by using program 'Shrmmtcl' for cantilever beams or 'Shr&mmt' for simply supported and overhanging beams. Material selection is by a code number. Nine structural steels, 18 aluminum alloys, and 6 types of wood are offered (see Figure 16.6(a), (b), (c), and (d)).

PROGRAM 'Shapes1'

STEEL BEAMS-W, S, & C

Input

		If US	If SI	Matl. Code
				#	Steel

Input field	If US	If SI	#	Steel
Units: Enter US or SI				
Max Bend Moment	lb-in.	Nmm	1	A 36
Max Shear Force	lb	N	2	A 242
Max Deflection	in.	mm	3	A 441
Matl: Enter Code #			4	A 529
Basic Deflection			5	A 572 G42
Based on either:			6	A 572 G50
I =1in.^4 or 10^6mm^4	in.	mm	7	A 572 G60
			8	A 572 G65
			9	A 588

Calcs

		If US	If SI
Allow Bend Stress	ERROR	psi	MPa
Allow Shear Stress	ERROR	psi	MPa

Output

		If US	If SI
Min Section Modulus	ERR	in.^3	10^3mm^3
Min Mmt of Inertia	ERR	in.^4	10^6mm^4
Min Web Area	ERR	in.^2	mm^2

Candidates

Beam Code	Beam Designation	Sect Mod	Mmt of Inertia	Web Area	
ERR	ERR	ERR	ERR	ERR	
ERR	ERR	ERR	ERR	ERR	
ERR	ERR	ERR	ERR	ERR	FOR BENDING STRESS
ERR	ERR	ERR	ERR	ERR	
ERR	ERR	ERR	ERR	ERR	

Beam Code	Beam Designation	Sect Mod	Mmt of Inertia	Web Area	
ERR	ERR	ERR	ERR	ERR	
ERR	ERR	ERR	ERR	ERR	
ERR	ERR	ERR	ERR	ERR	FOR DEFLECTION
ERR	ERR	ERR	ERR	ERR	
ERR	ERR	ERR	ERR	ERR	

(a) Input/output for program 'Shapes1'

Figure 16.6

PROGRAM 'Shapes2'

STEEL BEAMS-L SHAPES

Input

		If US	If SI	Matl. Code	
				#	Steel
Units: Enter US or SI		If US	If SI	1	A 36
Max Bend Moment		lb-in.	Nmm	2	A 242
Max Shear Force		lb	N	3	A 441
Max Deflection		in.	mm	4	A 529
Matl: Enter Code #				5	A 572 G42
Basic Deflection				6	A 572 G50
Based on either:				7	A 572 G60
I =1in.^4 or 10^6mm^4		in.	mm	8	A 572 G65
				9	A 588

Calcs

		If US	If SI
Allow Bend Stress	ERROR	psi	MPa
Allow Shear Stress	ERROR	psi	MPa

Output

		If US	If SI
Min Section Modulus	ERR	in.^3	10^3mm^3
Min Mmt of Inertia	ERR	in.^4	10^6mm^4
Min Shear Area	ERR	in.^2	mm^2

Candidates

Beam Code	Beam Designation	Sect Mod	Mmt of Inertia	Shear Area	Mass or Weight per Unit Length	
ERR	ERR	ERR	ERR	ERR	ERR	FOR BENDING STRESS
ERR	ERR	ERR	ERR	ERR	ERR	
ERR	ERR	ERR	ERR	ERR	ERR	
ERR	ERR	ERR	ERR	ERR	ERR	
ERR	ERR	ERR	ERR	ERR	ERR	
ERR	ERR	ERR	ERR	ERR	ERR	FOR DEFLECTION
ERR	ERR	ERR	ERR	ERR	ERR	
ERR	ERR	ERR	ERR	ERR	ERR	
ERR	ERR	ERR	ERR	ERR	ERR	
ERR	ERR	ERR	ERR	ERR	ERR	

(b) Input/output for program 'Shapes2'

Figure 16.6 (*Continued*)

PROGRAM 'Shapes2'

AL. ALLOY BEAMS-I & C

Input

		If US	If SI
Units: Enter US or SI			
Max Bend Moment		lb-in.	Nmm
Max Shear Force		lb	N
Max Deflection		in.	mm
Matl: Enter Code #			
Basic Deflection			
Based on either:			
I =1in.^4 or 10^6mm^4		in.	mm

Matl. Code #	Aluminum Alloy
10	1100-H12
11	1100-H18
12	2014-0
13	2014-T4
14	2014-T6
15	3003-0
16	3003-H12
17	3003-H18
18	5154-0
19	5154-H32
20	5154-H38
21	6061-0
22	6061-T4
23	6061-T6
24	7075-0
25	7075-T6
26	204.0-T4
27	356.0-T6

Calcs

		If US	If SI
Allow Bend Stress	ERROR	psi	MPa
Allow Shear Stress	ERROR	psi	MPa

Output

		If US	If SI
Min Section Modulus	ERR	in.^3	10^3mm^3
Min Mmt of Inertia	ERR	in.^4	10^6mm^4
Min Web Area	ERR	in.^2	mm^2

Candidates

Beam Code	Beam Designation	Sect Mod	Mmt of Inertia	Web Area	Mass or Weight per Unit Length	
ERR	ERR	ERR	ERR	ERR	ERR	FOR BENDING STRESS
ERR	ERR	ERR	ERR	ERR	ERR	
ERR	ERR	ERR	ERR	ERR	ERR	
ERR	ERR	ERR	ERR	ERR	ERR	
ERR	ERR	ERR	ERR	ERR	ERR	
ERR	ERR	ERR	ERR	ERR	ERR	FOR DEFLECTION
ERR	ERR	ERR	ERR	ERR	ERR	
ERR	ERR	ERR	ERR	ERR	ERR	
ERR	ERR	ERR	ERR	ERR	ERR	
ERR	ERR	ERR	ERR	ERR	ERR	

(c) Input/output for program 'Shapes2'

Figure 16.6 (*Continued*)

PROGRAM 'Shapes2'

TIMBER BEAMS

Input

					Matl. Code	
Units: Enter US or SI		If US	If SI		#	Timber
Max Bend Moment		lb-in.	Nmm		28	Redwood
Max Shear Force		lb	N		29	Douglas Fir
Max Deflection		in.	mm		30	Hemlock
Matl: Enter Code #					31	Pond. Pine
Basic Deflection					32	Sitka Spruce
Based on either:					33	Southn Pine
$I = 1\text{in.}^4$ or 10^6mm^4		in.	mm			

Calcs

		If US	If SI
Allow Bend Stress	ERROR	psi	MPa
Allow Shear Stress	ERROR	psi	MPa

Output

		If US	If SI
Min Section Modulus	ERR	in.^3	10^3mm^3
Min Mmt of Inertia	ERR	in.^4	10^6mm^4
Min Shear Area	ERR	in.^2	mm^2

Candidates

Beam Code	Beam Designation	Scct Mod	Mmt of Inertia	Shear Area	Mass or Weight per Unit Length	
ERR	ERR	ERR	ERR	ERR	ERR	FOR BENDING STRESS
ERR	ERR	ERR	ERR	ERR	ERR	
ERR	ERR	ERR	ERR	ERR	ERR	
ERR	ERR	ERR	ERR	ERR	ERR	
ERR	ERR	ERR	ERR	ERR	ERR	
ERR	ERR	ERR	ERR	ERR	ERR	FOR DEFLECTION
ERR	ERR	ERR	ERR	ERR	ERR	
ERR	ERR	ERR	ERR	ERR	ERR	
ERR	ERR	ERR	ERR	ERR	ERR	
ERR	ERR	ERR	ERR	ERR	ERR	

(d) Input/output for program 'Shapes2'

Figure 16.6   (Continued)

The basic deflection may be calculated by hand, using the principle of superposition if necessary, or by using program 'Deflect'. If U.S. customary units are being used, then $I = 1$ in.^{4}; if SI are used, then $I = 1 \times 10^6$ mm^4 for the *basic deflection* calculations.

The programs determine the allowable bending and shear stresses, the minimum section modulus, minimum moment of inertia, and minimum shear area required. They then select five candidate beams whose section modulii exceed the required value and five candidate beams whose moments of inertia exceed the required value, and then display the following:

1. Beam code, which is the listing number for the beam. The listing numbers are tabulated on page B of the screen.
2. Beam designation.
3. Section modulus.
4. Moment of inertia.
5. Shear area.
6. Weight or mass per unit length.

Page B of program 'Shapes1' contains a list of 100 steel W-, S-, and C-shapes. Page B of program 'Shapes2' contains lists of 75 steel L shapes, 35 aluminum alloy I- and C-shapes, and 33 timber sizes, in U.S. customary units and in SI units. The dimensions and elastic properties are tabulated for each.

These tables are rearranged in order of increasing section modulus on pages C of the program screens. After determining the minimum allowable section modulus for a set of input parameters, the programs search the data on pages C to find the first section modulus that is greater than the minimum allowable. This beam is the first of the candidates. Pages D are the same as pages C except that the section modulus entries are shifted down one row relative to the rest of the data. This is a simple way of making the programs choose the beam with the next higher section modulus for the second candidate. Repeating this process on pages E, F, and G gives candidates 3, 4, and 5.

The same approach is repeated for the moment of inertia, where pages I, J, K, L, and M are used.

The following examples demonstrate how to use these programs.

EXAMPLE PROBLEM 16.4

Select a W-, S-, or C-shape A36 structural steel cantilever beam with a span of 12 ft to carry a uniformly distributed load of 720 lb/ft and a concentrated load of 15,000 lb at its midpoint. The allowable maximum deflection is 0.50 in.

Solution

1. Determine the maximum shear force and maximum bending moment. These can be calculated readily by hand.

Since it is a cantilever beam, the maximum shear force is simply the total load.

$$V_{max} = (-720 \text{ lb/ft})(12 \text{ ft}) + (-15,000 \text{ lb}) = -23,640 \text{ lb}$$

The center of gravity of this total load is at the midpoint of the beam, i.e., 6 ft from the wall. Therefore, the maximum bending moment (which occurs at the wall for cantilever beams) is

$$M_{max} = (-23,640 \text{ lb})(6 \text{ ft}) = -148,040 \text{ lb-ft}.$$

We are going to need this as lb-in.:

$$M_{max} = (-148,040 \text{ lb-ft})(12 \text{ in./ft}) = 1.702 \times 10^6 \text{ lb-in.}$$

2. Determine the *basic deflection*. Because we are using U.S. customary units, the moment of inertia will be 1 in.^{4} for this calculation. We calculate the deflections using configuration 9 and then configuration 10; then we add the results. This is left to the student as an exercise.

For configuration 9, the resulting maximum deflection is -160.883 in. For configuration 10, the maximum deflection is -111.202 in.

Adding the two results gives

$$y_{max} = (-160.883 \text{ in.}) + (-111.202 \text{ in.}) = -272.1 \text{ in.}$$

PROGRAM 'Shapes1'

STEEL BEAMS-W, S, & C

Input

Units: Enter US or SI	US	If US	If SI
Max Bend Moment	1702100	lb-in.	Nmm
Max Shear Force	23640	lb	N
Max Deflection	-0.5	in.	mm
Matl: Enter Code #	1		
Basic Deflection			
Based on either:			
I =1in.^4 or 10^6mm^4	-272.1	in.	mm

Matl. Code	
#	Steel
1	A 36
2	A 242
3	A 441
4	A 529
5	A 572 G42
6	A 572 G50
7	A 572 G60
8	A 572 G65
9	A 588

Calcs

		If US	If SI
Allow Bend Stress	24000	psi	MPa
Allow Shear Stress	14500	psi	MPa

Output

		If US	If SI
Min Section Modulus	70.9208333	in.^3	10^3mm^3
Min Mmt of Inertia	544.2	in.^4	10^6mm^4
Min Web Area	1.63034483	in.^2	mm^2

Candidates

Beam Code	Beam Designation	Sect Mod	Mmt of Inertia	Web Area	
20	W14 × 53	77.8	541	5.1504	
29	W12 × 58	78	475	4.3884	
12	W16 × 50	81	659	6.1788	FOR BENDING STRESS
5	W21 × 44	81.6	843	7.231	
38	W10 × 77	85.9	455	5.618	

Beam Code	Beam Designation	Sect Mod	Mmt of Inertia	Web Area	
28	W12 × 72	97.4	597	5.2675	
8	W18 × 40	68.4	612	5.6385	
37	W10 × 100	112	623	7.548	FOR DEFLECTION
12	W16 × 50	81	659	6.1788	
36	W10 × 112	126	716	8.5768	

Figure 16.7   Input/output for Example Problem 16.4.

3. Open program 'Shapes1' and enter the values shown on the input/output sheet (Figure 16.7). The output shows that the following minimum values are needed:

$$\text{section modulus} = 70.92 \text{ in.}^3$$

$$\text{moment of inertia} = 544.2 \text{ in.}^4$$

$$\text{web area} = 1.630 \text{ in.}^2$$

Looking at the first five candidates (whose section modulii are all greater than 70.92 in.^{3}), two of them satisfy the moment of inertia and the web area requirements. These are the W16 $\times$ 50 and the W21 $\times$ 44.

Of the second five candidates (whose moments of inertia are all greater than 544.2 in.^{4}), only one, the W18 $\times$ 40 beam, does not meet the necessary section modulus.

So we have the following from which to choose:

W16 $\times$ 50	weighing 50 lb/ft
W21 $\times$ 44	weighing 44 lb/ft
W12 $\times$ 72	weighing 72 lb/ft
W10 $\times$ 100	weighing 100 lb/ft
W10 $\times$ 112	weighing 112 lb/ft

We select the W21 $\times$ 44 beam because it is lightest beam that satisfies all the requirements.

Answer
Select a W21 $\times$ 44 beam.

If you need to know more about this beam, its dimensions and elastic properties are tabulated on page B of your screen, code number 5.

EXAMPLE PROBLEM 16.5

A simply supported 2-m pine beam is to be used as a shelf to carry a total uniformly distributed load of 1000 kg, and two concentrated loads of 500 kg at 0.5 m from the end and 750 kg at 0.75 m from the same end. The maximum deflection is not to exceed 2.0 mm. What standard-size timber should be selected? Use an elastic modulus of 9000 MPa.

Solution
1. Establish consistent units.

$$\text{beam length} = 2 \text{ m} = 2000 \text{ mm}$$

$$\text{distributed load} = (-1000 \text{ kg})(9.81 \text{ m/s}^2) = -9810 \text{ N}$$

$$\omega = -9810 \text{ N/2 m} = -4905 \text{ N/m} = -4.905 \text{ N/mm}$$

$$\text{concentrated load 1} = (-500 \text{ kg})(9.81 \text{ m/s}^2)$$
$$= -4905 \text{ N} \quad \text{at } x = 0.50 \text{ m} = 500 \text{ mm}$$
$$\text{concentrated load 2} = (-750 \text{ kg})(9.81 \text{ m/s}^2)$$
$$= -7358 \text{ N} \quad \text{at } x = 0.75 \text{ m} = 750 \text{ mm}$$

2. Determine the maximum bending moment and the maximum shear force. Open program 'Shr&mmt.' The solutions are

$$\text{maximum shear force} = 13,173 \text{ N}$$
$$\text{maximum bending moment} = 7281 \text{ N} \cdot \text{m}$$

3. Determine the basic deflection. We use $I = 1 \times 10^6 \text{ mm}^4$ because this is a calculation in SI units. Open program 'Deflect'. The input/output sheet is shown in Figure 16.8(a) *Be careful* to use the correct units.

The maximum deflection (which is the basic deflection, because we used an *I*-value of $1 \times 10^6 \text{ mm}^4$) is -301.6 mm.

4. Print out candidate beam sizes. Open program 'Shapes2'. Move across the screen to **TIMBER BEAMS**.

The input/output is shown in Figure 16.8(b). Again, be careful of the units. The output shows that the following minimum values are required:

$$\text{section modulus} = 1055 \times 10^3 \text{ mm}^3$$
$$\text{moment of inertia} = 150.8 \times 10^4 \text{ mm}^4$$
$$\text{shear area} = 0.029 \times 10^6 \text{ mm}^2$$

Of the first set of five candidate beams, the first does not meet the minimum moment of inertia requirement, and the second does not meet the minimum shear area requirement.

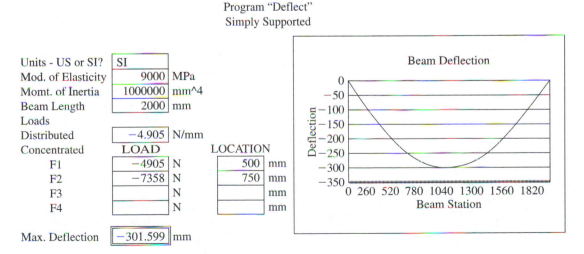

Figure 16.8   Example Problem 16.5.

PROGRAM 'Shapes2'

TIMBER BEAMS

Input

					Matl. Code	
Units: Enter US or SI	SI	If US		If SI	#	Timber
Max Bend Moment	7281000	lb-in.		Nmm	28	Redwood
Max Shear Force	13173	lb		N	29	Douglas Fir
Max Deflection	−2	in.		mm	30	Hemlock
Matl: Enter Code #	31				31	Pond. Pine
Basic Deflection					32	Sitka Spruce
Based on either:					33	Southn Pine
I =1in.^4 or 10^6mm^4	−301.6	in.		mm		

Calcs

		If US	If SI
Allow Bend Stress	6.9	psi	MPa
Allow Shear Stress	0.45	psi	MPa

Output

		If US	If SI
Min Section Modulus	1055.21739	in.^3	10^3mm^3
Min Mmt of Inertia	150.8	in.^4	10^6mm^4
Min Shear Area	29.2733333	in.^2	10^3mm^2

Candidates

Beam Code	Beam Designation	Sect Mod	Mmt of Inertia	Shear Area	Mass or Weight per Unit Length	
1223	200 × 200	1161	111	36.5	21.9	FOR BENDING STRESS
1217	100 × 300	1213	174	25.4	15.3	
1220	150 × 250	1355	163	33.7	20.2	
1224	200 × 250	1849	223	46	27.6	
1221	150 × 300	1989	290	40.9	24.6	
1220	150 × 250	1355	163	33.7	20.2	FOR DEFLECTION
1217	100 × 300	1213	174	25.4	15.3	
1224	200 × 250	1849	223	46	27.6	
1228	250 × 250	2333	281	58.1	34.8	
1221	150 × 300	1989	290	40.9	24.6	

(b) Input/output

Figure 16.8   (*Continued*)

Note: The units in the candidate table match those of the output. All you have to do is compare the numbers.

All in the second set except the second entry satisfy all the requirements. The lightest of the seven acceptable entries is the 150 × 250 beam, which occurs in both sets, so we select this as the best.

Answer

Choose a 150 × 250 beam. This is close to a 6 × 10 in inches.

EXAMPLE PROBLEM 16.6

Figure 16.9 (a) shows an 8-ft aluminum alloy 6061-0 cantilevered beam carrying a uniformly distributed load of 50 lb/ft, a download of 1500 lb at the free end, and an upload of 1000 lb at its midpoint. The tip deflection is not to exceed $\frac{9}{16}$ in. Select the beam shape and size. 6061-0 has an elastic modulus of 10×10^6 psi.

Solution

1. Determine the maximum shear force and the maximum moment of inertia. Open program 'Shrmmtcl.' We show the input/output sheet in Figure 16.9(b). Be careful to enter the upload as a positive value. Notice that by entering the distances in inches, we obtained the bending moment output in lb-in., which are the units we need for the 'Shapes2' program. We see that the maximum shear force is 1696 lb, and the maximum bending moment is 115,202 lb-in. You can also see the shear force and bending moment diagrams by moving to the right of the screen.

2. Determine the basic deflection. Since the units for this problem are U.S. customary, we will use $I = 1$ in^4. So $EI = 1 \times 10^7$ lb-in^2. We use beam configuration 9 (see Figure 15.13) for the concentrated loads.

For the download:

$$L = 8 \text{ ft} = 96 \text{ in.}, \qquad a = 0, \qquad F = -1500 \text{ lb}$$

$$C_1 = F/(6EI) = (-1500 \text{ lb})/(6 \times 10^7 \text{ lb-in.}^2) = -2.5 \times 10^{-5} \text{ in.}^{-2}$$

The maximum deflection is

$$y_{max} = C_1(L - a)^2(2L + a)$$
$$= (-2.5 \times 10^{-5} \text{ in.}^{-2})(96 \text{ in.})^2(2 \times 96 \text{ in.})$$
$$= -44.237 \text{ in.}$$

For the upload:

$$L = 96 \text{ in.}, \qquad a = 4 \text{ ft} = 48 \text{ in.}, \qquad F = +1000 \text{ lb}$$

$$C_1 = F/(6EI) = (1000 \text{ lb})/(6 \times 10^7 \text{ lb-in.}^2) = 1.667 \times 10^{-5} \text{ in.}^{-2}$$

The maximum deflection is

$$y_{max} = C_1(L - a)^2(2L + a)$$
$$= 1.667 \times 10^{-5} \text{ in.}^{-2})(96 \text{ in.} - 48 \text{ in.})^2(2 \times 96 \text{ in.} + 48 \text{ in.})$$
$$= +9.216 \text{ in.}$$

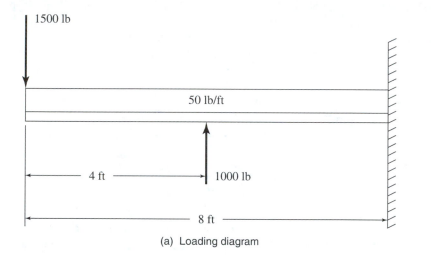

(a) Loading diagram

PROGRAM 'SHRMMTCL'

SHEAR FORCE AND BENDING MOMENT DIAGRAMS
CANTILEVER BEAM
PROGRAM SHRMMTCL

NOTE: DO NOT ERASE OR ENTER ANY VALUES IN
ANY SPACE SURROUNDED BY DOUBLE LINES.

UNITS	
CONCENTR'D LOAD	lb
DISTRIBUTED LOAD	lb/in.
DISTANCE	in.

BEAM LENGTH	WALL REACT	MAXIMUM	MAXIMUM
in.	RW	S.F.	B.M.
96	lb	lb	
	900	1696	115202

CONCENTRATED LOADS		DISTRIBUTED LOADS			DISTANCE FROM LEFT END OF BEAM	SHEAR FORCE	BENDING MOMENT
LOAD	X	LOAD	START	END	in.	lb	
lb	in.	lb/in.	in.	in.			
−1500	0	−4.167	0	96	0	−1500	0
1000	48				9.6	−1540	−14592.02
					19.2	−1580	−29568.06
					28.8	−1620	−44928.14
					38.4	−1660	−60672.25
−500					48	−700	−76800.38
					57.6	−740	−83712.55
					67.2	−780	−91008.75
					76.8	−820	−98688.98
					86.4	−860	−106753.2
					96	−900	−115201.5

(b) Input/output

Figure 16.9   Example Problem 16.9.

PROGRAM 'Shapes 2'
AL. ALLOY BEAMS-I & C

Input

			If SI		Matl. Code	Aluminum
Units: Enter US or SI	US	If US			#	Alloy
Max Bend Moment	115202	lb-in.	Nmm		10	1100-H12
Max Shear Force	1696	lb	N		11	1100-H18
Max Deflection	−0.5625	in.	mm		12	2014-0
Matl: Enter Code #	21				13	2014-T4
Basic Deflection					14	2014-T6
Based on either:					15	3003-0
I =1in.^4 or 10^6mm^4	−39.445	in.	mm		16	3003-H12
					17	3003-H18
					18	5154-0
					19	5154-H32
					20	5154-H38

Calcs

		If US	If SI		21	6061-0
Allow Bend Stress	4930	psi	MPa		22	6061-T4
Allow Shear Stress	2000	psi	MPa		23	6061-T6
					24	7075-0
					25	7075-T6
					26	204.0-T4
					27	356.0-T6

Output

		If US	If SI
Min Section Modulus	23.36754564	in.^3	10^3mm^3
Min Mmt of Inertia	70.12444444	in.^4	10^6mm^4
Min Web Area	0.848	in.^2	mm^2

Candidates Beam Code	Beam Designation	Sect Mod	Mmt of Inertia	Web Area	Mass or Weight per Unit Length	
187	10.00 × 6.00	26.42	132.09	2.5	8.646	FOR BENDING STRESS
209	12.00 × 4.00	26.63	159.76	3.48	8.274	
188	10.00 × 6.00	31.16	155.79	2.9	10.286	
210	12.00 × 5.00	39.95	239.69	4.2	11.822	
189	12.00 × 7.00	42.6	255.57	3.48	11.672	
206	9.00 × 4.00	17.4	78.31	2.61	6.97	FOR DEFLECTION
207	10.00 × 3.50	16.64	83.22	2.5	6.136	
186	9.00 × 5.50	22.67	102.02	2.43	8.361	
208	10.00 × 4.25	23.23	116.15	3.1	8.36	
187	10.00 × 6.00	26.42	132.09	2.5	8.646	

(c) Input/output

Figure 16.9 (*Continued*)

Using beam configuration 10 (see Figure 15.13) for the distributed load,

$$L = 96 \text{ in.,} \qquad \omega = -50 \text{ lb/ft} = -4.1667 \text{ lb/in.}$$

$$C_1 = \omega/(8EI) = (-4.1667 \text{ lb/in.})/(8 \times 10^7 \text{ lb-in.}^2) = -5.208 \times 10^{-8} \text{ in.}^{-3}$$

The maximum deflection is

$$y_{max} = C_1 L^4$$
$$= (-5.208 \times 10^{-8} \text{ in.}^{-3})(96 \text{ in.})^4$$
$$= -4.424 \text{ in.}$$

The total maximum basic deflection is

$$y_{max} = -44.237 \text{ in.} + 9.216 \text{ in.} - 4.424 \text{ in.} = -39.445 \text{ in.}$$

3. Select the beam. Open program 'Shapes2'. The input/output sheet is shown in Figure 16.9(c). The output shows the following minimum required values:

$$\text{section modulus} = 23.37 \text{ in.}^3$$
$$\text{moment of inertia} = 70.12 \text{ in.}^4$$
$$\text{web area} = 0.848 \text{ in.}^2$$

In the first set, all the candidates have moments of inertia that exceed the required minimum value of 70.12 in.^{4}, so they are all acceptable. Notice that there are two in the first set that both have the same designation, 10.00 × 6.00, but different code numbers. This is not an error. They both have the same overall size, but one is thicker than the other. The one with the 187 code number weighs 8.646 lb/ft, whereas the one with the 188 code number weighs 10.268 lb/ft. In the second set, only the last one exceeds the minimum required section modulus and so is the only acceptable candidate in that set.

So the acceptable beam sizes are

10.00 × 6.00	weighing 8.646 lb/ft
12.00 × 4.00	weighing 8.274 lb/ft
10.00 × 6.00	weighing 10.286 lb/ft
12.00 × 5.00	weighing 11.822 lb/ft
12.00 × 7.00	weighing 11.672 lb/ft

The lightest beam satisfying all the requirements is the 12.00 × 4.00.

Answer

Select a 12.00-in. × 4.00-in. channel beam. You can see the properties of this beam on screen page B, code number 209.

16.6   PROBLEMS FOR CHAPTER 16

Hand Calculations

1. An I-beam has a depth of 20.66 in. and a section modulus of 81.6 in^3. What is its moment of inertia?

2. An A529 structural steel W18 × 46 beam is to be used as a cantilever to support a 100,000-lb load at its free end. A W18 × 46

shape has a section modulus of 78.8 in.^{3}, a moment of inertia of 712 in.^{4}, and a web area of 6.502 in^2. The allowable bending stress for A529 structural steel is 27,700 psi, and the allowable shear stress is 16,800 psi. Its elastic modulus is 29 $\times$ 10^6 psi.

a. What is the longest safe length that can be used?

b. Is it safe in shear?

c. What is the maximum deflection if the beam is made to its longest safe length?

3. A 10-ft-long A36 structural steel pipe is to be used as a simply supported beam to support a 1000-lb load at its midpoint. The maximum allowable deflection is $-\frac{5}{8}$ in. Select the diameter and schedule of the lightest acceptable pipe from the following table.

Nominal size (in.)	Actual size (in.)	Schedule	I (in.^{4})	S (in.^{3})	Weight per ft (lb)
2.5	2.875	40	1.530	1.064	5.793
2.5	2.875	80	1.925	1.339	7.661
2.5	2.875	160	2.353	1.637	10.01
3.0	3.500	10S	1.822	1.041	4.33
3.0	3.500	40	3.020	1.724	7.58
3.5	4.000	5S	1.960	0.980	3.47
3.5	4.000	10S	2.756	1.378	4.97
4.0	4.500	10S	3.960	1.762	5.61

4. Based on the pipe you selected in Problem 3:

a. what will be the actual bending stress?

b. what will be the actual maximum deflection?

Hint: Do this by simple ratios.

Computer Calculations

5. Select a W-, S-, or C-shape A36 structural steel cantilever beam with a span of 3.5 m to carry a uniformly distributed load of 10 kN/m and a concentrated load of 3300 N at its midpoint. The allowable maximum deflection is -10 mm. See Figure 16.10.

6. The beam material in Problem 5 is changed to structural steel A588. What effect does this have on the optimum beam shape? Why?

7. A design change requires that the maximum deflection of the beam in Problem 5 be limited to -5 mm. Select the optimum beam shape.

8. A simply supported 6-ft Sitka spruce beam (Figure 16.11) is to be used to support a total uniformly distributed load of 7200 lb and two concentrated loads, one 5000 lb at 2 ft from the end and the other 6000 lb at 4 ft from the same end. The maximum deflection is not to exceed -0.10 in. What standard-size timber should be selected? Use an elastic modulus of 1.3 $\times$ 10^6 psi. (Don't forget that the shear area must be satisfied also.)

9. The Sitka spruce in Problem 8 is to be replaced by a beam of Douglas fir with an elastic modulus of 1.7 $\times$ 10^6 psi. Select the optimum beam size.

10. Figure 16.12 shows a 2.5-m aluminum alloy 6061-T6 cantilevered beam carrying a uniformly distributed load of 730 N/m, a download of 4500 N at its midpoint, and an upload of 6500 N at the free end. The tip deflection is not to exceed 6 mm upward. Select the beam shape and size. 6061-T6 has an elastic modulus of 69 GPa.

11. The 6061-T6 of Problem 10 is to be changed

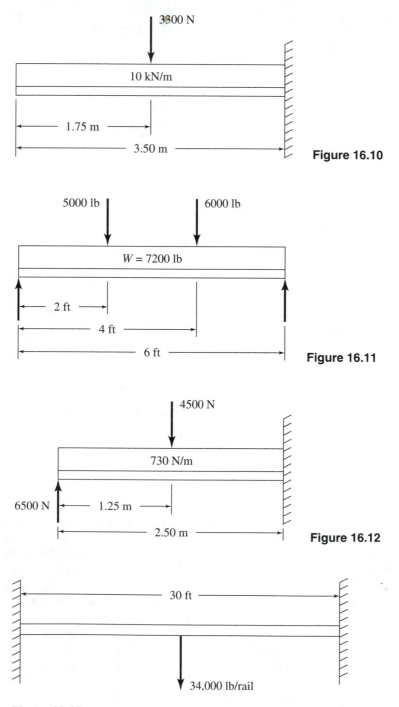

Figure 16.10

Figure 16.11

Figure 16.12

Figure 16.13

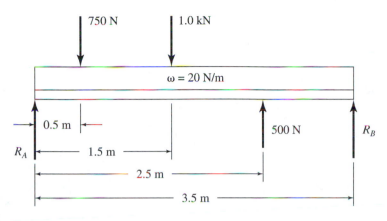

Figure 16.14

to aluminum alloy 2014-T4. Select the new optimum beam shape.

12. An overhead crane rides on two identical rails that are rigidly fixed in concrete at their ends. The span is 30 ft (Figure 16.13). The crane's maximum capacity is 34 tons, and the maximum deflection is limited to -0.25 in. The rail material is A572 Grade 42 structural steel. The buyer has requested a safety factor on the bending stress of 3.0 above the normally accepted allowable stress. Select a suitable W-, S-, or C-shape for the rails.

Note: The maximum shear force is $F/2$, and the maximum bending moment is $FL/8$ (see Chapter 18).

Hint: Calculate the bending moment; then multiply by 3.0 to allow for the safety factor.
  The design condition is when the crane is at midspan.

13. The rails in Problem 12 are to be replaced with structural steel A36 rails. Choose the optimum shape.

14. Referring to Problem 12, a design change requires that the maximum deflection be limited to -0.125 in. Choose the optimum shape.

15. Select A36 structural steel W-, S-, or C-shape beams for a balcony. The beams are

4 ft on centers and support a load of 500 lb/ft^2. The balcony is constructed as a propped cantilever, extending 12 ft from the wall. Maximum deflection is limited to $\frac{1}{360}$ of the span.

Hint: Determine the load on one beam, and express it as a uniformly distributed load.

Note: For a propped cantilever beam carrying a uniformly distributed load:

$$\text{maximum shear force} = 5\omega L/8$$

$$\text{maximum bending moment} = \omega L^2/8$$

These both occur at the wall.

16. A design change requires that the maximum deflection of the balcony in Problem 15 be limited to -0.10 in. Choose the optimum beam shape.

17. The simply supported Ponderosa pine beam shown in Figure 16.14 is limited to a maximum deflection of -6 mm. Select the lightest acceptable standard size. Use an elastic modulus of 7.6×10^3 MPa.

18. Select the optimum timber size if the maximum allowable deflection for Problem 17 is changed to -3 mm.

19. Select the optimum timber size if the maximum allowable deflection for Problem 17 is changed to -1.5 mm.

17

Statically Indeterminate Beams

■ *Objectives*

In this chapter you will learn how to determine moments and reactions for continuous beams having more than two supports, for propped cantilever beams, and for beams built in at both ends.

Many beams we may encounter in real situations cannot be solved by the methods discussed in previous chapters. Most of those came under the classification *statically determinate beams*, meaning that complete solutions could be obtained by using the equations for static equilibrium:

$$\sum F_x = 0$$
$$\sum F_y = 0$$
$$\sum M = 0$$

Indeterminate beams, such as those shown in Figure 13.2, cannot be solved using these equations alone. One very powerful means of analysis is the *finite-element method*, in which the body is analyzed as a group of many little structural elements. This, of course, requires the use of a computer and a program such as 'ALGOR SUPERSAP' by Algor Inc. We present spreadsheet programs for propped and built-in beams at the end of this chapter. First, however, we investigate two methods of analysis that provided accurate results before computers became commonplace. This is not a history lesson; it is important to understand the underlying principles from which these methods were derived. The two methods are the *three-moment equation* and the *method of superposition*.

17.1   THE THREE-MOMENT EQUATION

The three-moment equation is sometimes referred to as Clapeyron's three-moment equation.

Figure 17.1(a) shows part of a statically indeterminate beam. It has more than two simple supports and is called a **continuous** beam. We know that the sum of the reaction forces acting up must be equal to the sum of the loads acting down and that the algebraic sum of the moments about any point must be zero, but we cannot determine

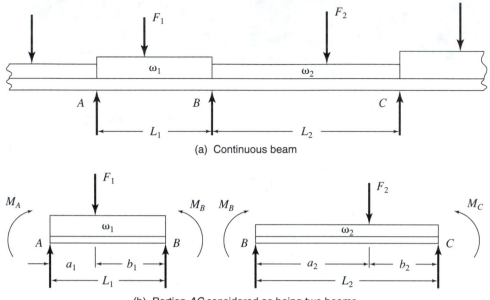

(a) Continuous beam

(b) Portion *AC* considered as being two beams

Figure 17.1

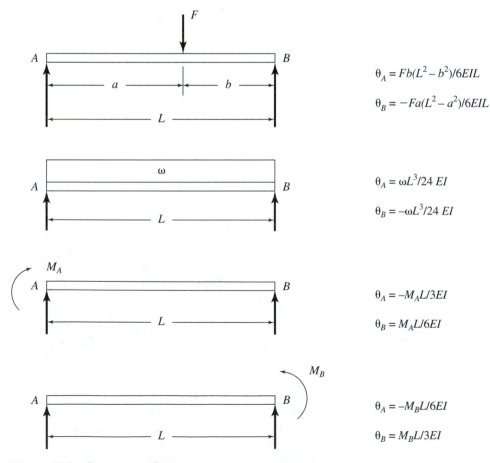

$\theta_A = Fb(L^2 - b^2)/6EIL$

$\theta_B = -Fa(L^2 - a^2)/6EIL$

$\theta_A = \omega L^3/24\ EI$

$\theta_B = -\omega L^3/24\ EI$

$\theta_A = -M_A L/3EI$

$\theta_B = M_A L/6EI$

$\theta_A = -M_B L/6EI$

$\theta_B = M_B L/3EI$

Figure 17.2   Slope equations.

the individual reactions by using the static equilibrium equations alone. We consider that portion of the beam in the region from support A to support C. If we divide that portion into two beams (see Figure 17.1(b)), then the clue to the solution is that the *slope* at B for the drawing on the left must be the same as the *slope* at B for the drawing on the right, because the beam is continuous. Since there are loads to the left of A and to the right of C, there will be moments M_A at A and M_C at C, in addition to the moment M_B at B.

We showed the derivation of the equation for the slope of a beam in Chapter 15. Using that equation,

$$\theta = dy/dx = \int M/(EI)dx + C$$

we can derive the expressions required for our present analysis, shown in Figure 17.2. (For example, see Using Calculus.)

USING CALCULUS

EXAMPLE

Derive the slope equation for a simply supported beam carrying a uniformly distributed load over its entire span, as in the second case of Figure 17.2.

Solution

The general slope equation is

$$dy/dx = \int M/(EI)\ dx + C$$

As before, we will derive the equation for the case of a *positive* load (upward acting) but be careful to enter downloads as negative in actual calculations.

The reactions are $R_A = R_B = -\omega L/2$; therefore, at any station x, R_A creates a bending moment of $-\omega Lx/2$. At station x there is also a load of ωx to the left whose center of gravity is at $x/2$. This creates a bending moment of $\omega x^2/2$, making the total bending moment at station x equal to $M = -\omega Lx/2 + \omega x^2/2$.

So

$$dy/dx = -\omega/(2EI) \int (Lx - x^2)\ dx + C$$

$$= -\omega(Lx^2/2 - x^3/3)/(2EI) + C$$

When $x = L/2$, $dy/dx = 0$. Therefore,

$$0 = -\omega(L^3/8 - L^3/24)/(2EI) + C$$

$$0 = -\omega L^3/(24EI) + C$$

Therefore, $C = \omega L^3/(24EI)$, and

$$dy/dx = -\omega(Lx^2/2 - x^3/3)/(2EI) + \omega L^3/(24EI)$$

$$\theta = dy/dx = -\omega(Lx^2/2 - x^3/3 - L^3/12)/(2EI)$$

This is the general slope equation for a simply supported beam carrying a uniformly distributed load over its entire span.

At the ends of the beam, $x = 0$ and $x = L$. At $x = 0$:

$$\theta_A = -\omega(-L^3/12)/(2EI)$$

$$= \omega L^3/(24EI)$$

(Notice that this is positive, but when we enter a download of $-\omega$, the slope at $x = 0$ will be negative, as it should be.) At $x = L$:

$$\theta_B = -\omega(L^3/2 - L^3/3 - L^3/12)/(2EI)$$

$$= -\omega L^3/(24EI)$$

Answer
The slopes at the ends of the beam are

$$\theta_A = \omega L^3/(24EI) \quad \text{and} \quad \theta_B = -\omega L^3/(24EI)$$

We are now going to calculate the slope at B by considering the left portion (Figure 17.1(b)). It consists of the sum of four slopes (see Figure 17.2).

1. Due to the concentrated load:
$$\theta_1 = -F_1 a_1(L_1^2 - a_1^2)/(6EIL_1)$$

2. Due to the distributed load:
$$\theta_2 = -\omega_1 L_1^3/(24EI)$$

3. Due to the moment at A:
$$\theta_3 = M_A L_1/(6EI)$$

4. Due to the moment at B:
$$\theta_4 = M_B L_1/(3EI)$$

The slope at B is

$$\theta_B = \theta_1 + \theta_2 + \theta_3 + \theta_4$$
$$= -F_1 a_1(L_1^2 - a_1^2)/(6EIL_1) - \omega_1 L_1^3/(24EI)$$
$$+ M_A L_1/(6EI) + M_B L_1/(3EI) \tag{17.1}$$

We now calculate the slope at B by considering the right portion (Figure 17.1(b)). Again, it consists of the sum of four slopes (see Figure 17.2).

1. Due to the concentrated load:
$$\theta_1 = F_2 b_2(L_2^2 - b_2^2)/(6EIL_2)$$

2. Due to the distributed load:
$$\theta_2 = \omega_2 L_2^3/(24EI)$$

3. Due to the moment at B:
$$\theta_3 = -M_B L_2/(3EI)$$

4. Due to the moment at C:
$$\theta_4 = -M_C L_2/(6EI)$$

The slope at B is

$$\theta_B = \theta_1 + \theta_2 + \theta_3 + \theta_4$$
$$= F_2 b_2(L_2^2 - b_2^2)/(6EIL_2) + \omega_2 L_2^3/(24EI)$$
$$- M_B L_2/(3EI) - M_C L_2/(6EI) \tag{17.2}$$

The slopes at B determined from both sides must be the same, so equations (17.1) and (17.2) are equal. Equating the two and dividing throughout by $(6EI)$ gives

$$\frac{-F_1a_1(L_1^2 - a_1^2)}{L_1} - \frac{\omega_1 L_1^3}{4} + M_A L_1 + 2M_B L_1$$
$$= \frac{F_2 b_2(L_2^2 - b_2^2)}{L_2} + \frac{\omega_2 L_2^3}{4} - 2M_B L_2 - M_C L_2$$

Rearranging, we have

$$M_A L_1 + 2M_B(L_1 + L_2) + M_C L_2 = \frac{F_1 a_1(L_1^2 - a_1^2)}{L_1} + \frac{F_2 b_2(L_2^2 - b_2^2)}{L_2}$$
$$+ \frac{\omega_1 L_1^3}{4} + \frac{\omega_2 L_2^3}{4} \tag{17.3}$$

Equation (17.3) is one of the more general forms of the three-moment equation.

EXAMPLE PROBLEM 17.1

Figure 17.3 shows a uniform beam with four supports carrying uniformly distributed loads of 360 lb/ft, 300 lb/ft, and 240 lb/ft. Determine the moments at B and C and the reactions at all four supports.

Solution

1. Determine the moments at B and C. Because there are no concentrated loads, the three-moment equation, equation (17.3), becomes

$$M_A L_1 + 2M_B(L_1 + L_2) + M_C L_2 = \omega_1 L_1^3/4 + \omega_2 L_2^3/4$$

We first apply this to the portion ABC.

$$M_A = 0 \quad \text{because there are no loads to the left of } A$$
$$L_1 = 4 \text{ ft}, \quad L_2 = 5 \text{ ft}$$
$$\omega_1 = -360 \text{ lb/ft}, \quad \omega_2 = -300 \text{ lb/ft}$$

So

$$2M_B(4 \text{ ft} + 5 \text{ ft}) + M_C(5 \text{ ft}) = \frac{(-360 \text{ lb/ft})(4 \text{ ft})^3}{4} + \frac{(-300 \text{ lb/ft})(5 \text{ ft})^3}{4}$$

$$(18 \text{ ft})M_B + (5 \text{ ft})M_C = -5760 \text{ lb-ft}^2 - 9375 \text{ lb-ft}^2$$

$$18M_B + 5M_C = -15{,}135 \text{ lb-ft} \tag{1}$$

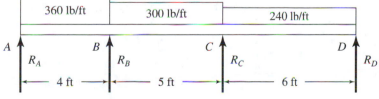

Figure 17.3　Loading for Example Problem 17.1.

Now we apply the three-moment equation to portion BCD. The equation becomes

$$M_B L_2 + 2M_C(L_2 + L_3) + M_D L_3 = \omega_2 L_2^3/4 + \omega_3 L_3^3/4$$

$$M_D = 0 \qquad \text{because there are no loads to the right of } D$$

$$L_2 = 5 \text{ ft}, \qquad L_3 = 6 \text{ ft}$$

$$\omega_2 = -300 \text{ lb/ft}, \qquad \omega_3 = -240 \text{ lb/ft}$$

So

$$M_B(5 \text{ ft}) + 2M_C(5 \text{ ft} + 6 \text{ ft}) = (-300 \text{ lb/ft})(5 \text{ ft})^3/4 + (-240 \text{ lb/ft})(6 \text{ ft})^3/4$$

$$(5 \text{ ft})M_B + (22 \text{ ft})M_C = -9375 \text{ lb-ft}^2 - 12{,}960 \text{ lb-ft}^2$$

$$5M_B + 22M_C = -22{,}335 \text{ lb-ft} \tag{2}$$

To solve the simultaneous equations (1) and (2), multiply (1) by 5 and (2) by 18. Then (1) becomes

$$90M_B + 25M_C = -75{,}675 \text{ lb-ft} \tag{1'}$$

and (2) becomes

$$90M_B + 396M_C = -402{,}030 \text{ lb-ft} \tag{2'}$$

Now subtract (1') from (2'):

$$(396 - 25)M_C = -402{,}030 \text{ lb-ft} - (-75{,}675 \text{ lb-ft})$$

$$371M_C = -326{,}355 \text{ lb-ft}$$

$$M_C = -879.7 \text{ lb-ft}$$

Substituting for M_C in equation (1) gives

$$18M_B + 5M_C = -15{,}135 \text{ lb-ft}$$

$$18M_B + 5(-879.7 \text{ lb-ft}) = -15{,}135 \text{ lb-ft}$$

$$18M_B - 4398 \text{ lb-ft} = -15{,}135 \text{ lb-ft}$$

$$18M_B = -15{,}135 \text{ lb-ft} + 4398 \text{ lb-ft} = -10{,}737 \text{ lb-ft}$$

$$M_B = -596.5 \text{ lb-ft}$$

The negative signs indicate that the beam is being bent upward at B and C.

2. Calculate the reactions.

Reaction at A, R_A: We take moments about B. Using the sign convention—clockwise moments are negative, counterclockwise moments are positive—the total moment at B consists of the algebraic sum of

a. The moment due to $R_A = -R_A(4 \text{ ft})$,
b. The moment due to the distributed load, $\omega_1 = (360 \text{ lb/ft})(4 \text{ ft})(4 \text{ ft}/2) = 2880$ lb ft, and
c. The moment at $B = M_B = -596.5$ lb-ft.

$$\sum M)_B = -R_A(4 \text{ ft}) + 2880 \text{ lb-ft} - 596.5 \text{ lb-ft} = 0$$

$$R_A = -(596.5 \text{ lb-ft} - 2880 \text{ lb-ft})/(4 \text{ ft})$$

$$= 571 \text{ lb}$$

Reaction at B, R_B: We take moments about C. The total moment at C consists of the algebraic sum of

a. The moment due to $R_A = -R_A(9 \text{ ft}) = -(571 \text{ lb})(9 \text{ ft}) = -5139 \text{ lb-ft}$,
b. The moment due to the distributed load, $\omega_1 = (360 \text{ lb/ft})(4 \text{ ft})(5 \text{ ft} + 4 \text{ ft}/2) = 10{,}080 \text{ lb-ft}$,

Note: $(360 \text{ lb/ft})(4 \text{ ft})$ is the weight, and $(5 \text{ ft} + 4 \text{ ft}/2) = 7 \text{ ft}$ is the distance.

c. The moment due to $R_B = -R_B(5 \text{ ft})$,
d. The moment due to the distributed load, $\omega_2 = (300 \text{ lb/ft})(5 \text{ ft})(5 \text{ ft}/2) = 3750 \text{ lb-ft}$,
e. The moment at $C = M_C = -879.7 \text{ lb-ft}$.

$$\sum M)_C = -5139 \text{ lb-ft} + 10{,}080 \text{ lb-ft} - R_B(5 \text{ ft})$$
$$+ 3750 \text{ lb-ft} - 879.7 \text{ lb-ft} = 0$$
$$-R_B(5 \text{ ft}) + 7811.3 \text{ lb-ft} = 0$$
$$R_B = 1562 \text{ lb}$$

Reaction at C, R_C: We take moments about D. The total moment at D consists of the algebraic sum of

a. The moment due to $R_A = -R_A(15 \text{ ft}) = -(571 \text{ lb})(15 \text{ ft}) = -8565 \text{ lb-ft}$,
b. The moment due to the distributed load, $\omega_1 = (360 \text{ lb/ft})(4 \text{ ft})(11 \text{ ft} + 4 \text{ ft}/2) = 18{,}720 \text{ lb-ft}$,
c. The moment due to $R_B = -R_B(11 \text{ ft}) = -(1562 \text{ lb})(11 \text{ ft}) = -17{,}182 \text{ lb-ft}$,
d. The moment due to the distributed load, $\omega_2 = (300 \text{ lb/ft})(5 \text{ ft})(6 \text{ ft} + 5 \text{ ft}/2) = 12{,}750 \text{ lb-ft}$,
e. The moment due to $R_C = -R_C(6 \text{ ft})$,
f. The moment due to the distributed load, $\omega_3 = (240 \text{ lb/ft})(6 \text{ ft})(6 \text{ ft}/2) = 4320 \text{ lb-ft}$.

$$\sum M)_D = -8565 \text{ lb-ft} + 18{,}720 \text{ lb-ft} - 17{,}182 \text{ lb-ft}$$
$$+ 12{,}750 \text{ lb-ft} - R_C(6 \text{ ft}) + 4320 \text{ lb-ft} = 0$$
$$- R_C(6 \text{ ft}) + 10{,}043 \text{ lb-ft} = 0$$
$$R_C = 1674 \text{ lb}$$

Reaction at D, R_D: We could determine the reaction at D by taking moments about any point, but we know the moment at C ($M_C = -879.7 \text{ lb-ft}$) and that the moment must be zero at D (because it is the end of the beam), then M_C minus the moments about C created by the distributed load ω_3 and the reaction R_D must be zero.

$$M_C - [-(240 \text{ lb/ft})(6 \text{ ft})(6 \text{ ft}/2) + R_D(6 \text{ ft})] = 0$$
$$-879.7 \text{ lb-ft} - [-4320 \text{ lb-ft} + R_D(6 \text{ ft})] = 0$$
$$-879.7 \text{ lb-ft} + [4320 \text{ lb-ft} - R_D(6 \text{ ft})] = 0$$
$$[3440.3 \text{ lb-ft} - R_D(6 \text{ ft})] = 0$$
$$R_D = 3440.3 \text{ lb-ft}/6 \text{ ft} = 573 \text{ lb-ft}$$

Answer

$$M_A = 0, \qquad\qquad M_B = -596.5 \text{ lb-ft}$$
$$M_C = -879.7 \text{ lb-ft}, \qquad M_D = 0$$
$$R_A = 571 \text{ lb}, \qquad R_B = 1562 \text{ lb}$$
$$R_C = 1674 \text{ lb}, \qquad R_D = 573 \text{ lb}$$

Check

The algebraic sum of the vertical forces should be zero. From left to right (Figure 17.3):

$$\sum F_y = 571 \text{ lb} + (-360 \text{ lb/ft})(4 \text{ ft}) + 1562 \text{ lb} + (-300 \text{ lb/ft})(5 \text{ ft})$$
$$+ 1674 \text{ lb} + (-240 \text{ lb/ft})(6 \text{ ft}) + 573 \text{ lb} = ?$$
$$= 571 \text{ lb} - 1440 \text{ lb} + 1562 \text{ lb} - 1500 \text{ lb} + 1674 \text{ lb}$$
$$- 1440 \text{ lb} + 573 \text{ lb}$$
$$= 0 \qquad \text{Check}$$

EXAMPLE PROBLEM 17.2

This is the same beam as in Example Problem 17.1, but it is extended beyond the support to the left. See Figure 17.4. It is now an overhanging beam, with a concentrated load at the left end. Calculate the moments at the four supports.

Solution

This time there *is* a moment at A, M_A. Remember the sign convention for bending moments discussed in Section 13.2: the bending moment is considered to be *positive* if the sections of the beam above the neutral axis are in *compression* and *negative* if they are in *tension.* The beam is being pushed down on both sides of the support A (see Figure 17.4), so the fibers, or molecules, above the neutral axis are in tension. The moment at A is therefore negative.

$$M_A = -(1000 \text{ lb} \times 3 \text{ ft}) = -3000 \text{ lb-ft}$$

Applying the three-moment equation to the portion *ABC* gives

$$M_A L_1 + 2M_B(L_1 + L_2) + M_C L_2 = \omega_1 L_1^3/4 + \omega_2 L_2^3/4$$
$$M_A = -3000 \text{ lb-ft}$$
$$L_1 = 4 \text{ ft}, \qquad L_2 = 5 \text{ ft}$$
$$\omega_1 = -360 \text{ lb/ft}, \qquad \omega_2 = -300 \text{ lb/ft}$$

So

$$(-3000 \text{ lb-ft})(4 \text{ ft}) + 2M_B(4 \text{ ft} + 5 \text{ ft}) + M_C(5 \text{ ft}) = (-360 \text{ lb/ft})(4 \text{ ft})^3/4$$
$$+ (-300 \text{ lb/ft})(5 \text{ ft})^3/4 -12,000 \text{ lb-ft}^2 + (18 \text{ ft})M_B + (5 \text{ ft})M_C =$$
$$-5760 \text{ lb-ft}^2 - 9375 \text{ lb-ft}^2$$
$$18M_B + 5M_C = -3135 \text{ lb-ft} \qquad\qquad (1)$$

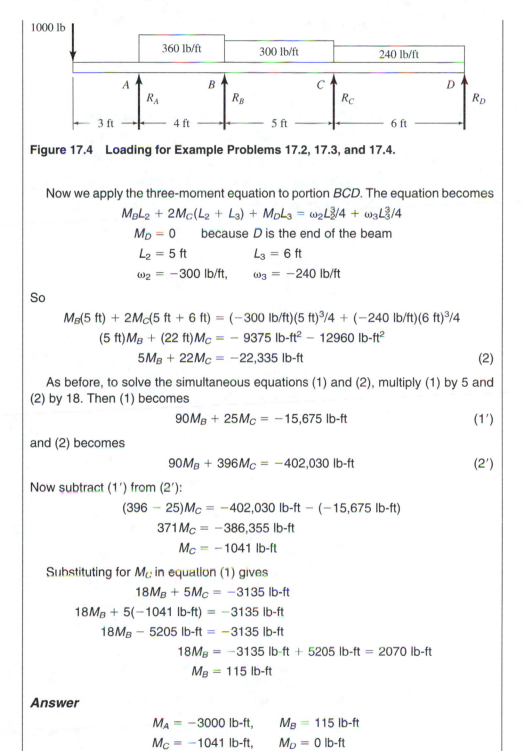

Figure 17.4 Loading for Example Problems 17.2, 17.3, and 17.4.

Now we apply the three-moment equation to portion BCD. The equation becomes

$$M_B L_2 + 2M_C(L_2 + L_3) + M_D L_3 = \omega_2 L_2^3/4 + \omega_3 L_3^3/4$$

$$M_D = 0 \qquad \text{because } D \text{ is the end of the beam}$$

$$L_2 = 5 \text{ ft} \qquad\qquad L_3 = 6 \text{ ft}$$

$$\omega_2 = -300 \text{ lb/ft}, \qquad \omega_3 = -240 \text{ lb/ft}$$

So

$$M_B(5 \text{ ft}) + 2M_C(5 \text{ ft} + 6 \text{ ft}) = (-300 \text{ lb/ft})(5 \text{ ft})^3/4 + (-240 \text{ lb/ft})(6 \text{ ft})^3/4$$

$$(5 \text{ ft})M_B + (22 \text{ ft})M_C = -9375 \text{ lb-ft}^2 - 12960 \text{ lb-ft}^2$$

$$5M_B + 22M_C = -22,335 \text{ lb-ft} \qquad (2)$$

As before, to solve the simultaneous equations (1) and (2), multiply (1) by 5 and (2) by 18. Then (1) becomes

$$90M_B + 25M_C = -15,675 \text{ lb-ft} \qquad (1')$$

and (2) becomes

$$90M_B + 396M_C = -402,030 \text{ lb-ft} \qquad (2')$$

Now subtract (1') from (2'):

$$(396 - 25)M_C = -402,030 \text{ lb-ft} - (-15,675 \text{ lb-ft})$$

$$371M_C = -386,355 \text{ lb-ft}$$

$$M_C = -1041 \text{ lb-ft}$$

Substituting for M_C in equation (1) gives

$$18M_B + 5M_C = -3135 \text{ lb-ft}$$

$$18M_B + 5(-1041 \text{ lb-ft}) = -3135 \text{ lb-ft}$$

$$18M_B - 5205 \text{ lb-ft} = -3135 \text{ lb-ft}$$

$$18M_B = -3135 \text{ lb-ft} + 5205 \text{ lb-ft} = 2070 \text{ lb-ft}$$

$$M_B = 115 \text{ lb-ft}$$

Answer

$$M_A = -3000 \text{ lb-ft}, \qquad M_B = 115 \text{ lb-ft}$$

$$M_C = -1041 \text{ lb-ft}, \qquad M_D = 0 \text{ lb-ft}$$

EXAMPLE PROBLEM 17.3

Determine the four reactions for the loading of Example Problem 17.2.

Solution

Refer again to Figure 17.4.

From Example Problem 17.2, we have the moments,

$$M_A = -3000 \text{ lb-ft}, \qquad M_B = 115 \text{ lb-ft}$$
$$M_C = -1041 \text{ lb-ft}, \qquad M_D = 0 \text{ lb-ft}$$

1. Find the reaction at A, R_A. We take moments about B.

The total moment at B consists of the algebraic sum of

a. The moment due to the 1000 lb concentrated load, $(1000 \text{ lb})(7 \text{ ft}) = 7000 \text{ lb-ft}$,

b. The moment due to $R_A = -R_A(4 \text{ ft})$,

c. The moment due to the distributed load, $\omega_1 = (360 \text{ lb/ft})(4 \text{ ft})(4 \text{ ft}/2) = 2880 \text{ lb-ft}$, and

d. The moment at $B = M_B = 115 \text{ lb-ft}$.

$$\sum M)_B = 7000 \text{ lb-ft} - R_A(4 \text{ ft}) + 2880 \text{ lb-ft} + 115 \text{ lb-ft} = 0$$
$$R_A = 9995 \text{ lb ft}/(4 \text{ ft})$$
$$= 2499 \text{ lb}$$

2. Find the reaction at B, R_B. We take moments about C. The total moment at C consists of the algebraic sum of

a. The moment due to the 1000-lb concentrated load, $(1000 \text{ lb})(12 \text{ ft}) = 12{,}000 \text{ lb-ft}$,

b. The moment due to $R_A = -R_A(9 \text{ ft}) = -(2499 \text{ lb})(9 \text{ ft}) = -22{,}489 \text{ lb-ft}$,

c. The moment due to the distributed load, $\omega_1 = (360 \text{ lb/ft})(4 \text{ ft})(5 \text{ ft} + 4 \text{ ft}/2) = 10{,}080 \text{ lb-ft}$,

d. The moment due to $R_B = -R_B(5 \text{ ft})$,

e. The moment due to the distributed load, $\omega_2 = (300 \text{ lb/ft})(5 \text{ ft})(5 \text{ ft}/2) = 3750 \text{ lb-ft}$,

f. The moment at $C = M_C = -1041 \text{ lb-ft}$.

$$\sum M)_C = 12{,}000 \text{ lb-ft} - 22{,}489 \text{ lb-ft} + 10{,}080 \text{ lb-ft}$$
$$- R_B(5 \text{ ft}) + 3750 \text{ lb-ft} - 1041 \text{ lb-ft} = 0$$
$$- R_B(5 \text{ ft}) + 2300 \text{ lb-ft} = 0$$
$$R_B = 460 \text{ lb}$$

3. Find the reaction at C, R_C. We take moments about D. The total moment at D consists of the algebraic sum of

a. The moment due to the 1000-lb concentrated load, $(1000 \text{ lb})(18 \text{ ft}) = 18{,}000 \text{ lb-ft}$,

b. The moment due to $R_A = -R_A(15 \text{ ft}) = -(2499 \text{ lb})(15 \text{ ft}) = -37{,}485 \text{ lb-ft}$,

c. The moment due to the distributed load, $\omega_1 = (360 \text{ lb/ft})(4 \text{ ft})(11 \text{ ft} + 4 \text{ ft}/2) = 18{,}720 \text{ lb-ft}$,

    d. The moment due to $R_B = -R_B(11 \text{ ft}) = -(460 \text{ lb})(11 \text{ ft}) = -5060$ lb-ft,

    e. The moment due to the distributed load, $\omega_2 = (300 \text{ lb/ft})(5 \text{ ft})(6 \text{ ft} + 5 \text{ ft}/2)$
       $= 12{,}750$ lb-ft,

    f. The moment due to $R_C = -R_C(6 \text{ ft})$,

    g. The moment due to the distributed load, $\omega_3 = (240 \text{ lb/ft}) (6 \text{ ft})(6 \text{ ft}/2) = 4320$
       lb-ft.

$$\sum M)_D = 18{,}000 \text{ lb-ft} - 37{,}485 \text{ lb-ft} + 18{,}720 \text{ lb-ft}$$
$$- 5060 \text{ lb-ft} + 12{,}750 \text{ lb-ft}$$
$$- R_C(6 \text{ ft}) + 4320 \text{ lb-ft} = 0$$
$$- R_C(6 \text{ ft}) + 11{,}245 \text{ lb-ft} = 0$$
$$R_C = 1874 \text{ lb}$$

4. Find the reaction at D, R_D. We know that the moment must be zero at D (because it is the end of the beam); therefore, M_C minus the moments about C created by the distributed load ω_3 and the reaction R_D must be zero.

$$M_C - [-(240 \text{ lb/ft})(6 \text{ ft})(6 \text{ ft}/2) + R_D(6 \text{ ft})] = 0$$
$$-1041 \text{ lb-ft} - [-4320 \text{ lb-ft} + R_D(6 \text{ ft})] = 0$$
$$-1041 \text{ lb-ft} + 4320 \text{ lb-ft} - R_D(6 \text{ ft})] = 0$$
$$3279 \text{ lb-ft} - R_D(6 \text{ ft})] = 0$$
$$R_D = 3279 \text{ lb-ft}/6 \text{ ft} = 547 \text{ lb-ft}$$

Answer

$$R_A = 2499 \text{ lb}, \qquad R_B = 460 \text{ lb}$$
$$R_C = 1874 \text{ lb}, \qquad R_D = 547 \text{ lb}$$

Check

The algebraic sum of the vertical forces should be zero. From left to right (Figure 17.4):

$$\sum F_y = -1000 \text{ lb} + 2499 \text{ lb} + (-360 \text{ lb/ft})(4 \text{ ft}) + 460 \text{ lb} + (-300 \text{ lb/ft})(5 \text{ ft})$$
$$+ 1874 \text{ lb} + (-240 \text{ lb/ft})(6 \text{ ft}) + 547 \text{ lb} = ?$$
$$\sum F_y = -1000 \text{ lb} + 2499 \text{ lb} - 1440 \text{ lb} + 460 \text{ lb}$$
$$- 1500 \text{ lb} + 1874 \text{ lb} - 1440 \text{ lb} + 547 \text{ lb}$$
$$\sum F_y = 0 \qquad \text{Check}$$

EXAMPLE PROBLEM 17.4

Sketch the shear force and bending moment diagrams for the loading shown in Figure 17.4.

Solution
From Example Problem 17.3,

$$R_A = 2499 \text{ lb}, \qquad R_B = 460 \text{ lb}$$
$$R_C = 1874 \text{ lb}, \qquad R_D = 547 \text{ lb}$$

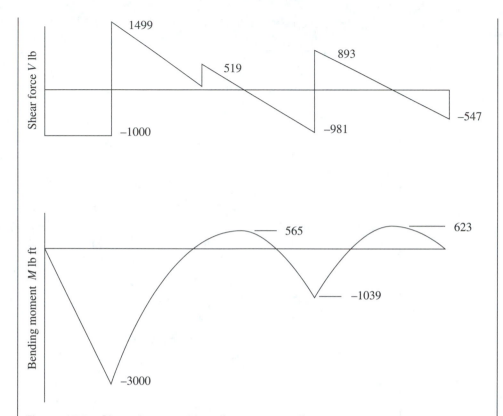

Figure 17.5 Shear force and bending moment diagrams.

1. Shear force diagram: Starting at the left end of the beam, the 1000-lb concentrated load creates a constant shear force of -1000 lb up to $x = 3$ ft (see Figure 17.5), where the reaction R_A (2499 lb) increases it to $(-1000$ lb $+ 2499$ lb$) = 1499$ lb.

The shear force then drops at a rate of 360 lb/ft for 4 ft, which is a drop of (360 lb/ft $\times$ 4 ft) $= 1440$ lb, resulting in a shear force of (1499 lb $-$ 1440 lb) $= 59$ lb at $x = 7$ ft.

The reaction R_B (460 lb) increases the shear force to 59 lb $+$ 460 lb $= 519$ lb at B, where it beings to fall again, this time at a rate of 300 lb/ft. The shear force reaches zero at 519 lb/(300 lb/ft) $= 1.73$ ft from B.—i.e., at $x = 8.73$ ft. It continues falling at the rate of 300 lb/ft to C ($x = 12$ ft), which is an additional 3.27 ft.

At C, then, $V = 0 - 300$ lb/ft $\times$ 3.27 ft $= -981$ lb, which is then increased by R_C (1874 lb) to $(-981$ lb $+ 1874$ lb$) = 893$ lb. The final distributed load decreases the shear force at a rate of 240 lb/ft. It reaches zero at 893 lb/(240 lb/ft) $= 3.72$ ft from C—i.e., at $x = 15.72$ ft. Continuing to drop at the rate of 240 lb/ft for the remaining 2.28 ft, the shear force reaches $0 - (240$ lb/ft $\times$ 2.28 ft$) = -547$ lb at D, where the reaction R_D (547 lb) returns it to zero.

2. Bending moment diagram: From the areas under the shear force diagram (see Figure 17.5),

$$x = 1 \text{ ft to 3 ft:} \quad A = -(1000 \text{ lb})(3 \text{ ft}) = -3000 \text{ lb-ft}$$

So the bending moment at $x = 3$ ft is -3000 lb-ft.

$$x = 3 \text{ ft to 7 ft:} \quad A = (1499 \text{ lb} + 59 \text{ lb})(4 \text{ ft})/2 = 3116 \text{ lb-ft}$$

So the bending moment at $x = 7$ ft is $(-3000 \text{ lb-ft} + 3116 \text{ lb-ft}) = 116$ lb-ft.

$$x = 7 \text{ ft to 8.73 ft (where } V = 0): \quad A = (519 \text{ lb})(8.73 \text{ ft} - 7 \text{ ft})/2 = 448.9 \text{ lb-ft}$$

So the bending moment at $x = 8.73$ ft is $(116 \text{ lb-ft} + 448.9 \text{ lb-ft}) = 564.9$ lb-ft.

$$x = 8.73 \text{ ft to 12 ft:} \quad A = -(981 \text{ lb})(12 \text{ ft} - 8.73 \text{ ft})/2 = -1603.9 \text{ lb-ft}$$

So the bending moment at $x = 12$ ft is $(564.9 \text{ lb-ft} - 1603.9 \text{ lb-ft}) = -1039$ lb-ft.

$$x = 12 \text{ ft to 15.72 ft (where } V = 0): \quad A = (893 \text{ lb})(15.72 \text{ ft} - 12 \text{ ft})/2 = 1661 \text{ lb-ft}$$

So the bending moment at $x = 15.72$ ft is $(-1039 \text{ lb-ft} + 1661 \text{ lb-ft}) = 622$ lb-ft.

$$x = 15.72 \text{ ft to 18 ft:} \quad A = (-547 \text{ lb})(18 \text{ ft} - 15.72 \text{ ft})/2 = 623.6 \text{ lb-ft}$$

So the bending moment at $x = 18$ ft is $(622 \text{ lb-ft} - 623.6 \text{ lb-ft}) = -1.6$ lb-ft.

This should be zero because it is the end of the beam. The -1.6 lb-ft is due to rounding errors.

Answer

We show the complete shear force and bending moment diagrams in Figure 17.5.

EXAMPLE PROBLEM 17.5

Determine the reactions and sketch the shear force and bending moment diagrams for the loading shown in Figure 17.6(a).

Solution

1. Determine the moments. In this case there is a distributed load, ω_1, and a concentrated load, F_2, so the three-moment equation (equation (17.3)) becomes

$$M_A L_1 + 2M_B(L_1 + L_2) + M_C L_2 = F_2 b(L_2^2 - b_2)/L_2 + \omega_1 L_1^3/4$$

where
$$L_1 = 6 \text{ m}, \quad L_2 = 4.5 \text{ m}, \quad b = 1.5 \text{ m} \quad (\text{see Figure 17.2})$$
$$F_2 = -110 \text{ kN}, \quad \omega_1 = -40 \text{ kN/m}$$
$$M_A = M_C = 0 \quad \text{because } A \text{ and } C \text{ are the ends of the beam}$$

Substituting these values into the equation gives

$$2M_B(6 \text{ m} + 4.5 \text{ m}) = (-110 \text{ kN})(1.5 \text{ m})[(4.5 \text{ m})^2 - (1.5 \text{ m})^2]/4.5 \text{ m}$$
$$+ (-40 \text{ kN/m})(6 \text{ m})^3/4$$
$$(21 \text{ m})M_B = -660 \text{ kN·m}^2 - 2160 \text{ kN·m}^2 = -2820 \text{ kN·m}^2$$
$$M_B = -134.3 \text{ kN·m}$$

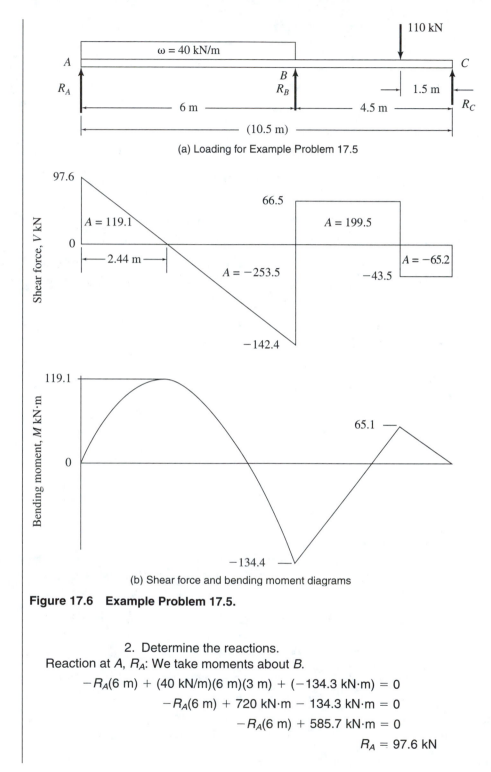

(a) Loading for Example Problem 17.5

(b) Shear force and bending moment diagrams

Figure 17.6   Example Problem 17.5.

2. Determine the reactions.

Reaction at A, R_A: We take moments about B.

$$-R_A(6 \text{ m}) + (40 \text{ kN/m})(6 \text{ m})(3 \text{ m}) + (-134.3 \text{ kN·m}) = 0$$

$$-R_A(6 \text{ m}) + 720 \text{ kN·m} - 134.3 \text{ kN·m} = 0$$

$$-R_A(6 \text{ m}) + 585.7 \text{ kN·m} = 0$$

$$R_A = 97.6 \text{ kN}$$

Reaction at B, R_B: We take moments about C.

$$-R_A(10.5 \text{ m}) + (40 \text{ kN/m})(6 \text{ m})(4.5 \text{ m} + 3 \text{ m}) - R_B(4.5 \text{ m})$$
$$+ (110 \text{ kN})(1.5 \text{ m}) = 0$$
$$-(97.6 \text{ kN})(10.5 \text{ m}) + 1800 \text{ kN·m} - R_B(4.5 \text{ m}) + 165 \text{ kN·m} = 0$$
$$-1025 \text{ kN·m} + 1965 \text{ kN·m} - R_B(4.5 \text{ m}) = 0$$
$$940.2 \text{ kN·m} - R_B(4.5 \text{ m}) = 0$$
$$R_B = 208.9 \text{ kN.}$$

Reaction at C, R_c: The moment at C must be zero because C is the end of the beam. The moment at B is -134.3 kN·m. Therefore, the moment at B minus the sum of the moments about B created by the concentrated load and the reaction R_c must equate to zero. Therefore,

$$-134.3 \text{ kN·m} - [-(110 \text{ kN})(4.5 \text{ m} - 1.5 \text{ m}) + R_C(4.5 \text{ m})] = 0$$
$$-134.3 \text{ kN·m} + 330 \text{ kN·m} - R_C(4.5 \text{ m}) = 0$$
$$195.7 \text{ kN·m} - R_C(4.5 \text{ m}) = 0$$
$$R_C = 195.7 \text{ kN·m/4.5 m}$$
$$= 43.5 \text{ kN}$$

Answer
$R_A = 97.6$ kN, $R_B = 208.9$ kN, $R_c = 43.5$ kN

Check
Algebraically adding the vertical forces:

$$97.6 \text{ kN} + (-40 \text{ kN/m})(6 \text{ m}) + 208.9 \text{ kN} + (-110 \text{ kN}) + 43.5 \text{ kN} = ?$$
$$97.6 \text{ kN} - 240 \text{ kN} + 208.9 \text{ kN} - 110 \text{ kN} + 43.5 \text{ kN} = 0 \qquad \text{Check}$$

3. Shear force diagram: See Figure 17.6(b). Starting at the left end of the beam, the reaction R_A creates a shear force of 97.6 kN, which then decreases at a rate of 40 kN/m due to the distributed load. The shear force reaches zero at $x = 97.9$ kN/(40 kN/m) = 2.44 m. It continues decreasing for another (6 m − 2.44 m) = 3.56 m at 40 kN/m to −(40 kN/m)(3.56 m) = −142.4 kN.

The reaction R_B (208.9 kN) increases it to (−142.4 kN + 208.9 kN) = 66.5 kN, which remains constant up to the concentrated load at $x = 9$ m.

The 110-kN load drops the shear to (66.5 kN − 110 kN) = −43.5 kN, which is then returned to zero by the reaction R_c.

4. Bending moment diagram: Refer again to Figure 17.6(b).

$$x = 0 \text{ to } 2.44 \text{ m}: \quad A = (97.6 \text{ kN})(2.44 \text{ m})/2 = 119.1 \text{ kN·m}$$

So the bending moment at $x = 2.44$ m is 119.1 kN·m.

$$x = 2.44 \text{ m to } 6 \text{ m}: \quad A = (-142.4 \text{ kN})(3.56 \text{ m})/2 = -253.5 \text{ kN·m}$$

So the bending moment at $x = 6$ m is (119.1 kN·m − 253.5 kN·m) = −134.4 kN·m.

$$x = 6 \text{ m to } 9 \text{ m}: \quad A = (66.5 \text{ kN})(3 \text{ m}) = 199.5 \text{ kN·m}$$

So the bending moment at $x = 9$ m is $(-134.4 \text{ kN·m} + 199.5 \text{ kN·m}) = 65.1$ kN·m.

$$x = 9 \text{ m to } 10.5 \text{ m:} \quad A = (-43.5 \text{ kN})(1.5 \text{ m}) = -65.3 \text{ kN·m}$$

So the bending moment at the end of the beam, which should be zero, is 65.1 kN·m − 65.3 kN·m = −0.2 kN·m, due to rounding errors.

Answer
We show the complete shear force and bending moment diagrams in Figure 17.6(b).

17.2   THE METHOD OF SUPERPOSITION

In the method of superposition we determine the deflection that *would exist at the support if the support were removed*; then we calculate what force would be required to bend the beam back to its undeflected position. As an example of this method, we determine the reaction at B for the loading of Figure 17.6(a).

EXAMPLE PROBLEM 17.6

Determine the reaction at B in Figure 17.6(a) using the method of superposition.

Solution
We calculated this reaction to be 208.9 kN in Example Problem 17.5, by Clapeyron's three-moment equation.

1. Remove the support at B.

Determine the deflection at B due to the distributed load. From configuration 6 of Figure 15.13:

$$C_1 = \omega/(24EI) \qquad C_2 = a(2L - a)$$
$$y = C_1 x(C_2^2 - 2C_2 x^2 + Lx^3)/L$$
$$\omega = -40 \text{ kN/m}, \qquad a = 6 \text{ m}, \qquad L = 10.5 \text{ m}$$

Therefore,

$$C_1 = (-40 \text{ kN/m})/(24EI) = (-1.667 \text{ kN/m})/(EI)$$
$$C_2 = (6 \text{ m})[2(10.5 \text{ m}) - 6 \text{ m}] = 90 \text{ m}^2$$
$$y = [(-1.667 \text{ kN/m})/(EI)](6 \text{ m})[(90 \text{ m}^2)^2 - 2(90 \text{ m}^2)(6 \text{ m})^2$$
$$+ (10.5 \text{ m})(6 \text{ m})^3]/10.5 \text{ m}$$
$$= [-10 \text{ kN}/(EI)](8100 \text{ m}^4 - 6480 \text{ m}^4 + 2268 \text{ m}^4)/10.5 \text{ m}$$
$$= [-10 \text{ kN}/(EI)] (3888 \text{ m}^4)/10.5 \text{ m}$$
$$= -3702.9 \text{ kN·m}^3/EI$$

Determine the deflection at B due to the concentrated load. From configuration 1 of Figure 15.13:

$$C_1 = F/(3EIL), \qquad y = C_1 bx(L^2 - b^2 - x^2)/2$$
$$L = 10.5 \text{ m}, \qquad b = 1.5 \text{ m}, \qquad x = 6 \text{ m}$$

$$C_1 = (-110 \text{ kN})/[3EI(10.5 \text{ m})] = (-3.492 \text{ kN/m})/EI$$

$$y = [(-3.492 \text{ kN/m})/(EI)](1.5 \text{ m})(6 \text{ m})[(10.5 \text{ m})^2 - (1.5 \text{ m})^2 - (6 \text{ m})^2]/2$$

$$= [-31.429 \text{ kN·m}/(EI)](72 \text{ m}^2)/2$$

$$= -1131.4 \text{ kN·m}^3/EI$$

Determine total deflection:

$$y = -3702.9 \text{ kN·m}^3/EI - 1131.4 \text{ kN·m}^3/EI$$

$$= -4834.3 \text{ kN·m}^3/EI$$

2. Determine the reaction at B. To do this we have to calculate what force, R_B, would be required to bend the beam back by an amount $+4834.3$ kN·m^3/EI. See Figure 17.7. From configuration 1 of Figure 15.13:

$$C_1 = F/3EIL, \qquad\qquad y = C_1bx(L^2 - b^2 - x^2)/2$$

$$F = R_B, \qquad L = 10.5 \text{ m}, \qquad b = 4.5 \text{ m}, \qquad x = 6 \text{ m}$$

$$C_1 = R_B/3EI(10.5 \text{ m}) = R_B/(31.5 \text{ m})(EI)$$

$$y = [R_B/(31.5 \text{ m})(EI)](4.5 \text{ m})(6 \text{ m})[(10.5 \text{ m})^2 - (4.5 \text{ m})^2 - (6 \text{ m})^2]/2$$

$$= [R_B(0.8571)(54)/2] \text{ m}^3/EI$$

$$= (23.14R_B \text{ m}^3)/EI$$

But we know that this is to be $+4834.3$ kN·m^3/EI. Therefore

$$(4834.3 \text{ kN·m}^3)/EI = 23.14R_B \text{ m}^3/EI$$

$$4834.3 \text{ kN·m}^3 = 23.14R_B \text{ m}^3$$

$$R_B = 4834.3 \text{ kN·m}^3/23.14 \text{ m}^3$$

$$= 208.9 \text{ kN}$$

which agrees exactly with the three-moment equation method.

Answer
$R_B = 208.9$ kN

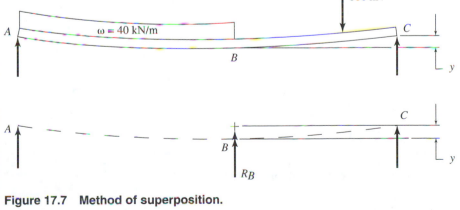

Figure 17.7   Method of superposition.

EXAMPLE PROBLEM 17.7

Apply the method of superposition to a propped cantilever. Determine the reaction at the support A for the loading shown in Figure 17.8(a).

Solution

1. Determine the deflection at A with the support removed. From Figure 15.13, configuration 10:

$$C_1 = \omega/8EI, \qquad y_{max} = C_1 L^4$$
$$\omega = -25 \text{ lb/in.}, \qquad L = 80 \text{ in.}$$
$$C_1 = (-25 \text{ lb/in.})/8EI = (3.125 \text{ lb/in.})/EI$$
$$y_{max} = [-(3.125 \text{ lb/in.})/EI](80 \text{ in.})^4$$
$$= -1.28 \times 10^8 \text{ lb-in.}^3/EI$$

2. What upward force, R_A, at A is required to deflect the beam an amount $+1.28 \times 10^8$ lb-in.$^3/EI$?

From Figure 15.13, configuration 9:

$$C_1 = F/6EI, \qquad\qquad y_{max} = C_1(L - a)^2(2L + a)$$
$$F = R_A \qquad L = 80 \text{ in.}, \qquad a = 0$$
$$C_1 = R_A/6EI$$

and with $a = 0$,

$$y_{max} = 2C_1 L^3$$
$$= 2(R_A/6EI)(80 \text{ in.})^3$$
$$= (1.707 \times 10^5 \text{ in.}^3)R_A/EI$$

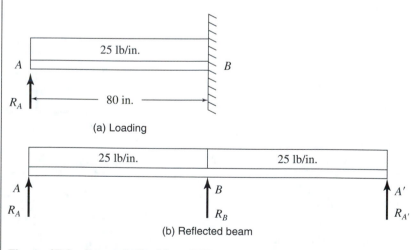

(a) Loading

(b) Reflected beam

Figure 17.8   Example Problem 17.7.

But $y_{max} = +1.28 \times 10^8$ lb-in.^{3}/EI. Therefore,

$$(1.707 \times 10^5 \text{ in.}^3)R_A/EI = 1.28 \times 10^8 \text{ lb-in.}^3/EI$$

$$(1.707 \times 10^5 \text{ in.}^3)R_A = 1.28 \times 10^8 \text{ lb-in.}^3$$

$$R_A = 1.28 \times 10^8 \text{ lb in.}^3/1.707 \times 10^5 \text{ in.}^3$$

$$= 750 \text{ lb}$$

Answer
$R_A = 750$ lb

We will now repeat this problem by using *reflection* and the three-moment equation. Since the slope of a cantilever beam is zero at the wall, we can imagine that it is *reflected* in the wall to form a continuous beam, as shown in Figure 17.8(b).

EXAMPLE PROBLEM 17.8

Apply the method of reflection to determine the reaction at the support A for the loading shown in Figure 17.8(a).

Solution
Figure 17.8(b) shows the reflected, equivalent continuous beam. It is as though the wall is a mirror, and A' is the reflection of A. We can do this because the slope of the beam at B is zero, so there is no discontinuity going from the actual beam to its reflection.

The three-moment equation, equation (17.3), becomes

$$M_A L_1 + 2M_B(L_1 + L_2) + M_{A'}L_2 = \omega_1 L_1^3/4 + \omega_2 L_2^3/4$$

$$M_A = M_{A'} = 0 \quad \text{because A and } A' \text{ are the ends of the beam}$$

$$L_1 = L_2 = 80 \text{ in.}, \quad \omega_1 = \omega_2 = -25 \text{ lb/in.}$$

Substituting these values into the equation gives

$$2M_B(80 \text{ in.} + 80 \text{ in.}) = (-25 \text{ lb/in.})(80 \text{ in.})^3/4 + (-25 \text{ lb/in.})(80 \text{ in.})^3/4$$

$$M_B(320 \text{ in.}) = -6.4 \times 10^6 \text{ lb-in.}^2$$

$$M_B = -2.0 \times 10^4 \text{ lb-in.}$$

Taking moments about B:

$$-R_A(80 \text{ in.}) + (25 \text{ lb/in.})(80 \text{ in.})(80 \text{ in.}/2) - 2.0 \times 10^4 \text{ lb-in.} = 0$$

$$-R_A(80 \text{ in.}) + 8.0 \times 10^4 \text{ lb-in.} - 2.0 \times 10^4 \text{ lb-in.} = 0$$

$$-R_A(80 \text{ in.}) + 6.0 \times 10^4 \text{ lb-in.} = 0$$

$$R_A = 6.0 \times 10^4 \text{ lb-in.}/(80 \text{ in.})$$

So $R_A = 750$ lb, as before

Answer
$R_A = 750$ lb

17.3   FIXED BEAMS

We show a fixed, or built-in, beam in Figure 17.9(a). This is replaced in Figure 17.9(b) by a simply supported beam with *fixed-end moments* (sometimes called *fixing moments*), M_A and M_B. The method of analysis uses the fact that the slope of the beam at both A and B must be zero.

Slope at A

Referring to Figure 17.2, the slope at A caused by the concentrated load, F, is

$$\theta_1 = Fb(L^2 - b^2)/6EIL$$

The slope at A caused by the distributed load is

$$\theta_2 = \omega L^3/24EI$$

The slope at A caused by the fixing moment M_A is

$$\theta_3 = -M_A L/3EI$$

and the slope at A caused by the fixing moment M_B is

$$\theta_4 = -M_B L/6EI$$

The sum of these must be zero:

$$Fb(L^2 - b^2)/(6EIL) + \omega L^3/24EI - M_A L/3EI - M_B L/6EI = 0$$

Multiply throughout by $6EI/L$:

$$Fb(L^2 - b^2)/L^2 + \omega L^2/4 - 2M_A - M_B = 0 \qquad (17.4)$$

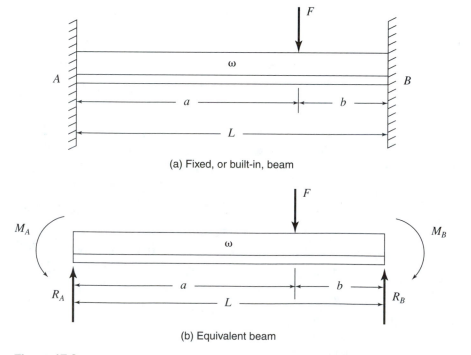

(a) Fixed, or built-in, beam

(b) Equivalent beam

Figure 17.9

Similarly, the slope at B must be zero.

The slope at B caused by the concentrated load, F, is

$$\theta_1 = -Fa(L^2 - a^2)/6EIL$$

The slope at B caused by the distributed load is

$$\theta_2 = -\omega L^3/24EI$$

The slope at B caused by the fixing moment M_A is

$$\theta_3 = M_AL/6EI$$

and the slope at B caused by the fixing moment, M_B, is

$$\theta_4 = M_BL/3EI$$

The sum of these must be zero:

$$-Fa(L^2 - a^2)/6EIL - \omega L^3/24EI + M_AL/6EI + M_BL/3EI = 0$$

Multiply throughout by 6EI/L:

$$-Fa(L^2 - a^2)/L^2 - \omega L^2/4 + M_A + 2M_B = 0 \qquad (17.5)$$

To solve the simultaneous equations (17.4) and (17.5), multiply equation (17.4) by 2 and add it to equation (17.5):

$$2Fb(L^2 - b^2)/L^2 + 2\omega L^2/4 - 4M_A - 2M_B = 0 \qquad (17.4a)$$

$$-Fa(L^2 - a^2)/L^2 - \omega L^2/4 + M_A + 2M_B = 0 \qquad (17.5)$$

$$F[2b(L^2 - b^2) - a(L^2 - a^2)]/L^2 + \omega L^2/4 - 3M_A = 0 \qquad (17.6)$$

The first term can be greatly simplified by noting that $L = a + b$. Hence, $L^2 = (a + b)^2 = a^2 + 2ab + b^2$

and

$$(L^2 - b^2) = a^2 + 2ab$$

Therefore, $$2b(L^2 - b^2) = 2b(a^2 + 2ab) = (2ba^2 + 4ab^2)$$

and $$(L^2 - a^2) = 2ab + b^2$$

so $$a(L^2 - a^2) = 2a^2b + ab^2$$

The term $2b (L^2 - b^2) - a (L^2 - a^2)$ becomes

$$(2ba^2 + 4ab^2) - (2a^2b + ab^2) = 3ab^2$$

So equation (17.6) may be expressed as

$$F(3ab^2)/L^2 + \omega L^2/4 - 3M_A = 0$$

or $$M_A = Fab^2/L^2 + \omega L^2/12 \qquad (17.7)$$

Returning to equation (17.4):

$$Fb(L^2 - b^2)/L^2 + \omega L^2/4 - 2M_A - M_B = 0 \qquad (17.4)$$

Again, $$(L^2 - b^2) = a^2 + 2ab$$

So $$b(L^2 - b^2) = a^2b + 2ab^2$$

Therefore, equation (17.4) may be written as

$$Fa^2b/L^2 + 2Fab^2/L^2 + \omega L^2/4 - 2M_A - M_B = 0$$

Now substitute for M_A from equation (17.7):

$$Fa^2b/L^2 + 2Fab^2/L^2 + \omega L^2/4 - 2(Fab^2/L^2 + \omega L^2/12) - M_B = 0$$
$$Fa^2b/L^2 + 2Fab^2/L^2 + \omega L^2/4 - 2Fab^2/L^2 - \omega L^2/6 - M_B = 0$$
$$Fa^2b/L^2 + \omega L^2/12 - M_B = 0$$

or

$$M_B = Fa^2b/L^2 + \omega L^2/12$$

$$(17.8)$$

Equations (17.7) and (17.8) define the fixing moments for the beam.

EXAMPLE PROBLEM 17.8

Plot the shear force and bending moment diagrams for the built-in beam shown in Figure 17.10(a).

Solution
1. Determine the fixing moments. For this case equations (17.7) and (17.8) become

$$M_A = Fab^2/L^2$$
$$M_B = Fa^2b/L^2$$

where

$$F = -30 \text{ kN}, \quad a = 3 \text{ m}, \quad b = 1 \text{ m}, \quad L = 4 \text{ m}$$
$$M_A = (-30 \text{ kN})(3 \text{ m})(1 \text{ m})^2/(4 \text{ m})^2 = -5.625 \text{ kN·m}$$
$$M_B = (-30 \text{ kN})(3 \text{ m})^2(1 \text{ m})/(4 \text{ m})^2 = -16.875 \text{ kN·m}$$

2. Determine the reactions. Taking moments about B (see Figure 17.10(b)):

$$-R_A(4 \text{ m}) + (30 \text{ kN})(1 \text{ m}) + 5.625 \text{ kN·m} - 16.875 \text{ kN·m} = 0$$
$$-R_A(4 \text{ m}) + 18.75 \text{ kN·m} = 0$$

Therefore,

$$R_A = 4.69 \text{ kN}$$

Taking moments about A:

$$R_B(4 \text{ m}) - (30 \text{ kN})(3 \text{ m}) + 5.625 \text{ kN·m} - 16.875 \text{ kN·m} = 0$$
$$R_B(4 \text{ m}) - 101.5 \text{ kN·m} = 0$$

Therefore,

$$R_B = 25.31 \text{ kN}$$

Note: We *cannot* write $(3 \text{ m})R_A = (1 \text{ m})R_B$ as we could if it were a simply supported beam, because the fixing moments affect the reactions. If the loading were symmetrical, such as a uniformly distributed load or a concentrated load at midspan, then the fixing moments would be equal, and each reaction would support one-half of the total load.

Answer

$$M_A = -5.625 \text{ kN·m}, \quad M_B = -16.875 \text{ kN·m}$$
$$R_A = 4.69 \text{ kN}, \quad R_B = 25.31 \text{ kN}$$

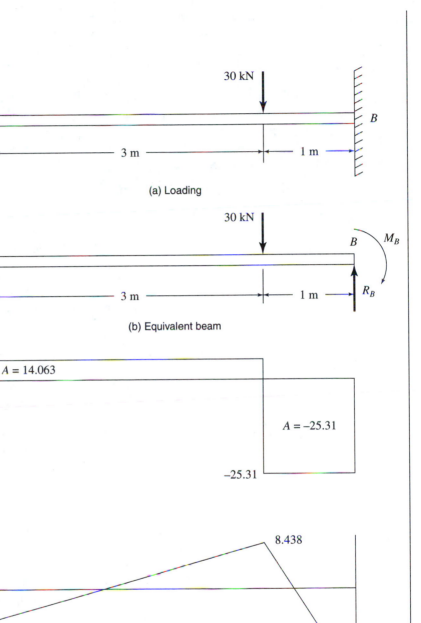

(a) Loading

(b) Equivalent beam

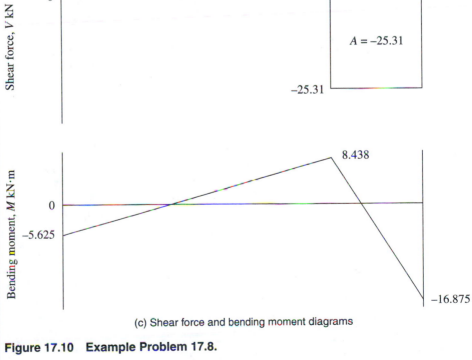

(c) Shear force and bending moment diagrams

Figure 17.10   Example Problem 17.8.

We know R_A, F, and R_B, so plotting the shear force diagram is simply a matter of drawing two rectangles. Notice, however, that the bending moment diagram *does not start at zero*. There is a moment of -5.625 kN·m at A, which is superimposed on the whole bending moment diagram. For example, the area under the shear force diagram from $x = 0$ to $x = 3$ m is 14.063 kN·m, but the moment at $x = 3$ m is *not* 14.063 kN·m; it is (14.063 kN·m $-$ 5.625 kN·m) = 8.438 kN·m.

We show the shear force and bending moment diagrams in Figure 17.10(c).

17.4   COMPUTER ANALYSES

17.4.1   Program 'Propbeam'

The program 'Propbeam' calculates the reactions, shear force diagram, and bending moment diagram for a continuous beam having three supports. It may be overhanging at both ends. The loading (see Figure 17.11) is defined as follows:

F_0 concentrated load on the left overhang

ω_0 distributed load on the left overhang

F_1 concentrated load on the first bay

ω_1 distributed load on the first bay

F_2 concentrated load on the second bay

ω_2 distributed load on the second bay

F_3 concentrated load on the right overhang

ω_3 distributed load on the right overhang

All distances are measured from the left end of the beam.

The program uses the three-moment equation to determine the moments at the supports; then we calculate the reactions, shear forces, and bending moments.

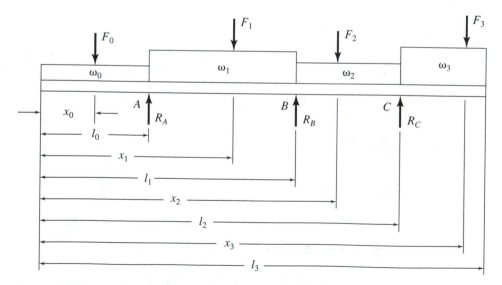

Figure 17.11   Continuous beam with three supports.

EXAMPLE PROBLEM 17.9

Calculate the reactions, and plot the shear force and bending moment diagrams for the loading shown in Figure 17.12(a).

Solution

Open program 'Propbeam'. The input values are (see Figure 17.12(a)):

$$F_0 = 0 \qquad F_1 = -160 \qquad F_2 = -80 \qquad\qquad F_3 = -150$$

$$x_0 = 0 \qquad x_1 = 1.8 \qquad x_2 = 3.7 \qquad\qquad x_3 = 4.7$$

$$\omega_0 = -20 \qquad \omega_1 = 0 \qquad \omega_2 = -45 \qquad\qquad \omega_3 = 0$$

$$l_0 = 1 \qquad l_1 = 2.5 \qquad l_2 = 4 \qquad \text{beam length} = 5$$

We show the input/output sheet in Figure 17.12(b). Be careful to enter the loads in the correct bays.

Answer

The reactions are shown to be

$$R_A = 97.18 \text{ kN}, \qquad R_B = 66.73 \text{ kN}, \qquad R_C = 313.59 \text{ kN}$$

and the maximum shear force and bending moments are

$$V_{max} = 163.59 \text{ kN} \qquad M_{max} = 105.00 \text{ kN·m}$$

Figure 17.12(c) gives the shear force and bending moment diagrams shown on the screen.

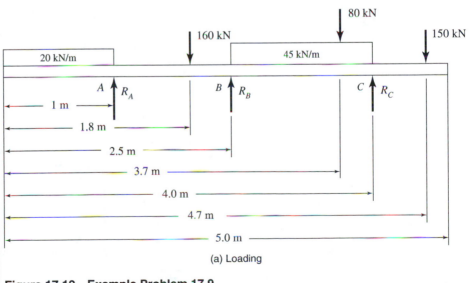

(a) Loading

Figure 17.12   Example Problem 17.9.

552

PROGRAM 'PROPBEAM'

SHEAR FORCE AND BENDING MOMENT DIAGRAM
FOR PROPPED CONTINUOUS BEAMS 'PROPBEAM'

NOTE: DO NOT ERASE OR ENTER ANY VALUES IN
ANY SPACE SURROUNDED BY DOUBLE LINES.

UNITS

CONCENTR'D LOAD	N
DISTRIBUTED LOAD	kN/m
DISTANCE	m

MAXIMUM SHEAR FORCE	MAXIMUM BENDING MOMENT
kN	kNm
163.59	105.00

SUPPORTS

I0		I1		I2	
m		m		m	
1		2.5		4	

BEAM LENGTH

m	
5	

SUPPORT REACTIONS

	RA		RB		RC	
kN		kN		kN	
	97.18		66.73		313.59

CONCENTRATED LOADS

	LOAD	X
	kN	m
F0		
F1	−160	1.8
F2	−80	3.7
F3	−150	4.7

−390

DISTRIBUTED LOADS

	LOAD
	kN/m
w0	−20
w1	
w2	−45
w3	

DISTANCE FROM LEFT END OF BEAM	SHEAR FORCE	BENDING MOMENT
m	kN	kNm
0	0	0
0.5	−10	−2
1	77	−10
1.5	77	29
2	−83	35
2.5	−83	−6
3	−39	−20
3.5	−61	−45
4	−164	−105
4.5	150	−30
5	−0	0

(b) Input/output

Figure 17.12 *(Continued)*

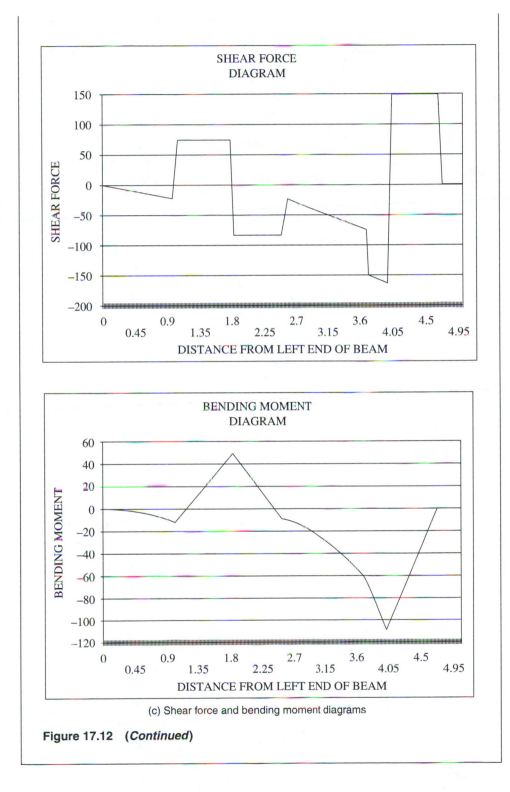

(c) Shear force and bending moment diagrams

Figure 17.12 (*Continued*)

Note: Pages C and E give detailed listings of the shear force and bending moment, respectively. The values are shown for every $\frac{1}{100}$ of beam length.

If the concentrated load, F_3, were increased sufficiently, all the weight would be taken off the central support; i.e., R_B would become zero. To what would F_3 have to be increased to achieve this?

17.4.2 Program 'Propclr'

Program 'Propclr' is for propped cantilever beams. Using the method of reflection and the three-moment equation, it calculates the reactions at the support and at the wall and plots the shear force and bending moment diagrams.

EXAMPLE PROBLEM 17.10

Figure 17.13(a) shows a cantilever beam with a prop 3 ft from the free end. It carries two distributed loads and two concentrated loads. Determine the reactions at the prop and at the wall and plot the shear force and bending moment diagrams.

Solution
Open program 'Propclr'. Referring to Figure 17.13(a), the required input values are

$$F_0 = -250, \qquad F_1 = -500, \qquad \omega_0 = -80, \qquad \omega_1 = 100$$
$$x_0 = 0, \qquad x_1 = 6, \qquad l_0 = 3, \qquad L = 8$$

We show the input/output sheet in Figure 17.13(b).

Answer
The reactions are

$$R_A = 1114.5 \text{ lb}, \qquad R_W = 375.5 \text{ lb}$$

and the maximum shear force and bending moments are

$$V_{max} = 620.5 \text{ lb-ft}, \qquad M_{max} = 1090.46 \text{ lb-ft}$$

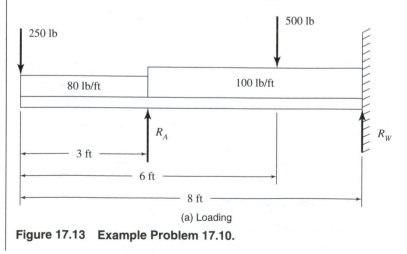

(a) Loading

Figure 17.13   Example Problem 17.10.

PROGRAM 'PROPCLR'

SHEAR FORCE AND BENDING MOMENT DIAGRAM
FOR PROPPED CANTILEVER BEAMS 'PROPCLR'

NOTE: DO NOT ERASE OR ENTER ANY VALUES IN
ANY SPACE SURROUNDED BY DOUBLE LINES.

NOTE: ALL DISTANCES MEASURED FROM THE FREE END.

SUPPORT	BEAM LENGTH
I0	L
ft	ft
3	8

CONCENTRATED LOADS

	LOAD	X
	lb	ft
F0	−250	0
F1	−500	6

SUPPORT REACTIONS

	RA	RW
	lb	lb
	1114.5	375.5

DISTRIBUTED LOADS

	LOAD
	lb/ft
w0	−80
w1	−100

UNITS

CONCENTR'D LOAD	lb
DISTRIBUTED LOAD	lb/ft
DISTANCE	ft

MAXIMUM SHEAR FORCE	MAXIMUM BENDING MOMENT
lb	lbft
620.5	1090.464

DISTANCE FROM LEFT END OF BEAM	SHEAR FORCE	BENDING MOMENT
ft	lb	lbft
0	−250	0
0.8	−314	−226
1.6	−378	−502
2.4	−442	−830
3.2	604	−987
4	524	−535
4.8	444	−148
5.6	364	176
6.4	−216	235
7.2	−296	31
8	−376	−237.5

(b) Input/output

Figure 17.13 *(Continued)*

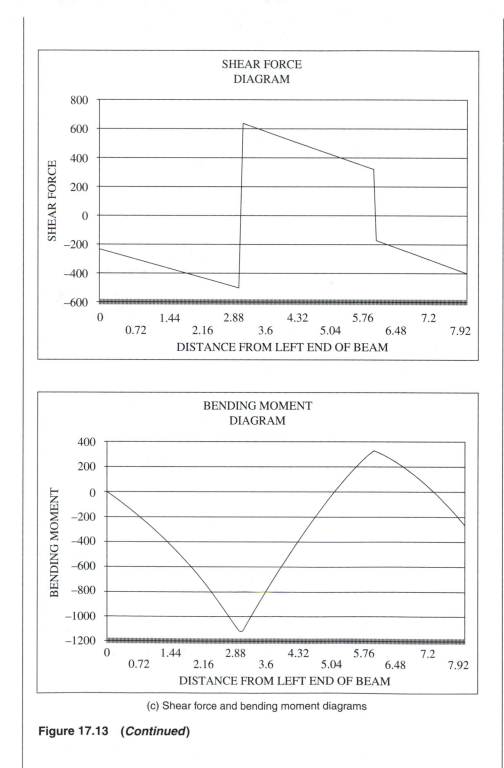

(c) Shear force and bending moment diagrams

Figure 17.13   (*Continued*)

Figure 17.13(c) gives the shear force and bending moment diagrams shown on the screen.

Note: Pages C and E give detailed listings of the shear force and bending moment, respectively. The values are shown for every $\frac{1}{100}$ of beam length.

Suppose, with the same loading, we move the prop to the end of the beam. What would be the force on the prop, and what would the shear force and bending moment diagrams look like?

To do this, change F_0 to a positive value (upward force) in increments until $R_A = 0$. Then the answer is this new F_0 *plus* 250 lb.

17.4.3   Program 'Fixbeam'

Program 'Fixbeam' is for fixed or built-in beams. We calculate the fixing moments from equations (17.7) and (17.8), then the wall reactions, R_A and R_B, and then the shear forces and bending moments. Only four inputs are required. These are the beam length, the concentrated load and its location, and the distributed load.

EXAMPLE PROBLEM 17.11

Figure 17.14(a) shows a built-in beam carrying a distributed load and a concentrated load that is not at midspan. Determine the reactions at the walls and plot the shear force and bending moment diagrams.

Solution
Open program 'Fixbeam.' Referring to Figure 17.14(a), the required input values are:

$$L = 15, \qquad F = -650, \qquad x = 5, \qquad \omega = -30$$

We show the input/output sheet in Figure 17.14(b).

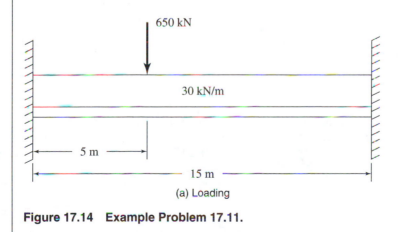

(a) Loading

Figure 17.14   Example Problem 17.11.

PROGRAM 'FIXBEAM'

SHEAR FORCE AND BENDING MOMENT DIAGRAM FOR BUILT-IN BEAM 'FIXBEAM'

NOTE: DO NOT ERASE OR ENTER ANY VALUES IN
ANY SPACE SURROUNDED BY DOUBLE LINES.

NOTE: ALL DISTANCES MEASURED FROM THE FREE END.

BEAM LENGTH

	L
m	15

SUPPORT REACTIONS

	RA		RB
kN	706.48	kN	393.519

CONCENTRATED LOAD

	LOAD		X
kN	-650	m	5

F

DISTRIBUTED LOAD

	LOAD
kN/m	-30

w

UNITS

CONCENTR'D LOAD	kN
DISTRIBUTED LOAD	kN/m
DISTANCE	m

MAXIMUM SHEAR FORCE		MAXIMUM BENDING MOMENT	
kN		kNm	
	706.48		1284.72

DISTANCE FROM LEFT END OF BEAM	SHEAR FORCE	BENDING MOMENT
m	kN	kNm
0	706.48	-2006.94
1.5	661.48	-980.97
3	616.48	-22.50
4.5	571.48	868.47
6	-123.52	1041.94
7.5	-168.52	822.92
9	-213.52	536.39
10.5	-258.52	182.36
12	-303.52	-239.17
13.5	-348.52	-728.19
15	-393.52	-1284.72

(b) Input/output

Figure 17.14 *(Continued)*

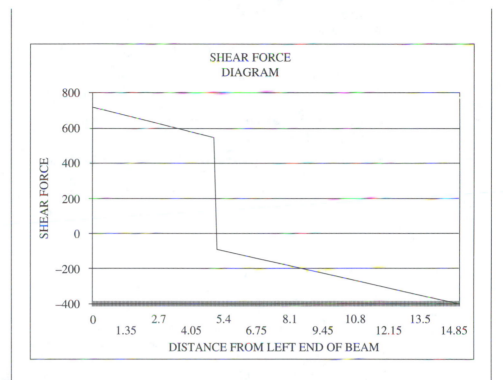

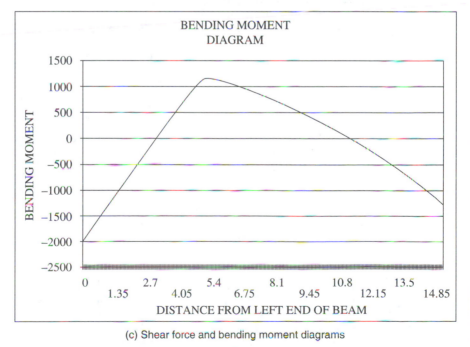

(c) Shear force and bending moment diagrams

Figure 17.14   (*Continued*)

Answer
The reactions are

$$R_A = 706.4 \text{ kN}, \qquad R_B = 393.5 \text{ kN}$$

and the maximum shear force and bending moments are

$$V_{max} = 706.5 \text{ kN}, \qquad M_{max} = 1284.7 \text{ kN·m}$$

Figure 17.14(c) gives the shear force and bending moment diagrams that appear on the screen.

Note: Pages C and E give detailed listings of the shear force and bending moment, respectively.

17.5   PROBLEMS FOR CHAPTER 17

Hand Calculations

1. Calculate the support reactions and plot the shear force and bending moment diagrams for the loading shown in Figure 17.15.

2. Calculate the support reactions and plot the shear force and bending moment diagrams for the loading shown in Figure 17.16.

3. Calculate the support reactions and plot the shear force and bending moment diagrams for the loading shown in Figure 17.17.

4. Calculate the support reactions and plot the shear force and bending moment diagrams for the loading shown in Figure 17.18.

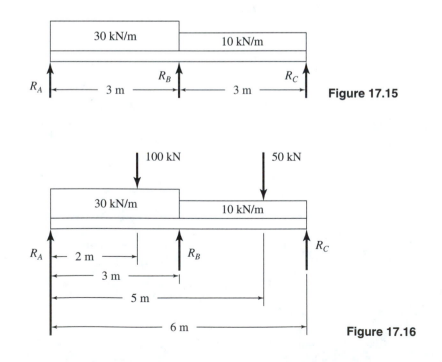

Figure 17.15

Figure 17.16

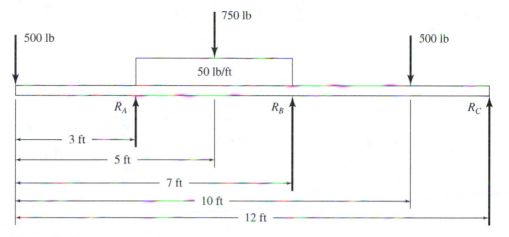

Figure 17.17

Figure 17.18

5. Calculate the support reactions and plot the shear force and bending moment diagrams for the loading shown in Figure 17.19.

6. Calculate the support reactions and plot the shear force and bending moment diagrams for the loading shown in Figure 17.20.

7. Calculate the support reactions and plot the shear force and bending moment diagrams for the loading shown in Figure 17.21.

8. Calculate the support reactions and plot the shear force and bending moment diagrams for the loading shown in Figure 17.22.

9. Calculate the support reactions and plot the shear force and bending moment diagrams for the loading shown in Figure 17.23.

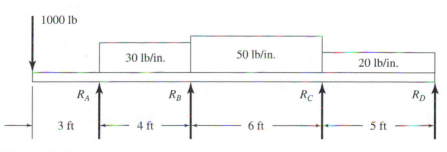

Figure 17.19

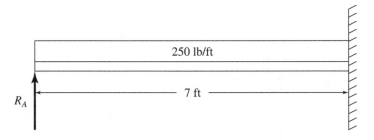

Figure 17.20

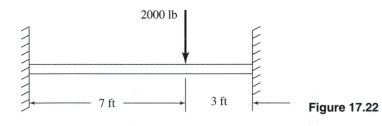

Figure 17.21

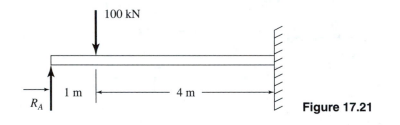

Figure 17.22

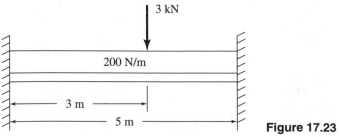

Figure 17.23

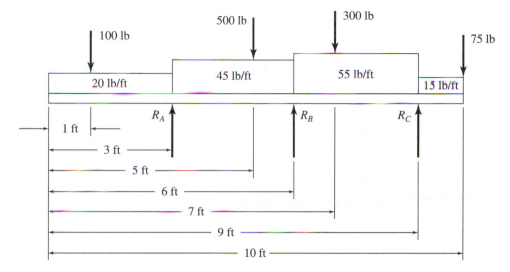

Figure 17.24

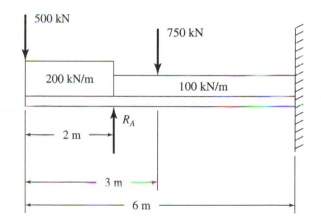

Figure 17.25

Computer Calculations

10. Repeat Problem 1 using the program 'Prop-beam'.

11. Repeat Problem 2 using the program 'Prop-beam'.

12. Repeat Problem 3 using the program 'Prop-beam'.

13. Determine the support reactions and plot the shear force and bending moment diagrams for the loading shown in Figure 17.24. Use the program 'Propbeam'.

14. Repeat Problem 6 using the program 'Prop-clr'.

15. Repeat Problem 7 using the program 'Prop-clr'.

16. Determine the support reactions and plot the shear force and bending moment diagrams for the loading shown in Figure 17.25. Use the program 'Propclr'.

17. What would be the load on the prop in Problem 16 if it were moved to the left end of the beam?

18. Repeat Problem 8 using the program 'Fixbeam'.

19. Repeat Problem 9 using the program 'Fixbeam'.

20. Reverse the direction of the 3-kN load in Problem 19 and replot the shear force and bending moment diagrams.

Combined Loadings

■ *Objectives*

In this chapter you will learn how to calculate the stresses—and, particularly, the maximum stresses—in members subjected to a combination of loadings. For example, a bolt is passed through clearance holes in two pieces of metal, a nut is attached, and then the bolt head is held while the nut is tightened. This action tends to stretch the bolt (puts it in tension) and tends to shear it by twisting. This combined loading of tension and shear can result in maximum values of tensile and shear stresses that would be greater than those calculated individually. The presence of the shear force can increase the maximum tensile stress, and the presence of the tensile force can increase the maximum shear stress. Since combined loadings occur in many practical cases, it is important to be able to recognize and treat these cases correctly.

Rather than develop an all-encompassing equation, which perhaps can be confusing, we will consider various types of combined loadings and show how they relate to what is called Mohr's circle.

18.1 PURE TENSION

Before considering the effects of combined loadings, it is important to realize that *shear stresses* exist even in members subjected only to pure tension.

Figure 18.1(a) shows a bar under pure tension, like a test specimen in a stress-strain experiment. What we have referred to so far as *tensile stress* and *compressive stress* are actually the *normal* (at right angles to the force) tensile and normal compressive stresses. We denote normal stresses with the symbol s_n, so that normal stress in the y-direction is s_{ny}, as shown in Figure 18.1(b). In this case the force equals 1000 N, and the cross-sectional area normal to the force is A = (10 mm)(10 mm) = 100 mm². Therefore, the normal stress is

$$s_{ny} = \text{force/area} = 1000 \text{ N}/100 \text{ mm}^2 = 10 \text{ N/mm}^2 \qquad (10 \text{ MPa})$$

We are going to show that on any plane that is not normal to or parallel to the force, there exist *both tensile and shear stresses* and that the maximum shear stress is equal to one-half of the *normal* tensile stress.

Let us first calculate the stresses on the slant plane AB (shown in Figure 18.1(c)) using specific numbers; then we repeat the calculation using symbols to derive a more general solution. The plane is sloped at 30° to the horizontal.

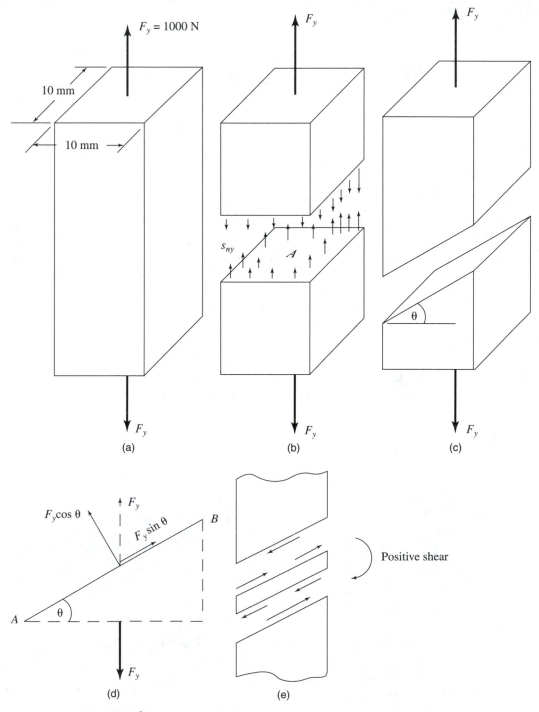

Figure 18.1   Pure tension.

Step 1

Resolve the force on the slant face into forces normal to and along the plane. The vertical downward force, F_y (= 1000 N) acting on plane AB must be balanced by an upward force of the same magnitude (see Figure 18.1(d)), which can be resolved into a force F_n normal to AB and a force F_s parallel to AB.

The force normal to AB is

$$F_n = (1000 \text{ N})(\cos 30°) = 866 \text{ N}$$

and that parallel to AB is

$$F_s = (1000 \text{ N})(\sin 30°) = 500 \text{ N}$$

Note: We are using the sign convention for shear that *shear is positive if it tends to rotate the element in a clockwise direction*. This couple is resisted by a counterclockwise couple, which means that on upper faces, positive shear is from left to right, and on lower faces it is from right to left. See Figure 18.1(e).

Step 2

Determine the slant plane area. The area of plane AB is equal to the length of AB times the depth into the page, which is 10 mm.

The length of AB is 10 mm/cos 30° = 11.547 mm, so

$$A_{AB} = (11.547 \text{ mm})(10 \text{ mm}) = 115.47 \text{ mm}^2$$

Step 3

Calculate the normal stress on the slant plane, which we denote as $s_{n\theta}$. The normal stress on the face AB is

$$s_{n\theta} = F_n/A_{AB} = 866 \text{ N}/115.47 \text{ mm}^2 = 7.50 \text{ N/mm}^2 \qquad (= 7.50 \text{ MPa})$$

Step 4

Calculate the shear stress on the slant plane, which we denote as $s_{s\theta}$. The shear stress on face AB is

$$s_{s\theta} = F_s/A_{AB} = 500 \text{ N}/115.47 \text{ mm}^2 = 4.33 \text{ N/mm}^2 \qquad (= 4.33 \text{ MPa})$$

Step 5

Calculate the ratio of the normal stress on the slant plane to the normal tensile stress.

$$s_{s\theta}/s_{ny} = 7.50 \text{ MPa}/10 \text{ MPa} = 0.750$$

Step 6

Calculate the ratio of the shear stress on the slant plane to the normal tensile stress.

$$s_{s\theta}/s_{ny} = 4.33 \text{ MPa}/10 \text{ MPa} = 0.433$$

We see from this that a considerable shear stress is developed on the slant plane, even though the member is subjected only to pure tension.

We are now going to repeat this calculation, but for any axial force F_y and any slant plane angle θ.

Step 1

Resolve the force on the slant face into forces normal to and along the plane. The vertical downward force, F_y, acting on plane AB must be balanced by an upward force of the same magnitude (see Figure 18.1(d)), which can be resolved into a force F_n normal to AB and a force F_s parallel to AB.

The force normal to AB is

$$F_n = F_y \cos\theta$$

and that parallel to AB is

$$F_s = F_y \sin\theta$$

Step 2

Determine the slant plane area. The cross-sectional area normal to the force is A, so the slant area is

$$A_{AB} = A/\cos\theta$$

Step 3

Calculate the normal stress on the slant plane. The normal stress on the face AB is

$$s_{n\theta} = F_n A_{AB} = F_y \cos\theta/(A/\cos\theta) = F_y \cos^2\theta/A$$

Step 4

Calculate the shear stress on the slant plane. The shear stress on face AB is

$$s_{s\theta} = F_s/A_{AB} = F_y \sin\theta/(A/\cos\theta) = F_y \sin\theta \cos\theta/A$$

Step 5

Calculate the ratio of the normal stress on the slant plane to the normal stress. The normal stress is

$$s_{ny} = F_y/A$$

Therefore,

$$s_{n\theta}/s_{ny} = (F_y \cos^2\theta/A)/(F_y/A)$$
$$s_{n\theta}/s_{ny} = \cos^2\theta \tag{18.1}$$

Step 6

Calculate the ratio of the shear stress on the slant plane to the normal tensile stress.

$$s_{s\theta}/s_{ny} = (F_y \sin\theta \cos\theta/A)/(F_y/A)$$
$$s_{s\theta}/s_{ny} = \sin\theta \cos\theta \tag{18.2}$$

Plotting these ratios as functions of the angle θ (see Figure 18.2) shows that the shear stress reaches a maximum value at $\theta = 45°$ and that this value is one-half the normal tensile stress. Notice also that at $\theta = 45°$ the tensile stress on the slant plane is equal to the shear stress; i.e., it is one-half of the normal tensile stress.

18.1.1   Mohr's Circle for Pure Tension

Mohr's circle (sometimes called Culman's circle) was developed in 1882 by Professor Otto Mohr, a German engineer. It is a plot of shear stress against normal stress. Before

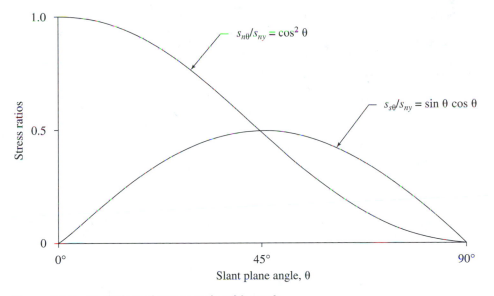

Figure 18.2   Variation of stress ratio with angle.

the advent of computers, it was a convenient method of graphically determining shear and normal stresses on any plane as well as principal planes (to be discussed later). We include it here because it demonstrates quite clearly how normal and shear stresses within a member vary with the angle of the plane being considered.

We showed above that the ratio of tensile stress on a plane at angle θ to the normal tensile stress is

$$s_{n\theta}/s_{ny} = \cos^2\theta \qquad (18.1)$$

From trigonometry we have the relationship $\cos^2\theta = \frac{1}{2} + (\cos 2\theta)/2$, so equation (18.1) may be expressed as

$$s_{n\theta}/s_{ny} = \frac{1}{2} + (\cos 2\theta)/2 \qquad (18.1(a))$$

Also from trigonometry, $\sin\theta\cos\theta = (\sin 2\theta)/2$, so equation (18.2) becomes

$$s_{s\theta}/s_{ny} = (\sin 2\theta)/2 \qquad (18.2(a))$$

We are now going to plot $s_{s\theta}/s_{ny}$ against $s_{n\theta}/s_{ny}$, with $s_{s\theta}/s_{ny}$ as the ordinate (y-axis) and $s_{n\theta}/s_{ny}$ as the abscissa (x-axis). First, look at Figure 18.3(a), which is a right-angle triangle with one angle equal to 2θ and hypotenuse equal to $\frac{1}{2}$. The vertical side is equal to $\frac{1}{2}\sin 2\theta$, and the horizontal side is $\frac{1}{2}\cos 2\theta$. But $\frac{1}{2}\sin 2\theta$ is $s_{s\theta}/s_{ny}$, and $\frac{1}{2}\cos 2\theta$ is $(s_{n\theta}/s_{ny} - \frac{1}{2})$.

If we now draw the axes of the graph (see Figure 18.3(b)) and, starting at zero, draw a horizontal line of length $\frac{1}{2}$, followed by a sloping line at angle 2θ whose length is also $\frac{1}{2}$, then the final point is at $x = \frac{1}{2} + (\frac{1}{2}\cos 2\theta)$, which is $s_{n\theta}/s_{ny}$, and $y = \frac{1}{2}\sin 2\theta$, which is $s_{s\theta}/s_{ny}$. Swinging the angle 2θ through 360° completes Mohr's circle for pure tension (Figure 18.3(c)).

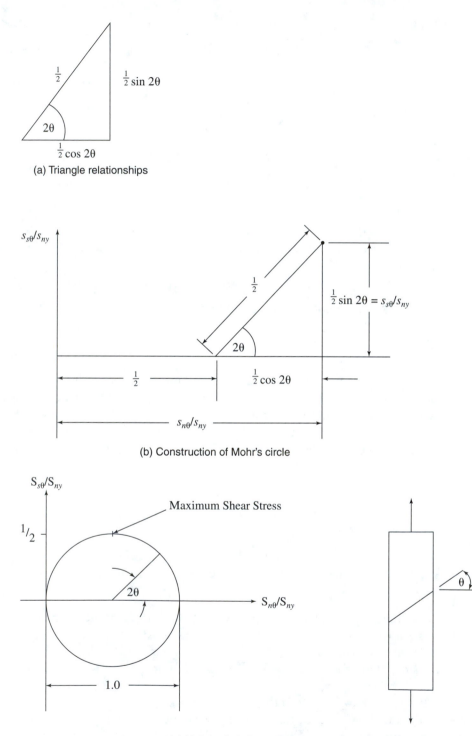

(a) Triangle relationships

(b) Construction of Mohr's circle

(c) Mohr's circle for pure tension

Figure 18.3

We see the following from Figure 18.3(c):

1. When $2\theta = 0$, $\theta = 0$, which means that the plane is normal to the applied force, $s_{n\theta}/s_{ny} = 1.0$, and $s_{s\theta}/s_{ny} = 0$. So for a member in pure tension, a plane normal to the applied force experiences the maximum tensile stress and no shear stress.
2. When $2\theta = 90°$ $\theta = 45°$, the shear stress is at its maximum value, and both the shear stress and the tensile stress on the plane are equal to one-half the normal tensile stress.
3. When $2\theta = 180°$ $\theta = 90°$, which means that the plane is parallel to the applied force and both the tensile stress and the shear stress are zero.

The same holds for pure compression, but of course, the signs are reversed.

18.1.2   Scarf Joints

A **scarf joint** is a joint cut at an angle to the axis of the pieces being connected. Figure 18.3(c) shows a scarf joint connecting two pieces under tension. If they were connected with a butt joint, the adhesive would have to be capable of withstanding the full tensile stress. The scarf reduces this requirement, although it does introduce a shear stress. Mohr's circle shows that we can select a scarf angle that would satisfy both the tensile and shear stress limits of the adhesive.

18.2   COMBINED TENSION-TENSION (STRESS FIELDS NORMAL TO EACH OTHER)

The approach to determining the stresses on any plane in the member is similar to that for pure tension, but this time we have to include the force F_x; see Figure 18.4(a). Figure 18.4(a) shows a square block of side length l, having tensile forces F_x and F_y on the vertical and horizontal faces, respectively. There are no forces on the faces parallel to the plane of the page. We consider a small triangular block inside the member that has its slant face at an angle θ to the horizontal. Figure 18.4(b) shows the triangular block taken out as a free body. We choose the length AB to be one unit (1 in. or 1 mm), and the depth of the block into the page is one unit also.

Now the force F_x creates a normal stress of $s_{nx} = F_x/l^2$ on all vertical planes in the body. Similarly, the force F_y creates a normal stress of $s_{ny} = F_y/l^2$ on all horizontal planes in the body.

So the face BC in Figure 18.4(b) is subjected to a normal stress of $s_{nx} (= F_x/l^2)$. Because the force on the face, call it f_x, is stress times area,

$$f_x = (s_{nx})(\text{area of BC})$$

The length of BC is AB sin θ, but AB = 1, so BC = sin θ, and the area is BC = sin θ because the depth of the element is also unity. The force acting on face BC is, therefore,

$$f_x = (s_{nx})(\sin \theta) \tag{18.3}$$

In the same way, the force on face AC is

$$f_y = (s_{ny})(\text{area of AC})$$

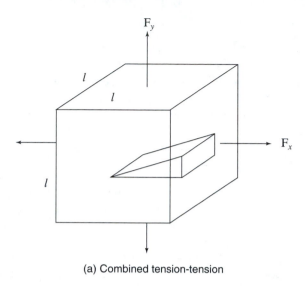

(a) Combined tension-tension

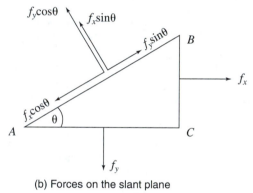

(b) Forces on the slant plane   **Figure 18.4**

The area of AC is $\cos \theta$, so

$$f_y = (s_{ny})(\cos \theta) \tag{18.4}$$

For equilibrium, there are forces $f_y \cos \theta$ and $f_x \sin \theta$ normal to the plane AB (see Figure 18.4(b)) and $f_x \cos \theta$ and $f_y \sin \theta$ parallel to the plane. So the force normal to the plane is

$$f_{n\theta} = f_y \cos \theta + f_x \sin \theta$$

and the force along the plane is

$$f_{s\theta} = f_y \sin \theta - f_x \cos \theta$$

Substituting for f_x and for f_y from equations (18.3) and (18.4) gives

$$f_{n\theta} = (s_{ny})(\cos^2\theta) + (s_{nx})(\sin^2\theta)$$

and

$$f_{s\theta} = (s_{ny})(\cos \theta \sin \theta) - (s_{nx})(\sin \theta \cos \theta)$$

If we now divide these by the slant area AB, we will get the stresses, but area AB is unity, so

$$s_{n\theta} = (s_{ny})(\cos^2\theta) + (s_{nx})(\sin^2\theta) \tag{18.5}$$

and

$$s_{s\theta} = (s_{ny} - s_{nx})(\sin\theta\cos\theta) \tag{18.6}$$

Notice that we resolved and added *forces* in the preceding analysis. You cannot resolve stresses.

We know from Figure 18.2 that the maximum value of $(\sin\theta\cos\theta)$ is $\frac{1}{2}$. It follows, then, that the maximum value of s_s is

$$s_{s\theta)\,max} = (s_{ny} - s_{nx})/2$$

If the stresses s_{nx} and s_{ny} are equal and of the same sign—i.e., both tensile or both compressive—then equation (18.6) shows that there are no shear forces on any plane in the member. Also, equation (18.5) shows that under this condition, $s_{n\theta} = s_{ny} = s_{nx}$ on any plane, since, from trigonometry, $\cos^2\theta + \sin^2\theta = 1$.

18.2.1   Mohr's Circle for Combined Tension-Tension (Stress Fields Normal to Each Other)

In the previous section we derived the expressions for normal and shear stresses on any plane within the body as

$$s_{n\theta} = (s_{ny})(\cos^2\theta) + (s_{nx})(\sin^2\theta) \tag{18.5}$$

and

$$s_{s\theta} = (s_{ny} - s_{nx})(\sin\theta\cos\theta) \tag{18.6}$$

where s_{nx} is the normal tensile stress in the x-direction and s_{ny} is the normal tensile stress in the y-direction. θ is the angle of the plane to the x-direction.

Using the trigonometric identities,

$$\cos^2\theta = \tfrac{1}{2} + (\cos 2\theta)/2$$
$$\sin^2\theta = \tfrac{1}{2} - (\cos 2\theta)/2$$

equation (18.5) may be written as

$$s_{n\theta} = (s_{ny})[\tfrac{1}{2} + (\cos 2\theta)/2] + (s_{nx})[\tfrac{1}{2} - (\cos 2\theta)/2]$$
$$= (s_{nx} + s_{ny})/2 + (s_{ny} - s_{nx})(\cos 2\theta)/2 \tag{18.5(a)}$$

And using the identity $\sin\theta\cos\theta = (\sin 2\theta)/2$, equation (18.6) becomes

$$s_{s\theta} = (s_{ny} - s_{nx})(\sin 2\theta)/2 \tag{18.6(a)}$$

Figure 18.5(a) shows a right-angle triangle with one angle of 2θ and having a hypotenuse equal to $(s_{ny} - s_{nx})/2$. The vertical side will be equal to $(s_{ny} - s_{nx})(\sin 2\theta)/2$, and the horizontal side will be $(s_{ny} - s_{nx})(\cos 2\theta)/2$. But $(s_{ny} - s_{nx})(\sin 2\theta)/2 = s_{s\theta}$, and $(s_{ny} - s_{nx})(\cos 2\theta)/2 = [s_{n\theta} - (s_{nx} + s_{ny})/2]$.

If we now draw the axes of the graph (see Figure 18.5(b)) and, starting at zero, draw a horizontal line equal to $(s_{ny} + s_{nx})/2$, followed by a sloping line at angle 2θ whose length is $(s_{ny} - s_{nx})/2$, then the final point is at $x = [(s_{ny} + s_{nx})/2 + (s_{ny} - s_{nx})(\cos 2\theta)/2]$, which is $s_{n\theta}$, and $y = (s_{ny} - s_{nx})(\sin 2\theta)/2$, which is $s_{s\theta}$. Swinging the angle 2θ through 360° completes Mohr's circle for the case of tension-tension (Figure 18.5(b)). The diagram shows the following very clearly:

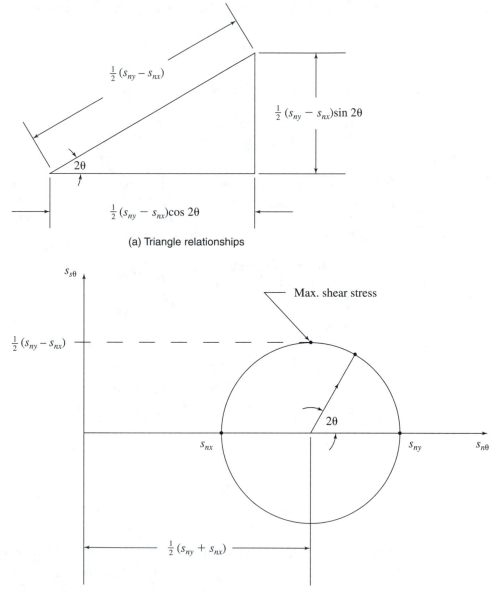

(a) Triangle relationships

(b) Mohr's circle for tension-tension

Figure 18.5

1. When $2\theta = 0$, $\theta = 0$, which means that the plane is normal to the applied force F_y, $s_{n\theta} = s_{ny}$, and $s_{s\theta} = 0$. This tells us that the lateral force, F_x, has no effect whatsoever on a plane normal to F_y.
2. When $2\theta = 90°$ ($\theta = 45°$), the shear stress is at its maximum value of $(s_{ny} - s_{nx})/2$ and the tensile stress on the plane is equal to $(s_{ny} + s_{nx})/2$, which is the average of the two normal tensile stresses.

3. When $2\theta = 180°$, $\theta = 90°$, which means that the plane is normal to the applied force F_x, $s_{n\theta} = s_{nx}$, and the shear stress is zero.

Combined compression-compression or tension-compression cases are handled the same way except that compressive forces are negative.

EXAMPLE PROBLEM 18.1

A rectangular sheet of metal, 0.10 in. thick, has sides of 5.0 in. and 8.0 in. The shorter side is subjected to a tensile force of 1000 lb, and the longer side is subjected to a tensile force of 5000 lb.

   a. What maximum shear stress is developed in the sheet?
   b. What is the angle between the maximum shear plane and the shorter side?
   c. What is the tensile force on this plane?

Solution

Because the applied loads are at right angles to each other, we know from the preceding analysis that maximum shear occurs at 45° to the lines of action of the forces. So the answer to (b) is 45°.

Take one of the loads, say the 1000-lb load, as F_x. (See Figure 18.6.) This is acting on a face whose area is $A_x = (5.0 \text{ in.})(0.1 \text{ in.}) = 0.50 \text{ in}^2$. So the normal stress in the x-direction is

$$s_{nx} = F_x/A_x = (1000 \text{ lb})/0.50 \text{ in.}^2 = 2000 \text{ psi}$$

Similarly,

$$F_y = 500 \text{ lb}, \quad \text{and} \quad A_y = (8.0 \text{ in.})(0.10 \text{ in.}) = 0.80 \text{ in.}^2$$
$$s_{ny} = F_y/A_y = (500 \text{ lb})/0.80 \text{ in.}^2 = 625 \text{ psi}$$

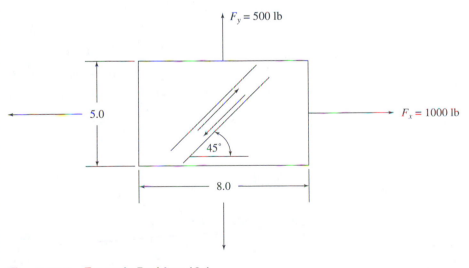

Figure 18.6   Example Problem 18.1.

a. The maximum shear stress is

$$S_{s\theta)max} = (S_{ny} - S_{nx})/2 = (625 \text{ psi} - 2000 \text{ psi})/2 = -687.5 \text{ psi}$$

The negative sign indicates that it runs from right to left on the upper face of the 45° plane (see Figure 18.6). It should be understood that we are referring to *all* planes in the sheet that are at 45° to the applied loads.

b. As we said earlier, the answer is 45°.

c. The tensile stress on this plane is the average of the two normal stresses:

$$S_{n\theta} = (S_{ny} + S_{nx})/2 = (625 \text{ psi} + 2000 \text{ psi})/2 = 1312.5 \text{ psi}$$

Answers

(a) = −687.5 psi, (b) 45°, and (c) 1312.5 psi.

EXAMPLE PROBLEM 18.2

Repeat Example Problem 18.1 but with the 1000-lb load being compressive.

Solution

F_y and s_{ny} will be the same as in the previous example.

$$S_{ny} = 625 \text{ psi}$$

F_x and s_{nx} will have the same magnitudes as in the previous example but will be negative.

$$S_{nx} = -2000 \text{ psi}$$

The equations for the stresses are

$$S_{n\theta} = (S_{nx} + S_{ny})/2 + (S_{ny} - S_{nx})(\cos 2\theta)/2 \qquad (18.5(\text{a}))$$
$$S_{s\theta} = (S_{ny} - S_{nx})(\sin 2\theta)/2 \qquad (18.6(\text{a}))$$

We know that the maximum value of sin 2θ is 1.0, and it occurs at θ = 45°, so

$$S_{s\theta)max} = (S_{ny} - S_{nx})/2 = [625 \text{ psi} - (-2000 \text{ psi})]/2 = 1312.5 \text{ psi}$$

When θ = 45°, cos 2θ = cos 90° = 0, so the tensile stress on the 45° plane is

$$S_{n\theta} = (S_{nx} + S_{ny})/2 = (-2000 \text{ psi} + 625 \text{ psi})/2 = -687.5 \text{ psi}$$

The negative sign indicates that the stress normal to the 45° plane is compressive, not tensile.

Answer

(a) = 1312.5 psi, (b) 45°, and (c) −687.5 psi.

Both the shear stress and the stress normal to the 45° plane act in opposite directions to those of Example Problem 18.1. Notice that the shear stress has been increased substantially.

18.3   COMBINED TENSION AND SHEAR

Just as tensile and compressive forces can create shear stresses, so can shear forces create tensile and compressive stresses. Figure 18.7(a) shows a square, ABCD, with a shear force, F_{sy}, on its right edge. Obviously this square element is not in equilibrium and requires a matching downward force, $-F_{sy}$, to prevent it moving in the y-direction. (See Figure 18.7(b).) But these two forces create a couple that must be balanced by an equal and opposite couple. Forces F_{sx} and $-F_{sx}$, shown in Figure 18.7(c), create this balancing couple. We see from this that all four faces must be subjected to shear and that the magnitude of the shear force on each face must be equal.

Consider now the plane BD (Figure 18.7(d)). The resultant of the forces acting on faces BC and CD is a force of magnitude $\sqrt{F_{sx}^2 + F_{sy}^2}$, acting upward to the right at 45°. But because the magnitude of F_{sx} is equal to the magnitude of F_{sy}, the resultant is $\sqrt{2F_s^2} = F_s\sqrt{2}$, where F_s is the magnitude of the shear force. There is an equal and opposite force acting down to the left. The result is that the face BD is under pure tension.

The area of face BD is the area of one of the sides divided by cos 45°:

$$A_{BD} = A_{BC}/\cos 45° = A_{BC}\sqrt{2} \qquad \text{because } \cos 45° = 1/\sqrt{2}$$

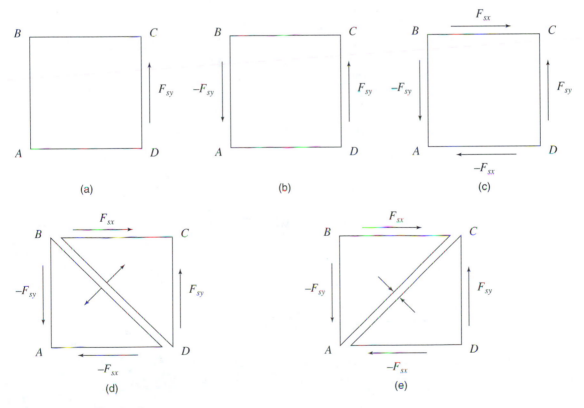

(a)   (b)   (c)

(d)   (e)

Figure 18.7   Simple shear.

The applied shear stress is $s_s = F_s/A_{BC}$ and the normal (tensile) stress on plane BD is

is
$$s_n 45° = F_s\sqrt{2}/A_{BD} = F_s\sqrt{2}/A_{BC}\sqrt{2} = F_s/A_{BC}$$

Therefore,
$$s_n 45° = s_s$$

This tells us that the normal stress on the 45° face is equal to the applied shear stress.

We see that the same is true for plane AC (Figure 18.7(e), except that the stress is pure compression.

Turning now to the case of combined tension and shear, we show (Figure 18.8(a)) a body subjected to tensile forces F_x, F_y, and shear force F_s.

If the body is a cube of side 1, then the stresses are
$$s_{nx} = F_x/l^2, \qquad s_{ny} = F_y/l^2, \qquad s_s = F_s/l^2$$

As in the previous analysis, we take a small triangular element ABC whose hypotenuse is unity, depth is unity, and angle between the hypotenuse and the x-direction is θ. See Figure 18.8(b). The forces acting on the faces of this element are equal to the stresses times the face areas.

The face area BC is $A_{BC} = \sin\theta$. The face area AC is $A_{AC} = \cos\theta$.

So,
$$f_x = s_{nx}A_{BC} = s_{nx}\sin\theta \qquad (18.7)$$
$$f_y = s_{ny}A_{AC} = s_{ny}\cos\theta \qquad (18.8)$$
$$f_{sx} = s_s A_{AC} = s_s\cos\theta \qquad (18.9)$$
$$f_{sy} = s_s A_{BC} = s_s\sin\theta \qquad (18.10)$$

Notice that although the shear *stresses* are equal on faces AB and AC, the shear *forces* are not, except when $\theta = 45°$. We now resolve the forces normal to and along face AB. See Figure 18.8 (b).

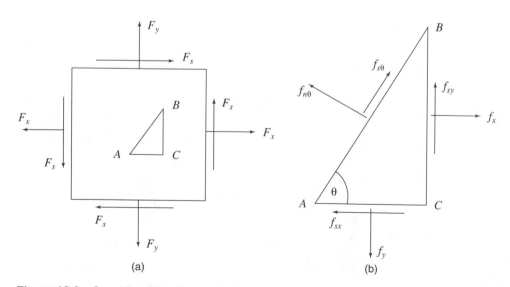

(a)   (b)

Figure 18.8   Combined tension and shear.

Force components normal to AB are as follows:

$$\text{Due to } f_x: \quad f_{nx} = f_x \sin\theta$$
$$\text{Due to } f_y: \quad f_{ny} = f_y \cos\theta$$
$$\text{Due to } f_{sx}: \quad f_{nsx} = f_{sx} \sin\theta$$
$$\text{Due to } f_{sy}: \quad f_{nsy} = f_{sy} \cos\theta$$

Adding,
$$f_{n\theta} = f_x \sin\theta + f_y \cos\theta + f_{sx} \sin\theta + f_{sy} \cos\theta$$

Substituting for f_x, f_y, f_{sx}, and f_{sy} from equations (18.7), (18.8), (18.9), and (18.10) gives

$$f_{n\theta} = s_{nx} \sin^2\theta + s_{ny}\cos^2\theta + s_s\cos\theta\sin\theta + s_s\sin\theta\cos\theta$$
$$f_{n\theta} = s_{nx} \sin^2\theta + s_{ny}\cos^2\theta + 2s_s\cos\theta\sin\theta$$

Force components along AB are as follows:

$$\text{Due to } f_x: \quad f_{sx} = -f_x \cos\theta$$
$$\text{Due to } f_y: \quad f_{sy} = f_y \sin\theta$$
$$\text{Due to } f_{sx}: \quad f_{ssx} = -f_{sx} \cos\theta$$
$$\text{Due to } f_{sy}: \quad f_{ssy} = f_{sy} \sin\theta$$

Adding,
$$f_{s\theta} = -f_x \cos\theta + f_y \sin\theta - f_{sx} \cos\theta + f_{sy} \sin\theta$$

Substituting for f_x, f_y, f_{sx}, and f_{sy} from equations (18.7), (18.8), (18.9), and (18.10) gives

$$f_{s\theta} = -s_{nx} \sin\theta\cos\theta + s_{ny}\cos\theta\sin\theta - s_s\cos^2\theta + s_s\sin^2\theta$$
$$= (s_{ny} - s_{nx})\sin\theta\cos\theta - 2s_s(\cos^2\theta - \sin^2\theta)$$

Because the area of face AB is unity, $s_{n\theta} = f_{n\theta}$ and $s_{s\theta} = f_{s\theta}$.

$$s_{n\theta} = s_{nx}\sin^2\theta + s_{ny}\cos^2\theta + 2s_s\cos\theta\sin\theta \qquad (18.11)$$
$$s_{s\theta} = (s_{ny} - s_{nx})\sin\theta\cos\theta - 2s_s(\cos^2\theta - \sin^2\theta) \qquad (18.12)$$

Using the trigonometric identities

$$\sin^2\theta = \tfrac{1}{2} - \cos 2\theta$$
$$\cos^2\theta = \tfrac{1}{2} + \cos 2\theta \qquad \text{and}$$
$$\sin\theta\cos\theta = (\sin 2\theta)/2$$

equations (18.11) and (18.12) may be written as

$$s_{n\theta} = (s_{nx} + s_{ny})/2 + [(s_{ny} - s_{nx})/2]\cos 2\theta + s_s\sin 2\theta \qquad (18.11(a))$$
$$s_{s\theta} = [(s_{ny} - s_{nx})/2]\sin 2\theta - s_s\cos 2\theta \qquad (18.12(a))$$

Equations (18.11(a)) and (18.12(a)) give the normal and shear stresses, respectively, on a plane inclined at angle θ to the x-direction.

In the earlier example of tension-tension, the maximum normal stresses were on planes normal to the applied loads, and the maximum shear stresses were on planes at 45° to these. The present case of combined tension and shear is a little more complicated. There *are* planes of maximum normal stress and maximum shear stress, but their angles depend on the applied stresses. As an example, we will calculate these maximum values for a specific case.

580 CHAPTER 18 COMBINED LOADINGS

EXAMPLE PROBLEM 18.3

A bolt having a cross-sectional area of 0.50 in.^{2} is subjected to a tensile load of 5000 lb and a shear force of 3500 lb. Determine the maximum normal and maximum shear stresses.

Solution

At this stage in our studies we don't know at what angles these maxima occur. One way to solve the problem is to plot graphs of $s_{n\theta}$ and $s_{s\theta}$ as functions of θ and read off the maximum values.

For the present case,

$$F_x = 0, \quad F_y = 5000 \text{ lb}, \quad F_s = 3500 \text{ lb}, \quad A = 0.50 \text{ in.}^2$$

So the stresses are

$$s_{nx} = 0, \quad s_{ny} = (5000 \text{ lb})/0.50 \text{ in.}^2 = 10{,}000 \text{ psi}$$
$$s_s = (3500 \text{ lb})/0.50 \text{ in.}^2 = 7000 \text{ psi}$$

Inserting these into equations (18.11(a)) and (18.12(a)) gives

$$s_{n\theta} = (s_{nx} + s_{ny})/2 + [(s_{ny} - s_{nx})/2]\cos 2\theta + s_s\sin 2\theta$$
$$= s_{ny}/2 + (s_{ny}/2)\cos 2\theta + s_s\sin 2\theta$$
$$= 5000 \text{ psi} + (5000 \text{ psi})\cos 2\theta + (7000 \text{ psi})\sin 2\theta$$
$$= (5000 \text{ psi})(1 + \cos 2\theta) + (7000 \text{ psi})\sin 2\theta \tag{1}$$
$$s_{s\theta} = [(s_{ny} - s_{nx})/2]\sin 2\theta - s_s\cos 2\theta$$
$$= (s_{ny}/2)\sin 2\theta - s_s\cos 2\theta$$
$$= (5000 \text{ psi})\sin 2\theta - (7000 \text{ psi})\cos 2\theta \tag{2}$$

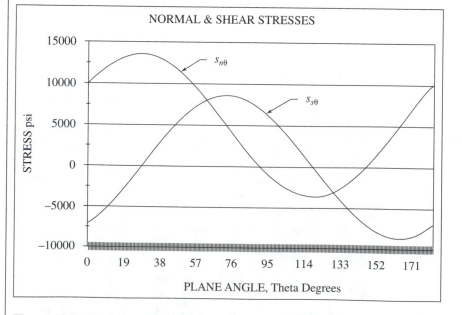

Figure 18.9 Variation of stresses with plane angle for Example Problem 18.3.

We show plots of $s_{n\theta}$ and $s_{s\theta}$ as functions of angle θ (for θ ranging from 0 to 180°) in Figure 18.9.

Answer
The maximum value of $s_{n\theta}$ is 13,600 psi, occurring at $\theta = 27.2°$ and 117.2°. The maximum value of $s_{s\theta}$ is 8600 psi, occurring at $\theta = 72.2°$ and 162.2°.

The results of Example Problem 18.3 show the following important facts:

1. The maximum normal stress (13,600 psi) is greater than the applied normal stress (10,000 psi).
2. The maximum shear stress (8600 psi) is greater than the applied shear stress (7000 psi).
3. The minimum normal stress (-3600 psi—the negative sign indicates a compressive stress) is on a plane that is at 90° to the maximum normal stress.
4. The minimum shear stress occurs on a plane that is at 90° to the maximum shear stress.
5. The maximum normal stress and the maximum shear stress occur on planes that are 45° apart.
6. The shear is zero on planes of maximum normal stress.

This example shows that if we had designed the bolt for the applied stresses, it would not have been strong enough. There are planes in the bolt where the stresses are greater than the *apparent* stresses.

The planes of maximum normal stress are called **principal planes**, and the stresses on them are the **principal stresses**.

18.4   PRINCIPAL PLANES AND PRINCIPAL STRESSES

In Example Problem 18.3 we determined the principal planes and principal stresses graphically. Finding the principal planes analytically requires the use of calculus. If you are familiar with calculus, it is simply a matter of differentiating equation (18.11(a)) with respect to θ and setting the result to zero. See Using Calculus.

USING CALCULUS

Finding the principal planes.
   From equation (18.11(a)),

$$s_{n\theta} = (s_{nx} + s_{ny})/2 + [(s_{ny} - s_{nx})/2]\cos 2\theta + s_s\sin 2\theta$$

Differentiating with respect to θ gives

$$d(s_{n\theta})/d\theta = -(s_{ny} - s_{nx})\sin 2\theta + 2s_s\cos 2\theta$$

Setting this equal to zero gives the maximum and the minimum values of $s_{n\theta}$.

$$0 = -(s_{ny} - s_{nx})\sin 2\theta + 2s_s\cos 2\theta$$

$$(s_{ny} - s_{nx})\sin 2\theta = 2s_s\cos 2\theta$$

$$\sin 2\theta/\cos 2\theta = \tan 2\theta = 2s_s/(s_{ny} - s_{nx}) \tag{1}$$

This is equation (18.13).

   This result gives two values of 2θ that differ by 180°, or two values of θ, say θ_1 and θ_2, that differ by 90°. On one of these planes $s_{n\theta}$ is a maximum and on the other it is a minimum. Let us say that the maximum value of $s_{n\theta}$ is s_1, and it occurs on plane θ_1. Similarly, s_2, the minimum value, occurs on plane θ_2. Now in general, if $\tan \alpha = a/b$, then

$$\sin \alpha = a/(a^2 + b^2)^{1/2} \quad \text{and}$$

$$\cos \alpha = b/(a^2 + b^2)^{1/2}$$

These arise simply from the definitions of the trigonometric functions and the Pythagorean theorem.
In equation (1),

$$a = 2s_s, \qquad b = (s_{ny} - s_{nx})$$

So, $$(a^2 + b^2)^{1/2} = [4s_s^2 + (s_{ny} - s_{nx})^2]^{1/2}$$

If $\alpha = 2\theta_1$, then $$\sin 2\theta_1 = 2s_s/[4s_s^2 + (s_{ny} - s_{nx})^2]^{1/2} \tag{2}$$

and $$\cos 2\theta_1 = (s_{ny} - s_{nx})/[4s_s^2 + (s_{ny} - s_{nx})^2]^{1/2} \tag{3}$$

If $\alpha = 2\theta_2$, then $$\sin 2\theta_2 = -2s_s/[4s_s^2 + (s_{ny} - s_{nx})^2]^{1/2} \tag{4}$$

and $$\cos 2\theta_2 = -(s_{ny} - s_{nx})/[4s_s^2 + (s_{ny} - s_{nx})^2]^{1/2} \tag{5}$$

This is because $2\theta_2 = 2\theta_1 + 180°$,

so $$\sin 2\theta_2 = \sin (2\theta_1 + 180°) = -\sin 2\theta_1$$

and $$\cos 2\theta_2 = \cos (2\theta_1 + 180°) = -\cos 2\theta_1$$

   Substituting equations (2) and (3) into equation (18.11(a)),

$$s_1 = (s_{nx} + s_{ny})/2 + [(s_{ny} - s_{nx})/2](s_{ny} - s_{nx})/[4s_s^2 + (s_{ny} - s_{nx})^2]^{1/2}$$

$$+ s_s2s_s/[4s_s^2 + (s_{ny} - s_{nx})^2]^{1/2}$$

$$= (s_{nx} + s_{ny})/2 + \tfrac{1}{2}[(s_{ny} - s_{nx})^2 + 4s_s^2]/[4s_s^2 + (s_{ny} - s_{nx})^2]^{1/2}$$

$$= (s_{nx} + s_{ny})/2 + \tfrac{1}{2}[(s_{ny} - s_{nx})^2 + 4s_s^2]^{1/2}$$

$$= (s_{nx} + s_{ny})/2 + [\{(s_{ny} - s_{nx})/2\}^2 + s_s^2]^{1/2} \tag{6}$$

Similarly,

$$s_2 = (s_{nx} + s_{ny})/2 - \tfrac{1}{2}[(s_{ny} - s_{nx})^2 + 4s_s^2]^{1/2} \tag{7}$$

   Equation (6) gives the maximum normal stress. It occurs on plane θ_1, expressed by equation (2) or (3). Similarly, equation (7) gives the minimum normal stress. It occurs on plane θ_2, expressed by equation (4) or (5).

   Turning now to the shear stress on any plane, θ, we had the equation [equation (18.12(a))]

$$s_{s\theta} = [(s_{ny} - s_{nx})/2]\sin 2\theta - s_s\cos 2\theta \tag{8}$$

Let us see what happens to the shear stress on the planes where the normal stress is a maximum or a minimum, i.e., at θ_1 and θ_2. When the normal stress is a maximum:

Substituting equations (2) and (3) into equation (8) gives

$$s_{s\theta} = [(s_{ny} - s_{nx})/2]2s_s/[4s_s^2 + (s_{ny} - s_{nx})^2]^{1/2} - s_s(s_{ny} - s_{nx})/[4s_s^2 + (s_{ny} - s_{nx})^2]^{1/2}$$

$$s_{s\theta} = [s_s(s_{ny} - s_{nx}) - s_s(s_{ny} - s_{nx})]/[4s_s^2 + (s_{ny} - s_{nx})^2]^{1/2}$$

So, $s_{s\theta} = 0$.

When the normal stress is a minimum, we can substitute equations (4) and (5) into equation (8):

$$s_{s\theta} = -[(s_{ny} - s_{nx})/2]2s_s/[4s_s^2 + (s_{ny} - s_{nx})^2]^{1/2}$$

$$+ s_s(s_{ny} - s_{nx})/[4s_s^2 + (s_{ny} - s_{nx})^2]^{1/2}$$

$$s_{s\theta} = [-s_s(s_{ny} - s_{nx}) + s_s(s_{ny} - s_{nx})]/[4s_s^2 + (s_{ny} - s_{nx})^2]^{1/2}$$

So, $s_{s\theta} = 0$.

We see that the *shear stress is zero on planes where the normal stress is maximum or minimum*. These planes are called the **principal planes** and the normal stresses on them are the **principal stresses**.

Finding the magnitude of the maximum shear stress

For maximum shear stress $d(s_{s\theta})/d\theta = 0$. Differentiating equation (8) with respect to θ gives

$$d(s_{s\theta})/d\theta = [(s_{ny} - s_{nx})/2]2\cos 2\theta + 2s_s\sin 2\theta = 0, \qquad \text{from which}$$

$$s_s\sin 2\theta = -[(s_{ny} - s_{nx})/2]\cos 2\theta$$

$$\sin 2\theta/\cos 2\theta = \tan 2\theta = -(s_{ny} - s_{nx})/2s_s \qquad (9)$$

Again, this equation gives two values of θ, say θ_3 and θ_4, that differ by 90°.

From equation (1), $\tan 2\theta_1 = 2s_s/(s_{ny} - s_{nx})$, and from equation (9),

$$\tan 2\theta_3 = -(s_{ny} - s_{nx})/2s_s$$

Therefore, $\qquad \tan 2\theta_1 \tan 2\theta_3 = -1$

This can be so only if $2\theta_1$ and $2\theta_3$ differ by 90°, so θ_1 and θ_3 differ by 45°.

Consequently, *maximum shear stress occurs on planes that are inclined at 45° to the principal planes*.

As before, we can determine $\sin 2\theta$ and $\cos 2\theta$ from equation (9): We had, if $\tan \alpha = a/b$,

$$\sin \alpha = a/(a^2 + b^2)^{1/2} \quad \text{and}$$

$$\cos \alpha = b/(a^2 + b^2)^{1/2}$$

This time, $a = (s_{ny} - s_{nx})$ and $b = 2s_s$, so

$$\sin 2\theta = (s_{ny} - s_{nx})/[(s_{ny} - s_{nx})^2 + 4s_s^2]^{1/2} \qquad \text{and}$$

$$\cos 2\theta = 2s_s/[(s_{ny} - s_{nx})^2 + 4s_s^2]^{1/2}$$

To find the maximum value of the shear stress, we substitute for sin 2θ and cos 2θ from these equations into equation (8):

$$
\begin{aligned}
s_{smax} &= [(s_{ny} - s_{nx})/2](s_{ny} - s_{nx})/[(s_{ny} - s_{nx})^2 + 4s_s^2]^{1/2} \\
&\quad - s_s \times 2s_s/[(s_{ny} - s_{nx})^2 + 4s_s^2]^{1/2} \\
&= [(s_{ny} - s_{nx})^2/2] + 2s_s^2]/[(s_{ny} - s_{nx})^2 + 4s_s^2]^{1/2} \\
&= (\tfrac{1}{2})[(s_{ny} - s_{nx})^2 + 4s_s^2]^{1/2} \\
&= [\{(s_{ny} - s_{nx})/2\}^2 + s_s^2]^{1/2}
\end{aligned}
\tag{10}
$$

The principal planes are found to be given by the equation

$$\tan 2\theta = 2s_s/(s_{ny} - s_{nx}) \tag{18.13}$$

This gives two values of 2θ that differ by 180°, or two values of θ that differ by 90°. The normal stress, $s_{n\theta}$, is maximum on one of these planes and minimum on the other. The maximum value is

$$s_{n\theta)max} = (s_{nx} + s_{ny})/2 + \sqrt{[(s_{ny} - s_{nx})/2]^2 + s_s^2} \tag{18.14(a)}$$

and the minimum value is

$$s_{n\theta)min} = (s_{nx} + s_{ny})/2 - \sqrt{[(s_{ny} - s_{nx})/2]^2 + s_s^2} \tag{18.14(b)}$$

Maximum shear stress occurs on planes that are at 45° to the principal planes. The maximum shear stress is

$$s_{s\theta)max} = \pm \sqrt{[(s_{ny} - s_{nx})/2]^2 + s_s^2} \tag{18.15}$$

The sign indicates the direction of the shear.

EXAMPLE PROBLEM 18.4

Repeat Example Problem 18.3 analytically.

Solution
The stresses were

$$s_{nx} = 0, \qquad s_{ny} = 10{,}000 \text{ psi}, \qquad s_s = 7000 \text{ psi}$$

Use equation (18.13) to find the principal planes:

$$
\begin{aligned}
\tan 2\theta &= 2s_s/(s_{ny} - s_{nx}) \\
&= 2(7000 \text{ psi})/(10{,}000 \text{ psi} - 0) \\
&= 1.4 \\
2\theta &= 54.5° \quad \text{or} \quad 234.5° \\
\theta &= 27.2° \quad \text{or} \quad 117.2°
\end{aligned}
$$

From equation (18.14(a)) the maximum value of normal stress is

$$s_{n\theta)max} = (s_{nx} + s_{ny})/2 + \sqrt{[(s_{ny} - s_{nx})/2]^2 + s_s^2}$$

$$= s_{ny}/2 + \sqrt{(s_{ny}/2)^2 + s_s^2}$$

$$= 10,000\ psi/2 + \sqrt{(10,000\ psi/2)^2 + (7000\ psi)^2}$$

$$= 5000\ psi + \sqrt{(25 \times 10^6 + 49 \times 10^6)\ psi}$$

$$= 5000\ psi + 8602\ psi$$

$$= 13,600\ psi$$

From equation (18.14(b)), the minimum value of normal stress is

$$s_{n\theta)min} = (s_{nx} + s_{ny})/2 - \sqrt{[(s_{ny} - s_{nx})/2]^2 + s_s^2}$$

$$= s_{ny}/2 - \sqrt{(s_{ny}/2)^2 + s_s^2}$$

$$= 5000\ psi - 8602\ psi$$

$$= -3600\ psi$$

Maximum shear stress occurs on planes that are at 45° to the principal planes. These are, therefore,

$$\theta = 27.2° + 45° = 72.2° \quad \text{and} \quad \theta = 117.2° + 45° = 162.2°$$

The maximum shear stress is

$$s_{s\theta)max} = \pm \sqrt{[(s_{ny} - s_{nx})/2]^2 + s_s^2} \qquad \text{from equation} \qquad (18.15)$$

$$= \pm \sqrt{(s_{ny}/2)^2 + s_s^2}$$

$$= \pm \sqrt{[(10,000\ psi/2)^2 + (7000\ psi)^2]}$$

$$= \pm \sqrt{25 \times 10^6 + 49 \times 10^6}\ psi$$

$$= \pm 8600\ psi$$

Answer
The maximum normal stress is a tensile stress of 13,600 psi on a plane at 27.2° to the x-direction. The minimum normal stress is a compressive stress of 3600 psi on a plane at 117.2° to the x-direction. The maximum shear stress is 8600 psi on planes at 72.2° and 162.2° to the x-direction.

18.5   COMBINED TORSION AND BENDING

The same basic equations for maximum normal stress and maximum shear stress apply. The maximum normal stress is

$$s_{n\theta)max} = (s_{nx} + s_{ny})/2 + \sqrt{[(s_{ny} - s_{nx})/2]^2 + s_s^2} \qquad (18.14(a))$$

and the maximum shear stress is

$$s_{s\theta)max} = \pm \sqrt{[(s_{ny} - s_{nx})/2]^2 + s_s^2} \qquad (18.15)$$

In the case of combined torsion and bending, the applied maximum shear stress is given by

$$s_s = T/Z_p \qquad \text{(Section 12.3)}$$

Here T is the torque and Z_p is the polar section modulus. Bending causes a maximum tensile stress of

$$s_{nx} = M/S \qquad \text{(Section 16.1.4)}$$

M is the maximum bending moment and S is the section modulus.

EXAMPLE PROBLEM 18.5

A 70-mm-diameter shaft is to transmit a torque of 500 N·m to a pump. The shaft is to be driven by a belt that will create a maximum bending moment in the shaft of 300 N·m. The design is to be based on a safety factor of 4. If the shaft material's allowable tensile and shear stresses are 60 MPa and 45 MPa, respectively, is the shaft diameter satisfactory?

Solution
1. Tensile stress due to bending:

$$s_{nx} = M/S$$
$$M = 300 \text{N·m} = 300 \times 10^3 \text{ N·mm}$$

The section modulus of a round shaft of diameter d is

$$S = \pi d^3/32$$
$$d = 70 \text{ mm}$$

Therefore,

$$S = \pi(70 \text{ mm})^3/32 = 33.7 \times 10^3 \text{ mm}^3$$

So
$$s_{nx} = M/S = 300 \times 10^3 \text{ N·mm}/33.7 \times 10^3 \text{ mm}^3$$
$$= 8.90 \text{ MPa}$$

2. Shear stress due to torsion:

$$s_s = T/Z_p$$

where Z_p, the polar section modulus, equals $\pi d^3/16 = 67.3 \times 10^3 \text{ mm}^3$

$$T = 500 \text{ N·m} = 500 \times 10^3 \text{ N·mm}$$

So
$$s_s = 500 \times 10^3 \text{ N·mm}/67.3 \times 10^3 \text{ mm}^3$$
$$= 7.43 \text{ MPa}$$

3. Principal stress (maximum tensile stress): From equation (18.14 (a))

$$s_{n\theta)\text{max}} = (s_{nx} + s_{ny})/2 + \sqrt{[(s_{ny} - s_{nx})/2]^2 + s_s^2}$$
$$s_{nx} = 8.90 \text{ MPa}, \qquad s_{ny} = 0, \qquad s_s = 7.43 \text{ MPa}$$
$$s_{n\theta)\text{max}} = s_{nx}/2 + \sqrt{(-s_{nx}/2)^2 + s_s^2}$$

$$= 8.90 \text{ MPa}/2 + \sqrt{(-8.90 \text{ MPa}/2)^2 + (7.43 \text{ MPa})^2}$$
$$= 4.45 \text{ MPa} + 8.66 \text{ MPa}$$
$$= 13.1 \text{ MPa}$$

4. Maximum shear stress: From equation (18.15)

$$s_{s(t)max} = \pm \sqrt{[(s_{ny} - s_{nx})/2]^2 + s_s^2}$$
$$= \pm \sqrt{(-s_{nx}/2)^2 + s_s^2}$$
$$= \pm 8.66 \text{ MPa}$$

5. Allowable stresses: Including the safety factor of 4, the allowable tensile stress is 60 MPa/4 = 15 MPa, and the allowable shear stress is 45 MPa/4 = 11.25 MPa. These both exceed the maximum stresses, so the design is acceptable.

Answer
The design is satisfactory.

EXAMPLE PROBLEM 18.6

A shaft is to be designed to transmit 35 hp at 280 rpm to a gear box. The shaft is supported on bearings 6 ft apart, and power is transmitted to it by means of a belt drive midway between the bearings. The drive-belt pull on the shaft is 850 lb. The shaft material is AISI 1080 OQT700, and a safety factor of 4, based on the yield strength in shear and yield strength in tension, is to be used. The yield strength in shear of AISI 1080 OQT700 is 71 ksi, and in tension it is 141 ksi. What diameter shaft is required? See Figure 18.10 (a).

Solution
Let the shaft diameter be d inches.

1. Calculate the torque.

$$T = 63,000(hp/rpm)$$

where T, the torque, has the units lb-in.

$$T = (63,000)(35)/280 = 7875 \text{ lb-in.}$$

2. Calculate the shear stress.

$$s_s = T/Z_p$$

where Z_p is the polar section modulus, which is $\pi d^3/16$ in^3.
So
$$s_s = 7875 \text{ lb-in.}/(\pi d^3/16 \text{ in.}^3) = 40,107/d^3 \text{ psi}$$

3. Calculate the maximum bending moment. The shear force (Figure 18.10(b)) starts at 850 lb/2 = 425 lb and remains constant to mid span. The area under the

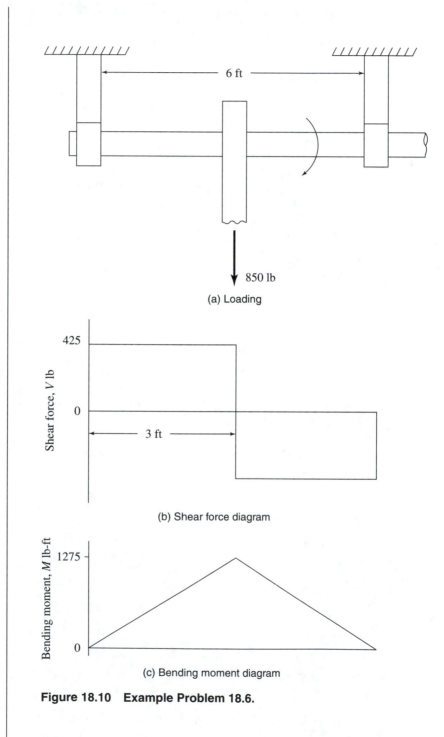

(a) Loading

(b) Shear force diagram

(c) Bending moment diagram

Figure 18.10   Example Problem 18.6.

shear force diagram, from the left end to the midsection, is (425 lb) (3 ft) = 1275 lb-ft, which is the maximum bending moment (Figure 18.10(c)).

$$M = 1275 \text{ lb-ft} = 15,300 \text{ lb-in.}$$

4. Calculate the bending stress.

$$s_b = M/S$$

where S is the section modulus, which is $\pi d^3/32$ in^3.

So $\qquad s_b = 15,300 \text{ lb-in.}/(\pi d^3/32 \text{ in.}^3) = 155,850/d^3 \text{ psi}$

This is the tensile stress, s_{nx}.

5. Calculate the maximum shear stress. From equation (18.15),

$$
\begin{aligned}
s_{s\theta)max} &= \pm \sqrt{[(s_{ny} - s_{nx})/2]^2 + s_s^2} \\
&= \pm \sqrt{(-s_{nx}/2)^2 + s_s^2} \\
&= \pm \sqrt{[-(155,850/d^3 \text{ psi})/2]^2 + (40,107/d^3 \text{ psi})^2} \\
&= 87,638/d^3 \text{ psi}
\end{aligned}
$$

6. Calculate the diameter to satisfy the shear requirement. The yield shear strength for the material is 71,000 psi. With a safety factor of 4, the allowable shear stress is 71,000 psi/4 = 17,750 psi. Therefore,

$$87,638/d^3 \text{ psi} = 17,750 \text{ psi}$$

So $\qquad\qquad\qquad d^3 = 87,638/17,750 = 4.937 \text{ in.}^3$

or $\qquad\qquad\qquad d = 1.70 \text{ in.}$

7. Calculate the maximum tensile stress. From equation (18.14 (a),

$$
\begin{aligned}
s_{n\theta)max} &= (s_{nx} + s_{ny})/2 + \sqrt{[(s_{ny} - s_{nx})/2]^2 + s_s^2} \\
&= s_{nx}/2 + \sqrt{(-s_{nx}/2)^2 + s_s^2} \\
&= (155,850/d^3 \text{ psi})/2 + 87,638/d^3 \text{ psi} \\
&= 165,563/d^3 \text{ psi}
\end{aligned}
$$

8. Calculate the diameter to satisfy the tensile requirement. The yield strength for the material is 141,000 psi. With a safety factor of 4, the allowable tensile stress is 141,000 psi/4 = 35,250 psi. Therefore,

$$165,563/d^3 \text{ psi} = 35,250 \text{ psi}$$

So $\qquad\qquad\qquad d^3 = 165,563/35,250 = 4.697 \text{ in.}^3$

or $\qquad\qquad\qquad d = 1.675 \text{ in.}$

compared with 1.70 in. for shear.

Answer

The shaft diameter should be at least 1.70 in.

18.6 COMBINED BENDING AND TENSION OR COMPRESSION

Shear stresses created due to bending are maximum on the neutral axis, falling to zero on the beam surface. The bending stresses, tensile and compressive, are zero on the neutral axis, increasing to maximum values on the beam surface. In general, the bending stresses are far greater than the shear stresses, so it is the combination of *bending* stresses and tensile, or compressive, stresses that is usually the design criterion.

In the absence of shear, equation (18.14(a)),

$$s_{n\theta)max} = (s_{nx} + s_{ny})/2 + \sqrt{[(s_{ny} - s_{nx})/2]^2 + s_s^2}$$

becomes

$$s_{n\theta)max} = (s_{nx} + s_{ny})/2 + (s_{ny} - s_{nx})/2,$$
$$= s_{ny}$$

It is our choice which we designate x and which we choose to be y. This equation simply tells us that the maximum normal stress is the maximum applied stress, so it is the algebraic sum of the bending and tensile, or compressive, stresses. With regard to maximum shear stress, equation (18.15),

$$s_{s\theta)max} = \pm \sqrt{[(s_{ny} - s_{nx})/2]^2 + s_s^2}$$

becomes

$$s_{s\theta)max} = (s_{ny} - s_{nx})/2$$

If the bending and tensile, or compressive, stresses act in the same direction, a shear stress is developed that is equal to one-half of their algebraic sum. This is the same result we found in Section 18.1.

Remember that bending stresses are positive (tensile) on one surface and negative (compressive) on the opposite surface. In combined loading situations we must be careful to use the correct signs.

18.7 ECCENTRIC LOADS

EXAMPLE PROBLEM 18.7

Determine the maximum tensile and compressive stresses in the 1-in. schedule 160 pipe shown in Figure 18.11(a). A 1-in. schedule 160 pipe has an outside diameter of 1.315 in. and an inside diameter of 0.815 in.

Solution
1. Resolve the force into x- and y-components. (See Figure 18.11(b).)

$$F_x = (500 \text{ lb})\cos 30° = 433 \text{ lb}$$
$$F_y = (500 \text{ lb})\sin 30° = 250 \text{ lb}$$

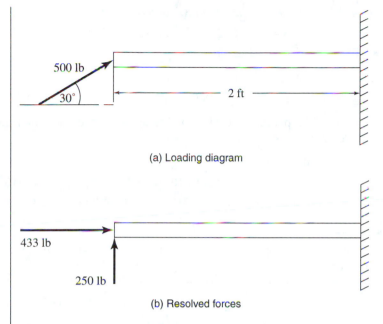

(a) Loading diagram

(b) Resolved forces

Figure 18.11   Example Problem 18.7.

2. Determine the normal (compressive) stress. The cross-sectional area of the pipe is

$$A = \pi[(1.315 \text{ in.})^2 - (0.815 \text{ in.})^2]/4 = 0.8364 \text{ in.}^2$$
$$s_{nx} = F_x/A = 433 \text{ lb}/0.8364 \text{ in.}^2$$
$$= 518 \text{ psi}$$

3. The bending stress is calculated from

$$s_b = M/S$$

where M is the maximum bending moment and S is the section modulus.

The pipe is cantilevered from the wall, so the maximum bending moment is at the wall and has the value

$$M = (250 \text{ lb})(2 \text{ ft} \times 12 \text{ in./ft}) = 6000 \text{ lb-in.}$$

This moment will create compressive stresses on the upper surface and tensile stresses on the lower.

For a pipe, the section modulus is

$$S = \pi(d_o^4 - d_i^4)/(32d_o)$$

where d_o and d_i are the outside and inside diameters.

$$S = \pi[(1.315 \text{ in.})^4 - (0.815 \text{ in.})^4]/[(32)(1.315 \text{ in.})]$$
$$= 0.1903 \text{ in.}^3$$

The bending stress is

$$s_b = M/S = 6000 \text{ lb-in.}/0.1903 \text{ in.}^3 = 31{,}530 \text{ psi}$$

This is −31,530 psi (compressive) on the upper surface and +31,530 psi (tensile) on the lower surface.

4. The total stresses are

on the upper surface: 518 psi − 31,530 psi = −31,010 psi

on the lower surface: 518 psi + 31,530 psi = 32,050 psi

Answer

The stress on the upper surface is −31,010 psi. The stress on the lower surface is 32,050 psi.

Eccentric loads, loads that are offset, create combined stresses. For example, the hacksaw blade shown in Figure 18.12 is under tension to keep it from bending while sawing through metal. The tensile force in the blade is tending to *compress* and *bend* the top section of the frame. Figure 18.13 shows a clamp. The force F is putting the clamp in *tension* and tending to *bend* it.

18.8   COMPUTER ANALYSES

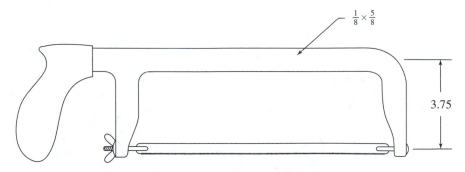

Figure 18.12   Hack saw: Example Problem 18.8.

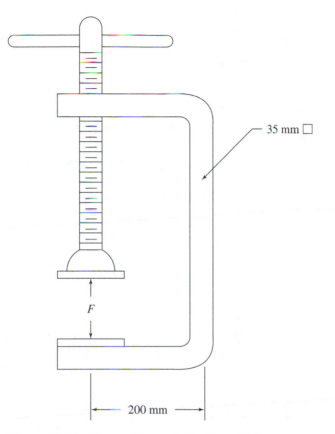

Figure 18.13   Clamp: Example Problem 18.8.

EXAMPLE PROBLEM 18.8

The hacksaw blade shown in Figure 18.12 is under a tensile stress of 2000 psi. The blade is $\frac{1}{2}$ in. deep and $\frac{1}{32}$ in. thick. The top section of the frame is $\frac{5}{8}$ in. $\times \frac{1}{8}$ in., and the distance from its centerline to the blade centerline is 3.75 in. Determine the maximum tensile and compressive stresses in the top section of the frame.

Solution
1. Calculate the eccentric force. The blade's cross-sectional area is

$$A_b = (0.50 \text{ in.})(0.03125 \text{ in.}) = 0.0156 \text{ in.}^2$$

Force = stress × area, so

$$F_x = (2000 \text{ psi})(0.0156 \text{ in.}^2) = 31.25 \text{ lb}$$

2. Calculate the compressive stress on the frame's top section. The cross-sectional area is

$$A_f = (0.625 \text{ in.})(0.125 \text{ in.}) = 0.0781 \text{ in.}^2$$

Stress = force/area, so

$$s_{nx} = -31.25 \text{ lb}/0.0781 \text{ in.}^2 = -400 \text{ psi}$$

3. Calculate the bending stress. The bending moment is

$$M = (31.25 \text{ lb})(3.75 \text{ in.}) = 117 \text{ lb-in.}$$

The moment of inertia is

$$I = bd^3/12, \qquad \text{where } b \text{ is the width and } d \text{ is the depth}$$

$$b = 0.125 \text{ in.}, \qquad d = 0.625 \text{ in.}$$

$$I = (0.125 \text{ in.})(0.625 \text{ in.})^3/12 = 2.54 \times 10^{-3} \text{ in.}^4$$

The bending stress is

$$s_b = Mc/I$$

where c is the distance from the neutral axis to the outer edge. Because $c = 0.625$ in./2 = 0.3125 in.,

$$s_b = (117 \text{ lb-in.})(0.3125 \text{ in.})/(2.54 \times 10^{-3} \text{ in.}^4)$$
$$= 14{,}395 \text{ psi}$$

This will be compressive on the lower surface of the top section and tensile on the upper. So on the lower surface the total stress is -400 psi $- 14{,}395$ psi $= -14{,}795$ psi, and on the upper surface the total stress is -400 psi $+ 14{,}395$ psi $= + 13{,}995$ psi.

Answer
The stress on the lower surface is $-14{,}800$ psi. The stress on the upper surface is 14,000 psi.

EXAMPLE PROBLEM 18.9

The allowable tensile stress for the clamp shown in Figure 18.13 is 110 MPa. What maximum force F can be applied safely?

Solution
The total stress consists of the sum of the tensile and the bending stresses.

1. Tensile stress: The cross-sectional area of the back of the frame is

$$A_f = (35 \text{ mm})(35 \text{ mm}) = 1225 \text{ mm}^2$$

So the tensile stress is

$$s_{ny} = F/1225 \text{ mm}^2 = (8.2 \times 10^{-4} \text{ mm}^{-2})F$$

2. Bending stress: The bending moment is

$$M = (200 \text{ mm})F$$

The moment of inertia of the frame back is

$$I = bd^3/12 = b^4/12$$

for a square section, where b is the length of the side. Therefore,

$$I = (35 \text{ mm})^4/12 = 1.251 \times 10^5 \text{ mm}^4$$

$$c = b/2 = 35 \text{ mm}/2 = 17.5 \text{ mm}$$

The bending stress is

$$s_b = Mc/I = (200 \text{ mm})F(17.5 \text{ mm})/(1.251 \times 10^5 \text{ mm}^4)$$

$$s_b = (27.98 \times 10^{-3} \text{ mm}^{-2})F$$

3. Total tensile stress: The total stress is

$$(8.2 \times 10^{-4} \text{ mm}^{-2})F + (27.98 \times 10^{-3} \text{ mm}^{-2})F = 0.0288F \text{ mm}^{-2}$$

4. Allowable maximum force: The maximum allowable stress is 110 MPa; therefore,

$$0.0288F \text{ mm}^{-2} = 110 \text{ N/mm}^2$$

$$F = (110 \text{ N/mm}^2)/(0.0288 \text{ mm}^{-2})$$

$$= 3819 \text{ N}$$

Answer
The maximum allowable force is 3820 N.

18.8.1   Program 'Combload'

This program determines the principal stresses, principal plane angles, and maximum shear stress and its angle for combined loadings. It also creates a plot of normal and shear stresses as functions of plane angle. Zero angle is defined as the x-direction. The combined loadings may be direct, bending, and/or torsional, in any or all combinations.

Figure 18.14 shows the input/output sheet for the program.

1. Starting in the top left corner, enter US or SI (uppercase letters) for the type of units.
2. If there are direct tensile, compressive, and/or shear forces, enter Y in the box marked "Direct? (Y or N)." If there are no direct forces, enter N or leave the box blank.
3. If there are bending loads, enter Y in the "Bending? (Y or N)" box. If not, enter N or leave it blank.
4. If there are torsional loads, enter Y in the box marked "Torsion? (Y or N)." If not, enter N or leave it blank.
5. If there are torsional loads, you must indicate whether the input will be in the form of torque or power and rpm. Enter Y in the appropriate box.
6. If there are direct forces, enter their values in lb or N in the boxes marked F_x, F_y,

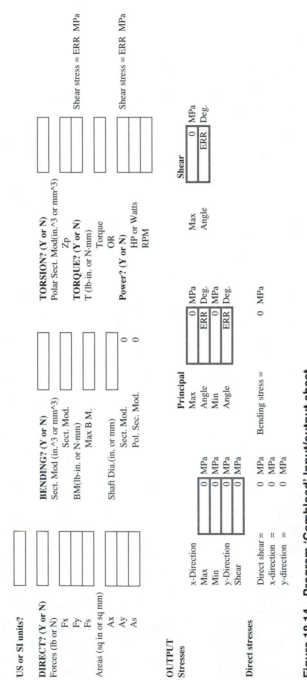

Figure 18.14 Program 'Combload' Input/output sheet.

F_s. Also enter the areas (in in.^{2} or mm^2) associated with those forces in the boxes marked A_x, A_y, A_s.

7. Under *Bending*, enter the section modulus and the maximum bending moment.

Note: If the member is a round shaft, you can enter its diameter, and the section modulus will be calculated for you. *This must then be typed into the section modulus box.*

8. Under *Torsion*, enter the polar section modulus. This will be calculated if you enter the shaft diameter, but *it must be then typed into the polar section modulus box*. Either torque is entered or power and rpm are entered to complete the input.

The output is divided into three parts, giving direct stresses, principal stresses and principal angles, and maximum shear stress and its plane angle. A graph showing the normal and shear stresses plotted against plane angle is displayed on the screen.

The following examples should clarify the use of this program.

EXAMPLE PROBLEM 18.10

Repeat Example Problem 18.3 using program 'Combload'. The question was: A bolt, having a cross-sectional area of 0.50 in.^{2}, is subjected to a tensile load of 5000 lb and a shear force of 3500 lb. Determine the maximum normal and maximum shear stresses.

Solution

Although we assume that the shear stress is the result of torsion, we are given the shear force directly, so the only input values are in the Direct column. We show the input/output sheet in Figure 18.15.

1. Open program 'Combload'.
2. Enter US in the units box.
3. Enter Y in the Direct? box.
4. We may choose to call the direct force F_x or F_y. Remember, however, that the principal planes are measured from the *x*-direction. To be consistent with Example Problem 18.3, make $F_x = 0$ and $F_y = 5000$. $F_s = 3500$, and the areas are 0.50.

Answer

The output shows that the maximum principal stress is 13,600 psi (tensile), the minimum principal stress is 3600 psi (compressive), and the maximum shear stress is 8600 psi. The principal planes are at 27.2° and 117.2°, and the maximum shear stress occurs on a plane at an angle of 72.2° to the *x*-direction.

A plot of the stresses against plane angle is displayed on the screen.

COMBINED LOADINGS
Program 'COMBLOAD'

US or SI units? [US]

DIRECT? (Y or N) [Y]
Forces (lb or N)
Fx [0]
Fy [5000]
Fs [3500]
Areas (sq in or sq mm)
Ax [0]
Ay [0.5]
As [0.5]

BENDING? (Y or N) [N]
Sect. Mod (in.^3 or mm^3)   Sect. Mod. [0]
BM(lb-in. or N·mm)   Max B M. [0]
Shaft Dia.(in. or mm) [0]
Sect. Mod. [0]
Pol. Sec. Mod. [0]

TORSION? (Y or N)  [N]
Polar Sect. Mod(in.^3 or mm^3)   Zp [0]
TORQUE? (Y or N) [N]
T (lb-in. or N·mm)   Torque [0]
OR
Power? (Y or N) [Y]
HP or Watts [0]
RPM [0]

OUTPUT
Stresses
x-Direction
Max [0] psi
Min [0] psi
y-Direction [10000] psi
Shear [7000] psi

Direct stresses
Direct shear = 7000 psi   Bending stress = 0 psi
x-direction = 0 psi
y-direction = 10000 psi

Principal
Max [13602.325] psi
Angle [27.231161] Deg.
Min [−3602.325] psi
Angle [117.23116] Deg.

Shear
Max [8602.33] psi
Angle [72.23] Deg.

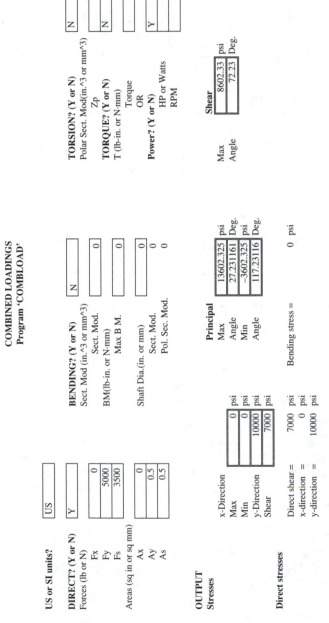

Figure 18.15   Program 'Combload' Input/output sheet for Example Problem 18.10.

EXAMPLE PROBLEM 18.11

Repeat Example Problem 18.5 using program 'Combload'. The question was: A 70-mm-diameter shaft is to transmit a torque of 500 N·m to a pump. The shaft is to be driven by a belt that will create a maximum bending moment in the shaft of 300 N·m. The design is to be based on a safety factor of 4. If the shaft material allowable tensile and shear stresses are 60 MPa and 45 MPa, respectively, is the shaft diameter satisfactory?

Solution

In this problem we are given bending moment and the torque. There are no direct loads. We show the input/output sheet in Figure 18.16 (a).

1. Open program 'Combload.'
2. Enter SI for units.
3. Enter Y for bending and Y for torsion. Also enter Y for torque.
4. Since the member is a round shaft, enter its diameter (70) in the Shaft Dia. box. The modulii appear beneath the diameter. They are Sect. Mod. = 33,673.9 mm^3 and Pol. Sec. Mod = 67,347.9 mm^3. Type these values into their appropriate boxes.
5. Enter the maximum bending moment of 3E05 and the torque of 5E05.

Answer

The output shows

maximum principal stress = 13.1 MPa   (tensile) at $-29.5°$

minimum principal stress = -8.7 MPa   (compressive) at 60.5°

maximum shear stress = 8.7 MPa   at 15.5°

bending stress alone = 8.9 MPa

shear stress due to torque alone = 7.4 MPa

Figure 18.16(b) shows the stress distributions, which can be found on the screen.

COMBINED LOADINGS
Program 'COMBLOAD'

US or SI units? [SI]

DIRECT? (Y or N) [N]
Forces (lb or N)
Fx [0]
Fy [0]
Fs [0]
Areas (sq in or sq mm)
Ax [0]
Ay [0]
As [0]

BENDING? (Y or N) [Y]
Sect. Mod (in.^3 or mm^3) [33674]
Sect. Mod.
BM(lb-in. or N·mm) [300000]
Max B M.
Shaft Dia.(in. or mm) [70]
Sect. Mod. 33673.946
Pol. Sec. Mod. 67347.892

TORSION? (Y or N) [Y]
Polar Sect. Mod(in.^3 or mm^3)
Zp [67348]
TORQUE? (Y or N) [Y]
T (lb-in. or N·mm)
Torque [500000]
OR
Power? (Y or N) [N]
HP or Watts [0]
RPM [0]

Shear stress = 7.4241254 MPa
Shear stress = #DIV/0! MPa

OUTPUT
Stresses

x-Direction
Max 8.9089505 MPa
Min −8.908951 MPa
y-Direction 0 MPa
Shear 7.4241254 MPa

Principal
Max 13.112419 MPa
Angle −29.51812 Deg.
Min −8.657944 MPa
Angle 60.481878 Deg.

Bending stress = 8.9089505 MPa

Shear
Max 8.66 MPa
Angle 15.48 Deg.

Direct stresses
Direct shear = 0 MPa
x-direction = 0 MPa
y-direction = 0 MPa

(a) Program 'Combload' Input/output sheet

Figure 18.16   Example Problem 18.11

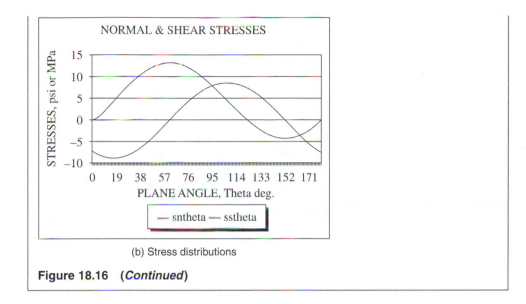

(b) Stress distributions

Figure 18.16 (*Continued*)

EXAMPLE PROBLEM 18.12

A 2.0-in.-diameter shaft is to deliver power at 430 rpm to a gear box for a large industrial mixer. It is possible that the shaft could be subjected to a maximum bending moment of 20,000 lb in. The shaft material is AISI 1141 OQT900, and a safety factor of 4 is to be used on the tensile and the shear yield strengths. What maximum power can be delivered safely?

Note: AISI 1141 OQT900 has a tensile yield strength of 129 ksi and a yield strength in shear of 64 ksi.

Solution
The maximum allowable principal stress is 129×10^3 psi/4 = 32,250 psi. The maximum allowable shear stress is 64×10^3 psi/4 = 16,000 psi. See Figure 18.17 for the input/output sheet.

1. Open program 'Combload.'
2. The units are US.
3. Enter Y into the boxes for *Bending, Torsion,* and *Power.*
4. Enter 2.0 for the shaft diameter, and copy (type) the section modulus and the polar section modulus values into their appropriate boxes.
5. Enter 20000 into the maximum bending moment box and 430 for the rpm.
6. Try a horsepower value, say, 150.

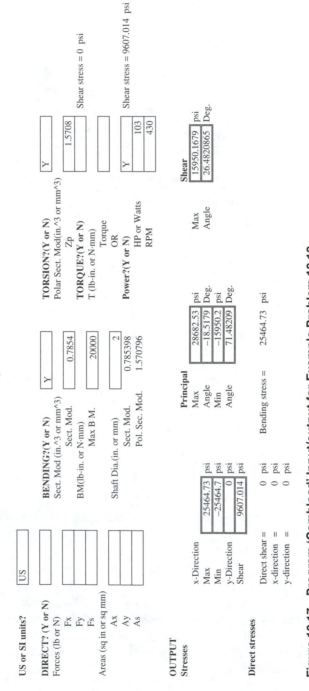

COMBINED LOADINGS
Program 'COMBLOAD'

US or SI units? US

DIRECT? (Y or N)
Forces (lb or N)
Fx
Fy
Fs
Areas (sq in or sq mm)
Ax
Ay
As

BENDING?(Y or N) Y
Sect. Mod (in.^3 or mm^3)
Sect. Mod. 0.7854
BM(lb-in. or N·mm)
Max B M. 20000
Shaft Dia.(in. or mm) 2
Sect. Mod. 0.785398
Pol. Sec. Mod. 1.570796

TORSION?(Y or N) Y
Polar Sect. Mod.(in.^3 or mm^3)
Zp 1.5708
TORQUE?(Y or N)
T (lb-in. or N·mm) Torque
OR
Power?(Y or N) Y
HP or Watts 103
RPM 430

Shear stress = 0 psi

Shear stress = 9607.014 psi

OUTPUT
Stresses

x-Direction
Max 25464.73 psi
Min −25464.7 psi
y-Direction 0 psi
Shear 9607.014 psi

Principal
Max 28682.53 psi
Angle −18.5179 Deg.
Min −15950.2 psi
Angle 71.48209 Deg.

Shear
Max 15950.1679 psi
Angle 26.4820865 Deg.

Direct stresses
Direct shear = 0 psi
x-direction = 0 psi
y-direction = 0 psi

Bending stress = 25464.73 psi

Figure 18.17 Program 'Combload' Input/output for Example Problem 18.12.

The results show

> maximum principal stress = 31,649 psi, which is *less* than the
> maximum allowable of 32,250 psi.
> maximum shear stress = 18,917 psi, which is *greater* than the
> maximum allowable of 16,000 psi

We must, therefore, reduce the horsepower to bring the shear stress to less than 16,000 psi. After a few tries you will "zero in" on a horsepower of 103, as shown in Figure 18.17, that results in a maximum shear stress of just less than 16,000 psi.

Answer
The maximum power that can be delivered safely is 103 hp.

18.9   PROBLEMS FOR CHAPTER 18

Hand Calculations

1. A piece of metal has a compressive stress of 300 MPa in the *x*-direction and a compressive stress of 500 MPa in the *y*-direction.
   a. Calculate the maximum shear stress.
   b. At what angle to the *x*-direction is the plane of maximum shear?
   c. What is the magnitude of the normal stress on this plane?

2. In a certain loading, there is no applied force in the *x*-direction, and the tensile stress in the *y*-direction is twice the shear stress. If $s_{n\theta}$ and $s_{s\theta}$ represent the normal and shear stresses on a plane at angle θ to the *x*-direction and s_s represents the applied shear stress, show that

$$(s_{n\theta} - s_{s\theta})/s_s = (4 \cos^2\theta - 1)$$

3. For the same type of loading as in Problem 2, at what plane angle are the normal and shear stresses equal?

4. Brittle materials tend to be weak in tension and will break along the plane of maximum normal stress when twisted. If you twist a piece of blackboard chalk, which is a brittle material, at what angle would you expect it to fail?

Hint: Use equation (18.13) with $s_{nx} = s_{ny} = 0$.

5. A $\frac{3}{4}$-in.-diameter rod is subjected to a direct shear force of 4000 lb and an axial tensile force of 7500 lb. What is the maximum shear stress, and at what angle is its plane to the axis of the rod?

6. A 3-in.-diameter solid shaft is subjected to a torque of 3450 lb-ft. What maximum bending moment can be applied to the shaft if the

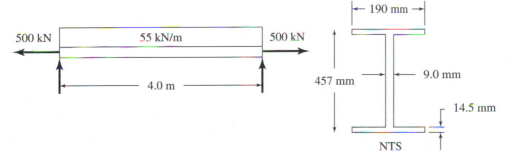

Figure 18.18

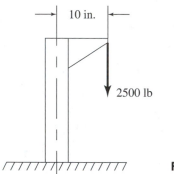

10 in.

2500 lb

Figure 18.19

maximum principal stress is limited to 20,000 psi and the maximum shear stress is limited to 15,000 psi?

7. A solid circular shaft, 2 m long, is to deliver 25 kW at 250 rpm. It is subjected to a bending moment created by a drive belt positioned at its midspan. The belt pull is 3.5 kN. The shaft material is AISI 1141 OQT900 steel, which has a yield strength of 889 MPa in tension and 445 MPa in shear. Using a factor of safety of 6, what is the minimum safe shaft diameter?

8. Figure 18.18 shows a 4.0-m-long, W460 × 74 simply supported beam, carrying a uni-

formly distributed load of 55 kN/m. It is also subjected to an axial tensile force of 500 kN. Calculate the maximum tensile stress and the maximum compressive stress in the beam.

9. Figure 18.19 shows a 4-in. nominal 80x steel pipe firmly concreted into the ground. A 2500-lb load is suspended as shown. Determine the maximum tensile and compressive stresses in the pipe.

Note: A 4-in. schedule 80x pipe has an outside diameter of 4.500 in. and an inside diameter of 3.826 in.

Computer Calculations

10. Repeat Problem 5.

11. Leaving the axial tensile force at 7500 lb, change the direct shear force in Problem 5 so that the plane of maximum shear stress is at 60° to the axis of the rod. What is the magnitude of this shear force?

12. Leaving the direct shear force at 4000 lb, change the axial tensile force in Problem 5 so that the plane of maximum shear stress is at 60° to the axis of the rod. What is the magnitude of this tensile force?

13. What axial tensile force would be necessary in Problem 5 to make the plane of maximum shear stress be at 45° to the axis of the rod?

14. Repeat Problem 6.

15. With the bending moment found in Problem 14, what maximum torque may be applied to

the shaft if the maximum principal stress is limited to 24,000 psi and the maximum shear stress is limited to 13,400 psi?

16. Repeat Problem 7.

17. A design change in Problem 7 requires the maximum rotational speed to be reduced to 200 rpm. Determine the new shaft diameter to the nearest millimeter. (Do not forget to change Z_p with each try.)

18. Repeat Problem 8.

19. Repeat Problem 9.

20. Write a computer program to calculate and plot the normal and shear stresses for combined axial tension and torque for a hollow circular shaft. The input variables are torque, tensile load, outside diameter, and inside diameter. Use either U.S. customary or SI units.

19

Struts and Columns

■ *Objectives*

In many cases in the design of structures, particularly lightweight ones such as aircraft and spacecraft, the criterion is not ultimate or yield strength but structural stability. In this chapter you learn about struts and columns and when they would be likely to buckle under compressive loads. We generally think of a column as a vertical structure, perhaps as part of a building, and a strut as a compression member in any orientation. The analysis is the same for both.

Suppose we cut a 1-in.-long piece from a 4-in.-diameter bar for a compression test specimen. As we increase the load, the specimen will compress, until eventually it will fail by cracking. If, on the other hand, our test specimen had been a 10-in.-long, $\frac{1}{4}$in.-diameter rod, it would have withstood a certain amount of compression but eventually would have bent and failed by buckling.

The questions to be answered are as follows:

1. When is a member considered to be slender enough to buckle under a compressive load?
2. What maximum load will a member support before buckling?
3. What effects do the end constraints have?

19.1 BASIC COLUMN THEORY

19.1.1 Euler's Column Equation

Figure 19.1 shows a strut just beginning to bend under a compressive load F_{cr}. Its maximum deflection is y_{max}. There is a bending moment, M, acting on the strut, where

$$M_{max} = F_{cr}y_{max} \tag{19.1}$$

In order to bend the strut, there must be a lateral force distribution along the strut, but we do not know what this distribution looks like. (See Using Calculus for a more detailed development of the theory.) Let us suppose that it is a uniformly distributed load. Then from Figure 15.13, configuration 5,

$$y_{max} = 5C_1L^4/16$$

where

$$C_1 = \omega/(24EI)$$

So

$$y_{max} = (5\omega L^4/16)/(24EI) \tag{19.2}$$

The maximum bending moment created by a uniformly distributed load is $M_{max} = \omega L^2/8$. (The first half of the shear force diagram is a triangle going from $\omega L/2$ to zero at $L/2$. Its area is $M_{max} = \frac{1}{2}(\omega L/2)(L/2) = \omega L^2/8$.)

Writing equation (19.2) as

$$y_{max} = (5L^2/2)(\omega L^2/8)/(24EI)$$

gives

$$y_{max} = (5L^2 M_{max})/(48EI)$$

Rearranging,

$$M_{max}/y_{max} = (48/5)EI/L^2$$

But from equation (19.1), $M_{max}/y_{max} = F_{cr}$; therefore,

$$F_{cr} = \frac{48}{5} EI/L^2 \tag{19.3}$$

Because $\frac{48}{5}$ is approximately equal to π^2,

$$F_{cr} = \pi^2 EI/L^2 \tag{19.4}$$

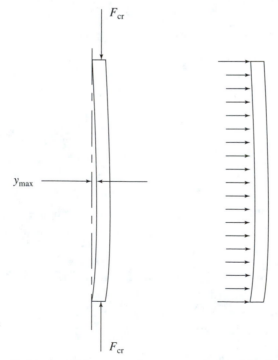

Strut under axial compression   Assumed loading   **Figure 19.1**

Equation (19.3) gives an approximate relationship based on the assumption of a uniformly distributed load. Equation (19.4) is the correct relationship and is known as **Euler's column equation**, first published in 1744. F_{cr} is the *Euler critical load*. This is the *maximum* load that can be withstood in compression. Any load greater than this value will cause a continually increasing deflection to failure.

USING CALCULUS

Development of Euler's Column Equation

Figure 19.2 shows a long, slender member loaded in compression. In this particular case, the solution to the differential equation to be developed is simplified by taking the midpoint of the member as the origin, O. The member length is $2a = L$.

At station x, where the deflection is y, the moment created by the right-end load, F, is Fy (ccw), so for rotational equilibrium, a bending moment $M = -Fy$ exists within the member. We saw in Chapter 15 that the radius of curvature is related to the bending moment through the equation

$$d^2y/dx^2 = M/EI$$

So
$$d^2y/dx^2 = -Fy/EI$$

or
$$d^2y/dx^2 + Fy/EI = 0$$

If we let $F/EI = \mu^2$ (which will assist in the solution), then we have the differential equation

$$d^2y/dx^2 + \mu^2y = 0$$

The solution to this equation is

$$y = C \cos \mu x + D \sin \mu x \tag{1}$$

You can verify this by differentiation:

$$dy/dx = -C\mu \sin \mu x + D\mu \cos \mu x$$
$$d^2y/dx^2 = -C\mu^2\cos \mu x - D\mu^2\sin \mu x$$

But
$$\mu^2y = C\mu^2\cos \mu x + D\mu^2\sin \mu x$$

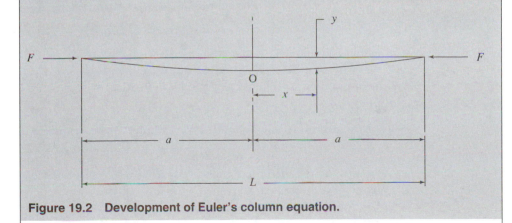

Figure 19.2   Development of Euler's column equation.

Therefore, $d^2y/dx^2 + \mu^2y = 0$.

The deflection, y, is zero at $x = a$ and at $x = -a$. Substituting $y = 0$ and $x = a$ in equation (1) gives

$$y = C \cos \mu x + D \sin \mu x$$

$$0 = C \cos \mu a + D \sin \mu a \qquad (2)$$

and substituting $y = 0$ and $x = -a$ gives

$$0 = C \cos (-\mu a) + D \sin (-\mu a)$$

But because $\cos (-\mu a) = \cos \mu a$ and $\sin (-\mu a) = -\sin \mu a$, we have

$$0 = C \cos \mu a - D \sin \mu a \qquad (3)$$

Adding equations (2) and (3) results in $C \cos \mu a = 0$, whereas subtracting (3) from (2) results in $D \sin \mu a = 0$.

These two equations must be satisfied simultaneously.

1. One solution is $C = D = 0$, but this is trivial since it means that the deflection is zero all along the member (see equation (1)).

2. Another solution is $C = 0$ and $\sin \mu a = 0$. This results in $\mu a = 0$ or $\mu a = \pi$ or $\mu a = 2\pi$ or $\mu a = n\pi$, where n is any integer. So $\mu^2 = F/EI = 0$ or π^2/a^2 or $4\pi^2/a^2$, etc., making $F = 0$ or $F = \pi^2 EI/a^2$ or $F = 4\pi^2 EI/a^2$, etc.

3. The third possible solution is $D = 0$ and $\cos \mu a = 0$. This results in $\mu a = \pi/2$ or $\mu a = 3\pi/2$ or $\mu a = n'\pi/2$, where n' is any odd integer. Therefore, $F = \pi^2 EI/4a^2$ or $F = 9\pi^2 EI/4a^2$, etc.

Apart from the trivial solution $F = 0$, we see that the smallest force satisfying the problem is

$$F = \pi^2 EI/4a^2$$

But $2a = L$; therefore, $F = \pi^2 EI/L^2$. This is the critical load.

19.1.2   Column Stress

If the cross-sectional area of the strut or column is A, then from equation (19.4) the critical stress is

$$s_{cr} = F_{cr}/A = (\pi^2 EI/L^2)/A \qquad (19.5)$$

I, the moment of inertia, and A, the cross-sectional area, are both functions of the size and shape of the cross section of the strut. We combine these into one term called the **radius of gyration**, r. Radius of gyration is defined as

$$r = \sqrt{I/A} \qquad (19.6)$$

Substituting this into equation (19.5) gives **Euler's stress equation**:

$$s_{cr} = \pi^2 E/(L/r)^2 \qquad (19.7)$$

The term L/r is called the **slenderness ratio**.

We see that the critical stress—i.e., the level of stress that can be sustained before the strut fails by buckling—depends on the square of the slenderness ratio and on the modulus of elasticity, E. Because the slenderness ratio is a function of I and I depends on the axis being considered if the cross section is not circular, then generally the slenderness ratio can have a maximum and a minimum value. We see from equation (19.7) that the largest value of slenderness ratio will result in the smallest (worst-case) critical stress. We can see that we *should* use the largest value by considering a simple demonstration. Take a 12-in. wooden or plastic ruler and push on each end. It will begin to bend about an axis through its width, not its thickness. Remembering that $I = bd^3/12$ for a rectangular cross section, you can see that it is bending about the axis of *least* moment of inertia. This is also the axis of least radius of gyration, because $r = \sqrt{I/A}$, and the axis of maximum slenderness ratio, L/r.

EXAMPLE PROBLEM 19.1

Determine the slenderness ratio of a 50-mm-diameter rod that is 1.50 m long.

Solution

$$r = \sqrt{I/A}$$
$$I = \pi d^4/64$$
$$A = \pi d^2/4$$
$$r = \sqrt{(\pi d^4/64)/(\pi d^2/4)}$$
$$= \sqrt{d^2/16}$$
$$r = d/4 \tag{19.8}$$
$$d = 50 \text{ mm}, \quad \text{so } r = 12.5 \text{ mm}$$
$$\text{slenderness ratio} = L/r = 1500 \text{ mm}/12.5 \text{ mm} = 120$$

Answer
The slenderness ratio is 120.

EXAMPLE PROBLEM 19.2

If the rod in Example Problem 19.1 were made of A242 structural steel, could it withstand an axial compressive stress of 15.0 MPa without buckling?

Solution
The critical stress is $s_{cr} = \pi^2 E/(L/r)^2$. E for structural steel is 200×10^3 MPa; therefore,

$$s_{cr} = \pi^2(200 \times 10^3 \text{ MPa})/(120)^2$$
$$= 13.9 \text{ MPa}, \quad \text{which is less than 15.0 MPa}$$

Answer
No, the rod could not withstand an axial compressive stress of 15.0 MPa without buckling.

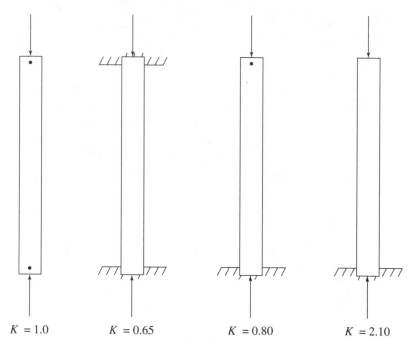

$K = 1.0$ $K = 0.65$ $K = 0.80$ $K = 2.10$

Figure 19.3 End constraints.

19.1.3 End Constraints

Euler's equations (19.4) and (19.7) were derived for struts or columns with pin joints or hinges at their ends. To allow for other types of end constraints (see Figure 19.3), we define an **effective length**,

$$L_e = KL \tag{19.9}$$

where the factor K is given in the following table.

End Constraints	K
1. Both ends pinned or hinged	1.00
2. Both ends fixed	0.65
3. One end pinned, one end fixed	0.80
4. One end fixed, one end free	2.10

Note: These are the recommended values that have been experimentally verified. The corresponding theoretical values are

$\quad$ (1) $\quad K = 1.0,$ $\quad\quad$ (2) $\quad K = 0.5,$ $\quad\quad$ (3) $\quad K = 0.707,$ $\quad\quad$ (4) $\quad K = 2.0$

19.2 ALLOWABLE LOAD

Obviously you would not design a column for its critical load, and as in all engineering design work it is essential to include design factors. For *long* columns (we discuss what we mean by *long* in the next section), the American Institute of Steel Construction (AISC) recommends a design factor of 1.92, and the Aluminum Association recommends 1.95. For mechanical engineering designs where there may be some uncertainty about the straightness of a strut, possible inhomogeneity of the material, and the fact that the load may not be aligned exactly with the strut axis, a design factor of 3 is recommended.

EXAMPLE PROBLEM 19.3

A 12-ft concrete column is to support an axial load of 50,000 lb. Both ends of the column are fixed. What would you recommend as the minimum diameter? Use a design factor of 2.0. Depending on the type of cement and the type and size of the aggregate, the elastic modulus for concrete can vary from 2.7×10^6 psi to 5.1×10^6 psi. Take $E = 3.0 \times 10^6$ for this problem.

Solution

The effective length is $L_e = KL$. For both ends fixed, $K = 0.65$. Therefore, $L_e = (0.65)(144 \text{ in.}) = 93.6 \text{ in.}$

From equation (19.4), the critical load is $F_{cr} = \pi^2 EI/L^2$, where we must replace L with L_e because the ends are not pin-jointed and also include a design factor of 2.0:

$$50,000 \text{ lb} = (\pi^2 EI/L_e^2)/2.0$$

(In effect, we are designing a column whose critical load is two times the applied load.) So,

$$50,000 \text{ lb} = \pi^2(3 \times 10^6 \text{ psi})I/[(93.6 \text{ in.})^2(2.0)]$$

or

$$I = (50,000 \text{ lb})(93.6 \text{ in.})^2(2.0)/[\pi^2(3 \times 10^6 \text{ psi})]$$

$$= 29.59 \text{ in.}^4$$

For a solid circular shape, $I = \pi d^4/64$, or

$$d^4 = 64I/\pi$$

So,

$$d^4 = 64(29.59 \text{ in.}^4)/\pi$$

$$d^4 = 602.8 \text{ in.}^4$$

$$d = 4.96 \text{ in.}$$

Answer

Recommend a column diameter of at least 5.0 in.

19.3 TRANSITION SLENDERNESS RATIO

Experimental results have confirmed Euler's equation for long slender columns down to a certain slenderness ratio that depends on the mechanical properties of the material. For lower slenderness ratios, J. B. Johnson proposed the equation

$$s_{cr} = F_{cr}/A = [1 - (L_e/r)^2/(2C_c^2)]s_y \qquad (19.10)$$

where

$$C_c = \sqrt{2\pi^2 E / s_y}$$

$$s_y = \text{yield stress of the material}$$

Equation (19.10) shows that as the slenderness ratio approaches zero—i.e., for short, thick members, the critical stress is simply the yield stress of the material. The member will fail because it has been compressed to its yield point.

As the slenderness ratio increases, the Johnson formula results get closer and closer to the Euler equation results, until they agree at a certain slenderness ratio, called the *transition slenderness ratio*. We can determine the value of the transition slenderness ratio by equating the two equations:

$$s_{cr} = [1 - (L_e/r)^2/(2C_c^2)]s_y \tag{19.10}$$

$$s_{cr} = \pi^2 E/(L_e/r)^2 \tag{19.7}$$

Equating the two:

$$[1 - (L_e/r)_{tr}^2/(2C_c^2)]s_y = \pi^2 E/(L_e/r)_{tr}^2$$

where $(L_e/r)_{tr}$ is the transition slenderness ratio. Multiplying both sides by $(L_e/r)_{tr}^2/s_y$ yields

$$(L_e/r)_{tr}^2 - (L_e/r)_{tr}^4/(2C_c^2) - \pi^2 E/s_y = 0$$

Now multiply throughout by $-2C_c^2$:

$$(L_e/r)_{tr}^4 - (2C_c^2)(L_e/r)_{tr}^2 + (2C_c^2)(\pi^2 E)/s_y = 0 \tag{19.11}$$

The general solution of a quadratic equation of the form

$$ax^2 + bx + c = 0$$

is

$$x = [-b \pm \sqrt{(b^2 - 4ac)}]/(2a) \tag{19.12}$$

Equation (19.11) is a quadratic equation in $(L_e/r)_{tr}^2$, where $a = 1$, $b = -(2C_c^2)$, and $c = (2C_c^2)(\pi^2 E)/s_y$. Notice that if $b^2 = 4ac$ in equation (19.12), then the square-root term will be zero, and x will equal $-b/(2a)$.

In our case, $b^2 = 4C_c^4$, and $4ac = 8(C_c^2)(\pi^2 E)/s_y$. But, $C_c^2 = 2\pi^2 E/s_y$, so $\pi^2 E/s_y = C_c^2/2$, and the term $4ac$ may be written as $8(C_c^2)(C_c^2/2)$ or $4C_c^4$. But this is the same as b^2, so $x = -b/(2a)$, which for the present case is

$$(L_e/r)_{tr}^2 = -(-2C_c^2)/2 = C_c^2$$

so

$$(L_e/r)_{tr} = C_c = \sqrt{2\pi^2 E/s_y} \tag{19.13}$$

C_c is called the *column constant*. Before calculating the critical load or critical stress for a column or strut, we must first determine whether the slenderness ratio is greater than or less than the column constant. If it is greater, we use the Euler equation; if less, the Johnson formula.

EXAMPLE PROBLEM 19.4

Go back to Example Problem 19.3 and check if the Euler equation was the correct equation to use. Take $s_y = 12{,}000$ psi.

Solution

We have to determine whether the slenderness ratio is greater than or less than the column constant.

In Example Problem 19.3 we found that:

$$I = 29.59 \text{ in.}^4, \qquad d = 4.96 \text{ in.}, \qquad L_e = 93.6 \text{ in.}$$

The column cross-sectional area is

$$A = \pi d^2/4 = \pi(4.96)^2/4 = 19.32 \text{ in.}^2$$

The radius of gyration is

$$r = \sqrt{I/A} = \sqrt{29.59 \text{ in.}^4/19.32 \text{ in.}^2} = 1.238 \text{ in.}$$

The slenderness ratio is

$$L_e/r = 93.6 \text{ in.}/1.238 \text{ in.} = 75.6$$

The column constant is

$$C_c = \sqrt{2\pi^2 E/s_y} = \sqrt{2\pi^2(3.0 \times 10^6 \text{ psi})/12{,}000 \text{ psi}}$$
$$= 70.3$$

Answer

The slenderness ratio is greater than the column constant; therefore, the Euler equation is the correct equation to use.

EXAMPLE PROBLEM 19.5

A connecting rod in an internal combustion engine is 8.75 in. long and has the cross section shown in Figure 19.4(a). The maximum axial compressive force on the rod is 8700 lb. The rod is to be made of a chromium steel that has a yield strength of 135 ksi and an elastic modulus of 29×10^6 psi. A design factor of 3.0 is required. Is the design acceptable?

Solution

1. Determine the *minimum* moment of inertia for the section.

About the *x-x* axis: To determine the moment of inertia about the *x-x* axis (see Figure 19.4(b)), we can calculate *I* for the complete rectangle, then subtract the moments of inertia of the two small rectangles. For the complete rectangle,

$$I = bd^3/12 \qquad \text{where } b = 0.75 \text{ in.}, \ d = 1.0 \text{ in.}$$
$$= (0.75 \text{ in.})(1.0 \text{ in.})^3/12 = 0.0625 \text{ in.}^4$$

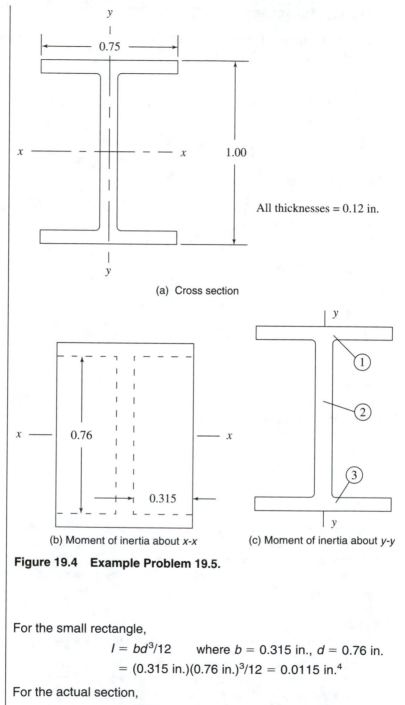

(a) Cross section

(b) Moment of inertia about *x-x*

(c) Moment of inertia about *y-y*

Figure 19.4   Example Problem 19.5.

For the small rectangle,

$$I = bd^3/12 \quad \text{where } b = 0.315 \text{ in., } d = 0.76 \text{ in.}$$
$$= (0.315 \text{ in.})(0.76 \text{ in.})^3/12 = 0.0115 \text{ in.}^4$$

For the actual section,

$$I = 0.0625 \text{ in.}^4 - 2(0.0115 \text{ in.}^4) = 0.0395 \text{ in.}^4$$

About the *y-y* axis: If we use the same method as before, we will have to shift the axes of the two small rectangles, because the axis *y-y* does not pass through their centroids. Probably it is easier to divide the cross section into three parts, as in Figure 19.4(c), where the y-y axis passes through the centroids of each part.

$$I = I_1 + I_2 + I_3 = I_1 + 2 \times I_2$$
$$I_1 = (0.76 \text{ in.})(0.12 \text{ in.})^3/12 = 1.094 \times 10^{-4} \text{ in.}^4$$
$$I_2 = (0.12 \text{ in.})(0.75 \text{ in.})^3/12 = 4.219 \times 10^{-3} \text{ in.}^4$$

So
$$I = 1.094 \times 10^{-4} \text{ in.}^4 + 2 \times 4.219 \times 10^{-3} \text{ in.}^4$$
$$= 8.55 \times 10^{-3} \text{ in.}^4$$

The smallest value of I for the section is, therefore,
$$I = 8.55 \times 10^{-3} \text{ in.}^4$$

2. Determine the cross-sectional area.
$$A = (0.12 \text{ in.})(0.76 \text{ in.}) + 2 \times (0.12 \text{ in.})(0.75 \text{ in.})$$
$$= 0.2712 \text{ in.}^2$$

3. Determine the radius of gyration.
$$r = \sqrt{I/A} = \sqrt{8.55 \times 10^{-3} \text{ in.}^4/0.2712 \text{ in.}^2}$$
$$= 0.1776 \text{ in.}$$

4. Determine the slenderness ratio.
$$\text{slenderness ratio} = L_e/r$$

The ends are pin-jointed, so $K = 1.0$; therefore,
$$\text{slenderness ratio} = L/r = 8.75 \text{ in.}/0.1776 \text{ in.} = 49.28$$

5. Determine the column constant.
$$C_c = \sqrt{2\pi^2 E/s_y} = \sqrt{2\pi^2(29 \times 10^6 \text{ psi})/135{,}000 \text{ psi}}$$
$$= 65.12$$

6. Select the column formula. Because the slenderness ratio is less than the column constant, the Euler equation is not applicable. From equation (19.10), the Johnson formula is
$$s_{cr} = [1 - (L_e/r)^2/(2C_c^2)]s_y$$

where $C_c = 65.12$, $L_e/r = 49.28$, and $S_y = 135{,}500 \text{ psi}$. So
$$s_{cr} = [1 - (49.28)^2/(2 \times 65.12^2)] (135{,}000 \text{ psi})$$
$$= 96{,}340 \text{ psi}$$

Including a design factor of 3.0, the actual stress must not exceed 96,340 psi/3.0 = 32,110 psi.

7. Calculate the actual stress.
$$F = 8700 \text{ lb}, \qquad A = 0.2712 \text{ in.}^2$$
$$s = F/A = 8700 \text{ lb}/0.2712 \text{ in.}^2 = 32{,}080 \text{ psi}$$

Answer
The actual stress is less than the critical stress; therefore, the design is acceptable.

19.4   AISC COLUMN CODE

The American Institute of Steel Construction gives the following equations for structural steel columns.

1. For $L_e/r \leq C_c$:

$$s_{cr} = [1 - (L_e/r)^2/(2C_c^2)]s_y/N \qquad (19.14(a))$$

where N is a safety factor to allow for possible misalignment, uncertainties about the yield stress, etc.

$$N = 1.67 + 0.375 \, (L_e/r)/C_c - 0.125[(L_e/r)/C_c]^3 \qquad (19.14(b))$$

The limiting values of N are

a. As (L_e/r) approaches zero (short columns), N approaches 1.67,
b. When $(L_e/r) = C_c$, $N = 1.67 + 0.375 - 0.125 = 1.92$.

2. For $L_e/r > C_c$ (max $L/r = 200$),

$$s_{cr} = 149 \times 10^6/(L_e/r)^2 \text{ psi} \qquad (19.15(a))$$

or

$$= 1.027 \times 10^6/(L_e/r)^2 \text{ MPa} \qquad (19.15(b))$$

EXAMPLE PROBLEM 19.6

What axial load can be supported safely by a 20-ft-long, A36 structural steel W14 $\times$ 74 column if the ends are pinned? For W14 $\times$ 74, the minimum radius of gyration is 2.48 in., and the cross-sectional area is 21.8 in². For A36 structural steel, $E = 29 \times 10^6$ psi and $s_y = 36,000$ psi.

Solution

1. Calculate the column constant.

$$C_c = \sqrt{2\pi^2 E/s_y} = \sqrt{2\pi^2(29 \times 10^6 \text{ psi})/36,000 \text{ psi}}$$

$$= 126.1$$

2. Calculate the slenderness ratio. The ends are pin-jointed; therefore, $K = 1$ and $L_e = L$.

$$\text{slenderness ratio} = L/r = 240 \text{ in.}/2.48 \text{ in.} = 96.77$$

3. Select the appropriate equation. Because $L_e/r < C_c$, the equation to use is

$$s_{cr} = [1 - (L_e/r)^2/(2C_c^2)]s_y/N$$

where

$$N = 1.67 + 0.375(L_e/r)/C_c - 0.125[(L_e/r)/C_c]^3$$

$$= 1.67 + 0.375(96.77)/126.1 - 0.125(96.77/126.1)^3$$

$$= 1.67 + 0.29 - 0.06$$

$$= 1.90$$

4. Calculate the critical stress:

$$s_{cr} = [1 - (L_e/r)^2/(2C_c^2)]s_y/N$$
$$= [1 - (96.77)^2/(2 \times 126.1^2)]36{,}000 \text{ psi}/1.90$$
$$= 13{,}370 \text{ psi}$$

Note: The AISC formula contains the design factor, N, so s_{cr} is the maximum *allowable* stress.

5. Calculate the maximum allowable load.

$$F = As_{cr} = (21.8 \text{ in.}^2)(13{,}370 \text{ psi}) = 291{,}400 \text{ lb}$$

Answer
The column can safely support an axial load of 291,000 lb.

The problem of finding the lightest column to support a given load is a little more difficult, because we do not know A, I, or r.

EXAMPLE PROBLEM 19.7

Select the most efficient (lightest) *W*-shape to support an axial load of 2000 kN. The column, which has fixed ends, is 4.5 m long and is made of A242 structural steel. The material yield strength is 345 MPa, and its elastic modulus is 200×10^3 MPa.

Solution

1. Calculate the column constant.

$$C_c = \sqrt{2\pi^2 E/s_y} = \sqrt{2\pi^2(200 \times 10^3 \text{ MPa})/345 \text{ MPa}}$$
$$= 107.0$$

2. Calculate the effective length. Since the ends are fixed, $K = 0.65$, and $L_e = (0.65)(4.5 \text{ m}) = 2.925 \text{ m}$.

3. Determine the smallest possible cross-sectional area. If L_e/r is small, then N approaches 1.67 from equation (19.14(b)). The allowable stress would then be $s_y/N = 345 \text{ MPa}/1.67 = 207 \text{ MPa}$. So the minimum possible area is

$$A_{min} = F/s_{allow} = 2000 \times 10^3 \text{ N}/207 \text{ MPa} = 9662 \text{ mm}^2$$

4. The following table lists common W-shapes whose cross-sectional areas are greater than 9662 mm². It also includes the minimum radius of gyration.

Section	Area (mm²)	r (mm)
W920 × 238	30,300	63.7
W840 × 329	41,900	91.4
W610 × 125	15,900	49.7
W530 × 92	11,800	45.0
W460 × 82	10,500	42.2
W410 × 149	19,000	63.8
W410 × 100	12,700	62.4
W360 × 237	30,100	101.6
W360 × 196	25,000	95.5
W360 × 147	18,800	94.2
W360 × 110	14,100	62.9
W360 × 79	10,100	48.7
W310 × 202	25,700	80.3
W310 × 143	18,200	78.6
W310 × 107	13,600	77.3
W310 × 86	11,000	63.6
W250 × 167	21,200	68.1
W250 × 149	19,000	67.4
W250 × 115	14,600	66.3
W250 × 80	10,200	64.9
W200 × 100	12,710	53.9

5. Reduce the candidates. The radii of gyration in the table range from 42.2 mm to 101.6 mm. This range, when considered in conjunction with the known variables, reduces the range of viable areas.

Ideally, the actual stress would be equal to the allowable stress:

$$F_{actual}/A = s_{allow} = [1 - (L_e/r)^2/(2C_c^2)]s_y/N$$

Rearranging gives

$$A = F_{actual}N/[1 - (L_e/r)^2/(2C_c^2)]s_y$$
$$= F_{actual}N/[1 - 0.5(L_e/C_c)^2/r^2]s_y \qquad (19.16)$$

From equation (19.14 (b)):

$$N = 1.67 + 0.375 (L_e/r)/C_c - 0.125[(L_e/r)/C_c]^3$$

which can be written as

$$N = 1.67 + 0.375 (L_e/C_c)/r - 0.125[(L_e/C_c)^3/r^3]$$

Substituting this into equation (19.16) yields

$$A = F_{actual}\{1.67 + 0.375 (L_e/C_c)/r$$
$$- 0.125[(L_e/C_c)^3/r^3]\}/[1 - 0.5 (L_e/C_c)^2/r^2]s_y \quad (19.17)$$

The maximum value of A will be when r is its minimum, and the minimum value of A will be when r is its maximum, which in the present problem are 42.2 mm and 101.6 mm, respectively.

Also, $F_{actual} = 2 \times 10^6$ N, $L_e/C_c = 2925$ mm/$107 = 27.34$ mm, and $s_y = 345$ MPa. Therefore,

$$A_{max} = (2 \times 10^6 \text{ N})\{1.67 + 0.375 (27.34 \text{ mm})/(42.2 \text{ mm}) - 0.125$$
$$\times [(27.34 \text{ mm})^3/(42.2 \text{ mm})^3]\}/[1 - 0.5(27.34 \text{ mm})^2/(42.2 \text{ mm})^2](345 \text{ MPa})$$
$$= 13{,}785 \text{ mm}^2$$

$$A_{min} = (2 \times 10^6 \text{ N})\{1.67 + 0.375 (27.34 \text{ mm})/(101.6 \text{ mm}) - 0.125$$
$$\times [(27.34 \text{ mm})^3/(101.6 \text{ mm})^3]\}/[1 - 0.5(27.34 \text{ mm})^2/(101.6 \text{ mm})^2](345 \text{ MPa})$$
$$= 10{,}637 \text{ mm}^2$$

These limits reduce our table of candidates to the following:

Section	Area (mm²)	r (mm)
W530 × 92	11,800	45.0
W410 × 100	12,700	62.4
W360 × 79	10,100	48.7
W310 × 107	13,600	77.3
W310 × 86	11,000	63.6
W200 × 100	12,710	53.9

Using equation (19.17) we now calculate the minimum area for each of the *r*-values in the table:

Section	Area (mm²)	r (mm)	Min. Area (mm²)
W530 × 92	11,800	45.0	13,294
W410 × 100	12,700	62.4	11,695
W360 × 79	10,100	48.7	12,789
W310 × 107	13,600	77.3	11,113
W310 × 86	11,000	63.6	11,633
W200 × 100	12,710	53.9	12,267

6. Select the best candidate. This shows that the possible candidates are

W410 × 100

W310 × 107

W200 × 100

Of the final candidates, either the W410 × 100 or the W200 × 100 would be the lightest, but the W410 × 100 has a higher radius of gyration and the same area (to within 0.08%), so this would be the better choice.

Answer

Select a W410 × 100 shape.

> **Check**
> The selected shape has $A = 12{,}700$ mm² and $r = 62.4$ mm.
>
> $$s_{allow} = [1 - (L_e/r)^2/(2C_c^2)]s_y/N$$
>
> where
>
> $$N = 1.67 + 0.375\,(L_e/r)/C_c - 0.125[(L_e/r)/C_c]^3$$
> $$L_e = 2925 \text{ mm}, \qquad C_c = 107, \qquad s_y = 345 \text{ MPa}$$
> $$L_e/r = 2925 \text{ mm}/62.4 \text{ mm} = 46.88$$
> $$(L_e/r)/C_c = 46.88/107 = 0.438$$
> $$N = 1.67 + 0.375(0.438) - 0.125[0.438]^3$$
> $$= 1.824$$
> $$s_{allow} = [1 - (46.88)^2/(2 \times 107^2)]345 \text{ MPa}/1.824$$
> $$s_{allow} = 171.0 \text{ MPa}$$
>
> The actual stress is 2×10^6 N/12,700 mm = 157.5 MPa.
> So the actual stress is less than the allowable stress.

19.5   TIMBER COLUMNS

The National Lumber Manufacturers Association recommends the following for wooden columns.

1. For $L_e/r < C_c$:

   s_{allow} = the allowable compressive stress of a short block loaded parallel to the grain

2. For $L_e/r \geq C_c$:

$$s_{allow} = 3.619E/(L_e/r)^2 \tag{19.18}$$

Note: $3.619 = \pi^2/2.727$, so equation (19.18) is Euler's equation with a design factor of 2.727.

EXAMPLE PROBLEM 19.8

A 7-ft-long ponderosa pine 4 × 6 column has one end cemented into the ground and the upper end free. What maximum axial load can it support safely? $E = 1.1 \times 10^6$ psi, and the allowable compressive stress parallel to the grain is 800 psi.

Solution
1. Calculate the effective length. With one end fixed and the other end free, $K = 2.10$.

Therefore, $\qquad\qquad\qquad L_e = (84 \text{ in})(2.10) = 176.4$ in.

2. Calculate the minimum radius of gyration. Generally, for a rectangular cross section having sides b and d, where $d \le b$,

$$I_{min} = bd^3/12, \quad \text{and} \quad A = bd$$
$$r_{min} = \sqrt{(I_{min}/A)} = \sqrt{(bd^3/12)/(bd)}$$
$$= \sqrt{d^2/12}$$
$$= 0.2887d \tag{19.19}$$

A nominal 4×6 is 3.5 in. $\times$ 5.5 in., so $d = 3.5$ in.

$$r_{min} = 1.010 \text{ in.}$$

3. Calculate the effective slenderness ratio.

effective slenderness ratio $L_e/r = 176.4$ in./1.010 in. $= 174.7$

4. Calculate the column constant.

$$C_c = \sqrt{2\pi^2 E/s_y} = \sqrt{2\pi^2(1.1 \times 10^6 \text{ psi})/800 \text{ psi}}$$
$$= 164.7$$

5. Calculate the allowable stress. $L_e/r \ge C_c$, so the appropriate equation is

$$s_{allow} = 3.619E/(L_e/r)^2$$
$$E = 1.1 \times 10^6 \text{ psi}, \quad L_e/r = 174.7$$

So

$$s_{allow} = 3.619 \, (1.1 \times 10^6 \text{ psi})/(174.7)^2$$
$$= 130.4 \text{ psi}$$

6. Calculate the allowable load.

$$F_{allow} = s_{allow}A$$
$$A = (3.5 \text{ in.})(5.5 \text{ in.}) = 19.25 \text{ in.}^2$$
$$F_{allow} = (130.4 \text{ psi})(19.25 \text{ in}^2) = 2510 \text{ lb}$$

Answer
The column can support a maximum load of 2500 lb safely.

EXAMPLE PROBLEM 19.9

We have decided to use a 4×4 instead of a 4×6 for the same conditions as in Example Problem 19.8. What maximum axial load can the column support safely?

Solution
The minimum edge dimension is the same, $d = 3.5$ in. Therefore, the minimum radius of gyration will be the same, which results in the same allowable stress of 130.4 psi.

The new cross-sectional area is (3.5 in.)(3.5 in.) $= 12.25$ in.^{2} The allowable load is (130.4 psi)(12.25 in.^{2}) $= 1597$ lb.

Answer
The column can support a maximum load of 1600 lb safely.

19.6   ECCENTRIC LOADS, INITIAL CURVATURE, AND STRESS REVERSAL

The analyses presented so far in this chapter are limited to compressive loads that act essentially through the centroid of each section of the column. If the load is not axially aligned or the column is slightly bowed, there will be a bending moment on the member in addition to the compressive load. The maximum bending moment will be $M_{max} = Fe$, where F is the end load and e is the maximum lateral distance from the line of action of the load to the column axis. See Figure 19.5. The maximum stress caused by this moment is $M_{max}\,c/I$, or M_{max}/S, which will be a compressive stress on one side of the axis and a tensile stress on the other. It is possible that the sum of the directly applied compressive stress and the tensile stress due to the bending moment can result in tension stresses in regions of the column. We show this in Figure 19.6.

This stress reversal could be important in the case of a concrete or brick column, whose tensile strengths are virtually zero. This condition arises only when the eccentricity exceeds a certain amount, which is a function of the cross section of the column. The region in which the load may be applied without causing a stress reversal is called the **kern.**

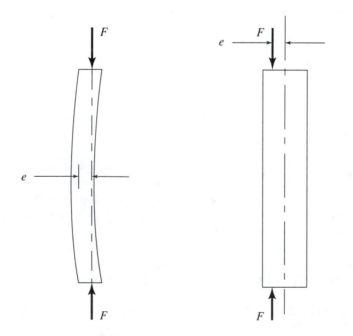

Figure 19.5   Nonaxial loadings.

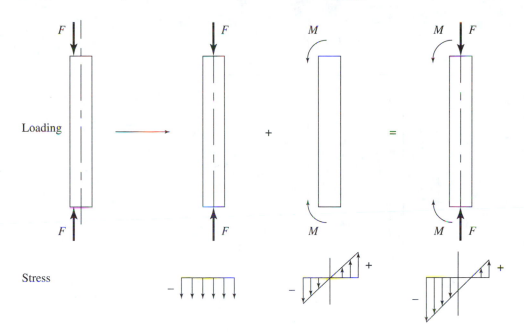

Figure 19.6 Stress reversal: Creation of tension by compression.

EXAMPLE PROBLEM 19.10

Determine the kern for a solid circular cross section.

Solution

Let the load be F located a distance e from the center of the section (Figure 19.7). If the diameter of the section is d, then

$$A = \pi d^2/4$$

so the direct compressive stress is $F/A = 4F/(\pi d^2)$.

The bending moment is $M = Fe$, which creates a bending stress of

$$s_b = \pm M/S = \pm Fe/S$$

This will be a compressive stress on one side and a tensile stress on the other. For a solid circular cross section, $S = \pi d^3/32$. Therefore,

$$s_b = \pm Fe/(\pi d^3/32) = \pm 32Fe/(\pi d^3)$$

The algebraic sum of the direct compressive stress and the tensile stress due to bending is

$$4F/(\pi d^2) - 32Fe/\pi d^3$$

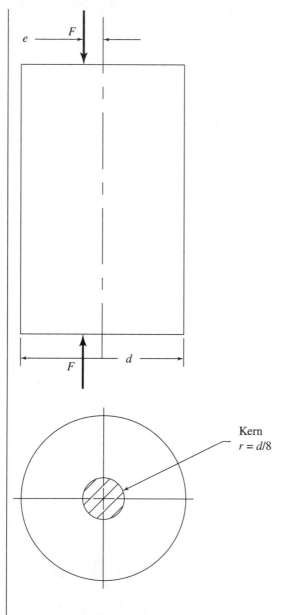

Figure 19.7   Kern of a solid circular section.

This will be zero on the edge of the kern. Therefore,

$$4F/\pi d^2 = 32Fe/\pi d^3$$

from which $e = d/8$.

   Since e is the outermost radius of the kern, the kern is a circular area whose radius is $d/8$. See Figure 19.7.

   To put this in perspective, a 4-in.-diameter column would have to have the center of gravity of its load no more than $\frac{1}{2}$ in. from its center to avoid any tensile stresses.

19.7 COMPUTER ANALYSES

19.7.1 Program 'Column'

Program 'Column' selects the most economical steel column shapes for the following input variables:

1. The type of units, either US or SI
2. The maximum load, in either lb or N
3. The column length, in either in. or mm
4. The end restraints, entered as the K factor
5. The type of structural steel, entered as a code number

The program is based on the AISC Column Code and contains the properties of numerous W-, S-, C-, and L-shapes. The output lists the best (in terms of the lightest) W-shape, S-shape, C-shape, and L-shape for the given inputs. You should pay particular attention to the possibility of an offset load when selecting the shape. W- and S-shapes are probably better choices than C- or L-shapes from this point of view.

Figure 19.8 shows the input/output sheet for the program. Notice that the output is showing "Check US SI." This is because the block marked "Units: Enter US or SI" is empty. This block must contain US or SI in *uppercase letters*. The maximum load and column length must be entered in their correct units, and the K factor is 0.65, 0.8, 1.0, or 2.1, depending on the end constraints. Select the material from the table at the right, and enter the code number. There are nine structural steels from which to choose.

All output values are in double-line cells. *Do not enter anything into these locations*. The first set of output variables are yield stress, column constant, and effective length. These are determined directly from the input values and the type of steel.

In the output section marked "OPTIMUM SHAPES," "Column #" refers to a code number for the particular shape. You can find this code number on sheet B of your screen, where the mechanical properties of each shape are tabulated. For example, if you need to know the flange width of a shape you have selected, you can find it on sheet B.

The shape designation is listed for each type, as are the cross-sectional area and minimum radius of gyration. The effective slenderness ratio, L_e/r, and the design factor, N, are also tabulated. If $L_e/r \le C_c$, equation (19.14(b)) is used to calculate the design factor, N, and the allowable stress is calculated from equation (19.14(a)). If $L_e/r > C_c$, $N = 1.92$, either equation (19.15(a)) or (19.15(b)), depending on the system of units being employed, is used to determine the allowable stress. Notice, however, that the allowable stress is in both cases a function of r, which is not known a priori. The program determines the minimum acceptable cross-sectional area by using equation (19.17) for each shape entered on sheet B. The closest, larger value to the tabulated area is then selected as the most economical shape for each of the W, S, C, and L columns tabulated.

PROGRAM 'Column'

INPUT

STEEL BEAMS-W, S, C, & L

Units: Enter US or SI	If US		If SI
Max Load	0	lb	N
Column Length	0	in.	mm
End Fixity, K Factor	0		
Matl: Enter Code #	0		

Matl. Code #	Steel
1	A 36
2	A 242
3	A 441
4	A 529
5	A 572 G42
6	A 572 G50
7	A 572 G60
8	A 572 G65
9	A 588

OUTPUT

ERROR	#VALUE!		
	If US		If SI
Yield Stress	psi		MPa
Column Const, Cc			
Effective Length, Le	0	in.	mm

OPTIMUM SHAPES

	W SHAPE		S SHAPE		C SHAPE		L SHAPE	
Column #	Check US	SI	Check US	SI	Check US	SI	Check US	SI
Designation	***		***		***		***	
Section Area	***		***		***		***	
Rad. of Gyration	***		***		***		***	
Le/r	***		***		***		***	
N	***		***		***		***	
Allowable Stress	***		***		***		***	
Actual Stress	***		***		***		***	

Figure 19.8 Input/output sheet for program 'Column'.

PROGRAM 'Column'

INPUT

STEEL BEAMS-W, S, C, & L

Units: Enter US or SI	SI	If US	If SI
Max Load	200000	lb	N
Column Length	3750	in.	mm
End Fixity, K Factor	0.65		
Matl: Enter Code #	3		

Matl. Code #	Steel
1	A 36
2	A 242
3	A 441
4	A 529
5	A 572 G42
6	A 572 G50
7	A 572 G60
8	A 572 G65
9	A 588

OUTPUT

		If US	If SI
Yield Stress	317	psi	MPa
Column Const, Cc	111.595		
Effective Length, Le	2437.5	in.	mm

OPTIMUM SHAPES

	W SHAPE	S SHAPE	C SHAPE	L SHAPE
Column #	1048	1069	1082	1131
Designation	W150 × 17.9	S200 × 27.4	C250 × 30	L89 × 89 × 11.1
Section Area	2290	3490	3790	1850
Rad. of Gyration	23.3	21.1	17.57	27.2
Le/r	104.61	115.52	138.73	89.61
N	1.92	1.92	1.92	1.90
Allowable Stress	92.77	76.96	53.36	112.85
Actual Stress	87.34	57.31	52.77	108.11

Figure 19.9 Input/output for Example Problem 19.11.

EXAMPLE PROBLEM 19.11

Determine the most efficient W- or S-shape for a 3.75-m structural steel column to carry an axial compressive load of 200 kN. Both ends of the column are firmly fixed, and the steel is to be ASTM A441.

Solution

1. Open program 'Column.'
2. Enter the input values as shown in Figure 19.9. Be careful to enter the load in newtons, i.e., 200,000, and the length in mm, 3750. The K factor for both ends fixed is 0.65, and the material code is 3.
3. The output gives the optimum W-shape as W150 × 17.9 and the optimum S-shape as S200 × 27.4. Because the second number is the mass in kg/m, the W-shape is 35% lighter than the S-shape.
4. Select the W150 × 17.9 shape.

The actual stress is 87 MPa, and the allowable stress is 93 MPa.

Answer

Select a W150 × 17.9 shape for the column.

EXAMPLE PROBLEM 19.12

You are to design a balcony to be supported by ten 12-ft columns. The total weight to be supported is 300 tons, but a safety factor of 2 is to be included on the weight. The steel is to be ASTM A36, and the ends of the columns are firmly fixed. Select the column shape using the AISC Column Code.

Solution

1. Open program 'Column'.
2. Enter the input values as shown in Figure 19.10.

$$\text{maximum total load} = 600,000 \text{ lb}$$

$$\text{maximum design load} = 2 \times 600,000 \text{ lb} = 1,200,000 \text{ lb}$$

$$\text{maximum load/column} = 1,200,000 \text{ lb}/10 \text{ columns} = 120,000 \text{ lb}$$

$$\text{column length} = 12 \text{ ft} = 144 \text{ in.}$$

$$K = 0.65$$

$$\text{material code} = 1$$

3. The output shows that the W8 × 24 shape is the most efficient, although we must check the weight of the L8 × 6 × 1/2 column. Go to sheet B, and find column code number 152. We see that it weighs 23 lb/ft, which is 1 lb/ft lighter than the W-

PROGRAM 'Column'

INPUT

STEEL BEAMS-W, S, C, & L

Units: Enter US or SI	US	If US	If SI		Matl. Code #	Steel
					1	A 36
Max Load	120000	lb	N		2	A 242
Column Length	144	in.	mm		3	A 441
End Fixity, K Factor	0.65				4	A 529
Matl: Enter Code #	1				5	A 572 G42
					6	A 572 G50
					7	A 572 G60
					8	A 572 G65
					9	A 588

OUTPUT

		If US	If SI
Yield Stress	36000	psi	MPa
Column Const, Cc	126.101		
Effective Length, Le	93.6	in.	mm

OPTIMUM SHAPES

	W SHAPE	S SHAPE	C SHAPE	L SHAPE
Column #	45	65	77	152
Designation	W8 × 24	S12 × 35	C15 × 40	L8 × 6 × 1/2
Section Area	7.08	10.3	11.8	6.75
Rad. of Gyration	1.61	0.98	0.886	1.79
Le/r	58.14	95.51	105.64	52.29
N	1.83	1.90	1.91	1.81
Allowable Stress	17604.19	13536.00	12248.82	18143.51
Actual Stress	16949.15	11650.49	10169.49	17777.78

Figure 19.10 Input/output sheet for Example Problem 19.12.

shape. Looking at the allowable and actual stresses we see that the L-shape is stressed to 98% of allowable, whereas the W-shape is stressed to 96%. The better choice is the W-shape. Also, end attachments are easier with a W-shape.

Answer

Select W8 × 24 columns.

19.8   PROBLEMS FOR CHAPTER 19

Hand Calculations

1. Determine the slenderness ratio of a cylinder having an outside diameter of 50 mm and an inside diameter of 40 mm. The cylinder is 1.60 m long.

2. The cylinder in the previous question has one end fixed and the other end free. It is made of aluminum alloy 3003-H18, which has a yield strength of 186 MPa and an elastic modulus of 69×10^3 MPa. Using a design factor of 3, what maximum axial compressive load can be withstood safely?

3. Design a 1.5-ft-long, circular-cross-section linkage rod that is to be subjected to an axial compressive load of 4000 lb. Its ends are pin-jointed, and it is to be made of AISI 1040 WQT 1300, which has a yield strength of 63 ksi and a modulus of elasticity of 30×10^6 psi. Use a design factor of 3.

4. Select the lightest standard-cross-section, 6-ft-long hemlock No. 2 post to support an axial compressive load of 1200 lb. It is firmly fixed at one end and pinned at the other. Its compressive strength parallel to the grain is 800 psi, and its elastic modulus is 1400 ksi. Use the National Lumber Manufacturers Association recommended method.

5. Show that the kern for a cylinder with an outside diameter of d_o and an inside diameter of d_i is a disk whose radius is $(d_o^2 + d_i^2)/(8d_o)$.

6. A 4-in. × 4-in square cross-section brick column is to support a compressive load whose center of gravity will be off center by 2 in. If the tensile strength of brick is zero, is it safe?

7. A 12-ft-long, *slender* column, pinned at each end, is to carry an axial compressive load of 350 lb. Determine the moments of inertia necessary for each.
   a. ASTM A36 structural steel ($s_y = 36,000$ psi, $E = 29 \times 10^6$ psi)
   b. Aluminum alloy 6061-T4 ($s_y = 21,000$ psi, $E = 10 \times 10^6$ psi)
   c. AISI 301 annealed stainless steel ($s_y = 40,000$ psi, $E = 28 \times 10^6$ psi)

8. I-sections are efficient shapes for struts. Show that the minimum radius of gyration for an I-section whose flange width is *b,* overall depth is *d,* and thickness of both the flange and the web is *t* can be expressed as

$$r = \sqrt{[2b^3 + (d - 2t)t^2]/[12(2b + d - 2t)]}$$

Computer Calculations

9. Using the equation developed in Problem 8, write a computer program to calculate the minimum radius of gyration of an I-section, given b, d, and t.

10. A compression member of a truss is to safely carry an axial load of 100 kN. It is 3.75 m long and has pinned ends. Select the most effi-cient L-shape, using the AISC code, if the material is ASTM A36 structural steel.

11. If the ends of the members in Problem 10 are fixed, what is the most efficient L-shape?

12. If the material in Problem 11 is changed to structural steel A441, what is the most efficient L-shape?

13. If the strut length in Problem 12 is reduced to 3 m, what is the most efficient L-shape?

14. A 10-ft-long S-shape column is to support an axial load of 60,000 lb. One end is fixed, and the other is pinned. It is to be made of ASTM A441 structural steel. Select the most economical section. Would you select a different S-shape if the material were to be ASTM A36 structural steel? If so, what?

15. The column length in Problem 14 is increased to 12 ft. If the material is A36, select the most economical W-, S-, or C-shape.

16. If both ends were fixed in Problem 15, what would be the most economical W-, S-, or C-shape?

17. If the maximum load were increased to 75,000 lb in Problem 16, what would be the most economical W-, S-, or C-shape?

18. Select the best W-shape columns to support a grandstand weighing 50 tons. There are to be sixteen 15-ft-long columns with one end fixed and the other end pinned. Include a safety factor of 2 on the weight, and use ASTM A242 structural steel.

19. A design change in Problem 18 requires the columns to be increased in length from 15 ft to 16 ft. Select the best W-shape.

20. A further design change requires the grandstand weight in Problem 19 to be increased to 60 tons. Select the best W-shape.

A MATHEMATICAL NOTES

A.1 AREAS OF FIGURES

A.1.1 Circle

A circle of radius r has an area of $A = \pi r^2$, where $\pi = 3.1415926\ldots$, a number that never ends. In engineering it is much easier to measure the diameter, d, of an object than to measure its radius, so instead of r we use $d/2$. Then

$$A = \pi(d/2)^2 = \pi d^2/4$$

A.1.2 Triangle

The area of a triangle is one-half the base times the height. The height is measured at right angles to the base, from the base to the peak. You can see this by studying Figure A.1, where triangle ABC has height h and base b. To show that the area of the triangle is one-half the base times the height, complete rectangle $ADEC$ around the triangle. The area of this rectangle is bh. The triangle is made up of triangle ABF (whose area is one-half the area of $ADBF$) plus triangle FBC (whose area is one-half the area of $FBEC$). So the area of the whole triangle ABC is one-half the area $ADEC$, i.e., the area of the triangle is $bh/2$.

A.1.3 Trapezoid

A trapezoid is a plane figure (a two-dimensional figure) with four sides, only two of which are parallel. Figure A.2 shows a trapezoid $ABCD$ in which the distance between

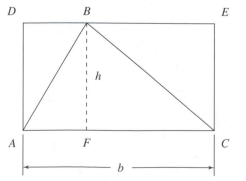

Figure A.1 Area of a triangle.

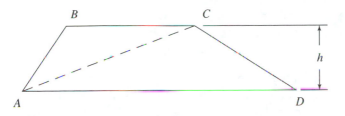

Figure A.2 Area of a trapezoid.

the parallel sides AD and BC is h. Draw the diagonal AC. The area of triangle ABC is $(BC)h/2$. The area of triangle ACD is $(AD)h/2$. The area of the trapezoid is the sum of these areas:

$$A = (BC)h/2 + (AD)h/2 = h(BC + AD)/2$$

So, the area of a trapezoid is equal to the average of the lengths of the parallel sides times the distance between them.

A.2 TRIGONOMETRY

A.2.1 Sine, Cosine, and Tangent

Figure A.3 shows *right-angled* triangle ABC with sides a, b, and c and opposite the angles A, B, and C, respectively. Angle C is the right angle. The sine of angle A, written as sin A, is equal to a/c. Side c, which is the side opposite the right angle, is called the **hypotenuse**, and side a, which is opposite angle A, is called the **opposite side**. So

$$\text{sine of an angle} = \text{opposite/hypotenuse}$$

Similar triangles are triangles that have the same shape (identical angles) but are different in size. Suppose there is another triangle that is similar to ABC, with sides $2a$, $2b$, and $2c$. Then,

$$\sin A = 2a/2c = a/c, \qquad \text{as before}$$

We see from this that the sine of an angle is a characteristic of the angle and is in no way affected by the size of the triangle.

For example, $\sin 30° = 0.5000$, and $\sin 45° = 0.7071$.

Referring again to Figure A.3, the cosine of angle A, written as cos A, is equal to b/c. Side b is called the **adjacent side** (adjacent means "near to"). So

$$\text{cosine of an angle} = \text{adjacent/hypotenuse}$$

Figure A.3 Right-angle tri-angle.

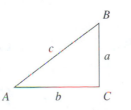

This is also a function of the angle only.

Similarly, the tangent of angle A, written as tan A, is equal to a/b. So

$$\text{tangent of an angle} = \text{opposite/adjacent}$$

Notice that because the hypotenuse must be longer than either of the other two sides, sine and cosine values are always less than or equal to 1.0. The tangent, on the other hand, can have any value.

A.2.2   Trigonometric Relationships

The sum of the interior angles of a triangle equals 180°, so in the right-angled triangle ABC of Figure A.3, angle A plus angle B = 90°, or B = (90° − A).
Now sin A = a/c and cos B = a/c; therefore,

$$\sin A = \cos (90° − A)$$

Similarly, cos A = b/c, and sin B = b/c; therefore,

$$\cos A = \sin (90° − A)$$

The relationship between sine, cosine, and tangent is

$$\sin A/\cos A = (a/c)/(b/c) = a/b = \tan A$$

$$\tan A = \sin A/\cos A$$

From the Pythagorean theorem

$$a^2 + b^2 = c^2$$

Dividing through by c^2:

$$(a/c)^2 + (b/c)^2 = 1$$

Therefore, $\sin^2 A + \cos^2 A = 1$.

For example, sin 30° = 0.5000. What are cos 30° and tan 30°?

$$\sin^2 30° + \cos^2 30° = 1.0$$

$$\cos^2 30° = 1.0 − \sin^2 30° = 1.0 − 0.5000^2 = 0.75$$

Therefore, cos 30° = $\sqrt{0.75}$ = 0.8660.

$$\tan 30° = \sin 30°/\cos 30° = 0.5000/0.8660 = 0.5774$$

A.2.3   Sin (α + β) and Cos (α + β)

Figure A.4 shows *any* triangle, ABC. Drop a perpendicular from B to meet AC at D. Call angle ABD α and angle DBC β. We are going to show that

$$\sin (α + β) = \sin α \cos β + \cos α \sin β$$

Using the sine rule (see Section A.2.5),

$$[\sin (α + β)]/AC = \sin C/AB$$

But sin C = BD/BC; therefore,

$$\sin (α + β) = AC(BD/BC)/AB$$

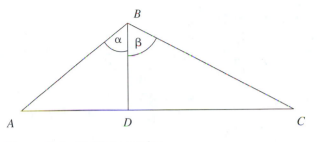

Figure A.4   Multiple angles.

But $AC = AD + DC$ (see Figure A.4); therefore,

$$\sin (\alpha + \beta) = (AD + DC)BD/(AB)(BC)$$
$$= (AD/AB)(BD/BC) + (BD/AB)(DC/BC)$$

or, $\sin (\alpha + \beta) = \sin \alpha \cos \beta + \cos \alpha \sin \beta$

An important relationship resulting from this is the following: If $\beta = \alpha$, then

$$\sin 2\alpha = 2 \sin \alpha \cos \alpha$$

We are now going to show that:

$$\cos (\alpha + \beta) = \cos \alpha \cos \beta - \sin \alpha \sin \beta$$

Using the cosine rule (see Section A.2.4), from Figure A.4;

$$(AC)^2 = (AB)^2 + (BC)^2 - 2(AB)(BC) \cos (\alpha + \beta)$$

Therefore, $2(AB)(BC) \cos (\alpha + \beta) = (AB)^2 + (BC)^2 - (AC)^2$

But $(AC)^2 = (AD + DC)^2 = (AD)^2 + 2(AD)(DC) + (DC)^2$

So $2(AB)(BC) \cos (\alpha + \beta) = (AB)^2 + (BC)^2 - (AD)^2$
$$- 2(AD)(DC) - (DC)^2$$

Rearranging,

$$2(AB)(BC) \cos (\alpha + \beta) = [(AB)^2 - (AD)^2] + [(BC)^2 - (DC)^2] - 2(AD)(DC)$$

From Figure A.4 we see that

$$[(AB)^2 - (AD)^2] = (BD)^2, \quad \text{and} \quad [(BC)^2 - (DC)^2] = (BD)^2$$

Also, therefore,

$$2(AB)(BC)\cos (\alpha + \beta) = 2(BD)^2 - 2(AD)(DC)$$

from which

$$\cos (\alpha + \beta) = (BD/AB)(BD/BC) - (AD/AB)(DC/BC)$$

Finally,

$$\cos (\alpha + \beta) = \cos \alpha \cos \beta - \sin \alpha \sin \beta$$

It follows from this that

$$\cos 2\alpha = \cos^2\alpha - \sin^2\alpha$$

Since $\cos^2 \alpha = 1 - \sin^2 \alpha$,

$$\cos 2\alpha = 1 - 2 \sin^2\alpha \quad \text{and} \quad \cos 2\alpha = 2 \cos^2\alpha - 1$$

A.2.4   The Cosine Rule

The Pythagorean theorem, $a^2 + b^2 = c^2$, is true only for right-angled triangles. The extension to this for any triangle is called the **cosine rule.** Figure A.5 shows a triangle ABC that is *not* a right-angled triangle. Call the angles A, B, and C and the sides opposite them a, b, and c, respectively. Drop a perpendicular from B onto b to meet b at point D. We now have two right-angled triangles, ABD and DBC.

Using the Pythagorean theorem,

$$a^2 = (BD)^2 + (DC)^2$$

But

$$DC = b - AD$$

so

$$(DC)^2 - (b - AD)^2 = b^2 - 2b(AD) + (AD)^2$$

Therefore,

$$a^2 = BD^2 + b^2 - 2b(AD) + (AD)^2 \qquad (A.1)$$

Also

$$c^2 = (AD)^2 + (BD)^2$$

or

$$(BD)^2 = c^2 - (AD)^2$$

Substituting for BD^2 in equation (A.1),

$$a^2 = c^2 - (AD)^2 + b^2 - 2b(AD) + (AD)^2$$

or

$$a^2 = c^2 + b^2 - 2b(AD)$$

Now, looking at Figure A.5 we see that $AD = c \cos A$; therefore,

$$a^2 = b^2 + c^2 - 2bc \cos A$$

A.2.5   The Sine Rule

Refer again to Figure A.5:

$$\sin A = BD/c, \quad \text{and} \quad \sin C = BD/a$$

Therefore,

$$\sin A/\sin C = (BD/c)/(BD/a) = a/c$$

so

$$(\sin A)/a = (\sin C)/c$$

This can be extended to

$$(\sin A)/a = (\sin B)/b = (\sin C)/c$$

which may also be expressed as

$$a/\sin A = b/\sin B = c/\sin C$$

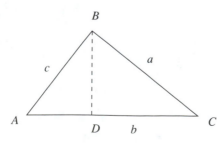

Figure A.5   Cosine rule and sine rule.

A.3 THE SIGMA NOTATION

The sigma notation (Σ, sigma, is the Greek letter for S, inferring "sum") is a shorthand symbol for the sum of a series. For example, if $S_1 = 1 + 2 + 3 + 4 + 5$, we may write:

$$S_1 = \sum_{N=1}^{5} N$$

This says put $N = 1$, then 2, then 3, then 4, then 5, and add.

Suppose $S_2 = 2 + 4 + 6 + 8 + 10$. We can write this as

$$S_2 = \sum_{N=1}^{5} 2N$$

which says put $N = 1$, so that we have $2 \times 1 = 2$, then put $N = 2$, so we have $2 \times 2 = 4$, etc., and then add.

We can write $S_2 = 2(1 + 2 + 3 + 4 + 5) = 2S_1$, from which we see that:

$$\sum_{N=1}^{5} 2N = 2 \sum_{N=1}^{5} N$$

or, in general,

$$\sum_{N=1}^{r} pN = p \sum_{N=1}^{r} N$$

where p is a constant.

Not all sums are in such a neat series, though. Suppose we want to express $S_3 = 7 + 3 + 2 + 9$ in sigma notation. Let $L_1 = 7, L_2 = 3, L_3 = 2$, and $L_4 = 9$. Then the expression

$$S_3 = \sum_{n=1}^{4} L_n$$

tells us to put $n = 1$, making L_1, which equals 7, then put $n = 2$, making L_2, which equals 3, and so on; then add.

B Computer Program Listings

The programs are written for Microsoft Excel 97. They are listed in alphabetical order.

PROGRAM 'CENTROID'   (See Chapter 11.)

'Centroid' calculates the *centroid location* and the *moments of inertia* of any shape that can be divided into any or all of the following:

1. Three rectangles
2. Three right triangles
3. Three circles
4. Two semicircles
5. Two quarter-circles

PROGRAM 'COLUMN'   (See Chapter 19.)

'Column' selects the most economical beam for W, S, C, and L steel shapes given the load, column length, end fixity factor, and type of steel.

PROGRAM 'COMBLOAD'   (See Chapter 18.)

'Combload' determines the stresses and the principal planes for any combination of the following loadings:

1. Direct: Enter the force components and the areas.
2. Bending: If the member has a circular cross section, enter the diameter and the bending moment. If it is not circular, enter the section modulus and the bending moment.
3. Torsion: Enter the torque or the power and rpm. If the member has a circular cross section, enter the diameter; if not circular, enter the polar section modulus.

A plot of stresses versus angle is included.

PROGRAM 'CONNECT'   (See Chapter 9.)

'Connect' calculates the strength and efficiency of bolted, riveted, and welded joints. For bolted and riveted joints, it also identifies the limiting mode (shear, tension, or bearing) and the shear force distribution on a group of fasteners if it is eccentrically loaded.

The program determines the strength and efficiency of both lap and butt joints, and in the case of bolted connections, either friction-type or bearing-type joints may be specified.

PROGRAM 'CONVERT'   (See Chapter 1.)

'Convert' is a basic program for the conversion from SI to U.S. and from U.S. to SI units.

PROGRAM 'COUPLING'   (See Chapter 12.)

'Coupling' is for the calculation of the maximum allowable torque and power that can be safely transmitted by a flange coupling. The input data include shaft diameter and rpm, key, web, and hub dimensions and allowable stresses, and bolt details. The program defines the limiting part and limiting mode.

PROGRAM 'DEFLECT'   (See Chapter 15.)

'Deflect' calculates the deflection of a simply supported beam carrying a distributed load and/or up to four concentrated loads. A plot of deflection versus beam station is included.

PROGRAM 'DEFLECTION'   (See Chapter 15.)

This program is designed to handle all the Figure 15.13 configurations from 1 to 10. The input parameters are the same as those shown in the configurations.

PROGRAM 'DSTRESS'   (See Chapter 6.)

'Dstress' is for direct stress calculations. Both tensile/compressive and shear stress problems may be solved. The program includes stress concentration factors for steps, grooves, and holes.

PROGRAM 'FIXBEAM'   (See Chapter 17.)

'Fixbeam' calculates and plots the shear force and bending moment diagrams for beams with fixed (built-in) ends. One distributed and/or one concentrated load(s) may be entered.

PROGRAM 'FRICTION'   (See Chapter 5.)

Program 'Friction' is actually three programs on one screen. They are for solving problems of wedges, jackscrews, and belts.

PROGRAM 'MOVELOAD'   (See Chapter 13.)

Program 'Moveload' calculates the maximum shear and bending moment, and their locations, for loads moving across simply supported beams.

PROGRAM 'PROPBEAM'   (See Chapter 17.)

'Propbeam' calculates and plots the shear force and bending moment diagrams for continuous beams with three supports. Up to four distributed and/or four concentrated loads may be entered.

PROGRAM 'PROPCLR'   (See Chapter 17.)

'Propclr' calculates and plots the shear force and bending moment diagrams for cantilever beams with a support at any location along the beam. Up to two distributed and/or two concentrated loads may be entered.

PROGRAM 'PROPTAB'   (See Chapter 7.)

'Proptab' retrieves material property data for the most commonly used

1. Irons,
2. Steels,
3. Stainless steels,
4. Structural steels,
5. Aluminum alloys,
6. Nonferrous alloys,
7. Plastics, and
8. Wood.

PROGRAM 'RESULTANT'   (See Chapter 2.)

'Resultant' determines the resultant of up to 10 coplanar or 10 noncoplanar forces.

PROGRAM 'SHAPES1'   (See Chapter 16.)

'Shapes1' selects five candidate beams that satisfy the bending stress requirement and five candidate beams that satisfy the deflection requirement. The candidates are selected from a table of W-, S-, and C-shape steel beams. The operator has a choice of nine structural steels.

PROGRAM 'SHAPES2'   (See Chapter 16.)

'Shapes2' is similar to 'Shapes1' but is for steel L-shapes, aluminum alloy I- and C-shapes, and timber beams. There is a choice of 18 aluminum alloys and 6 types of wood.

PROGRAM 'SHR&MMT'   (See Chapter 13.)

'Shr&mmt' calculates and plots the shear force and bending moment diagrams for simply supported and for overhanging beams.

PROGRAM 'SHRMMTCL'   (See Chapter 13.)

'Shrmmtcl' calculates and plots the shear force and bending moment diagrams for cantilever beams.

PROGRAM 'SPCFRAME'   (See Chapter 4.)

'Spcframe' determines the forces in the members of a space frame. The input data consist of the projected lengths of the members and the components of the external forces.

PROGRAM 'TORQUE'   (See Chapter 12.)

'Torque' calculates the torque, power, and twist angle for solid and hollow shafts. It includes stress concentrations for steps, grooves, and key seats.

PROGRAM 'TRUSS'   (See Chapter 4.)

'Truss' calculates the forces in the members of single triangle trusses. The input data consist of the angles of the members and the external force and its angle.

PROGRAM 'TSTRESS'   (See Chapter 8.)

'Tstress' is for thermal stress calculations. It may be used to determine the stresses, force, and length changes for constrained, heated, or cooled materials in series or the stresses and forces for constrained, heated, or cooled materials in parallel. Up to 5 materials may be combined from a selection of 16.

PROGRAM 'VESSEL'   (See Chapter 12.)

'Vessel' calculates maximum hoop and longitudinal stresses in cylindrical pressure vessels and maximum tangential stresses in spherical pressure vessels.

C REPRESENTATIVE MECHANICAL PROPERTIES OF SELECTED ENGINEERING MATERIALS

For more data on a wide variety of materials, use program 'Proptab.'

Representative Properties of Structural Steels: U.S. Customary Units.

ASTM Designation	Ultimate Strength, psi Tensile	Shear	Yield Strength, psi Tensile	Shear	Allowable Tensile Stress, psi	Allowable Bearing Stress, psi	Allowable Shear Stress, psi
A36	58,000	47,560	36,000	18,000	21,600	69,600	
A242	67,000	54,940	46,000	23,000	27,600	80,400	
A307 bolts							10,000
A325 bolts							
Friction-type							17,500
Bearing-type							30,000
A490 bolts							
Friction-type							22,000
Bearing-type							40,000
A500 Grade A	45,000	36,900	33,000	16,500	19,800	54,000	
A500 Grade B	58,000	47,560	42,000	21,000	25,200	69,600	
A500 Grade C	62,000	50,840	46,000	23,000	27,600	74,400	
A501	58,000	47,560	36,000	18,000	21,600	69,600	
A502-1 rivets							17,500
A502-2 rivets							22,000
A514	100,000	82,000	90,000	45,000	50,000	120,000	
A572 Grade 42	60,000	49,200	42,000	21,000	25,200	72,000	
A572 Grade 50	65,000	53,300	50,000	25,000	30,000	78,000	
A572 Grade 60	75,000	61,500	60,000	30,000	36,000	90,000	
A572 Grade 65	80,000	65,600	65,000	32,500	39,000	96,000	

Elastic Modulus, psi	29,000,000
Modulus of Rigidity, psi	12,000,000

Representative Properties of Structural Steels: SI Units.

ASTM Designation	Ultimate Strength, MPa Tensile	Shear	Yield Strength, MPa Tensile	Shear	Allowable Tensile Stress, MPa	Allowable Bearing Stress, MPa	Allowable Shear Stress, MPa
A36	400	328	248	124	149	480	
A242	462	379	317	159	190	554	
A307 bolts							69
A325 bolts							
Friction-type							121
Bearing-type							207
A490 bolts							
Friction-type							152
Bearing-type							276
A500 Grade A	310	254	228	114	137	372	
A500 Grade B	400	328	290	145	174	480	
A500 Grade C	427	351	317	159	190	513	
A501	400	328	248	124	149	480	
A502-1 rivets							121
A502-2 rivets							152
A514	690	565	621	310	345	827	
A572 Grade 42	414	339	290	145	174	496	
A572 Grade 50	448	368	345	172	207	538	
A572 Grade 60	517	424	414	207	248	621	
A572 Grade 65	552	452	448	224	269	662	

Elastic Modulus, MPa: 200,000

Modulus of Rigidity, MPa: 83,000

Representative Properties of Carbon and Alloy Steels: U.S. Customary Units.

AISI Designation	Ultimate Strength, psi		Yield Strength, psi		Elastic Modulus, psi	Modulus of Rigidity, psi	Density lb/in.³
	Tensile	Shear	Tensile	Shear			
1020 Ann	57,000	46,740	43,000	21,500			
1020 HR	65,000	53,300	48,000	24,000	30,000,000	11,500,000	0.283
1020 CD	75,000	61,500	64,000	32,000	Approx. the same for all carbon & alloy steels.		
1040 Ann	75,000	61,500	51,000	25,500			
1040 HR	90,000	73,800	60,000	30,000			
1040 CD	97,000	79,540	82,000	41,000			
1040 WQT 700	127,000	104,140	93,000	46,500			
1040 WQT 900	118,000	96,760	90,000	45,000			
1040 WQT 1100	107,000	87,740	80,000	40,000			
1040 WQT 1300	87,000	71,340	63,000	31,500			
1080 Ann	89,000	72,980	54,000	27,000			
1080 OQT 700	189,000	154,980	141,000	70,500			
1080 OQT 900	179,000	146,780	129,000	64,500			
1080 OQT 1100	145,000	118,900	103,000	51,500			
1080 OQT 1300	117,000	95,940	70,000	35,000			
1141 Ann	87,000	71,340	51,000	25,500			
1141 CD	112,000	91,840	95,000	47,500			
1141 OQT 700	193,000	158,260	172,000	86,000			
1141 OQT 900	146,000	119,720	129,000	64,500			
1141 OQT 1100	116,000	95,120	97,000	48,500			
1141 OQT 1300	94,000	77,080	68,000	34,000			
4140 Ann	95,000	77,900	60,000	30,000			
4140 OQT 700	231,000	189,420	212,000	106,000			
4140 OQT 900	187,000	153,340	173,000	86,500			
4140 OQT 1100	147,000	120,540	131,000	65,500			
4140 OQT 1300	118,000	96,760	101,000	50,500			
5160 Ann	105,000	86,100	40,000	20,000			
5160 OQT 700	263,000	215,660	238,000	119,000			
5160 OQT 900	196,000	160,720	179,000	89,500			
5160 OQT 1100	149,000	122,180	132,000	66,000			
5160 OQT 1300	115,000	94,300	103,000	51,500			

Ann = annealed   HR = hot-rolled   CD = cold-drawn   WQT = water-quenched, tempered   OQT = oil-quenched, tempered.

Representative Properties of Carbon and Alloy Steels: SI Units.

AISI Designation	Ultimate Strength, MPa		Yield Strength, MPa		Elastic Modulus, MPa	Modulus of Rigidity, MPa	Density kg/cu.m
	Tensile	Shear	Tensile	Shear			
1020 Ann	393	322	296	148			
1020 HR	448	368	331	165	207,000	80,000	7680
1020 CD	517	424	441	221	Approx. the same for all carbon & alloy steels.		
1040 Ann	517	424	352	176			
1040 HR	621	509	414	207			
1040 CD	669	548	565	283			
1040 WQT 700	876	718	641	321			
1040 WQT 900	814	667	621	310			
1040 WQT 1100	738	605	552	276			
1040 WQT 1300	600	492	434	217			
1080 Ann	614	503	372	186			
1080 OQT 700	1303	1069	972	486			
1080 OQT 900	1234	1012	889	445			
1080 OQT 1100	1000	820	710	355			
1080 OQT 1300	807	662	483	241			
1141 Ann	600	492	352	176			
1141 CD	772	633	655	328			
1141 OQT 700	1331	1091	1186	593			
1141 OQT 900	1007	825	889	445			
1141 OQT 1100	800	656	669	334			
1141 OQT 1300	648	531	469	234			
4140 Ann	655	537	414	207			
4140 OQT 700	1593	1306	1462	731			
4140 OQT 900	1289	1057	1193	596			
4140 OQT 1100	1014	831	903	452			
4140 OQT 1300	814	667	696	348			
5160 Ann	724	594	276	138			
5160 OQT 700	1813	1487	1641	821			
5160 OQT 900	1351	1108	1234	617			
5160 OQT 1100	1027	842	910	455			
5160 OQT 1300	793	650	710	355			

Ann = annealed   HR = hot-rolled   CD = cold-drawn   WQT = water-quenched, tempered   OQT = oil-quenched, tempered.

Representative Properties of Stainless Steels: U.S. Customary Units.

AISI Designation	Ultimate Strength, psi		Yield Strength, psi		Elastic Modulus, psi
	Tensile	Shear	Tensile	Shear	
301 Ann	110,000	90,200	40,000	20,000	28,000,000
301 FH	185,000	151,700	140,000	70,000	28,000,000
430 Ann	75,000	61,500	40,000	20,000	29,000,000
430 FH	90,000	73,800	80,000	40,000	29,000,000
501 Ann	70,000	57,400	30,000	15,000	29,000,000
501 OQT1000	175,000	143,500	135,000	67,500	29,000,000

Ann = annealed   FH = full hard

Representative Properties of Stainless Steels: SI Units.

AISI Designation	Ultimate Strength, MPa		Yield Strength, MPa		Elastic Modulus, MPa
	Tensile	Shear	Tensile	Shear	
301 Ann	758	622	276	138	193,000
301 FH	1276	1046	965	483	193,000
430 Ann	517	424	276	138	200,000
430 FH	621	509	552	276	200,000
501 Ann	483	396	207	103	200,000
501 OQT1000	1207	989	931	465	200,000

Ann = annealed   FH = full hard

Representative Properties of Wrought Aluminum Alloys; U.S. Customary Units.

Aluminium Assoc. Designation	Ultimate Strength, psi		Yield Strength, psi		Allowable Tensile Stress, psi	Allowable Bearing Stress, psi	Allowable Shear Stress, psi	Elastic Modulus, psi
	Tensile	Shear	Tensile	Shear				
1100-H12	16,000	10,000	15,000	7,500	8,160	9,750		10,000,000
1100-H14 rivets							4,000	
1100-H18	24,000	13,000	22,000	11,000	12,240	14,300		10,000,000
2014-0	27,000	18,000	14,000	7,000	8,540	9,100		10,600,000
2014-T4	62,000	38,000	42,000	21,000	25,620	27,300		10,600,000
2014-T6	70,000	42,000	60,000	30,000	35,700	39,000		10,600,000
2017-T64 rivets							14,500	
2024-T4 bolts							16,000	
3003-0	16,000	11,000	6,000	3,000	3,660	3,900		10,000,000
3003-H12	19,000	12,000	18,000	9,000	9,690	11,700		10,000,000
3003-H18	29,000	16,000	27,000	13,500	14,790	17,550		10,000,000
5154-0	35,000	22,000	17,000	8,500	10,370	11,050		10,200,000
5154-H32	39,000	22,000	30,000	15,000	18,300	19,500		10,200,000
5154-H38	48,000	28,000	39,000	19,500	23,790	25,350		10,200,000
6053-T61 rivets							8,500	
6061-0	18,000	12,000	8,000	4,000	4,880	5,200		10,000,000
6061-T4	35,000	24,000	21,000	10,500	12,810	13,650		10,000,000
6061-T6	45,000	30,000	40,000	20,000	22,950	26,000		10,000,000
6061-T6 rivets							11,000	
6061-T6 bolts							12,000	
7075-0	33,000	22,000	15,000	7,500	9,150	9,750		10,400,000
7075-T6	83,000	48,000	73,000	36,500	42,330	47,450		10,400,000
7075-T73 bolts							17,000	

Representative Properties of Wrought Aluminum Alloys: SI Units.

Aluminum Assoc. Designation	Ultimate Strength, MPa		Yield Strength, MPa		Allowable Tensile Stress, MPa	Allowable Bearing Stress, MPa	Allowable Shear Stress, MPa	Elastic Modulus, MPa
	Tensile	Shear	Tensile	Shear				
1100-H12	110	69	103	52	56	67	0	69,000
1100-H14 rivets							28	69,000
1100-H18	165	90	152	76	84	99		
2014-0	186	124	97	48	59	63		73,000
2014-T4	427	262	290	145	177	188		73,000
2014-T6	483	290	414	207	246	269		73,000
2017-T64 rivets							100	
2024-T4 bolts							110	
3003-0	110	76	41	21	25	27		69,000
3003-H12	131	83	124	62	67	81		69,000
3003-H18	200	110	186	93	102	121		69,000
5154-0	241	152	117	59	72	76		70,000
5154-H32	269	152	207	103	126	134		70,000
5154-H38	331	193	269	134	164	175		70,000
6053-T61 rivets							59	
6061-0	124	83	55	28	34	36		69,000
6061-T4	241	165	145	72	88	94		69,000
6061-T6	310	207	276	138	158	179		69,000
6061-T6 rivets							76	
6061-T6 bolts							83	
7075-0	228	152	103	52	63	67		72,000
7075-T6	572	331	503	252	292	327		72,000
7075-T73 bolts							117	

Representative Properties of Various Engineering Materials: U.S. Customary Units.

| Material | Ultimate Strength, psi | | | Yield Strength | Elastic |
	Tensile	Shear	Compressive	Tensile, psi	Modulus, psi
Concrete					
Low strength	—	200	1,000	—	2,500,000
Medium strength	—	800	4,000	—	3,600,000
High strength	—	2,000	10,000	—	4,500,000
Brass					
C36,000 — soft	49,000	—	—	18,000	16,000,000
C36,000 — hard	68,000	—	—	45,000	16,000,000
Bronze					
C54,400 — soft	68,000	—	—	57,000	17,000,000
C54,400 — hard	75,000	—	—	63,000	17,000,000
Cast iron — ductile					
A536 — 60-40-18	60,000	57,000	108,000	40,000	24,000,000
A536 — 120-90-2	120,000	—	180,000	90,000	23,000,000
Cast iron — Gray					
A48 grade 20	20,000	32,000	80,000	—	12,200,000
A48 grade 60	55,000	72,000	170,000	—	21,500,000
Cast iron — malleable					
A220 — 45008	65,000	49,000	240,000	45,000	26,000,000
A220 — 80002	95,000	75,000	240,000	80,000	27,000,000
Copper					
C14500 — soft	32,000	—	—	10,000	17,000,000
C14500 — hard	48,000	—	—	44,000	17,000,000
C17,000 — soft	175,000	—	—	150,000	19,000,000
C17,000 — hard	215,000	—	—	200,000	19,000,000
C26,800 — soft	46,000	8,000	—	14,000	15,000,000
C26,800 — hard	74,000	34,200	—	60,000	15,000,000
C71,500 — soft	44,000	11,400	—	20,000	16,500,000
C71,500 — hard	75,000	38,800	—	68,000	16,500,000
Magnesium — cast	40,000	21,000	—	19,000	6,500,000
Monel	80,000	56,000	—	40,000	26,000,000
Titanium alloy	170,000	100,000	—	155,000	16,500,000
Zinc — cast	58,000	—	—	47,000	12,000,000

Representative Properties of Various Engineering Materials: SI Units.

Material	Ultimate Strength, MPa			Yield Strength Tensile, MPa	Elastic Modulus, MPa
	Tensile	Shear	Compressive		
Concrete					
Low strength	—	1	7	—	17,238
Medium strength	—	6	28	—	24,822
High strength	—	14	69	—	31,028
Brass					
C36,000 — soft	338	—	—	124	110,320
C36,000 — hard	469	—	—	310	110,320
Bronze					
C54,400 — soft	469	—	—	393	117,215
C54,400 — hard	517	—	—	434	117,215
Cast iron — ductile					
A536 — 60-40-18	414	393	745	276	165,480
A536 — 120-90-2	827	—	1,241	621	158,585
Cast iron — gray					
A48 Grade 20	138	221	552	—	84,119
A48 Grade 60	379	496	1,172	—	148,243
Cast iron — Malleable					
A220 — 45008	448	338	1,655	310	179,270
A220 — 80002	655	517	1,655	552	186,165
Copper					
C14500 — soft	221	—	—	69	117,215
C14500 — hard	331	—	—	303	117,215
C17,000 — soft	1,207	—	—	1,034	131,005
C17,000 — hard	1,482	—	—	1,379	131,005
C26,800 — soft	317	55	—	97	103,425
C26,800 — hard	510	236	—	414	103,425
C71,500 — soft	303	79	—	138	113,768
C71,500 — hard	517	268	—	469	113,768
Magnesium — cast	276	145	—	131	44,818
Monel	552	386	—	276	179,270
Titanium alloy	1,172	690	—	1,069	113,768
Zinc — cast	400	—	—	324	82,740

Representative Properties of Timber: U.S. Customary Units.

Type of Wood	Allowable Stress, psi			Elastic Modulus, psi
	Bending	Horizontal Shear	Compression Parallel to Grain	
Douglas fir	1,450	95	1,000	1,700,000
Hemlock	1,250	75	925	1,400,000
Ponderosa pine	1,000	65	800	1,100,000
Redwood	1,350	100	1,050	1,300,000
Sitka spruce	1,150	70	875	1,300,000
Southern pine	1,000	70	550	1,300,000

Representative Properties of Reinforced Plastics: U.S. Customary Units.

Type	Strength, psi		Modulus of Elasticity, psi	Density lb/in.^{3}
	Tensile	Flexural		
Epoxy — molded	15,000	30,000	3,000,000	0.069
Nylon 6/6	26,000	35,000	1,300,000	0.041
Phenolic	9,000	18,000	2,500,000	0.066
Polyimide	27,000	50,000	3,200,000	0.069
Polystyrene	12,000	17,000	800,000	0.042

Representative Properties of Timber: SI Units.

Type of Wood	Allowable Stress, MPa			Elastic Modulus, MPa
	Bending	Horizontal Shear	Compression Parallel to Grain	
Douglas fir	10.0	0.66	6.90	11,700
Hemlock	8.6	0.52	6.38	9,700
Ponderosa pine	6.9	0.45	5.52	7,600
Redwood	9.3	0.69	7.24	9,000
Sitka spruce	7.9	0.48	6.03	9,000
Southern pine	6.9	0.48	3.79	9,000

Representative Properties of Reinforced Plastics: SI Units.

Type	Strength, MPa		Modulus of Elasticity, MPa	Density kg/m^3
	Tensile	Flexural		
Epoxy — molded	103	207	20,700	1,910
Nylon 6/6	179	241	9,000	1,135
Phenolic	62	124	17,200	1,825
Polyimide	186	345	22,100	1,910
Polystyrene	83	117	5,500	1,165

D TABLE OF CONVERSION FACTORS: U.S. CUSTOMARY UNITS TO SI UNITS

Also see program 'Convert'.

Quantity	Units	Multiply by	To Obtain
Length	inch	0.0254	meter
	feet	0.3048	meter
	yard	0.9144	meter
	mile	1.609	kilometer
Angle	degree	0.01745	radian
Area	square inch	6.452×10^{-4}	square meter
	square foot	0.0929	square meter
Volume	cubic inch	1.639×10^{-5}	cubic meter
	cubic foot	2.832×10^{-2}	cubic meter
Mass	slug	14.59	kilogram
	pound-mass	0.4536	kilogram
Force	pound	4.448	newton
Density	pound-mass/ft^3 (cu. ft)	16.02	kg/m^3
Stress and pressure	pound/in.^{3}	6.895×10^{-3}	megapascal
	pound/ft^2	4.788×10^{-5}	megapascal
Speed	feet/second	0.3048	meter/second
	feet/minute	$5.08E \times 10^{-3}$	meter/second
	miles/hour	0.447	meter/second
	miles/hour	1.609	kilometer/hour
Acceleration	feet/second^2	0.3048	meter/second^2
Moment and torque	pound-feet	1.356	newton-meter
	pound-inch	0.113	newton-meter
Work and energy	foot-pound	1.356	joule
Power	foot-pound/second	1.356	watt
	foot-pound/minute	2.26×10^{-2}	watt
	horsepower	745.7	watt
Temperature	degrees Fahrenheit	(F −32)/1.8	degrees Celsius

E TABLES OF STRUCTURAL STEEL SHAPES AND STANDARD TIMBER SIZES

Also see programs 'Shapes1,' 'Shapes2,' and 'Column.'

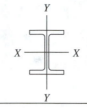

W-Shapes: U.S. Customary Units (Selected Listing)

				Flange		Axis X–X			Axis Y–Y		
			Web				Elastic Properties				
			Thickness	Width	Thickness	I	S	r	I	S	r
Designation	**Area** in.²	**Depth** in.	in.	in.	in.	in.⁴	in.³	in.	in.⁴	in.³	in.
W36 × 160	47.0	36.01	0.650	12.000	1.020	9750	542	14.4	295	49.1	2.50
W33 × 221	65.0	33.93	0.775	15.805	1.275	12800	757	14.1	840	106	3.59
W24 × 84	24.7	24.10	0.470	9.020	0.770	2370	196	9.79	94.4	20.9	1.95
W21 × 62	18.3	20.99	0.400	8.240	0.615	1330	127	8.54	57.5	13.9	1.77
× 44	13.0	20.66	0.350	6.500	0.450	843	81.6	8.06	20.7	6.36	1.26
W18 × 55	16.2	18.11	0.390	7.530	0.630	890	98.3	7.41	44.9	11.9	1.67
× 50	14.7	17.99	0.355	7.495	0.570	800	88.9	7.38	40.1	10.7	1.65
× 40	11.8	17.90	0.315	6.015	0.525	612	68.4	7.21	19.1	6.35	1.27
× 35	10.3	17.70	0.300	6.000	0.425	510	57.6	7.04	15.3	5.12	1.22
W16 × 100	29.4	16.97	0.585	10.425	0.985	1490	175	7.10	186	35.7	2.51
× 67	19.7	16.33	0.395	10.235	0.665	954	117	6.96	119	23.2	2.46
× 50	14.7	16.26	0.380	7.070	0.630	659	81.0	6.68	37.2	10.5	1.59
× 40	11.8	16.01	0.305	6.995	0.505	518	64.7	6.63	28.9	8.25	1.57
× 31	9.12	15.88	0.275	5.525	0.440	375	47.2	6.41	12.4	4.49	1.17
× 26	7.68	15.69	0.250	5.500	0.345	301	38.4	6.26	9.59	3.49	1.12
W14 × 159	46.7	14.98	0.745	15.565	1.190	1900	254	6.38	748	96.2	4.00
× 132	38.8	14.66	0.645	14.725	1.030	1530	209	6.28	548	74.5	3.76
× 99	29.1	14.16	0.485	14.565	0.780	1110	157	6.17	402	55.2	3.71
× 74	21.8	14.17	0.450	10.070	0.785	796	112	6.04	134	26.6	2.48
× 53	15.6	13.92	0.370	8.060	0.660	541	77.8	5.89	57.7	14.3	1.92
× 38	11.2	14.10	0.310	6.770	0.515	385	54.6	5.87	26.7	7.88	1.55
× 34	10.0	13.98	0.285	6.745	0.455	340	48.6	5.83	23.3	6.91	1.53
× 30	8.85	13.84	0.270	6.730	0.385	291	42.0	5.73	19.6	5.82	1.49
× 26	7.69	13.91	0.255	5.025	0.420	245	35.3	5.65	8.91	3.54	1.08
× 22	6.49	13.74	0.230	5.000	0.335	199	29.0	5.54	7.00	2.80	1.04
W12 × 136	39.9	13.41	0.790	12.400	1.250	1240	186	5.58	398	64.2	3.16
× 96	28.2	12.71	0.550	12.160	0.900	833	131	5.44	270	44.4	3.09
× 72	21.1	12.25	0.430	12.040	0.670	597	97.4	5.31	195	32.4	3.04
× 58	17.0	12.19	0.360	10.010	0.640	475	78.0	5.28	107	21.4	2.51
× 45	13.2	12.06	0.335	8.045	0.575	350	58.1	5.15	50.0	12.4	1.94
× 30	8.79	12.34	0.260	6.520	0.440	238	38.6	5.21	20.3	6.24	1.52
× 26	7.65	12.22	0.230	6.490	0.380	204	33.4	5.17	17.3	5.34	1.51
× 22	6.48	12.31	0.260	4.030	0.425	156	25.4	4.91	4.66	2.31	0.847
× 19	5.57	12.16	0.235	4.005	0.350	130	21.3	4.82	3.76	1.88	0.822
× 16	4.71	11.99	0.220	3.990	0.265	103	17.1	4.67	2.82	1.41	0.773
W10 × 112	32.9	11.36	0.755	10.415	1.250	716	126	4.66	236	45.3	2.68
× 100	29.4	11.10	0.680	10.340	1.120	623	112	4.60	207	40.0	2.65
× 77	22.6	10.60	0.530	10.190	0.870	455	85.9	4.49	154	30.1	2.60
× 54	15.8	10.09	0.370	10.030	0.615	303	60.0	4.37	103	20.6	2.56
× 45	13.3	10.10	0.350	8.020	0.620	248	49.1	4.32	53.4	13.3	2.01
× 39	11.5	9.92	0.315	7.985	0.530	209	42.1	4.27	45.0	11.3	1.98
× 22	6.49	10.17	0.240	5.750	0.360	118	23.2	4.27	11.4	3.97	1.33
W8 × 67	19.7	9.00	0.570	8.280	0.935	272	60.4	3.72	88.6	21.4	2.12
× 35	10.3	8.12	0.310	8.020	0.495	127	31.2	3.51	42.6	10.6	2.03
× 24	7.08	7.93	0.245	6.495	0.400	82.8	20.9	3.42	18.3	5.63	1.61
× 15	4.44	8.11	0.245	4.015	0.315	48.0	11.8	3.29	3.41	1.70	0.876
× 10	2.96	7.89	0.170	3.940	0.205	30.8	7.81	3.22	2.09	1.06	0.841
W6 × 25	7.34	6.38	0.320	6.080	0.455	53.4	16.7	2.70	17.1	5.61	1.52
× 12	3.55	6.03	0.230	4.000	0.280	22.1	7.31	2.49	2.99	1.50	0.918
× 9	2.68	5.90	0.170	3.940	0.215	16.4	5.56	2.47	2.19	1.11	0.905

W-Shapes: SI Units (Selected Listing)

				Flange		Axis X–X			Axis Y–Y		
			Web								
	Area	Depth	Thickness	Width	Thickness	I	S	r	I	S	r
Designation	mm²	mm	mm	mm	mm	10⁶mm⁴	10³mm³	mm	10⁶mm⁴	10³mm³	mm

Designation	Area mm²	Depth mm	Web Thickness mm	Width mm	Thickness mm	I 10^6mm⁴	S 10^3mm³	r mm	I 10^6mm⁴	S 10^3mm³	r mm
W920 × 238	30300	915	16.5	305	25.9	4060	8870	366	122.8	805	63.7
W840 × 329	41900	862	19.7	401	32.4	5330	12370	357	350	1746	91.4
W610 × 125	15900	612	11.9	229	19.6	986	3220	249	39.3	343	49.7
W530 × 92	11800	533	10.2	209	15.6	554	2080	217	23.9	229	45.0
× 65	8390	525	8.9	165	11.4	351	1337	205	8.62	104.5	32.1
W460 × 82	10500	460	9.9	191	16.0	370	1609	187.7	18.69	195.7	42.2
× 74	9480	457	9.0	190	14.5	333	1457	187.4	16.69	175.7	42.0
× 60	7610	455	8.0	153	13.3	255	1121	183.1	7.95	103.9	32.3
× 52	6650	450	7.6	152	10.8	212	942	178.5	6.37	83.8	30.9
W410 × 149	19000	431	14.9	265	25.0	620	2880	180.6	77.4	584	63.8
× 100	12700	415	10.0	260	16.9	397	1913	176.8	49.5	381	62.4
× 74	9480	413	9.7	180	16.0	274	1327	170.0	15.48	172.0	40.4
× 60	7610	407	7.7	178	12.8	216	1061	168.5	12.03	135.2	39.8
× 46.1	5880	403	7.0	140	11.2	156.1	775	162.9	5.16	73.7	29.6
× 38.7	4950	399	6.4	140	8.8	125.3	628	159.1	3.99	57.0	28.4
W360 × 237	30100	380	18.9	395	30.2	791	4160	162.1	311	1575	101.6
× 196	25000	372	16.4	374	26.2	637	3420	159.6	228	1219	95.5
× 147	18800	360	12.3	370	19.8	462	2570	156.8	167	903	94.2
× 110	14100	360	11.4	256	19.9	331	1839	153.2	55.8	436	62.9
× 79	10100	354	9.4	205	16.8	225	1271	149.3	24.0	234	48.7
× 57	7230	358	7.9	172	13.1	160.2	895	148.9	11.11	129.2	39.2
× 51	6450	355	7.2	171	11.6	141.5	797	148.1	9.70	113.5	38.8
× 44.6	5710	352	6.9	171	9.8	121.1	688	145.6	8.16	95.4	37.8
× 38.7	4960	353	6.5	128	10.7	102.0	578	143.4	3.71	58.0	27.3
× 32.7	4190	349	5.8	127	8.5	82.8	474	140.6	2.91	45.8	26.4
W310 × 202	25700	341	20.1	315	31.8	516	3026	141.7	165.7	1052	80.3
× 143	18200	323	14.0	309	22.9	347	2149	138.1	112.4	728	78.6
× 107	13600	311	10.9	306	17.0	248	1595	135.0	81.2	531	77.3
× 86	11000	310	9.1	254	16.3	197.7	1275	134.1	44.5	350	63.6
× 67	8520	306	8.5	204	14.6	145.7	952	130.8	20.8	204	49.4
× 44.6	5670	313	6.6	166	11.2	99.1	633	132.2	8.45	101.8	38.6
× 38.7	4940	310	5.8	165	9.7	84.9	548	131.1	7.20	87.3	38.2
× 32.7	4180	313	6.6	102	10.8	64.9	415	124.6	1.940	38.0	21.5
× 28.3	3590	309	6.0	102	8.9	54.1	350	122.8	1.565	30.7	20.9
× 23.8	3040	305	5.6	101	6.7	42.9	281	118.8	1.174	23.2	19.65
W250 × 167	21200	289	19.2	265	31.8	298	2060	118.6	98.2	741	68.1
× 149	19000	282	17.3	263	28.4	259	1837	116.8	86.2	656	67.4
× 115	14600	269	13.5	259	22.1	189.4	1408	113.9	64.1	495	66.3
× 80	10200	256	9.4	255	15.6	126.1	985	111.2	42.9	336	64.9
× 67	8580	257	8.9	204	15.7	103.2	803	109.7	22.2	218	50.9
× 58	7420	252	8.0	203	13.5	87.0	690	108.3	18.73	184.5	50.2
× 32.7	4190	258	6.1	146	9.1	49.1	381	108.3	4.75	65.1	33.7
W200 × 100	12710	229	14.5	210	23.7	113.2	989	94.4	36.9	351	53.9
× 52	6650	206	7.9	204	12.6	52.9	514	89.2	17.73	173.8	51.6
× 35.7	4570	201	6.2	165	10.2	34.5	343	86.9	7.62	92.4	40.8
× 22.3	2860	206	6.2	102	8.0	19.98	194.0	83.6	1.419	27.8	22.3
× 14.9	1910	200	4.3	100	5.2	12.82	128.2	81.9	0.870	17.40	21.3
W150 × 37.2	4740	162	8.1	154	11.6	22.2	274	68.4	7.12	92.5	38.8
× 17.9	2290	153	5.8	102	7.1	9.20	120.3	63.4	1.245	24.4	23.3
× 13.4	1730	150	4.3	100	5.5	6.83	91.1	62.8	0.912	18.24	23.0

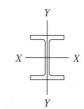

S-Shapes: U.S. Customary Units (Selected Listing)

Designation	Area in.^{2}	Depth in.	Web Thickness in.	Flange		Elastic Properties					
				Width in.	Thickness in.	Axis X-X			Axis Y-Y		
						I in.^{4}	S in.^{3}	r in.	I in.^{4}	S in.^{3}	r in.
S24 × 100	29.3	24.00	0.745	7.245	0.870	2390	199	9.02	47.7	13.2	1.27
× 90	26.5	24.00	0.625	7.125	0.870	2250	187	9.21	44.9	12.6	1.30
× 80	23.5	24.00	0.500	7.000	0.870	2100	175	9.47	42.2	12.1	1.34
S20 × 96	28.2	20.30	0.800	7.200	0.920	1670	165	7.71	50.2	13.9	1.33
× 86	25.3	20.30	0.660	7.060	0.920	1580	155	7.89	46.8	13.3	1.36
× 75	22.0	20.00	0.635	6.385	0.795	1280	128	7.62	29.8	9.32	1.16
× 66	19.4	20.00	0.505	6.255	0.795	1190	119	7.83	27.7	8.85	1.19
S18 × 70	20.6	18.00	0.711	6.251	0.691	926	103	6.71	24.1	7.72	1.08
× 54.7	16.1	18.00	0.461	6.001	0.691	804	89.4	7.07	20.8	6.94	1.14
S15 × 50	14.7	15.00	0.550	5.640	0.622	486	64.8	5.75	15.7	5.57	1.03
× 42.9	12.6	15.00	0.411	5.501	0.622	447	59.6	5.95	14.4	5.23	1.07
S12 × 50	14.7	12.00	0.687	5.477	0.659	305	50.8	4.55	15.7	5.74	1.03
× 40.8	12.0	12.00	0.462	5.252	0.659	272	45.4	4.77	13.6	5.16	1.06
× 35	10.3	12.00	0.428	5.078	0.544	229	38.2	4.72	9.87	3.89	0.980
× 31.8	9.35	12.00	0.350	5.000	0.544	218	36.4	4.83	9.36	3.74	1.00
S10 × 35	10.3	10.00	0.594	4.944	0.491	147	29.4	3.78	8.36	3.38	0.901
× 25.4	7.46	10.00	0.311	4.661	0.491	124	24.7	4.07	6.79	2.91	0.954
S8 × 23	6.77	8.00	0.441	4.171	0.426	64.9	16.2	3.10	4.31	2.07	0.798
× 18.4	5.41	8.00	0.271	4.001	0.426	57.6	14.4	3.26	3.73	1.86	0.831
S7 × 20	5.88	7.00	0.450	3.860	0.392	42.4	12.1	2.69	3.17	1.64	0.734
× 15.3	4.50	7.00	0.252	3.662	0.392	36.7	10.5	2.86	2.64	1.44	0.766
S6 × 17.25	5.07	6.00	0.465	3.565	0.359	26.3	8.77	2.28	2.31	1.30	0.675
× 12.5	3.67	6.00	0.232	3.332	0.359	22.1	7.37	2.45	1.82	1.09	0.705
S5 × 14.75	4.34	5.00	0.494	3.284	0.326	15.2	6.09	1.87	1.67	1.01	0.620
× 10	2.94	5.00	0.214	3.004	0.326	12.3	4.92	2.05	1.22	0.809	0.643

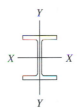

S-Shapes: SI Units (Selected Listing)

Designation	Area mm²	Depth mm	Web Thickness mm	Flange Width mm	Flange Thickness mm	Axis X-X I 10⁶mm⁴	Axis X-X S 10³mm³	Axis X-X r mm	Axis Y-Y I 10⁶mm⁴	Axis Y-Y S 10³mm³	Axis Y-Y r mm
S610 × 149	18900	610	18.9	184	22.1	995	3260	229	19.85	216	32.4
× 134	17100	610	15.9	181	22.1	937	3070	234	18.69	207	33.1
× 119	15160	610	12.7	178	22.1	874	2870	240	17.56	197.3	34.0
S510 × 143	18190	516	20.3	183	23.4	695	2690	195.5	20.89	228	33.9
× 128	16320	516	16.8	179	23.4	658	2550	201	19.48	218	34.5
× 112	14190	508	16.1	162	20.2	533	2100	193.8	12.40	153.1	29.6
× 98	12520	508	12.8	159	20.2	495	1949	198.8	11.53	145.0	30.3
S460 × 104	13290	457	18.1	159	17.6	385	1685	170.2	10.03	126.2	27.5
× 81.4	10390	457	11.7	152	17.6	335	1466	179.6	8.66	113.9	28.9
S380 × 74	9480	381	14.0	143	15.8	202	1060	146.0	6.53	91.3	26.2
× 64	8130	381	10.4	140	15.8	186.1	977	151.3	5.99	85.6	27.1
S310 × 74	9480	305	17.4	139	16.7	127.0	833	115.7	6.53	94.0	26.2
× 60.7	7740	305	11.7	133	16.7	113.2	742	120.9	5.66	85.1	27.0
× 52	6650	305	10.9	129	13.8	95.3	625	119.7	4.11	63.7	24.9
× 47.3	6030	305	8.9	127	13.8	90.7	595	122.6	3.90	61.4	25.4
S250 × 52	6650	254	15.1	126	12.5	61.2	482	95.9	3.48	55.2	22.9
× 37.8	4810	254	7.9	118	12.5	51.6	406	103.6	2.83	48.0	24.3
S200 × 34	4370	203	11.2	106	10.8	27.0	262	78.6	1.794	33.8	20.3
× 27.4	3490	203	6.9	102	10.8	24.0	236	82.9	1.553	30.5	21.1
S180 × 30	3790	178	11.4	98	10.0	17.65	198.3	68.2	1.319	26.9	18.66
× 22.8	2900	178	6.4	93	10.0	15.28	171.7	72.6	1.099	23.6	19.47
S150 × 25.7	3270	152	11.8	91	9.1	10.95	144.1	57.8	0.961	21.1	17.14
× 18.6	2370	152	5.9	85	9.1	9.20	121.1	62.3	0.758	17.84	17.88
S130 × 22.0	2800	127	12.5	83	8.3	6.33	99.7	47.5	0.695	16.75	15.75
× 15	1900	127	5.4	76	8.3	5.12	80.6	51.9	0.508	13.37	16.35

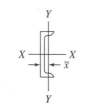

C-Shapes: U.S. Customary Units (Selected Listing)

| | | | Web | Flange | | Elastic Properties | | | | | | |
| | | | | | | Axis X-X | | | Axis Y-Y | | | |
Designation	Area in.²	Depth in.	Thickness in.	Width in.	Thickness in.	I in.⁴	S in.³	r in.	I in.⁴	S in.³	r in.	x̄ in.
C15 × 40	11.8	15.00	0.520	3.520	0.650	349	46.5	5.44	9.23	3.37	0.886	0.777
× 33.9	9.96	15.00	0.400	3.400	0.650	315	42.0	5.62	8.13	3.11	0.904	0.787
C12 × 30	8.82	12.00	0.510	3.170	0.501	162	27.0	4.29	5.14	2.06	0.763	0.674
× 25	7.35	12.00	0.387	3.047	0.501	144	24.1	4.43	4.47	1.88	0.780	0.674
× 20.7	6.09	12.00	0.282	2.942	0.501	129	21.5	4.61	3.88	1.73	0.799	0.698
C10 × 30	8.82	10.00	0.673	3.033	0.436	103	20.7	3.42	3.94	1.65	0.669	0.649
× 25	7.35	10.00	0.526	2.886	0.436	91.2	18.2	3.52	3.36	1.48	0.676	0.617
× 20	5.88	10.00	0.379	2.739	0.436	78.9	15.8	3.66	2.81	1.32	0.692	0.606
× 15.3	4.49	10.00	0.240	2.600	0.436	67.4	13.5	3.87	2.28	1.16	0.713	0.634
C9 × 20	5.88	9.00	0.448	2.648	0.413	60.9	13.5	3.22	2.42	1.17	0.642	0.583
× 15	4.41	9.00	0.285	2.485	0.413	51.0	11.3	3.40	1.93	1.01	0.661	0.586
× 13.4	3.94	9.00	0.233	2.433	0.413	47.9	10.6	3.48	1.76	0.962	0.669	0.601
C8 × 18.75	5.51	8.00	0.487	2.527	0.390	44.0	11.0	2.82	1.98	1.01	0.599	0.565
× 13.75	4.04	8.00	0.303	2.343	0.390	36.1	9.03	2.99	1.53	0.854	0.615	0.553
× 11.5	3.38	8.00	0.220	2.260	0.390	32.6	8.14	3.11	1.32	0.781	0.625	0.571
C7 × 14.75	4.33	7.00	0.419	2.299	0.366	27.2	7.78	2.51	1.38	0.779	0.564	0.532
× 12.25	3.60	7.00	0.314	2.194	0.366	24.2	6.93	2.60	1.17	0.703	0.571	0.525
× 9.8	2.87	7.00	0.210	2.090	0.366	21.3	6.08	2.72	0.968	0.625	0.581	0.540
C6 × 13	3.83	6.00	0.437	2.157	0.343	17.4	5.80	2.13	1.05	0.642	0.525	0.514
× 10.5	3.09	6.00	0.314	2.034	0.343	15.2	5.06	2.22	0.866	0.564	0.529	0.499
× 8.2	2.40	6.00	0.200	1.920	0.343	13.1	4.38	2.34	0.693	0.492	0.537	0.511
C5 × 9	2.64	5.00	0.325	1.885	0.320	8.90	3.56	1.83	0.632	0.450	0.489	0.478
× 6.7	1.97	5.00	0.190	1.750	0.320	7.49	3.00	1.95	0.479	0.378	0.493	0.484
C4 × 7.25	2.13	4.00	0.321	1.721	0.296	4.59	2.29	1.47	0.433	0.343	0.450	0.459
× 5.4	1.59	4.00	0.184	1.584	0.296	3.85	1.93	1.56	0.319	0.283	0.449	0.457

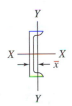

C-Shapes: SI Units (Selected Listing)

Designation	Area mm^2	Depth mm	Web Thickness mm	Flange Width mm	Flange Thickness mm	Axis X-X I 10^6mm^4	Axis X-X S 10^3mm^3	Axis X-X r mm	Axis Y-Y I 10^6mm^4	Axis Y-Y S 10^3mm^3	Axis Y-Y r mm	Axis Y-Y $\bar{x}$ mm
C380 × 60	7610	381	13.2	89.4	16.5	145.3	763	138.2	3.84	55.1	22.5	19.74
× 50.4	6430	381	10.2	86.4	16.5	131.1	688	142.8	3.38	50.9	22.9	19.99
C310 × 45	5690	305	13.0	80.5	12.7	67.4	442	108.8	2.14	33.8	19.39	17.12
× 37	4740	305	9.8	77.4	12.7	59.9	393	112.4	1.861	30.9	19.81	17.12
× 30.8	3930	305	7.2	74.7	12.7	53.7	352	116.9	1.615	28.3	20.27	17.73
C250 × 45	5690	254	17.1	77.0	11.1	42.9	338	86.8	1.640	27.1	16.98	16.48
× 37	4740	254	13.4	73.3	11.1	38.0	299	89.5	1.399	24.3	17.18	15.67
× 30	3790	254	9.6	69.6	11.1	32.8	258	93.0	1.170	21.6	17.57	15.39
× 22.8	2900	254	6.1	66.0	11.1	28.1	221	98.4	0.949	19.02	18.09	16.10
C230 × 30	3790	229	11.4	67.3	10.5	25.3	221	81.7	1.007	19.18	16.30	14.81
× 22	2850	229	7.2	63.1	10.5	21.2	185.2	86.2	0.803	16.65	16.79	14.88
× 19.9	2540	229	5.9	61.8	10.5	19.94	174.1	88.6	0.733	15.75	16.99	15.27
C200 × 27.9	3550	203	12.4	64.2	9.9	18.31	180.4	71.8	0.824	16.53	15.24	14.35
× 20.5	2610	203	7.7	59.5	9.9	15.03	148.1	75.9	0.637	14.02	15.62	14.05
× 17.1	2180	203	5.6	57.4	9.9	13.57	133.7	78.9	0.549	12.80	15.87	14.50
C180 × 22.0	2790	178	10.6	58.4	9.3	11.32	127.2	63.7	0.574	12.79	14.34	13.51
× 18.2	2320	178	8.0	55.7	9.3	10.07	113.1	65.9	0.487	11.50	14.49	13.34
× 14.6	1852	178	5.3	53.1	9.3	8.87	99.7	69.2	0.403	10.23	14.75	13.72
C150 × 19.3	2470	152	11.1	54.8	8.7	7.24	95.3	54.1	0.437	10.47	13.30	13.06
× 15.6	1994	152	8.0	51.7	8.7	6.33	83.3	56.3	0.360	9.22	13.44	12.67
× 12.2	1548	152	5.1	48.8	8.7	5.45	71.7	59.3	0.288	8.04	13.64	12.98
C130 × 13.4	1703	127	8.3	47.9	8.1	3.70	58.3	46.6	0.263	7.35	12.43	12.14
× 10.0	1271	127	4.8	44.4	8.1	3.12	49.1	49.5	0.199	6.20	12.51	12.29
C100 × 10.8	1374	102	8.2	43.7	7.5	1.911	37.5	37.3	0.180	5.62	11.45	11.66
× 8.0	1026	102	4.7	40.2	7.5	1.602	31.4	39.5	0.133	4.65	11.39	11.61

Angles, Equal Legs: U.S. Customary Units (Selected Listing)

Size and Thickness in.	Weight per Foot lb	Area in.²	Axis X-X or Y-Y				Axis Z-Z
			I in.⁴	S in.³	r in.	x or y in.	r in.
L8 × 8 × $1\frac{1}{8}$	56.9	16.7	98.0	17.5	2.42	2.41	1.56
× 1	51.0	15.0	89.0	15.8	2.44	2.37	1.56
× $\frac{7}{8}$	45.0	13.2	79.6	14.0	2.45	2.32	1.57
× $\frac{3}{4}$	38.9	11.4	69.7	12.2	2.47	2.28	1.58
× $\frac{5}{8}$	32.7	9.61	59.4	10.3	2.49	2.23	1.58
× $\frac{9}{16}$	29.6	8.68	54.1	9.34	2.50	2.21	1.59
× $\frac{1}{2}$	26.4	7.75	48.6	8.36	2.50	2.19	1.59
L6 × 6 × 1	37.4	11.0	35.5	8.57	1.80	1.86	1.17
× $\frac{7}{8}$	33.1	9.73	31.9	7.63	1.81	1.82	1.17
× $\frac{3}{4}$	28.7	8.44	28.2	6.66	1.83	1.78	1.17
× $\frac{5}{8}$	24.2	7.11	24.2	5.66	1.84	1.73	1.18
× $\frac{9}{16}$	21.9	6.43	22.1	5.14	1.85	1.71	1.18
× $\frac{1}{2}$	19.6	5.75	19.9	4.61	1.86	1.68	1.18
× $\frac{7}{16}$	17.2	5.06	17.7	4.08	1.87	1.66	1.19
× $\frac{3}{8}$	14.9	4.36	15.4	3.53	1.88	1.64	1.19
× $\frac{5}{16}$	12.4	3.65	13.0	2.97	1.89	1.62	1.20
L5 × 5 × $\frac{7}{8}$	27.2	7.98	17.8	5.17	1.49	1.57	0.973
× $\frac{3}{4}$	23.6	6.94	15.7	4.53	1.51	1.52	0.975
× $\frac{5}{8}$	20.0	5.86	13.6	3.86	1.52	1.48	0.978
× $\frac{1}{2}$	16.2	4.75	11.3	3.16	1.54	1.43	0.983
× $\frac{7}{16}$	14.3	4.18	10.0	2.79	1.55	1.41	0.986
× $\frac{3}{8}$	12.3	3.61	8.74	2.42	1.56	1.39	0.990
× $\frac{5}{16}$	10.3	3.03	7.42	2.04	1.57	1.37	0.994
L4 × 4 × $\frac{3}{4}$	18.5	5.44	7.67	2.81	1.19	1.27	0.778
× $\frac{5}{8}$	15.7	4.61	6.66	2.40	1.20	1.23	0.779
× $\frac{1}{2}$	12.8	3.75	5.56	1.97	1.22	1.18	0.782
× $\frac{7}{16}$	11.3	3.31	4.97	1.75	1.23	1.16	0.785
× $\frac{3}{8}$	9.8	2.86	4.36	1.52	1.23	1.14	0.788
× $\frac{5}{16}$	8.2	2.40	3.71	1.29	1.24	1.12	0.791
× $\frac{1}{4}$	6.6	1.94	3.04	1.05	1.25	1.09	0.795
L$3\frac{1}{2}$ × $3\frac{1}{2}$ × $\frac{1}{2}$	11.1	3.25	3.64	1.49	1.06	1.06	0.683
× $\frac{7}{16}$	9.8	2.87	3.26	1.32	1.07	1.04	0.684
× $\frac{3}{8}$	8.5	2.48	2.87	1.15	1.07	1.01	0.687
× $\frac{5}{16}$	7.2	2.09	2.45	0.976	1.08	0.990	0.690
× $\frac{1}{4}$	5.8	1.69	2.01	0.794	1.09	0.968	0.694
L3 × 3 × $\frac{1}{2}$	9.4	2.75	2.22	1.07	0.898	0.932	0.584
× $\frac{7}{16}$	8.3	2.43	1.99	0.954	0.905	0.910	0.585
× $\frac{3}{8}$	7.2	2.11	1.76	0.833	0.913	0.888	0.587
× $\frac{5}{16}$	6.1	1.78	1.51	0.707	0.922	0.869	0.589
× $\frac{1}{4}$	4.9	1.44	1.24	0.577	0.930	0.842	0.592
× $\frac{3}{16}$	3.71	1.09	0.962	0.441	0.939	0.820	0.596
L$2\frac{1}{2}$ × $2\frac{1}{2}$ × $\frac{1}{2}$	7.7	2.25	1.23	0.724	0.739	0.806	0.487
× $\frac{3}{8}$	5.9	1.73	0.984	0.566	0.753	0.762	0.487
× $\frac{5}{16}$	5.0	1.46	0.849	0.482	0.761	0.740	0.489
× $\frac{1}{4}$	4.1	1.19	0.703	0.394	0.769	0.717	0.491
× $\frac{3}{16}$	3.07	0.92	0.547	0.303	0.778	0.694	0.495

Angles, Equal Legs: SI Units (Selected Listing)

Size and Thickness mm	Mass per Meter kg	Area mm²	Elastic properties				Axis Z-Z
			Axis X-X or Y-Y				
			I × 10⁶ mm⁴	S × 10³ mm³	r mm	x or y mm	r mm
L203 × 203 × 28.6	84.6	10 770	40.8	287	61.5	61.2	39.6
× 25.4	75.9	9 680	37.0	259	62.0	60.2	39.6
× 22.2	67.0	8 520	33.1	229	62.2	58.9	39.9
× 19.0	57.9	7 350	29.0	200	62.7	57.9	40.1
× 15.9	48.7	6 200	24.7	169	63.2	56.6	40.1
× 14.3	44.0	5 600	22.5	153	63.5	56.1	40.4
× 12.7	39.3	5 000	20.2	137	63.5	55.6	40.4
L152 × 152 × 25.4	55.7	7 100	14.8	140	45.7	47.2	29.7
× 22.2	49.3	6 280	13.3	125	46.0	46.2	29.7
× 19.0	42.7	5 450	11.7	109	46.5	45.2	29.7
× 15.9	36.0	4 590	10.1	92.8	46.7	43.9	30.0
× 14.3	32.6	4 150	9.20	84.2	47.0	43.4	30.0
× 12.7	29.2	3 710	8.28	75.5	47.2	42.7	30.0
× 11.1	25.6	3 260	7.37	66.9	47.5	42.2	30.2
× 9.52	22.2	2 810	6.41	57.8	47.8	41.7	30.2
× 7.94	18.5	2 350	5.41	48.7	48.0	41.1	30.5
L127 × 127 × 22.2	40.5	5 150	7.41	84.7	37.8	39.9	24.7
× 19.0	35.1	4 480	6.53	74.2	38.4	38.6	24.8
× 15.9	29.8	3 780	5.66	63.3	38.6	37.6	24.8
× 12.7	24.1	3 060	4.70	51.8	39.1	36.3	25.0
× 11.1	21.3	2 700	4.16	45.7	39.4	35.8	25.0
× 9.52	18.3	2 330	3.64	39.7	39.6	35.3	25.1
× 7.94	15.3	1 950	3.09	33.4	39.9	34.8	25.3
L102 × 102 × 19.0	27.5	3 510	3.19	46.0	30.2	32.3	19.8
× 15.9	23.4	2 970	2.77	39.3	30.5	31.2	19.8
× 12.7	19.0	2 420	2.31	32.3	31.0	30.0	19.9
× 11.1	16.8	2 140	2.07	28.7	31.2	29.5	19.9
× 9.52	14.6	1 850	1.81	24.9	31.2	29.0	20.0
× 7.94	12.2	1 550	1.54	21.1	31.5	28.5	20.1
× 6.35	9.8	1 250	1.27	17.2	31.8	27.7	20.2
L89 × 89 × 12.7	16.5	2 100	1.52	24.4	26.9	26.9	17.3
× 11.1	14.6	1 850	1.36	21.6	27.2	26.4	17.4
× 9.52	12.6	1 600	1.19	18.8	27.2	25.7	17.4
× 7.94	10.7	1 350	1.02	16.0	27.4	25.1	17.5
× 6.35	8.6	1 090	0.837	13.0	27.7	24.6	17.6
L76 × 76 × 12.7	14.0	1 770	0.924	17.5	22.8	23.7	14.8
× 11.1	12.4	1 570	0.828	15.6	23.0	23.1	14.9
× 9.52	10.7	1 360	0.733	13.7	23.2	22.6	14.9
× 7.94	9.1	1 150	0.629	11.6	23.4	22.1	15.0
× 6.35	7.3	929	0.516	9.46	23.6	21.4	15.0
× 4.76	5.52	703	0.400	7.23	23.9	20.8	15.1
L64 × 64 × 12.7	11.4	1 450	0.512	11.9	18.8	20.5	12.4
× 9.52	8.9	1 120	0.410	9.38	19.1	19.4	12.4
× 7.94	7.4	942	0.353	7.90	19.3	18.8	12.4
× 6.35	6.1	768	0.293	6.46	19.5	18.2	12.5
× 4.76	4.57	594	0.228	4.97	19.8	17.6	12.6

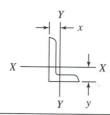

Angles, Unequal Legs: U.S. Customary Units (Selected Listing)

Size and Thickness in.	Weight per Foot lb	Area in.²	Axis X-X				Axis Y-Y			
			I in.⁴	S in.³	r in.	y in.	I in.⁴	S in.³	r in.	x in.
L9 × 4 × 1	40.8	12.0	97.0	17.6	2.84	3.50	12.0	4.00	1.00	1.00
× $\frac{3}{4}$	31.3	9.19	76.1	13.6	2.88	3.41	9.63	3.11	1.02	0.906
× $\frac{1}{2}$	21.3	6.25	53.2	9.34	2.92	3.31	6.92	2.17	1.05	0.810
L8 × 6 × 1	44.2	13.0	80.8	15.1	2.49	2.65	38.8	8.92	1.73	1.65
× $\frac{3}{4}$	33.8	9.94	63.4	11.7	2.53	2.56	30.7	6.92	1.76	1.56
× $\frac{1}{2}$	23.0	6.75	44.3	8.02	2.56	2.47	21.7	4.79	1.79	1.47
L7 × 4 × $\frac{7}{8}$	30.2	8.86	42.9	9.65	2.20	2.55	10.2	3.46	1.07	1.05
× $\frac{5}{8}$	22.1	6.48	32.4	7.14	2.24	2.46	7.84	2.58	1.10	0.963
× $\frac{3}{8}$	13.6	3.98	20.6	4.44	2.27	2.37	5.10	1.63	1.13	0.870
L6 × 4 × $\frac{7}{8}$	27.2	7.98	27.7	7.15	1.86	2.12	9.75	3.39	1.11	1.12
× $\frac{5}{8}$	20.0	5.86	21.1	5.31	1.90	2.03	7.52	2.54	1.13	1.03
× $\frac{3}{8}$	12.3	3.61	13.5	3.32	1.93	1.94	4.90	1.60	1.17	0.941
L6 × 3$\frac{1}{2}$ × $\frac{1}{2}$	15.3	4.50	16.6	4.24	1.92	2.08	4.25	1.59	0.972	0.833
× $\frac{3}{8}$	11.7	3.42	12.9	3.24	1.94	2.04	3.34	1.23	0.988	0.787
× $\frac{1}{4}$	7.9	2.31	8.86	2.21	1.96	1.99	2.34	0.847	1.01	0.740
L5 × 3 × $\frac{1}{2}$	12.8	3.75	9.45	2.91	1.59	1.75	2.58	1.15	0.829	0.750
× $\frac{3}{8}$	9.8	2.86	7.37	2.24	1.61	1.70	2.04	0.888	0.845	0.704
× $\frac{1}{4}$	6.6	1.94	5.11	1.53	1.62	1.66	1.44	0.614	0.861	0.657
L4 × 3 × $\frac{1}{2}$	11.1	3.25	5.05	1.89	1.25	1.33	2.42	1.12	0.864	0.827
× $\frac{3}{8}$	8.5	2.48	3.96	1.46	1.26	1.28	1.92	0.866	0.879	0.782
× $\frac{1}{4}$	5.8	1.69	2.77	1.00	1.28	1.24	1.36	0.599	0.896	0.736
L3$\frac{1}{2}$ × 2$\frac{1}{2}$ × $\frac{1}{2}$	9.4	2.75	3.24	1.41	1.09	1.20	1.36	0.760	0.704	0.705
× $\frac{3}{8}$	7.2	2.11	2.56	1.09	1.10	1.16	1.09	0.592	0.719	0.660
× $\frac{1}{4}$	4.9	1.44	1.80	0.755	1.12	1.11	0.777	0.412	0.735	0.614
L3 × 2 × $\frac{1}{2}$	7.7	2.25	1.92	1.00	0.924	1.08	0.672	0.474	0.546	0.583
× $\frac{3}{8}$	5.9	1.73	1.53	0.781	0.940	1.04	0.543	0.371	0.559	0.539
× $\frac{1}{4}$	4.1	1.19	1.09	0.542	0.957	0.993	0.392	0.260	0.574	0.493
L2$\frac{1}{2}$ × 2 × $\frac{3}{8}$	5.3	1.55	0.912	0.547	0.768	0.831	0.514	0.363	0.577	0.581
× $\frac{1}{4}$	3.62	1.06	0.654	0.381	0.784	0.787	0.372	0.254	0.592	0.537

Elastic Properties

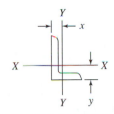

Angles, Unequal Legs: SI Units (Selected Listing)

Size and Thickness mm	Mass per Meter kg	Area mm²	Elastic Properties							
			Axis X-X				Axis Y-Y			
			I × 10⁶ mm⁴	S × 10³ mm³	r mm	y mm	I × 10⁶ mm⁴	S × 10³ mm³	r mm	x mm
L229 × 102 × 25.4	60.7	7740	40.4	288	72.1	88.9	4.99	65.5	25.4	25.4
× 19.0	46.6	5930	31.7	223	73.1	86.6	4.01	51.0	25.9	23.0
× 12.7	31.7	4030	22.1	153	74.2	84.1	2.88	35.6	26.7	20.6
L203 × 152 × 25.4	65.8	8390	33.6	247	63.2	67.3	15.9	146	43.9	41.9
× 19.0	50.3	6410	26.4	192	64.3	65.0	12.8	113	44.7	39.6
× 12.7	34.2	4350	18.4	131	65.0	62.7	9.03	78.5	45.5	37.3
L178 × 102 × 22.2	44.9	5720	17.9	158	55.9	64.8	4.25	56.7	27.2	26.7
× 15.9	32.9	4180	13.5	117	56.9	62.5	3.26	42.3	27.9	24.5
× 9.52	20.2	2570	8.57	72.8	57.7	60.2	2.12	26.7	28.7	22.1
L152 × 102 × 22.2	40.5	5150	11.5	117	47.2	53.8	4.06	55.6	28.2	28.5
× 15.9	29.8	3780	8.78	87.0	48.3	51.6	3.13	41.6	28.7	26.2
× 9.52	18.3	2330	5.62	54.4	49.0	49.3	2.04	26.2	29.7	23.9
L152 × 89 × 12.7	22.7	2900	6.91	69.5	48.8	52.8	1.77	26.1	24.7	21.2
× 9.52	17.4	2200	5.37	53.1	49.3	51.8	1.39	20.2	25.1	20.0
× 6.35	11.8	1490	3.69	36.2	49.8	50.5	0.974	13.9	25.7	18.8
L127 × 76 × 12.7	19.0	2420	3.93	47.7	40.4	44.4	1.07	18.8	21.1	19.0
× 9.52	14.6	1850	3.06	36.7	40.9	43.2	0.849	14.6	21.5	17.9
× 6.35	9.8	1250	2.13	25.1	41.1	42.2	0.599	10.1	21.9	16.7
L102 × 76 × 12.7	16.5	2100	2.10	31.0	31.8	33.8	1.01	18.4	21.9	21.0
× 9.52	12.6	1600	1.65	23.9	32.0	32.5	0.799	14.2	22.3	19.9
× 6.35	8.6	1090	1.15	16.4	32.5	31.5	0.566	9.82	22.8	18.7
L89 × 64 × 12.7	14.0	1770	1.35	23.1	27.7	30.5	0.566	12.5	17.9	17.9
× 9.52	10.7	1360	1.07	17.9	27.9	29.5	0.454	9.70	18.3	16.8
× 6.35	7.3	929	0.749	12.4	28.4	28.2	0.323	6.75	18.7	15.6
L76 × 51 × 12.7	11.5	1450	0.799	16.4	23.5	27.4	0.280	7.77	13.9	14.8
× 9.52	8.8	1120	0.637	12.8	23.9	26.4	0.226	6.08	14.2	13.7
× 6.35	6.1	768	0.454	8.88	24.3	25.2	0.163	4.26	14.6	12.5
L64 × 51 × 9.52	7.9	1000	0.380	8.96	19.5	21.1	0.214	5.95	14.7	14.8
× 6.35	5.39	684	0.272	6.24	19.9	20.0	0.155	4.16	15.0	13.6

Timber Sizes: U.S. Customary Units (Selected Listing)*

Nominal Size in.	American Standard Dressed Size in.	Area of Section in.^{2}	Weight per Foot lb	Moment of Inertia in.^{4}	Section Modulus in.^{3}
2 × 4	1.5 × 3.5	5.25	1.46	5.36	3.06
× 6	5.5	8.25	2.29	20.8	7.56
× 8	7.25	10.9	3.02	47.6	13.1
× 10	9.25	13.9	3.85	89.9	21.4
× 12	11.25	16.9	4.69	178	31.6
4 × 4	3.5 × 3.5	12.2	3.40	12.5	7.15
× 6	5.5	19.2	5.35	48.5	17.6
× 8	7.25	25.4	7.05	111	30.7
× 10	9.25	32.4	8.99	231	49.9
× 12	11.25	39.4	10.9	415	73.8
6 × 6	5.5 × 5.5	30.2	8.40	76.3	27.7
× 8	7.5	41.2	11.5	193	51.6
× 10	9.5	52.2	14.5	393	82.7
× 12	11.5	63.2	17.6	697	121
× 14	13.5	74.2	20.6	1128	167
8 × 8	7.5 × 7.5	56.2	15.6	263	70.3
× 10	9.5	71.2	19.8	536	113
× 12	11.5	86.2	24.0	951	165
× 14	13.5	101	28.1	1538	228
× 16	15.5	116	32.3	2327	300
10 × 10	9.5 × 9.5	90.2	25.1	679	143
× 12	11.5	109	30.3	1204	209
× 14	13.5	128	35.6	1948	289
× 16	15.5	147	40.9	2948	380
× 18	17.5	166	46.2	4243	485
× 20	19.5	185	51.5	5870	602
12 × 12	11.5 × 11.5	132	36.7	1458	253
× 14	13.5	155	43.1	2358	349
× 16	15.5	178	49.5	3569	460
× 18	17.5	201	55.9	5136	587
× 20	19.5	224	62.3	7106	729
× 22	21.5	247	68.7	9524	886
× 24	23.5	270	75.1	12440	1058

*All weights and properties calculated for dressed sizes. Weights based on 40 lb/ft^3.

Timber Sizes: SI Units (Selected Listing)*

Nominal Size mm	American Standard Dressed Size mm	Area of Section $\times 10^3$ mm^2	Mass per Meter kg	Moment of Inertia $\times 10^6$ mm^4	Section Modulus $\times 10^3$ mm^3
50 × 100	38 × 89	3.38	2.03	2.23	50.2
× 150	140	5.32	3.19	8.69	124
× 200	184	6.99	4.20	19.7	214
× 250	235	8.93	5.36	41.1	350
× 300	286	10.87	6.52	74.1	518
100 × 100	89 × 89	7.92	4.75	5.23	117
× 150	140	12.5	7.48	20.4	291
× 200	184	16.4	9.83	46.2	502
× 250	235	20.9	12.5	96.3	819
× 300	286	25.4	15.3	174	1 213
150 × 150	140 × 140	19.6	11.8	32.0	457
× 200	191	26.7	16.0	81.3	851
× 250	241	33.7	20.2	163	1 355
× 300	292	40.9	24.6	290	1 989
× 350	343	48.0	28.8	471	2 745
200 × 200	191 × 191	36.5	21.9	111	1 161
× 250	241	46.0	27.6	223	1 849
× 300	292	55.8	33.5	396	2 714
× 350	343	65.5	39.3	642	3 745
× 400	394	75.2	45.2	974	4 942
250 × 250	241 × 241	58.1	34.8	281	2 333
× 300	292	70.4	42.2	500	3 425
× 350	343	82.7	49.6	810	4 726
× 400	394	95.0	57.0	1 228	6 235
× 450	445	107	64.3	1 770	7 954
× 500	495	119	71.6	2 436	9 842
300 × 300	292 × 292	85.3	51.2	606	4 150
× 350	343	100	60.1	982	5 726
× 400	394	115	69.0	1 488	7 555
× 450	445	130	78.0	2 144	9 637
× 500	495	144	86.7	2 951	11 920
× 550	546	159	95.7	3 961	14 510
× 600	597	174	105	5 178.	17 350

*All masses and properties calculated for dressed sizes. Masses based on 600 kg/m^3.

REFERENCES

1. Giancoli, D. C. *Physics*. Upper Saddle River, N.J.: Prentice Hall, 1985.
2. Calter, P. *Technical Calculus*. Upper Saddle River, N.J.: Prentice Hall, 1988.
3. Funk, E. R., and L. J. Rieber. *Handbook of Welding*, Delmar Inc., Albany N.Y.: 1985.
4. Mott, R. L. *Applied Strength of Materials*, 3d. ed. Englewood Cliffs, N.J.: Prentice Hall, 1996.
5. Morrow, H. W. *Statics and Strength of Materials*, 2d ed. Upper Saddle River, N.J.: Prentice Hall, 1993.
6. Cheng, F. *Statics and Strength of Materials*, 2d ed. Glencoe/McGraw-Hill, Westerville, Ohio, 1997.
7. Spiegel, L., and G. F. Limbrunner. *Applied Statics and Strength of Materials*, 2d ed. Upper Saddle River, N.J.: Prentice Hall, 1995.
8. Bassin, M. G., S. M. Brodsky, and H. Wolkoff. *Statics and Strength of Materials*, 4th ed. Glencoe/Macmillan/McGraw-Hill, Westerville, Ohio, 1992.
9. *Structural Steel Shapes*. Pittsburgh, Pa.: United States Steel, 1980.
10. *Aluminum Standards and Data*, 11th ed. Washington, D.C.: Aluminum Association, 1993.
11. *Manual of Steel Construction*, 9th ed. Chicago, Il.: American Institute of Steel Construction, 1989.
12. Jensen, C., J. D. Helsel, and D. Short. *Engineering Drawing and Design*, 5th ed. Glencoe/McGraw-Hill, Westerville, Ohio, 1996.

Index

A

Acceleration of gravity, 4
Accuracy, 9
AISC Manual of Steel Construction, 206, 213, 214, 495, 610, 616
Allowable axial compressive load:
  for columns, 606, 610
Allowable axial compressive stress:
  for columns, 608, 616, 620
Allowable stress:
  for aluminum alloys, 496
  for bearing, 214
  for bolts and rivets, 213
  for shear, 175
  for steel beams, 495
  for steel columns, 608, 616
  for tension and compression, 172
  for timber columns, 620
  for timber construction, 497, 615
Allowable stress design, 174
Allowable torque. *See* Torque
Aluminum Association, 213, 496
Analysis of frames. *See* Frames
Angle of static friction. *See* Friction
Angle of twist, 289
Angle of wrap, 136
Areas of figures, 632
Aristotle, 4
ASTM American Society for Testing and Materials, 167
Atom, 1
Average web shear. *See* Stress
AWS American Welding Society, 203
Axially loaded machine parts, 605

B

Beam (s):
  bending deformation, 320
  bending moment, 349
  bending stress, 432
  composite, 439
  continuous, 528
  curvature, 452
  deflections, 451
    basic, 504
    double integration, 470
    equations, 473–78
    moment-area method, 451
    principle of superposition, 471
  design, 495
  fixed, 478, 546
  generalized equations, 368–78
  horizontal shear, 419
  propped cantilever, 477, 478
  reactions, 53
  selection, 500
  shear force, 327
  shear stresses, 419
  statically determinate, 319
  statically indeterminate, 319, 527
  types, 319
  vertical shear, 419
  weight, in design, 482, 503
Beam diagrams and formulas, 473–78
Beam deflections, 451
Beam selection table, Appendix E
Bearing stress (pressure). *See* Stress
Belt friction. *See* Friction

Bending stress. *See* Stress
Bending moment sign convention, 323
Bearing-type bolted connections. *See* Connections
Bolted connections. *See* Connections
Bolts:
   in connections. *See* Connections
   types of, 213, 214
Brass, 649, 650
Brittleness, 172, 174
Bronze, 649, 650

C

C shapes, dimensions and properties, 658, 659
Cantilever beam, 346
   method of reflection, 545
Carbon steel, 645, 646
Cast Iron, 649, 650
Center of gravity, 245
Centroid, 246
   of any shape, 255
   of composite shapes, 249
   of simple shapes, 249
Channel sections, design properties, 658, 659
Characteristics of a force. *See* force
Coefficient of linear expansion, 176
Coefficient of static friction. *See* Friction
Columns, 605
   allowable axial compressive loads and stresses, 606, 608, 610, 616, 620
   analysis (AISC), 610, 616
   basic theory, 605
   design, 617
   eccentric loads, 622
   effective length, 610
      table of factors, 610
   Euler buckling load, 606
   Johnson formula, 611
   slenderness ratio, 608, 611
   steel, 610
   timber, 620

Combined loading, 565
Combined stresses, 568
   axial and bending, 585
   normal and shear, 577
   torsion and bending, 585
Common bolts. *See* Connections
Components of force. *See* Force
Composite beams, 439
Computer programs:
   Centroid, 270
   Column, 625
   Combload, 595
   Connect, 220
   Convert, 10
   Coupling, 311
   Deflect, 483
   Deflection, 487
   Dstress, 159
   Fixbeam, 557
   Friction, 139
   Moveload, 409
   Propbeam, 550
   Propclr, 554
   Proptab, 185
   Resultant, 37
   Shapes1, 509
   Shapes2, 509
   Shr&mmt, 394
   Shrmmtcl, 394
   Spcframe, 111
   Torque, 309
   Truss, 108
   Tstress, 198
   Vessel, 239
Concentrated force. *See* Force
Concentrated load. *See* Load
Concrete, 649, 650
Connections, 203
   bolted and riveted, 209
   bearing type, 212
   friction type, 212
   high-strength, 212
   modes of failure, 211
   rivets and common bolts, 213, 214
   strength and efficiency, 205, 215

types of, 205, 209, 210
welded, 203
Continuous beams. *See* beams
Conversion factors, 652
Copper, 649, 650
Corner return, 206
Cosine rule, 636
Coulomb, Charles, 121
Coulomb friction. *See* Friction
Couples, 51
Couplings, shafts, 302
Culman's circle, 568

D

Deflection of beams. *See* Beams
Design of axially loaded members. *See*
 Columns
Design of beams. *See* Beams
Diameter, selection of, 506
Direct stress. *See* Stress
Direct stress formula, 145
Distributed load. *See* Load
Distributed force. *See* Force
Dressed size, 664, 665
Ductile Iron, 649, 650
Ductility, 172
Dynamic friction. *See* Kinetic friction

E

Eccentric loads on columns. *See*
 Columns
Eccentrically loaded members, 219,
 592, 622
Efficiency of a joint, 215
Effective length, 610
 k factor values, 610
Elasticity, 1
Elastic limit, 170
Elastic modulus. *See* Modulus of
 elasticity
Elastic range, 170
Elongation, 172
End return. *See* Corner return
Engineering Materials. *See* Mechanical
 properties of engineering
 materials

Equilibrant, 30
Equilibrium:
 concurrent force system, 35
 conditions for, 46, 79
 non-concurrent force system, 46
 parallel force system, 45
Euler buckling load. *See* Columns
Euler's column theory, 605

F

Factor of safety, 174
Fillet welds, 206
Fixed-end beams. *See* Beams, fixed
Flexure formula, 325
Force, 3, 13
 characteristics, 3
 components, 19
 concentrated, 43
 defined, 3
 distributed, 49
 internal resisting, 3
 resultants, 14
 systems, types, 13, 14
 units of, 2, 6
Force triangle, 15
Force triangle method, applied, 16, 17
Frames:
 space frames, 88
 two-dimensional, 83
Free-body diagram, 33
Friction, 121
 angle of static, 124
 applications, 121
 belt, 136
 coefficient of static, 122
 Coulomb, 121
 table of values, 123
 theory, 122
Friction type connections, 212

G

Gauge pressure, 231
General shear formula, 423
Gray Cast Iron, 649, 650
Groove welds. *See* Welds

H

High-strength bolt:
    connections using, 209
    strength, 213, 214
    designations, 213
Hooke's Law, 170
Hoop stress, 234
Horsepower, defined, 298

I

Internal resting force. *See* Force
Internal resisting moment. *See* Beams,
    Bending moment

J

Jackscrews, 133
Johnson formula. *See* Columns
Joints
    efficiency of, 205, 215
    method of, 98
Joule, 298

K

Kern, 622
Keyseats, 296
Kinetic friction, 122

L

Laws of equilibrium. *See* Equilibrium
Leg of weld, 206
Linear coefficient of thermal
    expansion, 176
    table of values, 176
Loads:
    concentrated, 43
    distributed, 49
Loads on beams. *See* Beams
Longitudinal stress, 234

M

Magnesium, 649, 650
Malleable Iron, 649, 650
Mass, 4

Mechanical properties of engineering
    materials, Appendix C
    carbon and alloy steel, 645, 646
    plastics, 651
    stainless steel, 647
    structural steel, 643, 644
    timber, 651
    various engineering materials, 649,
        650
    wrought aluminum, 648
Method of joints, 98
Method of sections, 104
Modulus of elasticity, 170
    in shear, 175
Modulus of rigidity, 175
Mohr, Otto, 568
Mohr's circle, 568
    for combined bending and tension
        or compression, 590
    for combined tension and shear, 577
    for combined tension-tension, 573
    for combined torsion and bending,
        585
    for pure tension, 568
Molecule, 1
Moment, 41
    arm, 42
    calculation of, 43
    sign convention for, 44
    internal resisting–*See* Beams,
        bending moment
    units, 41
Moment-area method, 451
Moment of inertia, 257
    any shape, 265
    composite shapes, 261
    polar, 268
    product of inertia, 267
    simple shapes, 258
    tabular format of calculation, 264
    transfer formula, 260
Monel, 649, 650
Moving loads, 381
    three axle vehicle, 386
    two axle vehicle, 381

N

Necking, 172
Neutral axis, neutral plane, 321
Newton, Sir Isaac, 4
Normal stress. *See* Stress
Nominal size, pipes, 499, 506
Nonferrous metals, 649, 650
Numerical accuracy, 9

O

Offset method for determination of
    yield strength, 172
Overhanging beam, 341, 364

P

Parallelogram law, 15
Percent elongation, 172
Pin reactions, 84
Plastics, 651
Poisson, Simeon, 175
Poisson's ratio, 175
Polar moment of inertia. *See* Moment
    of inertia
Power, 298
Precision, 9
Pressure vessels, 231
   joints in, 238
   stresses in, 234
   thin-walled, 234
   thick-walled, 236
Principal moments, 268
Principal planes, 581
Principal stress, 581
Properties of materials. See
    Mechanical properties of
    engineering materials
Proportional limit, 170
Propped cantilever, 477, 478
Pythagorean theorem, 636

Q

Quadratic formula, 612

R

Radial stress, 234
Radius of curvature, 452
Radius of gyration, 268
Reactions, 35
  calculating:
    for any number of concentrated
      loads, 56
    for any number of concentrated
      loads and distributed loads, 63
    for any number of distributed
      loads, 61
    for one concentrated load, 53
    for one distributed load, 57
    for two concentrated loads, 55
    for two distributed loads, 59
  general case:
    for M distributed loads, 69
    for N concentrated loads, 66
    for N concentrated and M
      distributed loads, 72
    for one concentrated load, 65
    for one distributed load, 68
References, 667
Reflection, method of, 545
Resolution, defined, 19
Resultant, defined, 14
  of collinear concurrent forces, 14
  of coplanar concurrent forces, 15, 21
  of noncoplanar concurrent forces,
    23, 27
Right triangle review, 633, 634
Rivets. *See* Connections:
  types of, 213
Root of weld, 205
Rounding numbers, 8
Rutherford, Ernest, 1

S

S shapes, dimensions and properties,
    656, 657
Safety factor, 172, 174
Section modulus, 497
Sections, method of, 104

Shafts:
  transmission of power by, 300, 302
Shear diagrams:
  procedure for, 327
Shear force:
  sign convention, 328
Shear in beams. *See* Beams
Shear strain. *See* Strain
Shear stress, 151
  torsional, 289
SI units, 2
Sigma notation, 637
Significant digit, defined, 8
Simple supports, 320
Sine rule, 636
Slenderness ratio. *See* Columns
Slopes of simple beams, 528
Stainless steel, 647
Static friction, 121
Strain:
  axial, 169
  shear, 288
Strength of a joint. *See* Connections,
      strength and efficiency
Stress:
  allowable, 172
  average web shear, 500
  bearing, 154
  bending, 439
  defined, 149
  direct, 145
  on inclined planes, 567, 568
  in members of two or more
      materials, 191, 194, 439
  normal, 565
  principal, 581
  shear, 151
  on mutually perpendicular planes, 577
  tangential, 234
  tensile and compressive, 145, 150
  thermal, 189
  thick-walled pressure vessels, 236
  thin-walled pressure vessels, 234
  torsional shear, 289
Stress concentration, 155, 293, 435
  factors for, 157, 293, 295, 437

Stresses in beams. *See* Beams
Stress-strain diagram, 171
Structural shapes, Appendix E
  angles
    equal legs, 660, 661
    unequal legs, 662, 663
  C shapes, 658, 659
  S shapes, 656, 657
  timber sizes, 664, 665
  W shapes, 654, 655
Structural steel, 643, 644
Struts, 605
Superposition, 542

T

Temperature stresses. *See* Stress,
      thermal
Tensile strength, 170
Tension, defined, 145
Tension and compression test, 167
Theorem of Three Moments, 527
Thermal deformation, 176
Thermal stress, 189
  materials in parallel, 191
  materials in series, 194
  single material, 189
Thick-wall pressure vessels. *See*
      Pressure Vessels
Thin-wall pressure vessels. *See* Pressure
      Vessels
Three moment equation, 527
Throat of weld, 206
Timber columns. *See* Columns
Timber sections, design properties, 651
Titanium, 649, 650
Torque, 285
Torsion, 285
  deformation, 287
Torsional shear stress, 289
Transfer formula. *See* Moment of
      inertia
Transmissibility of a force, 14
Triangle laws, 636
Trig functions, 633
Trusses:
  development of, 97